The Molecular Physics of Liquid Crystals

Based on lectures presented at a
NATO Advanced Study Institute, held in
Cambridge, August 21 to September 3, 1977

The Molecular Physics of Liquid Crystals

Edited by

G.R. LUCKHURST
Department of Chemistry,
University of Southampton

and

G.W. GRAY
Department of Chemistry,
University of Hull

1979

ACADEMIC PRESS

London New York San Francisco

A Subsidiary of Harcourt Brace Jovanovich, Publishers

ACADEMIC PRESS INC. (LONDON) LTD.
24/28 Oval Road
London NW1

United States Edition published by
ACADEMIC PRESS INC.
111 Fifth Avenue
New York, New York 10003

British Library Cataloguing in Publication Data
The Molecular Physics of Liquid Crystals. –
(NATO Advanced Study Institutes).
1. Liquid crystals – Congresses
I. Luckhurst, G R
II. Gray, George William III. Series
548′.9 QD923 78-52095
ISBN 0-12-458950-2

Printed in Great Britain by
Whitstable Litho Ltd., Whitstable, Kent

PREFACE

We know what a liquid crystal is - it is a state of matter occurring on the phase diagram between the crystal and the liquid phases. As such it is an unusually fascinating state because it combines properties of both phases. We know a great deal about the macroscopic behaviour of liquid crystals, indeed the continuum properties are particularly well understood. In contrast our knowledge, both experimental and theoretical, of liquid crystals at the molecular level is less advanced. This lack of understanding stems, in part, from the very wide spectrum of techniques required to probe the static and dynamic molecular properties of liquid crystals. The interpretation of the experimental results also presents considerable difficulties because of the inherent problem of developing molecular theories of dense liquids.

In an attempt to clarify this exceptionally complex situation we decided to organise a NATO Advanced Study Institute devoted to the molecular physics of liquid crystals. The participants at the Institute came from many scientific disciplines but shared a common interest in the molecular behaviour of liquid crystals. They met in Corpus Christi College, Cambridge, at the end of August, 1977 and for two weeks there followed a process of mutual education. Experimentalists attempted to coach each other in the particular subtleties and importance of their technique, whilst the theoreticians were determined to persuade us of the validity of their models of liquid crystals. In addition, students from many countries were able to benefit from such discourses and discussions. This book is one of the fruits of the Advanced Study Institute. It presents, possibly for the first time, a relatively complete and detailed account of the theories needed to understand the molecular behaviour of liquid crystals. In addition it describes, in depth, those techniques which provide the most searching tests of these theories.

Contrary to the theoretician's belief, the constituent molecules of liquid crystals are neither rigid nor cylindrically symmetric and this view is soon dispelled by Chapters 1 and 12 which describe the dramatic dependence of mesophase occurrence on molecular structure. The formation of a liquid crystal mesophase clearly depends on the anisotropy in the intermolecular potential but the question as to which particular contribution is dominant remains unanswered. The various possibilities are described in Chapter 2 and the theories based on these are dealt with in Chapters 4, 7 and 8, using the important theoretical foundations laid in Chapter 3. These theories are quite different and yet a definitive test of them is not yet possible; the experimental techniques which should eventually provide such a test are described in Chapters 10, 13, 15 and 17. If we consider, as we must, the complexity of the structure of the molecules forming liquid crystals then one of the few techniques capable of providing sufficient information to characterize the orientational order is NMR; this important technique is described in Chapters 15 and 16. The computer simulation experiments discussed in Chapter 9 have particular appeal for theoreticians since they are now able to determine the properties of a system for which they have defined the pair potential.

The cholesteric phase is simply a helical version of a nematic. However, the molecular origins of this twist have been the subject of considerable controversy which has yet to be resolved. Some hint of this is contained in Chapters 5 and 6 which describe opposing theories of the cholesteric phase.

The addition of a solute to a nematogen invariably depresses the nematic-isotropic transition and may even destroy the mesophase. Experimental studies of such solute-solvent interactions are described in Chapter 10. These are interpreted in the following chapter using models based on both long-range and short-range interactions, with apparently equal success.

Our knowledge and understanding of the molecular dynamics in liquid crystals is still less advanced than that of the static properties. However, an attempt has been made to understand molecular reorientation by modifying the Debye model for rotational diffusion; this extension is developed in Chapter 18. It is possible to test this model using various techniques and some of these are described in Chapters 19 and 20 although the most searching test may be provided by the molecular dynamics method presented in Chapter 9. Many other motions are possible and quasielastic incoherent neutron scattering, described in Chapter 20, provides a powerful technique for their study.

The rich polymorphism provided by the smectic phases is a source of considerable excitement. The occurrence of the various phases depends in a delicate manner on the molecular structure, as we see from Chapter

12. The molecular organisation in the more highly ordered smectic phases is well known largely as a result of the beautiful X-ray studies reviewed in Chapter 14.

It will be clear that our understanding of the molecular behaviour of liquid crystals has advanced but some problems still remain. The account given here illustrates the extent of our knowledge and indicates the experimental and theoretical techniques which may be brought to bear on those problems remaining. It is our hope that this text will lay the foundation and provide the stimulation for future studies of these problems.

Finally, we wish to express our gratitude to our colleagues all of whom have contributed so much to the subject, to the Advanced Study Institute and to the book. We also thank NATO for funding the Institute so generously. This book has been produced directly from type-written copy and its pleasing appearance owes much to the skill and patience of Kate Welfare. We are also indebted to Judy Sibley of Academic Press; her constant encouragement and advice has ensured that the time taken to produce this book was kept to a minimum.

University of Southampton, England — G. R. LUCKHURST
University of Hull, England — G. W. GRAY
June 1979

CONTENTS

CONTENTS

TO OUR FAMILIES

Chapter 1

LIQUID CRYSTALS AND MOLECULAR STRUCTURE: NEMATICS AND CHOLESTERICS

G. W. GRAY

Department of Chemistry, The University, Hull, HU6 7RX, England

Introduction

One of the fascinations of liquid crystals is that their study brings together scientifically minded people from a range of different areas of specialism who share a deep common interest in liquid crystals. The interaction is probably at its closest amongst chemists, physicists, and theoreticians and in witness of this we have seen the origination of this collected Volume on the Molecular Physics of Liquid Crystals. These interactions can of course be highly stimulating, but they can also lead to problems. For example, organic chemists are forced to think in theoretical and physical terms about the compounds they prepare, and conversely, physicists and theoreticians are forced to apply their thinking and experimentation to organic materials of which they may have little real understanding either with regard to their structures or their properties.

In my two Chapters on Liquid Crystals and Molecular Structure, it is my aim to generate at least some feeling amongst the non-chemists for the chemistry of the materials with which they are involved at the liquid crystal level. These Chapters will concentrate on the liquid crystals which we describe, somewhat loosely [1], as thermotropic. Through this term we are trying to convey that such liquid crystals are formed by thermal effects, i.e., by heating a solid or by cooling a liquid phase, and to distinguish them from lyotropic liquid crystals usually formed as a result of the action of a limited quantity of solvent on the solid, for example. We will therefore concentrate attention on molecular structural aspects of *thermotropic mesogens* where the term mesogen means a compound capable of forming a liquid crystal phase or mesophase.

We are all aware that the molecules of such mesogens are more or less elongated in shape. The molecules, when at rest, have usually more width than thickness, and so they are conveniently described as lath-like in shape. We must remember, of course, that in phases in which rotational motion about the long molecular axis can occur, e.g., in nematic, smectic A, smectic B phases, the volume swept out by the molecule will be more cylindrical or rod-like in nature. We also know that the molecules must possess some degree of rigidity if a liquid crystal phase is to be formed. A classical example is provided by the n-alkanoic acids (long chain fatty acids). Although these can be represented with an extended, elongated molecular conformation (1) for the dimers, the molecules are entirely flexible, and do not exhibit liquid crystal phases of the thermotropic kind. However, if double bonds are introduced into the

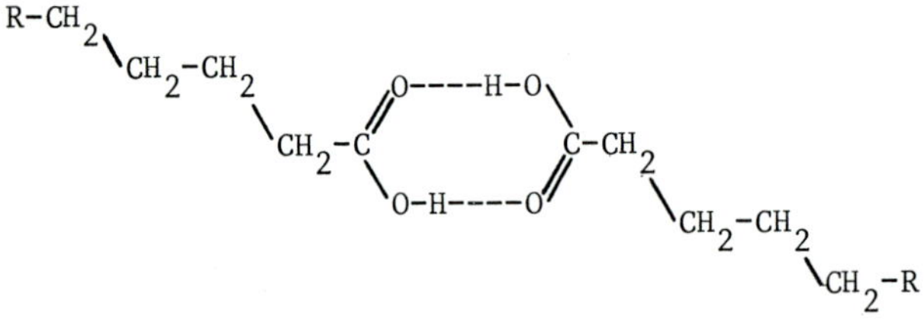

R-CH$_2$ \ CH$_2$-CH$_2$ \ CH$_2$-C(=O----H-O)(-O-H----O=)C-CH$_2$ \ CH$_2$-CH$_2$ \ CH$_2$-R (1)

R = $(CH_2)_nCH_3$ - a continuation of the chain

structure to give the alka-2,4-dienoic acids (2), the structure now has sufficient rigidity, due to restricted rotation about the double bonds of the sp^2 hybridised carbon atoms, and the compounds can form nematic phases [2]. The structure now has an extended semi-rigid core structure from A to B.

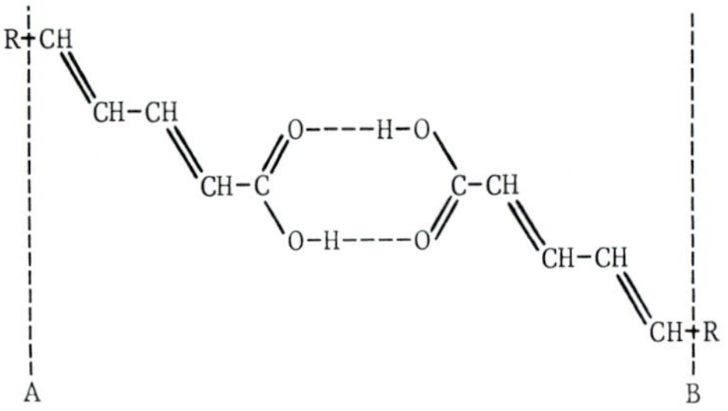

R-CH = CH-CH = CH-C(=O----H-O)(-O-H----O=)C-CH = CH-CH = CH-R (2)

A B

This requirement for at least a degree of rigidity associated with the molecules has led to molecular statistical descriptions of liquid crystal phases in terms of hard particles (rod-like, spherical, spherocylindrical etc.). Useful as these theories are, and valuable the conclusions that can be reached from their application to ordered, fluid phases of the nematic (cholesteric) and smectic kinds, their successes must be limited, because the molecules simply are not hard particles of a particular shape. Each type of molecule is endowed with almost a personality of its own, and as a consequence, even minor changes in molecular structure can have very large effects on the liquid crystal properties of a system.

For example, regarding the molecules as featureless hard particles cannot offer obvious explanations of observations such as the following:

(a)

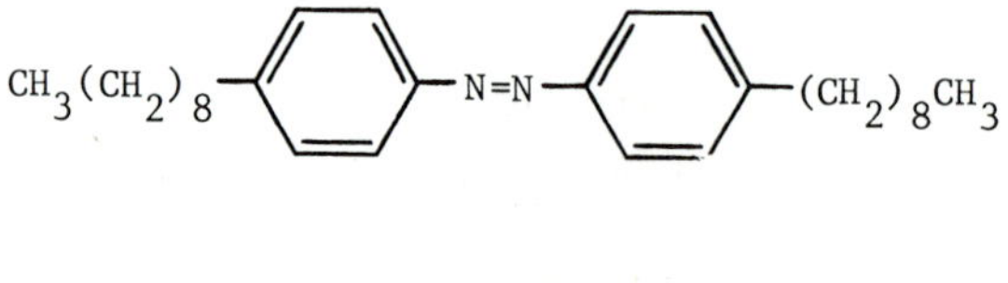

$CH_3(CH_2)_8$-C$_6$H$_4$-N=N-C$_6$H$_4$-$(CH_2)_8CH_3$ (3)

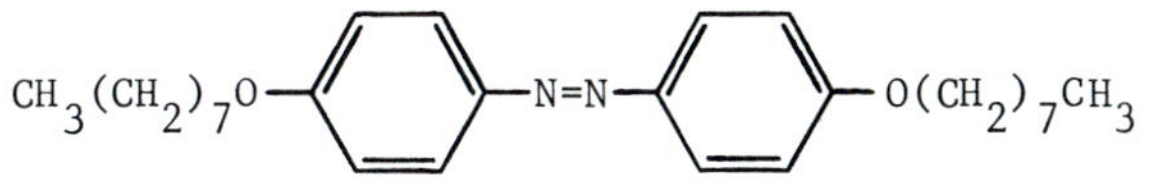

$CH_3(CH_2)_7O-C_6H_4-N{=}N-C_6H_4-O(CH_2)_7CH_3$ (4)

Compounds (3) and (4) represent very similar rod-like shapes with the same number of atoms in the terminal chains. Unless we say that the very small differences in bond angles and bond lengths relating to the C-O-ring and C-C-ring situations alter the rod shape significantly, how do we explain why compound (3) exhibits two smectic phases with transitions temperatures [3]: C-S_B, 37°; S_B-S_A, 40.5°; S_A-I, 53°, whereas compound (4) exhibits only nematic properties with the very different constants: C-N, 102°; N-I, 109°? The organic chemist would however anticipate such differences. He would, at least, recognise conjugative interactions between the O-alkyl functions and the rings in structure (4) and would expect a greater anisotropy of the polarisability and intermolecular attractions for this system, i.e., he would expect higher transition temperatures for compound (4). He would also expect that the intermolecular repulsions for the two systems would differ since a $-CH_2-$ and an -O-function differ electronically.

(b)

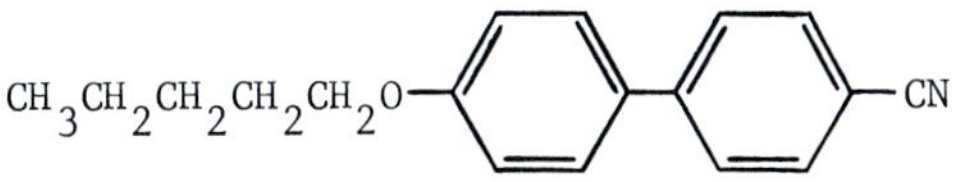

$CH_3CH_2CH_2CH_2CH_2O-C_6H_4-C_6H_4-CN$ (5)

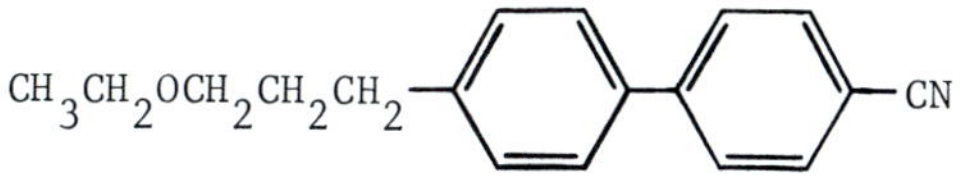

$CH_3CH_2OCH_2CH_2CH_2-C_6H_4-C_6H_4-CN$ (6)

Compounds (5) and (6), with identical chain lengths and rod-like shapes, differ only in the location of the oxygen, and yet compound (5) melts [4] at 53° and gives a nematic phase until 67.5°, whereas compound (6) melts at 36° and forms no nematic phase even on cooling to 0°, when recrystallisation occurs [5]. The organic chemist would, for the reasons just given, expect (5) to have a higher nematic isotropic (N-I) temperature than (6) because he looks at the detailed implications of the structural change at the molecular level.

(c)

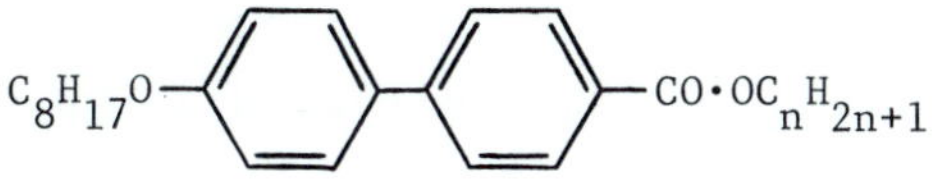

$C_8H_{17}O-C_6H_4-C_6H_4-CO{\cdot}OC_nH_{2n+1}$

When the isotropic liquids of these esters [6] are cooled, the following differences arise simply from the unit increases in the number of carbons in the terminal chain:

n	Phase types observed (on cooling)
2	I, S_A, S_B, S_E, C
3	I, S_A, S_B, C
4	I, S_A, S_C, C
5	I, S_A, C

That is at n = 4 we have the sudden injection of a smectic C phase and the complete elimination of S_B and S_E properties after n = 3.

(d)

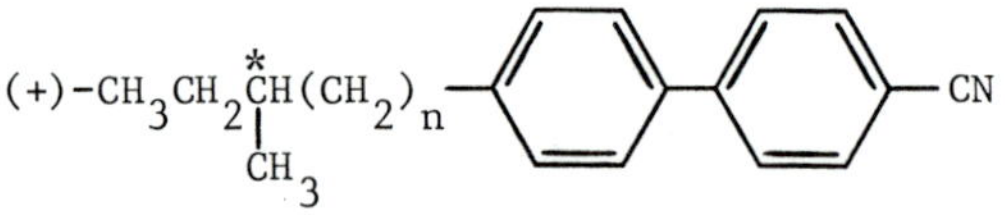

$$(+)\text{-}CH_3CH_2\overset{*}{C}H(CH_3)(CH_2)_n\text{-}C_6H_4\text{-}C_6H_4\text{-}CN \qquad (8)$$

As chiral molecules, all with the same absolute configuration (S) at the asymmetric carbon atom (asterisked), the materials with values of n from 1 to 3 are all cholesteric. However, if we mix the appropriate proportions of the compounds (8) with n = 1 and 2 or 2 and 3, we obtain a non-twisted nematic phase. The situation is that on moving the branching points from an even to an odd to an even point in the chain, the sense of the cholesteric helix changes from right-handed to left-handed and back to right-handed again [7].

(e)

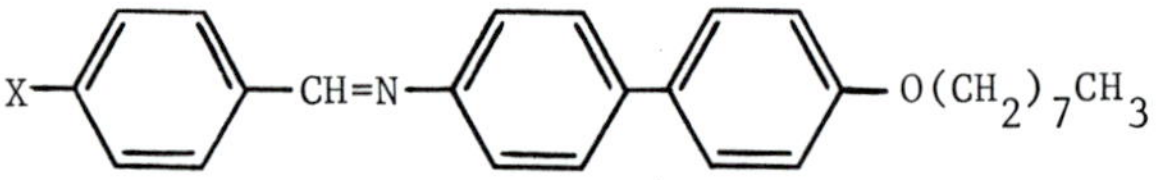

$$X\text{-}C_6H_4\text{-}CH{=}N\text{-}C_6H_4\text{-}C_6H_4\text{-}O(CH_2)_7CH_3 \qquad (9)$$

CH_3 and Cl substituents have effectively the same size, and so the molecules (9) with X = Cl *or* CH_3 have the same hard rod shape, but with X = Cl, the transition temperatures [8] are: C-N, 193°; N-I, 261.5° while with X=CH_3, the temperatures [8] are: C-S_A, 155°; S_A-N, 178°; N-I, 222°.

Such differences in behaviour cannot be classed as subtle; they are major and fundamental, affecting nematic, cholesteric and smectic properties. To begin to comprehend why they occur, we must think of the mesogen molecule as a dynamic entity consisting of a fairly rigid part, the core structure, and flexible parts such as alkyl chains. The entire molecule may be freely rotating about some axis which may not be along the most obvious line, unless due regard is paid to the precise stereochemistry of the molecule; moreover, parts of the molecule may be rotating and even bending in relation to other sections of the molecule. Superimposed on these effects, we have the electronic interactions which may occur within the molecule and which may affect or be affected by the rotational motions just discussed. All these factors influence the anisotropy of molecular properties which in turn determine the anisotropy of the intermolecular interactions (attractive and repulsive) which are responsible for the formation of the liquid crystal phase. In saying this, it is not to be inferred that molecular shape is unimportant. It is of fundamental importance, but it must be assessed in relation to other parameters and factors, and in many cases, in relation to a system that is capable of rotational motion.

Although these examples have impinged on all three liquid crystal phases, nematic, cholesteric and smectic, the discussion will now be divided into two sections relating first to Nematic and Cholesteric materials and second to Smectic materials.

Nematic (Cholesteric) Materials

In the nematic phase, the molecular centres of gravity are disordered as in a liquid, but we have a statistically parallel orientation of the long axes of the molecules along an axis (the director); there is no regular arrangement of the ends of the molecules. The nematic phase is

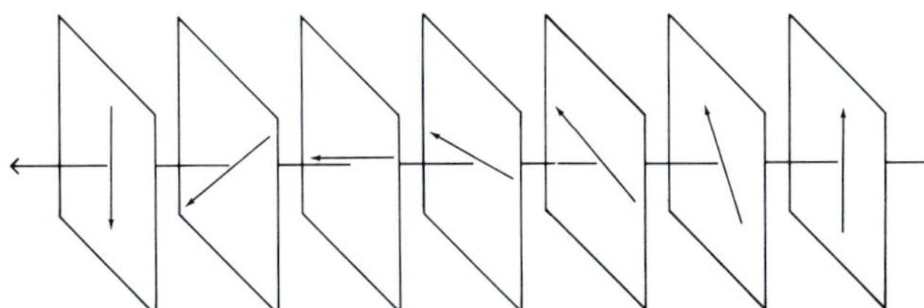

Fig. 1 *Simple representations of the molecular organisation in nematic and cholesteric phases.*

mobile and markedly affected by external fields and forces; the molecules therefore possess a high degree of mobility and are capable of rotational movement. Since a nematic phase is converted into a cholesteric phase by the addition of small amounts of optically active materials, and conversely, since two cholesteric materials of opposite helical senses give a nematic mesophase at some finite composition, the cholesteric phase may be regarded as a twisted nematic phase with a new director associated with the axis of the macromolecular cholesteric helix as in figure 1.

Molecular Structural Features Common to many Nematogens

The requirement that the molecule of a nematogen possesses a fairly rigid core structure is quite general [1, 9, 10]. This core structure, represented by the oblong in structure (10), usually has terminal substituents associated with it. These groups X and Y may be the same or different, and may vary from monatomic substituents such as Cl, through compact globular units such as NO_2 and NMe_2, to long chains such as $CH_3(CH_2)_n$ or $CH_3(CH_2)_nO$, i.e., alkyl or alkoxy groups.

X—[ ]—Y (10)

The rigidity of the central core structure is frequently achieved by using aromatic rings connected either directly as in 4-cyano-4'-n-pentylbiphenyl (11) and 4-cyano-4"-n-pentyl-*p*-terphenyl (12),

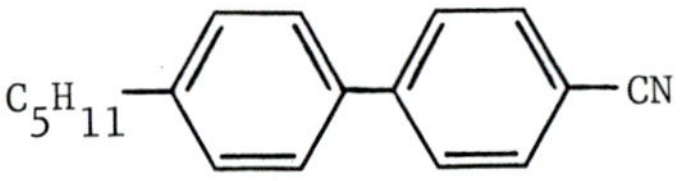

C_5H_{11}—⟨ ⟩—⟨ ⟩—CN (11)

C-N, 22.5°; N-I, 35°

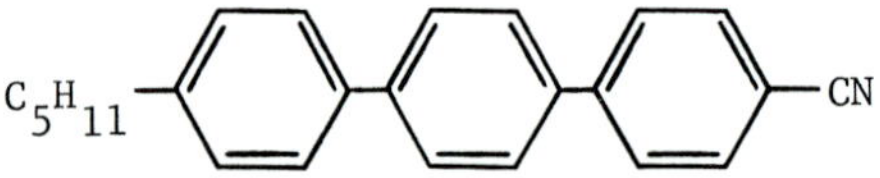

C_5H_{11}—⟨ ⟩—⟨ ⟩—⟨ ⟩—CN (12)

C-N, 130°; N-I, 239°

or through a linking unit which is represented as A▬B in (13)

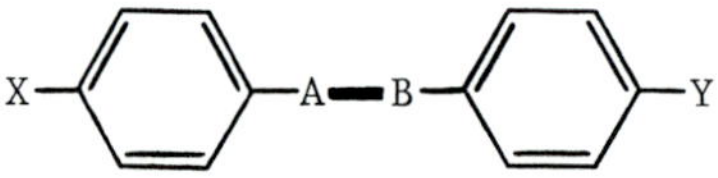

X—⟨ ⟩—A▬B—⟨ ⟩—Y (13)

Compounds (11) and (12) show [4], by the increase in melting temperature and the much greater (204°) increase in N-I temperature, the marked influence of extending the length of the rigid core from two to three *p*-phenylene rings. Naturally, the linking points of the rings are usually *para* to preserve the linearity of the molecule. The linking unit A▬B in (13) usually contains multiple bonds about which freedom of rotation is restricted, so preserving the rigidity and elongation of the molecules. Such multiple bonds can also conjugate with the phenylene rings, enhancing the anisotropic polarisability. Examples of common linking units are

-C≡C-; -CH=N-; -N=N(→O)-; -CH=N(→O)-; $(-CH=CH-)_n$; -CH=N-N=CH-

Also effective in this context is the carboxyl group, $-CO_2H$, and also $-CH=CH-CO_2H$; these can give cyclic dimers as shown in the hydrogen bonded structures (1) and (2). This obviously increases molecular length and maintains rigidity. The ester function too is effective since resonance interactions confer double bond character on the C-O link restricting rotational motions. The same arguments apply to the linking unit -CH=CH-CO·O-.

It should be noted that apart from the tolanes (14) which are linear,

$$X-C_6H_4-C\equiv C-C_6H_4-Y \qquad (14)$$

systems involving a linking unit with a double bond are not linear, e.g., the stilbene (15). Structure (15) represents the *trans*-isomer which is

$$X-C_6H_4-CH=CH-C_6H_4-Y \quad (trans) \qquad (15)$$

conducive to liquid crystal formation, whereas the *cis*-isomer (16) is not.

$$X-C_6H_4-CH=CH-C_6H_4-Y \quad (cis) \qquad (16)$$

The rapid decrease in the N-I transition temperature of compound (17) on exposure to light has been shown by NMR spectroscopy to be due to the accumulation of the non-linear *cis*-isomer through photochemical isomerisation [11]. After three days' exposure to ordinary laboratory light-

$$C_5H_{11}-C_6H_4-C_6H_4-CH=CH-C\equiv N \qquad (17)$$

trans-isomer: C-S_A, 80°; S_A-N, 99°; N-I, 147.1°

ing the constants were C-S_A, 79°; S_A-N, 90°; N-I, 125.5°. After one week's exposure, the ratio of *trans*- to *cis*-isomer was 2.6:1.

The importance of molecular elongation is thus stressed, as is the fact that extension of the linking unit from that in compound (18) to (19) greatly enhances the N-I transition temperature, i.e., increases the thermal stability of the nematic order from that of a monotropic phase in (18) to an enantiotropic phase in (19).

$$C_2H_5O-C_6H_4-CH=N-C_6H_4-OC_2H_5 \qquad (18)$$

C-I, 148° (I-N, 143°)

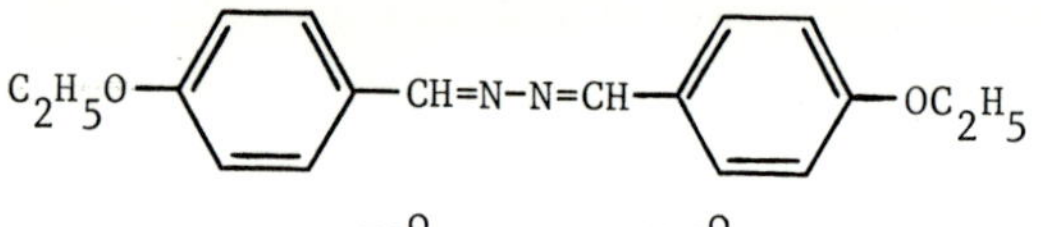

$$C_2H_5O-C_6H_4-CH=N-N=CH-C_6H_4-OC_2H_5 \qquad (19)$$

C-N, 172°; N-I, 199°

Increasing the number of rings and linking units has a similar effect resulting in a much greater thermal persistence of the phase for the more elongated system (20); when n = 1, the nematic phase is monotropic

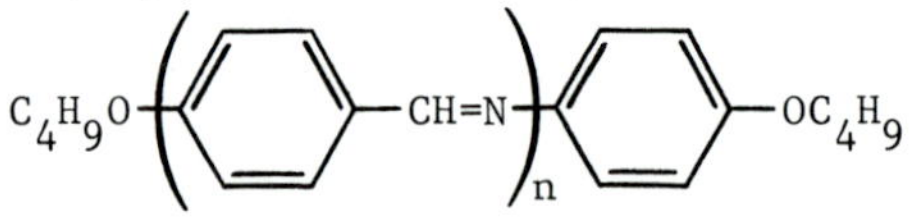

$$C_4H_9O\left(-C_6H_4-CH=N-\right)_n C_6H_4-OC_4H_9 \qquad (20)$$

(N-I, 121°), and when n = 2, the N-I is very high (297.5°).

For similar reasons, replacement of a single *p*-phenylene ring by a 4,4'-biphenylene or 2,6-naphthylene ring system strongly increases the N-I temperatures [12] as illustrated by compounds (21) to (23).

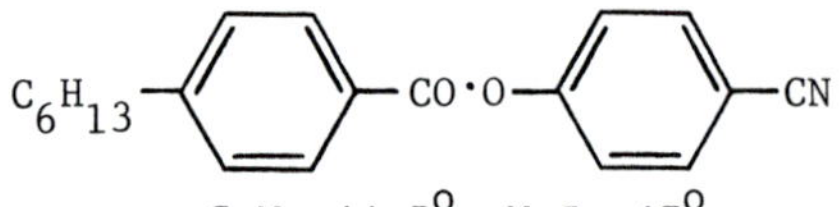

$$C_6H_{13}-C_6H_4-CO\cdot O-C_6H_4-CN \qquad (21)$$

C-N, 44.5°; N-I, 47°

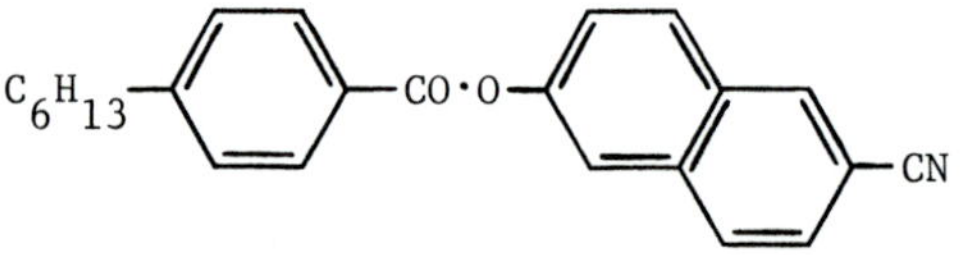

$$C_6H_{13}-C_6H_4-CO\cdot O-C_{10}H_6-CN \qquad (22)$$

C-N, 72.8°; N-I, 138.4°

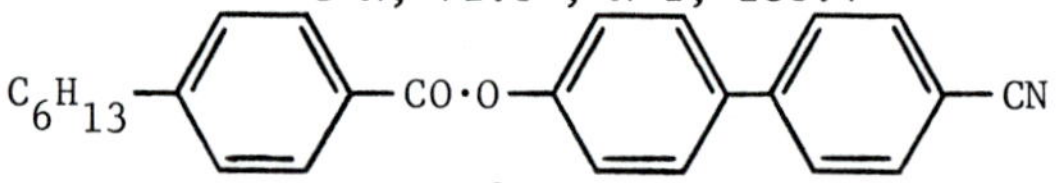

$$C_6H_{13}-C_6H_4-CO\cdot O-C_6H_4-C_6H_4-CN \qquad (23)$$

C-N, 91°; N-I, 229.6°

These examples emphasise the importance of extending the rigid core structure. The consequence is a more elongated molecule with a greater anisotropic polarisability. Conversely, changing the conjugated linking units to their flexible, saturated equivalents such as

$$-CH_2-CH_2-; \qquad -CH_2-O-; \qquad -O-CH_2-CH_2-O-$$

usually leads to non liquid crystalline materials or to mesogens with phases of low thermal persistence, if they link only two *p*-phenylene rings. When three rings are involved, and two are directly linked or are linked by a unit which preserves conjugation and molecular rigidity, a second linking unit can be of these more flexible types, and nematic phases of reasonable thermal persistence are obtained. This raises the interesting implication that if a considerable proportion of a lath-like molecule is rigid and packs parallel to neighbouring molecules, more flexible parts may be constrained to be in line with the more rigid parts. Evidence supporting this comes from the following observations [13]

The compound (24) has a flexible (CH_2-CH_2) linking unit. If the N-I

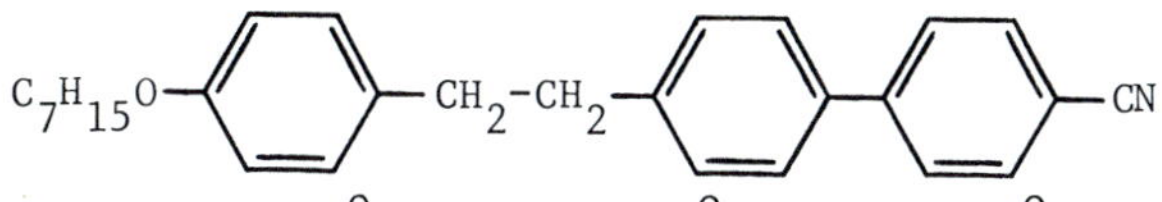

$C_7H_{15}O$–(ring)–CH_2-CH_2–(ring)–(ring)–CN (24)

C-S_A, 61.3°; S_A-N, 125.8°; N-I, 147.8°

transition temperatures for mixtures of compound (24) with compound (11) are plotted against percentage composition, the N-I line is not a straight line as normally found for binary systems, but is a curve. That is, compositions around the 50% region have considerably higher N-I temperatures than would be expected. This can be explained if we assume that in the pure compound (24), the breakdown of the nematic order is due to the onset of rotations about the CH_2-CH_2 link as the temperature is increased. At the lower N-I temperatures of the mixtures, these rotations may be less important, and compound (24) behaves as if it had a more rigid structure with a higher N-I value, probably around 180°. Such an N-I value would be more characteristic of a compound with three *p*-phenylene rings in a rigid molecule, e.g., a 4''-substituted 4-cyano-*p*-terphenyl [4].

If we use the N-I value for a mesogen as an index of the thermal persistence or thermal stability of that nematic phase, and if we consider molecules of the general structure

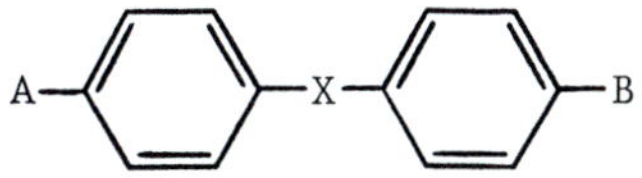

A–(ring)–X–(ring)–B

and draw information from various literature sources, [14-20] it is possible to construct an average order of efficiency for central groups X in increasing N-I transition temperatures. With minor exceptions, dependent upon the end groups A and B selected as constant for a given set of comparative data, the order is

X = –(ring)– > –(ring)– > -CH=CH- (*trans*) > -N=N-(↓O) > -CH=N-(↓O) >

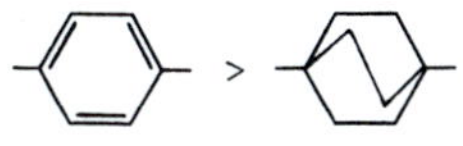

-C≡C- > -N=N- > -CH=N > CO·O- > none

Thus, stable mesophases are produced by substituted *p*-terphenyls and *trans*-stilbenes, whereas mesophases of lower thermal stability are formed by Schiff's bases, esters and biphenyls (X = none). Stereochemical considerations are important and affect the position in the order, e.g., stilbenes are planar, azoxybenzenes are slightly twisted and Schiff's bases [21] are considerably twisted. The relatively low position of the tolanes (X = -C≡C-) in the order is surprising, since crystalline diphenylacetylene [22] contains planar molecules. Possibly however, in the mesophase the rings become non-coplanar. The consequences could however be less serious than those in Schiff's bases, since the *cylinder* of electron density associated with the molecular orbitals of the -C≡C- linkage may still allow conjugative interactions to occur, even when the rings are non-coplanar.

The high position of the bicyclo-octane ring [23] in the order is interesting and shows that polarisable aromatic rings which permit conjugative interactions in the molecule are not absolutely essential. The bicyclo-octane ring can function quite effectively by maintaining the molecular linearity and rigidity. Dewar and co-workers [24] have extended their studies of non-aromatic rings as linking units using the system (25). Again using N-I transition temperatures as a basis for

MeO–C₆H₄–O·C(=O)–X–C(=O)·O–C₆H₄–OMe (25)

comparison, the effectiveness of various groups X in promoting nematic properties is:

X = [benzene ring] > [bicyclo-octene ring] > [cyclohexa-1,4-diene ring] > [bicyclo-octane ring] > [cyclohexa-1,3-diene ring] > [cyclohexane ring] > -CH=CH-

N-I(^{o}) 285 276 273 269 251 245 139

The results extend the previous series and show that all the rings are much more effective than a stilbene linkage. We now see that a bicyclo-octene ring is a little more effective than the bicyclo-octane ring, so that unsaturated linkages are of significance, although obviously not the only factor. This is also shown by the superiority of the two cyclohexadiene rings over the cyclohexane ring, although these rings are also more rigid than the cyclohexane ring; one would have anticipated that the cyclohexa-1,3-diene ring would have been superior to the cyclohexa-1,4-diene ring, since the former should permit extended conjugation within the molecule. However, the butadiene system is quite seriously twisted in the ring structure, and this unquestionably lowers the position of the cyclohexa-1,3-diene ring in the order. These results again emphasise the importance of a careful assessment of molecular shape, geometry, and conjugation on the anisotropic polarisability.

One must also keep a sense of balance about the extent to which simple rules of structure are applied. For instance, the cyclohexane ring is distinctly inferior to a benzene ring as a *central* linking unit, but we must not extend this conclusion to the situation where the ring is used more as a terminal function. For example, the N-I transition temperatures [25] of simple benzoate esters such as (26) are *lower* than those of corresponding or closely related (see the third example below) *trans*-cyclohexane carboxylic esters (27).

C_nH_{2n+1}–C₆H₄–CO·O–C₆H₄–R (26)

C_nH_{2n+1}–C₆H₁₀–CO·O–C₆H₄–R (27)

(26)				(27)			
n	R	C-N	N-I	n	R	C-N	N-I
5	OC_4H_9	49°	58°	5	OC_4H_9	49°	81°
4	CN	67.1°	42.6°	4	CN	54°	67.5°
6	C_5H_{11}	28°	(19°)	5	C_5H_{11}	36°	48°

( ) - monotropic transition

Similarly the results of Eidenschink *et al.*[26] on the cyclohexane analogues (29) of the 4-alkyl-4'-cyanobiphenyls (28) illustrate the point. Effects on C-N temperatures seem to be variable, but it is clear

C_nH_{2n+1}–C₆H₄–C₆H₄–CN (28)

C_nH_{2n+1}–C₆H₁₀–C₆H₄–CN (29)

(28)			(29)		
n	C-N	N-I	n	C-N	N-I
5	22.5°	35°	5	31°	55°
6	13.5°	27°	6	42°	47°
7	28.5°	42°	7	30°	59°

that using the cyclohexane ring in such a position can have advantageous effects on N-I temperatures. These results have also been confirmed [27] for a wider range of esters than those examined by Deutscher *et al.*[25]. Although the reasons for the enhancement of the N-I transition temperatures by the non-planar and somewhat flexible cyclohexane ring are not clear, it is a useful means of producing new mesogens. But the rules must not be stretched too far; if, for instance, we now replace the cyclohexane ring in such an environment by the bicyclo-octane ring, in the hope of achieving high N-I temperatures, we may be disappointed; in fact very low, monotropic N-I temperatures can be obtained [27].

Furthermore, if we examine the effect of changing a benzene ring to a cyclohexane ring at both extremities of a molecule, the overall result is a decrease in the N-I temperatures. For example, in the dimeric system (30), the benzene acids [28] have appreciably higher N-I values than the cyclohexane analogues [29]; the latter are however very much lower melting, and when we are interested in materials for display devices, the C-N value can be as important as the N-I value. Clearly

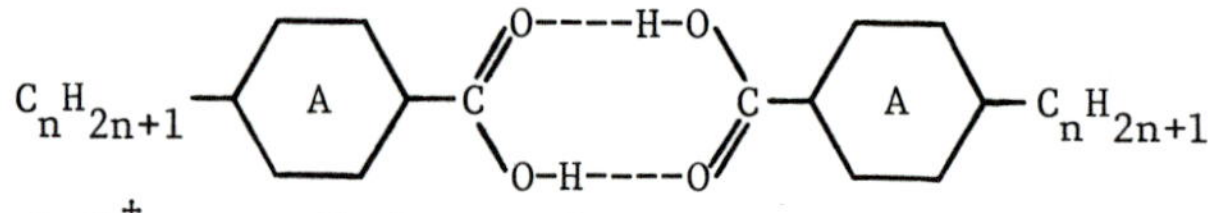

C_nH_{2n+1}–(A)–C(=O···H–O)(O–H···O=)C–(A)–C_nH_{2n+1} (30)

A	n	C-N†	N-I	A	n	C-N	N-I
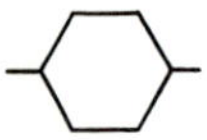 cyclohexane	4	41°	94°	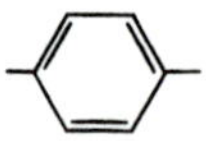 benzene	4	99.5°	113°
	5	54°	105°		5	88°	126.5°
	6*	32°	97°		6	97.5°	114.5°
	7*	39°	101°		7	101.5°	120°

† or S * also smectic

therefore, two non-planar and flexible cyclohexane rings at the ends of the core result in a less stable nematic phase.

Finally, heteroaromatic ring systems can also be employed in place of phenyl rings in the core structure. Less extensive systematic studies have been made in this area, and since a recent review has been given [10] for a range of heteroaromatic rings, we need simply to note the broad conclusions.

System (31) has been studied quite fully, and the effectiveness in

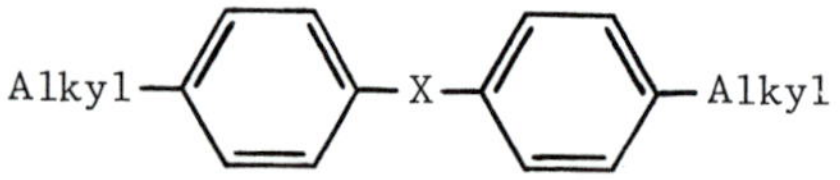

Alkyl–⟨phenyl⟩–X–⟨phenyl⟩–Alkyl (31)

enhancing N-I temperatures is

X = pyridazine > pyridine > pyrazine > pyrimidine > tetrazine

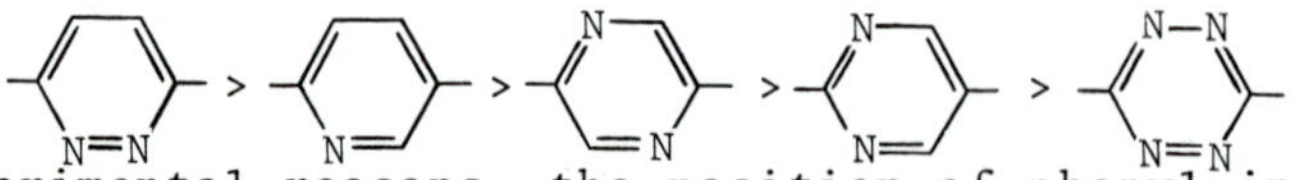

For experimental reasons, the position of phenyl in this nematic order could not be established. However, based on other studies of a range of compounds of the types

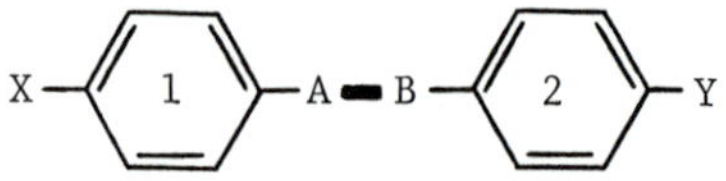

X–(1)–A=B–(2)–Y

and

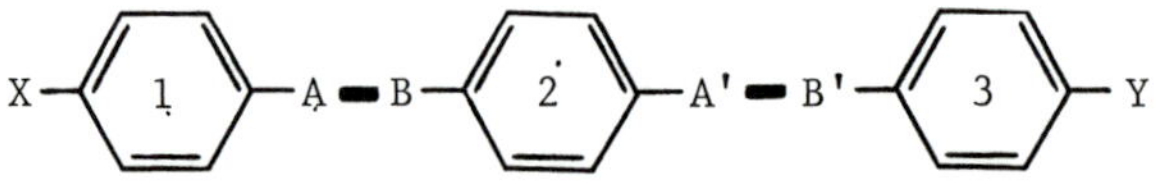

X–(1)–A=B–(2)–A'=B'–(3)–Y

the effect of replacing one of the rings (1, 2, or 3) by pyridyl, e.g.

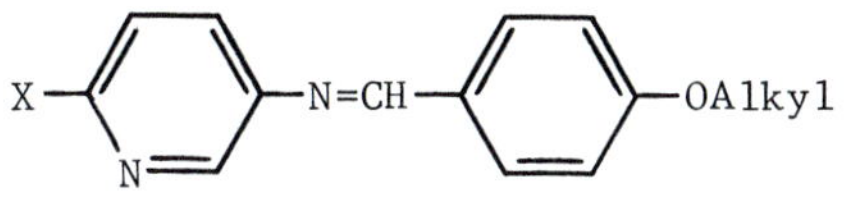

X—(pyridyl)—N=CH—(phenyl)—OAlkyl

or

$C_8H_{17}O$—(phenyl)—CH=N—(pyridyl)—CH_3

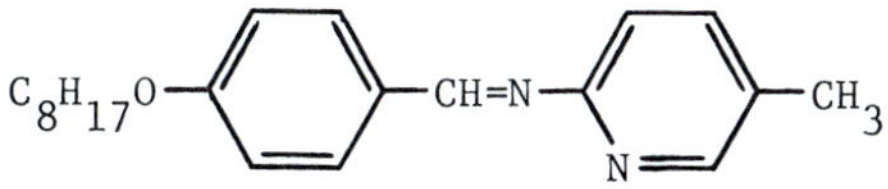

has been established, and N-I temperatures are generally lower. The effect can be appreciable; for example, the pyridyl analogue of MBBA has an N-I of 6.4°, *ca* 40° lower than that of MBBA. Often the smectic properties of the pyridyl systems are quite pronounced. It would seem therefore that phenyl should occupy a position in the order for system (31) higher than pyridyl.

The order of effectiveness of *terminal* heteroaromatic rings varies from system to system. For the compounds

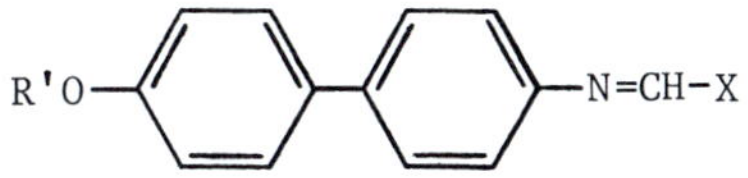

R'O—(phenyl)—(phenyl)—N=CH-X

it is

X = (pyridyl) > (pyridyl) > (phenyl) >> (pyridyl)

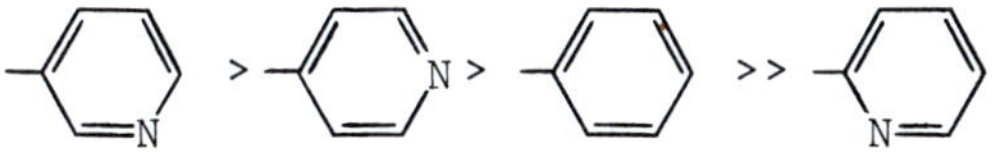

whereas for

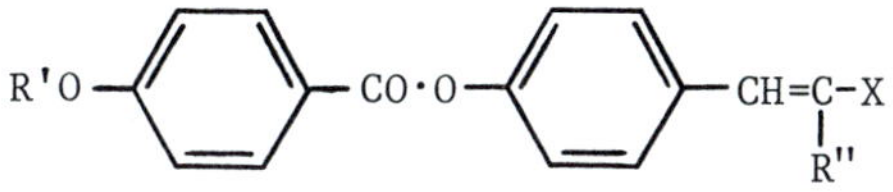

R'O—(phenyl)—CO·O—(phenyl)—CH=C(R'')-X

it is

X = (pyridyl) ≈ (pyridazinyl) > (phenyl) > (pyrazinyl) > (pyridyl) >> (pyridyl)

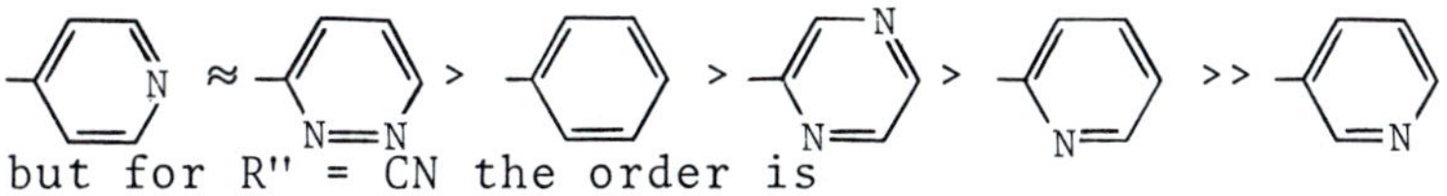

when R'' = H, but for R'' = CN the order is

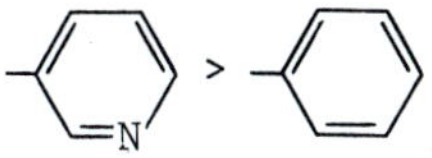

(pyridyl) > (phenyl)

Then again, in the system

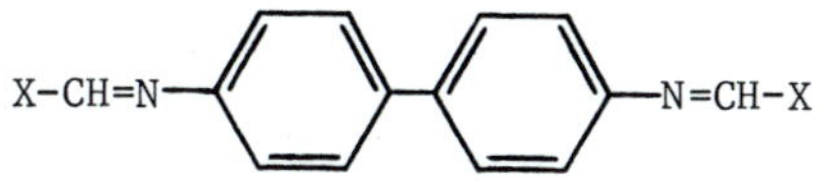

X-CH=N—(phenyl)—(phenyl)—N=CH-X

the order is

X = (pyridyl) > (phenyl) > (pyridyl) and (pyridyl)

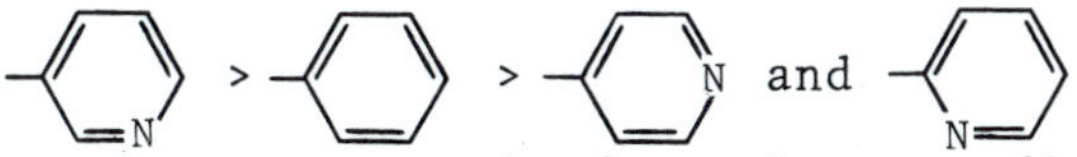

A clear picture does not result from these studies, but some understanding of the variations can be reached in terms of factors such as differences in conjugative interactions and twisting of the structure dependent on the location of the ring nitrogen relative to the Schiff's base or stilbene linkage.

Lateral Substituents

We will now consider the effects of broadening a molecule by introducing a substituent in place of a hydrogen in a position along the side of the core structure, e.g., the substituent X in

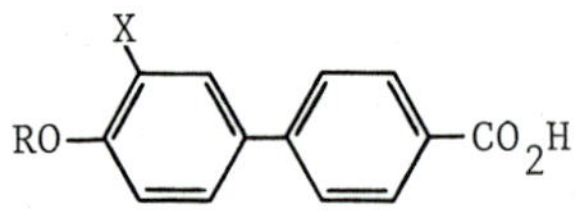

(32)

or

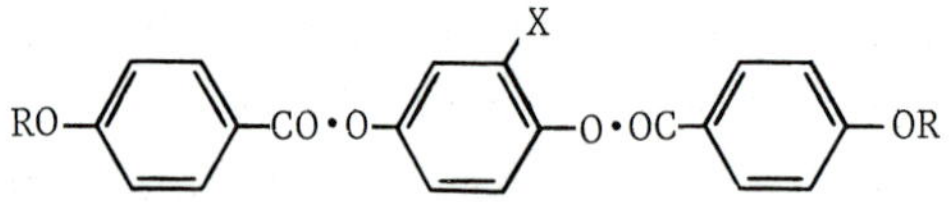

(33)

The situation for system (32), and other similarly substituted systems, is so well documented [30] that it need only be said that for a range of substituents X (F, Cl, Br, I, NO_2, CH_3) of varying size, polarisability and polarity, the N-I transition temperatures fall in proportion to the size of the substituent. The fact that the variable dipoles of the substituents do not interfere with this relationship is in accord with Maier-Saupe theory of the nematic phase (cf. Chapter 4).

So far we have considered the N-I transition temperature as a criterion of the effects of molecular structural change on the nematogenicity of the compound, and in the case of the lateral substituents, the effects have been rationalised in terms of a reduction in anisotropic polarisability and an increased axial separation of the molecules giving a reduction in intermolecular attractive forces. Studies of the enthalpies and entropies of N-I transitions have recently led some workers to consider whether this may not be an oversimplification. Young, Haller and Green [31] studied the system (34) and found a systematic decrease in

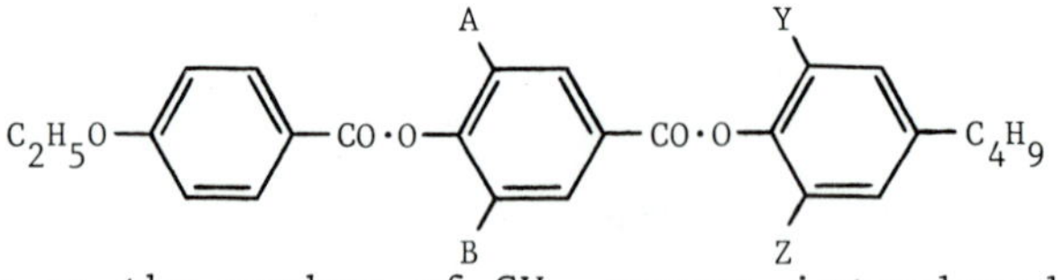

(34)

N-I temperatures as the number of CH_3 groups introduced at positions A, B, Y and Z was increased. Methyl groups at A and B *or* Y and Z gave about twice the effect of only one CH_3 group, whereas two methyl groups at A or B *and* Y or Z gave a smaller effect, as the two protruding groups need not be on opposite sides of the molecules. However, only in some instances were the observed enthalpies of transition for the CH_3-substituted compounds lower than that for the parent compound. In those cases where ΔH_{NI} increased on substitution, the decrease in transition temperature was attributed to an even larger increase in ΔS_{NI}. These authors did not conclude that the nematic phases of the methyl substituted compounds with lower N-I values necessarily had more ordered nematic phases;

they took a balanced view which is best summarised in their words. "There are a large number of degrees of freedom, both internal and external, that contribute to the entropy, and subtle changes in the restrictions that these experience in the two phases may result in a non-systematic variation of ΔS_{NI} with substitution. For example, the barriers to internal rotation about the ester linkages are undoubtedly influenced by the steric repulsions of both the methyl groups and the neighbouring molecules. Any non-systematic change in ΔS_{NI} with substitution could be rationalised on this basis."

If however it is assumed that changes in ΔS_{NI} are dominated by *intermolecular* degrees of freedom in the nematic phase, then one could postulate that the lateral CH_3 group(s) functions as a bulge in the repulsion envelope of the molecule enhancing positional correlation and short range order in the nematic phase (by an interlocking effect). Dewar and Griffin [24] reached a similar conclusion based on thermodynamic studies of the compounds (33) with R=CH_3 and X=H, F, Cl, Br, I, and CH_3; their data are shown in Table 1. The enthalpy values show no correlation

Table 1. *Thermodynamic data for the compounds (33).*

X	ΔH_{NI}/kJ mol^{-1}	ΔS_{NI}/J K^{-1} mol^{-1}	T_{NI}/^{o}C
H	1.71	2.97	301
F	1.69	3.06	278.5
Cl	1.94	3.69	252.4
CH_3	2.21	4.23	252.1
Br	2.04	3.96	241.1
I	2.06	4.16	222.9

with the transition temperatures, although some of the differences are sufficiently small to raise doubts as to their meaning. The entropy values do decrease steadily with increasing transition temperature, with the exception of X=CH_3, for which ΔS_{NI} is too high for T_{NI}. Based on these results, Dewar and Griffin concluded that the projecting substituents, far from decreasing intermolecular interactions by forcing the molecules apart, are increasing the order in the nematic phase by an interlocking effect on molecular mobility. They do not consider that there are any steric effects in this system to justify the cautionary comments quoted earlier from Young, Haller, and Green [31]. However, Griffin, Wertz, and Griffin [32] have since studied the case of X=CH_3 and CF_3 in the system (33) with R=n-C_4H_9. The results showed that T_{NI} decreases from CH_3 to CF_3 by 54° and that this decrease is accompanied by *decreases* in ΔH_{NI} of 0.59 kJ mol^{-1} and in ΔS_{NI} of 0.59 J K^{-1} mol^{-1}. The results are therefore inconsistent with the proposal of an interlocking effect and the authors proposed that either a steric effect between the CF_3 substituent and the ester function occurs in the nematic phase, but is less important in the isotropic liquid, or that the lone pairs of electrons on the CF_3 group give rise to repulsive interactions so that the intermolecular bonding loss is greater than the loss in positional order. In considering this result, it is clear that the size of the CF_3 group is too small to account for the large decrease in N-I temperature on the basis of separation of the long molecular axes {as proposed for system (32)}. However, molecular models show that with the CF_3 group present, there is steric strain in both the possible planar conformations, and this may be more extensive than we expect, if the views of Griffin *et al.* on repulsive interactions from CF_3 groups are correct. If the molecule is in fact twisted, then the large ΔT_{NI} would be explained (see later under steric effects).

It is certainly not wise to discount ΔS and ΔH values for transitions, and, indeed, the previous observations and those in Chapter 10 stress the need to consider these values, in addition to the transition temperatures, wherever possible. In doing so, however, all the factors that may affect ΔS_{NI} must be taken into account. An additional problem is that whereas an abundance of accurate transition temperatures exists for many compounds, rather few data on ΔH_{NI} and ΔS_{NI}, measured with a *sufficient* degree of accuracy, are available. It seems premature therefore to put too great an emphasis on theories based on such observations. We ought however to keep an open mind about the situation and see how it develops when a much larger range of enthalpy values is available on which to base more secure conclusions.

Furthermore, as will be seen later, regular trends in transition temperatures do occur when the systematic changes in molecular structure are connected with the ascent of homologous series of mesogens. Now, a much wider range of data on entropies of transition for homologous series exists than for any other group of materials, and yet regular trends in ΔS_{NI} do not always accompany the regularly changing transition temperatures [33]. A regularity may exist early in a series, but later, irregularities or even large jumps in the entropies are observed. At present, it seems reasonable therefore to continue to base our comparisons of mesogens on transition temperatures, for which the regular trends are most unlikely to be coincidental, but at the same time to keep a close watch on developments relating to enthalpies of transition, considerations of which may lead to alterations in our ideas as to why the transition temperatures are altered by structural changes in the molecules.

Other cases of lateral substitution worthy of mention are:

(a) when the substituent occupies an indentation in the molecular perimeter [10,30], the nematic phase has a *higher* transition temperature. This occurs in the dimeric 6-n-alkoxy-2-naphthoic acids (35), when a substituent X is introduced into the secluded 5-position. 5-Chloro-

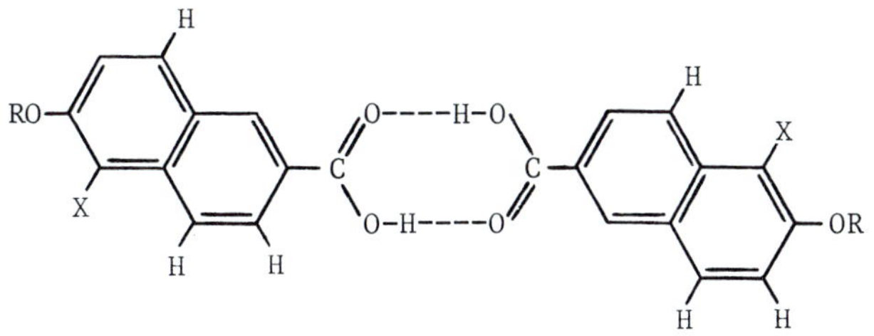

(35)

and 5-bromo-substituents give small, but definite *increases* in the N-I temperatures, whereas 5-iodo- protrudes sufficiently to give a decrease in N-I temperature;

(b) when the substituent is capable of intramolecular hydrogen bonding, this partly shields the broadening effect and can reduce rotational motions within the molecule, as in the hydroxy-substituted Schiff's base (36) [34] or related azo-compounds [35]. Again higher N-I temperatures occur compared with the non-hydroxy analogues;

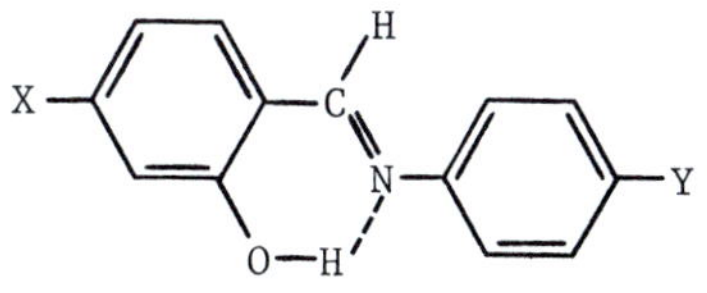

(36)

(c) if the substituent exerts a steric effect on the molecule such that the molecule adopts a less planar conformation, then much larger decreases in the N-I temperatures occur than those found when the same substituent occupies a position in a similar molecule, but exerts no steric effect. The simple rules of additivity of substituent effect, applicable when no steric effects are involved, can still be applied [10, 30, 36], provided the effect of each individual substituent in the same environment is known. System (37) illustrates some of these effects.

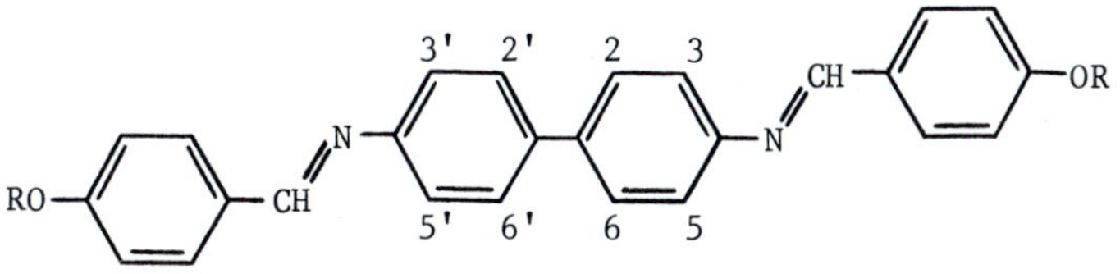

(37)

The rings of the terminal benzylidene groups are *not* coplanar [21] with those of the biphenyl ring system, and it is not known whether the two rings of the biphenyl structure are coplanar (as they are in the solid) in the nematic phase. However, substituents in the 2-, 2'- and in the equivalent 6- and 6'- positions will increase the twisted nature of the molecule by rotation about the 1,1'-bond. The results are given in Table 2.

Table 2. *Average results for system (37)* $(R = C_7\text{-}C_{10})$

Substituent(s) and position(s)	ΔT_{NI} (relative to subst.=H)	Comments
3-Chloro	48.4^{o}	No steric effect
3-Bromo	58.0^{o}	No steric effect
3,3'-Dichloro	94.2^{o}	Approx. additivity -
3,3'-Dibromo	112.4^{o}	no steric effect
2-Fluoro	29.4	All exert
2-Chloro	84.1^{o}	steric effects
2-Bromo	99.2^{o}	giving large
2-Iodo	122.6^{o}	decreases in
2-Methyl	85.4^{o}	T_{NI}
2,5-Dichloro	134.6^{o}	Approx. additivity -
2,5-Dibromo	157.7^{o}	steric effect operates

The effects of the same substituents in the 2-position are obviously much larger than those given in the 3-position, from which no steric effect operates, but an additivity of effect is shown for both the 3,3'- and 2,5-disubstituted compounds, i.e., irrespective of whether a steric effect operates or not.

When the N-I transition temperatures for a range of the 2-substituted Schiff's bases (37) are plotted against molecular breadth, a close approximation to a straight line is obtained, but the slope is much greater than that of the line obtained for the 3-substituted acids (32). The extra influence is the steric effect in the 2-substituted compounds, giving larger ΔT_{NI} values for a given substituent than in the case of the 3-substituted acids. A final point to be made relating to lateral substitution is that the effect of a given substituent in reducing N-I temperature is understandably greater if the molecular length is short.

Terminal Substituents

The nature of the terminal substituent(s) or end group(s) in the molecule of a mesogen has a profound influence on the liquid crystal properties of the compound [10,30]. An average order of terminal group efficiency in enhancing the N-I temperature is:

$$Ph > NHCOCH_3 > CN > OCH_3 > NO_2 > Cl > Br > N(CH_3)_2 > CH_3 > F > H$$

Replacement of the terminal ring hydrogen in a mesogen molecule by any of the commonly encountered substituents is therefore stabilising to the nematic order. As de Jeu has said [16] the effect should be greatest "if the *para*-substituent can easily be embedded in the conjugated system", because, by so doing it will contribute most to the axial polarisability of the molecule. Certainly the effect is greatest with groups which are themselves readily polarisable and/or are capable of giving strong conjugative interactions, and smallest with small substituents which conjugate weakly or with $N(CH_3)_2$ which probably lies out of the molecular plane.

Perhaps the terminal functions which attract most interest and study are those involving alkyl chains, i.e., the group $CH_3(CH_2)_n$ - or the alkoxy group, $CH_3(CH_2)_nO$-. The situation is very well documented for a large number of homologous series [9,10,30]. Provided adequate attention has been paid to the purity of the materials, it is found generally that:

(a) the N-I transition temperatures alternate in a regular manner, the degree of alternation diminishing as n is increased;

(b) the N-I temperatures for ethers containing the function $-O(CH_2)_{n-1}CH_3$ are always some 30-40° higher than those of the corresponding materials with the group $-(CH_2)_nCH_3$;

(c) the N-I temperatures for a given series lie on two smooth curves which may either fall or rise as the series is ascended. If the N-I temperatures are high, the curves are of the falling type (see figure 2); if the N-I temperatures are low, the curves rise. The cross over temperature between falling and rising curves would seem to be about 100°, in which region the curves are often rather flat;

(d) for a series of ethers, the even carbon chain members give the upper curve and the odd carbon chain members the lower curve. This situation is reversed for terminal alkyl groups; therefore the oxygen in an ether effectively replaces a CH_2 group.

If the alkyl chain adopts an extended, all *trans*-conformation, the

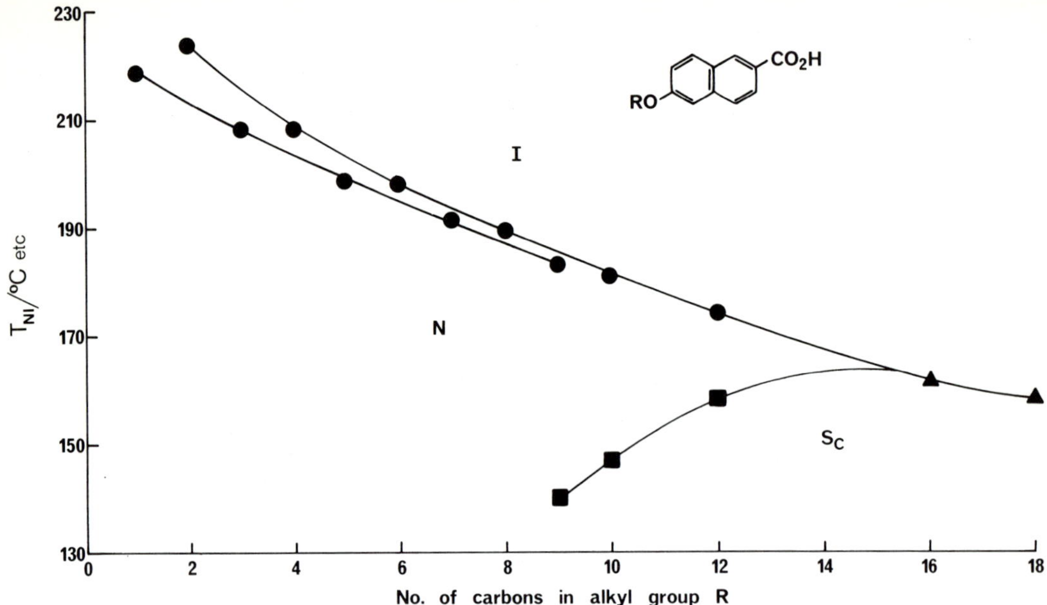

Fig. 2 *An example of alternating N-I temperatures constituting two falling curves.*

the axial polarisability increases about twice as much as that at right angles on passing from an even to an odd member of an alkyl-substituted series (38); on passing from an odd to an even number, the two polarisabilities increase about equally [40]. At comparable molecular

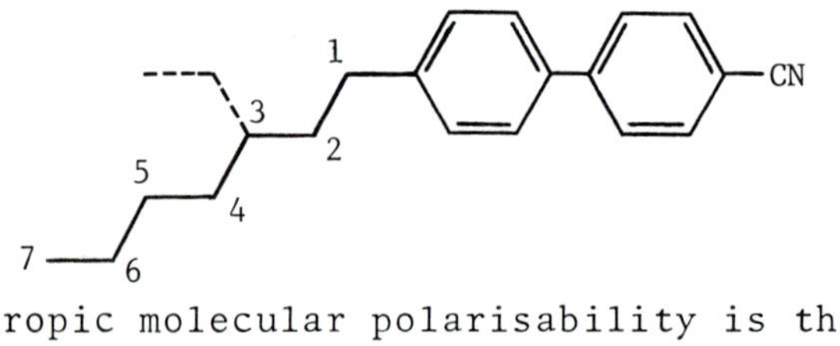

(38)

weights, the anisotropic molecular polarisability is therefore greater for odd than for even members and their N-I values are higher. The opposite situation obtains for ethers.

As we shall see in Chapters 15 and 20 the assumption of a regularly, extended conformation for a flexible alkyl chain is certainly unreasonable, but the regularity of alternation of T_{NI} values can be explained only if some preference for such a conformation in fact exists. The damping of the alternation as the series is ascended is explained by a statistical increase in the population of non-extended conformations giving a decreasing difference in the anisotropic polarisabilities for odd and even numbers.

Such arguments explain only the rising type of N-I curve, because of the increasing molecular polarisability along the series. Although the

Maier-Saupe theory [14] of the nematic state considers only attractive dispersion forces when considering the free energy differences (ΔA) between the isotropic and nematic phases, recent studies suggest that the entropy contribution due to steric intermolecular repulsions is also important, and an expression which considers both attractive and repulsive forces is

$$\Delta A = - a\overline{P}_2^2 - bT\overline{P}_2^2 - TS(\overline{P}_2),$$

where $\overline{P}_2$ is the order parameter, $-a\overline{P}_2^2$ is the internal energy from dispersion forces (a being approximately proportional to the square of the anisotropic molecular polarisability), $b\overline{P}_2^2$ is the packing entropy, and S is the orientational entropy. For a phase transition at constant volume ($\Delta A = 0$), T_{NI} is given by

$$T_{NI} = 2a/4.54k - 2b,$$

where k is Boltzmann's constant. b is in some way proportional to the effective molecular length:breadth ratio [40]; if reductions in this ratio due to chain bending are proportional [42] to temperature and molecular length, then b will decrease as the chain lengthens [40]. At lower temperatures, the effect of b will be smaller, and the change in T_{NI} with chain length will be determined mainly by changes in a. With higher T_{NI} values, chain motions will be more important, and the decrease in b may dominate and lead to a fall in T_{NI} with increase in chain length. A quantitative account of the influence of chain statistics on the N-I transition is given in reference [39] of Chapter 10.

For many years, one series has been considered exceptional [9,10,37]; as figure 3 shows, there are depressions in both N-I curves. However, the completion of studies on the homologous series of 4-n-alkyl- and 4-n-alkoxy-4'-cyanobiphenyls revealed [38] the results shown in figures 4 and 5. The depressions in figure 3 are reproduced as distinct minima for both series. Moreover, in series for which the N-I values alternate and rise, a maximum must eventually be reached and be followed by a subsequent fall. Certain series of esters have recently provided evidence for this behaviour [39].

The shapes of N-I curves are therefore regular, but can be complex, and perhaps for the majority of series for which the situation is simple, steadily rising or steadily falling curves, only portions of an overall T_{NI} against chain length trend of the types shown in figures 3-5 are seen.

Qualitative attempts to explain such trends involve [38] considerations of the conformations adopted by the growing alkyl chain as we shall see in Chapter 12. Consider the 4-n-alkyl-4'cyanobiphenyls [38]; the methyl and ethyl members exist as distinct conformers. From the standpoint of nematic order and/or conformer stability, it is best if the extended *trans*-conformations are adopted by the propyl and butyl members. A decrease in T_{NI} from C_1 to C_2 and from C_3 to C_4, with the T_{NI} values for C_1 and C_3 lying above the values for C_2 and C_4, respectively, would be expected. Continuation of the *trans*-conformation would give T_{NI} for C_5 a lower value than for C_3, but higher than for C_4. Rotation about the 2,3-bond could, however, allow the C_5 member to contain a predominance of a conformer shown by the dotted line in (38) which, though energetically less favourable, is especially favourable to the nematic order. The corresponding rotation for the C_4 member is not favourable, because of the freely rotating terminal CH_3 group. It is suggested that the presence of other similar conformers is responsible for the progressively increasing T_{NI} values further along the series, the relative population of the specially favourable conformations decreasing with temperature increase and causing the curve to level off slowly.

Further evidence that flexible alkyl chains may have a strong tendency to adopt statistical predominances of favourable, nematogenic conformations is given [4,43] by the pronounced alternation of the T_{NI} values for ω-phenylalkyl esters. If a regular, extended *trans*-arrangement of the $(CH_2)_n$ unit is favoured, the terminal phenyl ring falls alternately in and out of line with the rest of the molecule as n changes from even

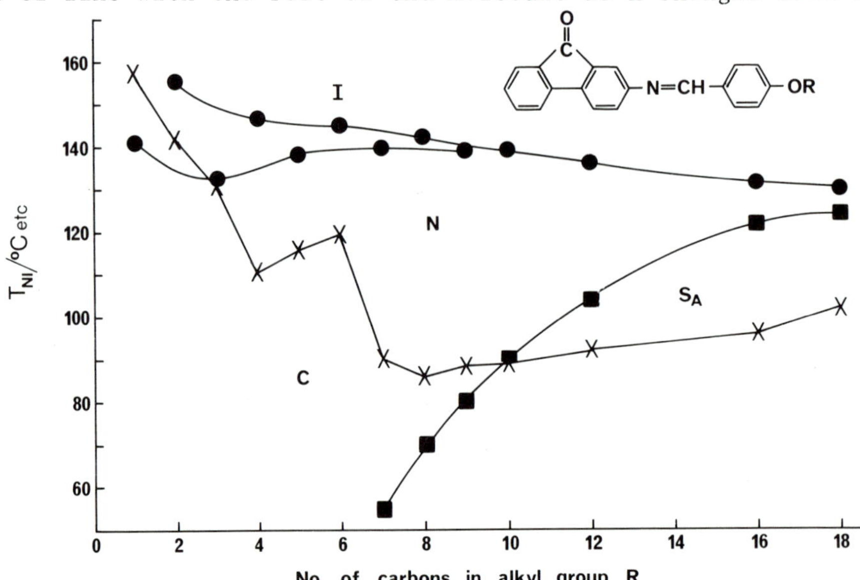

Fig. 3 *Transition temperature versus chain length for the series of 2-(4'n-alkoxybenzylidene)-aminofluorenones.*

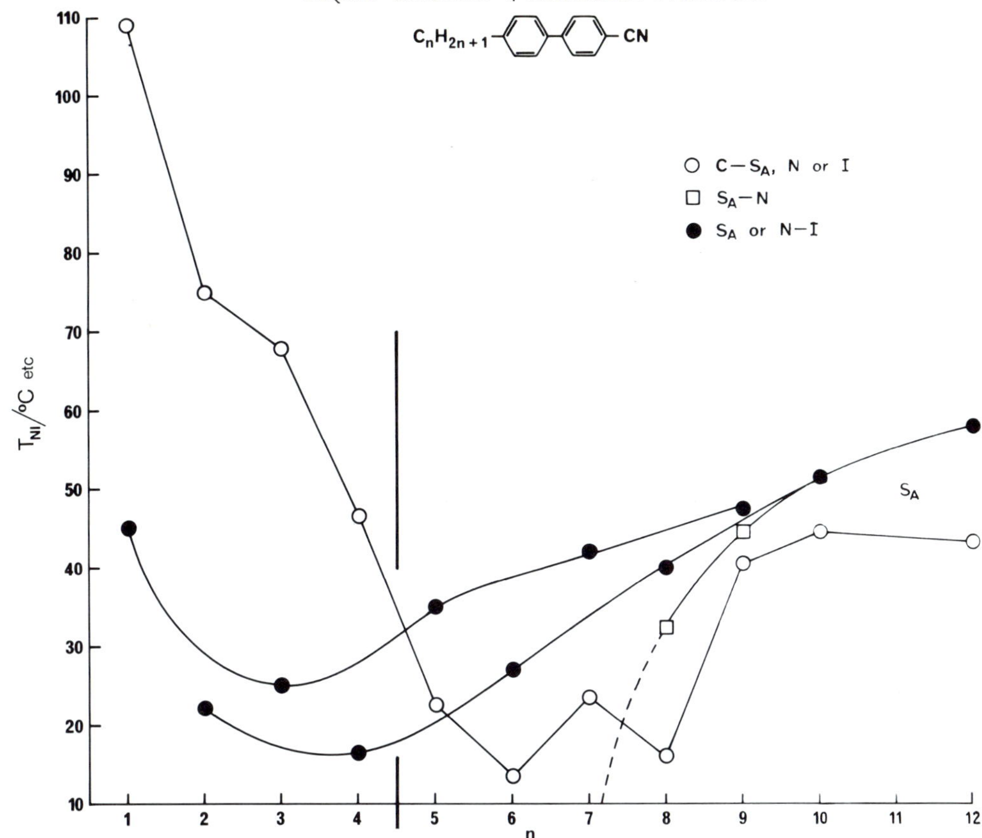

Fig. 4 *Transition temperatures for the 4-n-alkyl-4'-cyanobiphenyls.*

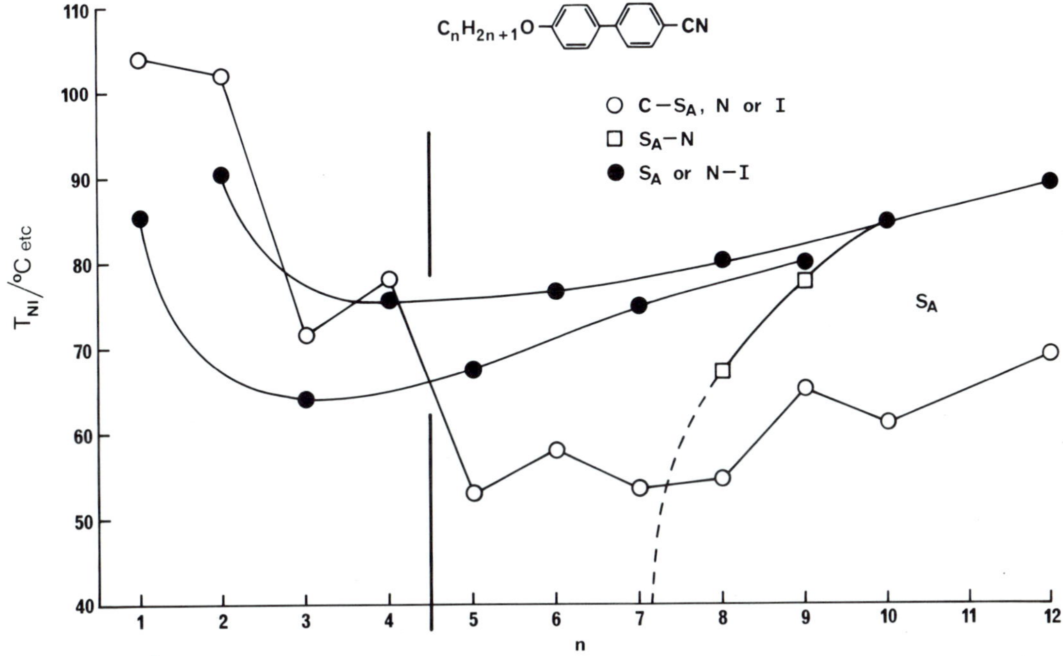

Fig. 5 *Transition temperatures for the 4-n-alkoxy-4'-cyanobiphenyls.*

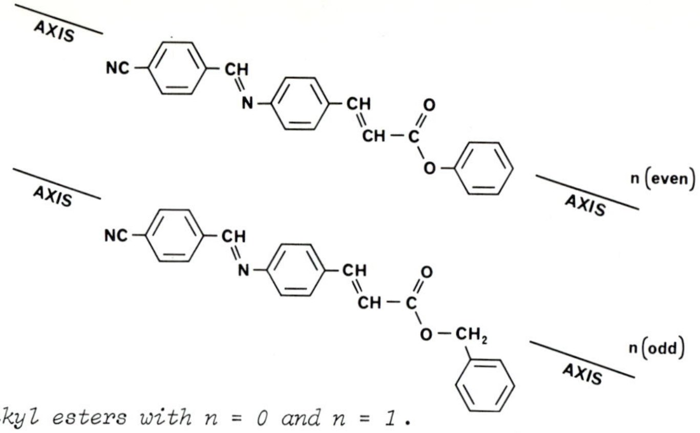

Fig. 6 *ω-Phenylalkyl esters with n = 0 and n = 1.*

(including zero) to odd as we see in figure 6. Very high alternations as shown in figure 7 would then arise; if all conformers were equally likely, no such clear distinction would arise, and a moderate alternation would be expected. Moreover, if the phenylalkyl ring is substituted in the 4-position, T_{NI} is raised when n is even, but is decreased when n is odd, because of the increased size of the protruding group relative to the unsubstituted system [4,53]. In addition, a 3-substituent decreases T_{NI} relative to the unsubstituted compound, as would be expected. The effect is sometimes smaller when n is odd [4,53] (and indeed in one case an increase in T_{NI} is observed), because, if a suitable population of one planar conformation (39) is adopted, the substituent increases the molecular length and counteracts the broadening effect of the other planar conformation resulting from rotation about the CH_2 to ring bond. With

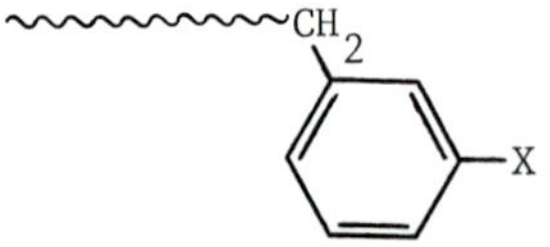

(39)

some types of structure, we must therefore be careful when we try to assess the likely effects of introducing lateral and terminal substituents; all the stereochemical factors must first be considered quite carefully.

The effects of branching of terminal alkyl chains in mesogens have been investigated systematically [43,44]. For instance, branching at the penultimate carbon of a n-alkyl chain gives a similar, but smaller effect on T_{NI} values to that just discussed. For the compounds (40) with X = CN, the N-I temperatures alternate much more than those for the

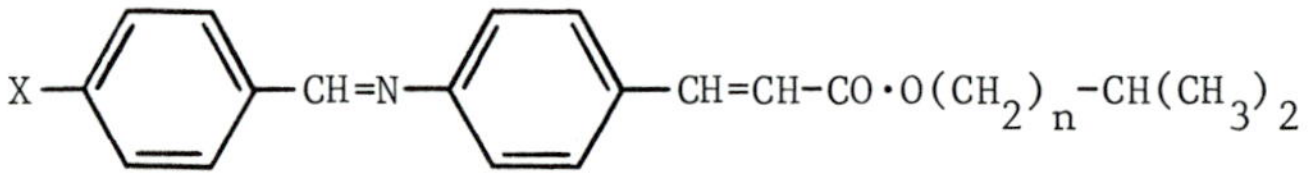

$$X-C_6H_4-CH{=}N-C_6H_4-CH{=}CH-CO\cdot O(CH_2)_n-CH(CH_3)_2 \quad (40)$$

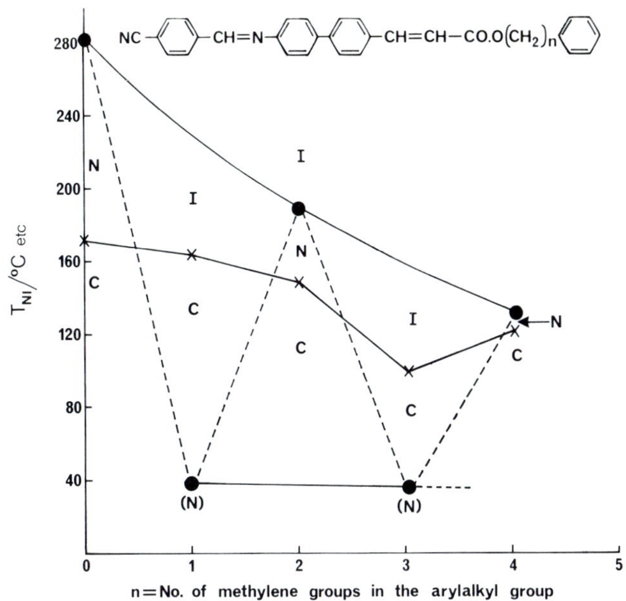

Fig. 7 *The high alternations of the N-I values result in an alternation between enantiotropic and monotropic N phases as n increases.*

corresponding n-alkyl series. The T_{NI} values are: n = 0, (134.5°); n = 1, 142.5°; n = 2, 110°; n = 3, 119.5°. The n value corresponding to a lower T_{NI} is now even, not odd, and this would be expected, since the protruding methyl group is on the second carbon from the end of the chain.

When a branching CH_3 group is moved progressively along a chain and finally onto the pentultimate carbon, consistent effects, exemplified by the following results, are obtained [43,44].

The 1-methyl group probably exerts a steric effect and has a

NC–⟨⟩–CH=N–⟨⟩–CH=CH–CO·OAlkyl

Alkyl	C-N or I	N -I
$CH_2CH_2CH_2CH_2CH_3$	73.5°	136.5°
$CH(CH_3)CH_2CH_2CH_2CH_3$	64.5°	(<20°)
$CH_2CH(CH_3)CH_2CH_2CH_3$	92.5°	112°
$CH_2CH_2CH(CH_3)CH_2CH_3$	82°	103°
$CH_2CH_2CH_2CH(CH_3)_2$	89°	119.5°

very large effect on T_{NI}, but in the 2-, 3-, or 4-position, the effect is much less, and in general diminishes as the branching position is moved towards the free end of the chain. Such results for ester alkyl groups have been confirmed for branching in alkoxy substituents [43] and in alkyl groups directly attached to aryl ring systems [45,46]. Since branched chain materials often have a lower melting point than the unbranched analogues, studies of branched chain mesogens have featured in the quest for low melting nematogens for applications [17,47].

Cholesteric Liquid Crystals

If we exclude from our consideration cholesteric liquid crystals derived from a range of different sterols and consider only (a) non-steryl materials which exhibit cholesteric phases because the molecules are nematogenic, but optically active or (b) cholesterogens in which only terminal groups in a *constant* steryl system are varied, then we have no need to define any new rules to relate the cholesteric properties to the molecular structure. Such materials are simply twisted nematics and the racemic modifications of type (a) cholesterogens have T_{NI} values identical to the T_{ChI} values of the individual optical isomers [48,49]. At least this is the situation for a considerable number of type (a) cholesterogens containing only *one* asymmetric centre, for which both the optical isomers and the racemic modifications have been studied. The situation is more complex when more than one asymmetric centre is present in the molecule. In one case [50], an ester molecule contained two chiral centres and had a Ch-I of 114° for the pure isomer, and an N-I of 118.5° for the racemic modification. The situation is different here because the racemic modification is now a physical mixture of two diastereoisomeric pairs of compounds.

The structural and geometric factors which influence T_{NI} values of nematogens are therefore exactly those that influence the T_{ChI} values of cholesterogens of types (a) and (b). For example, the cholesteric terminal group efficiency order is the same as for nematics [48], the T_{ChI} values for ω-phenylalkyl esters alternate very markedly [51], the T_{ChI} against chain length plots for homologous series of cholesterogens closely resemble those for T_{NI} temperatures, and the effects of lateral substituents on T_{ChI} values are similar to those on T_{NI} values.

In dealing with cholesterogens of the non-sterol type, a further important feature is the sense of the helix of the cholesteric phase that is produced. Recently we have completed a survey of a range of such materials [7]. Usually, but not always, the twisting power of the system diminishes as the chiral centre is moved away from the core structure, i.e., as n increases in system (41). Also if an oxygen re-

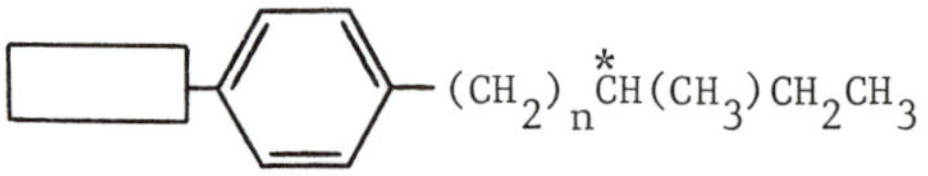

$$\square\!-\!C_6H_4\!-\!(CH_2)_n\overset{*}{C}H(CH_3)CH_2CH_3 \qquad (41)$$

places one of the CH_2 groups in $(CH_2)_n$, the twisting power always decreases. More important however is the consistent rule that if the asymmetric centre is S in absolute configuration, and the branching point is at an even number of atoms (E) from the core structure, the helix sense will be right handed (D), not left handed (L). We therefore have a simple rule of letters which combine as SED, SOL, REL, or ROD, where O refers to a branching point at an odd number of atoms from the core and R refers to the other absolute configuration of the asymmetric centre. This provides a valuable guiding rule which is also useful in predicting the behaviour of systems containing more than one chiral centre [7].

With regard to the effects of change in the molecular structure of the sterol skeleton on T_{ChI} values, helical pitch sense, and twisting power of sterol-type cholesterics, the complexities of the situation (recently reviewed [10]) are greater because of the subtleties of the stereochemistry of the fairly complex alicyclic system.

Concluding Comments on Nematics/Cholesterics

In the course of this account, some general, empirical rules relating the effects of molecular structural change to the nematic/cholesteric properties of mesogens, particularly the N- or Ch-I transition temperatures, have been developed and illustrated. These rules are of real value to practical chemists interested in the design of new materials with physical properties that are tailored to specific applications or to the requirements of physical chemists and physicists who wish to examine particular properties of mesophases. It is worth pointing out however, that our inability to predict with any certainty how melting points change with molecular structural change often frustrates the chemists' attempts to synthesise mesogens based on structures which could reasonably be expected to give liquid crystal properties. If the melting point turns out to be too high and supercooling of the melt is poor, no liquid crystal phase is formed.

For these reasons, any means that can help in assessing the likely potential of a structure to be mesogenic are of value in deciding upon the expenditure of synthetic effort which could ultimately be frustrated by a melting point problem. In this connection, recent studies by Destrade, Gasparoux, and Guillon [52] on the depolarised light scattering of liquids or solutions in an isotropic solvent are of interest. Using a polarised, monochromatic light beam and a liquid sample, one of the components of the intensity of light scattered in a direction at right angles to the incident beam is related to the optical anisotropy and to the orientational correlations of the molecules. An angular correlation function, J^A_{22}, may be deduced. By measuring the depolarised Rayleigh scattering and making comparisons over a number of suitably chosen materials as solutes, the contribution of each part of a molecule (central core, substituents) to the orientational correlations which lead to mesophase formation may be estimated. When J^A_{22} is 0, the molecules can be classed as isotropic, but the greater the positive value of J^A_{22} the more likely the compound is to be a mesogen. For example, core structures such as C_6H_5-CH=N-C_6H_5, C_6H_5-CO.O-C_6H_5, C_6H_5-C_6H_5, and C_6H_5-C≡C-C_6H_5 have J^A_{22} values >0.40 and give many mesogens with suitable end groups. On the other hand, C_6H_5-S-C_6H_5 has $J^A_{22} \approx 0$ and is not a potential core structure for mesophase formation, whereas C_6H_5-CO-C_6H_5 has $J^A_{22} \approx 0.34$ and is a possible core structure. At a given temperature, increases in the J^A_{22} values are observed when the core structure is substituted terminally, first at one end and then at the second end of the molecule, confirming earlier conclusions about the role of terminal groups in enhancing mesogenic properties. This technique seems particularly valuable, since it allows estimates of mesomorphic potential to be made using comparatively simple, and therefore often much more accessible, unsubstituted core structures.

References

1. G.W. Gray and P.A. Winsor in "*Liquid Crystals and Plastic Crystals*" (G.W. Gray and P.A. Winsor, eds.), Vol.1, Ellis Horwood, Chichester, England (1974).
2. K. Markau and W. Maier, *Chem. Ber.*, 95, 889 (1962).
3. W.H. de Jeu, *J. Phys. (Paris)*, **38**, 1265 (1977).
4. G.W. Gray, *J. Phys. (Paris)*, 36, 337 (1975); G.W. Gray, K.J. Harrison and J.A. Nash, *Electron. Lett.* 9, 130 (1973).
5. G.W. Gray and A. Mosley, unpublished results.
6. J.W. Goodby and G.W. Gray, *J. Phys. (Paris)*, 37, 18 (1976).
7. G.W. Gray and D.G. McDonnell, *Mol. Cryst. Liq. Cryst.*, 37, 189 (1976); *loc. cit.*, 34, 211 (1976).
8. G.W. Gray in *"Molecular Structure and the Properties of Liquid Crystals"*, Academic Press Inc., N.Y., p.131 (1962).
9. G.W. Gray,*"Molecular Structure and the Properties of Liquid Crystals"*, Academic Press Inc., N.Y. (1962).
10. G.W. Gray, *"Advances in Liquid Crystals"* (G.H. Brown, ed.) Vol.2, Academic Press Inc., N.Y. p.1. (1976).
11. D. Coates and G.W. Gray, *J. Chem. Soc. Chem. Commun.*, 514 (1975).
12. D. Coates and G.W. Gray, *Mol. Cryst. Liq. Cryst.*, 37, 249 (1976).
13. D. Coates and G.W. Gray, *J. Chem. Soc. Perkin II*, 7, 863 (1976).
14. G.W. Gray, K.J. Harrison, J.A. Nash, J. Constant, D.S. Hulme, J. Kirton, and E.P. Raynes, *Liquid Crystals and Ordered Fluids"*, (R.S. Porter and J.F. Johnson, eds.), Vol.2, p.617, Plenum Press, N.Y. (1973).
15. G.W. Gray, K.J. Harrison and J.A. Nash, *J. Chem. Soc. Chem. Commun.*, 431 (1974).
16. J. van der Veen, W.H. de Jeu, A.H. Grobben and J. Boven, *Mol. Cryst. Liq. Cryst.*, 17, 291 (1972); W.H. de Jeu and J. van der Veen, *Philips Res. Rep.* 27, 172 (1972); J. van der Veen, W.H. de Jeu, M.W.M. Wanninkhof and C.A.M. Tienhoven, *J. Phys. Chem.* 77, 2153 (1973); W.H. de Jeu and Th. W. Lathouwers, *Z. Naturforsch.*, A29, 905 (1974).
17. W.R. Young, A. Aviram, and R.J. Cox, *Angew. Chem.*, 83, 410 (1971); *I.B.M. Res. Rep.*, RC3559, (1974).

18. L. Verbit and R.L. Tuggey, *"Liquid Crystals and Ordered Fluids"*, (J.F. Johnson and R.S. Porter, eds.) Vol.2, p.307, Plenum Press, N.Y. 1973.
19. S. Malthéte, J. Leclerq, J. Gabart, J. Billard and J. Jacques, *C.R. Acad. Sci. Ser.* C, 273, 265 (1971).
20. L.E. Knaak, H.M. Rosenberg and M.P. Servê, *Mol. Cryst. Liq. Cryst.*, 17, 171 (1972).
21. H.B. Burgi and J.D. Dunitz, *Helv. chim. Acta*, 53, 1747 (1970); J. van der Veen and A.H. Grobben, *Mol. Cryst. Liq. Cryst.*, 15, 239 (1971).
22. J.M. Robertson and I. Woodward, *Proc. Roy. Soc. Ser.* A162, 568 (1937).
23. M.J.S. Dewar and R.S. Goldberg, *J. Org. Chem.*, 35, 2711 (1970); *J. Amer. Chem. Soc.*, 92, 1582 (1970).
24. M.J.S. Dewar, A.C. Griffin and R.M. Riddle in *"Liquid Crystals and Ordered Fluids"*, (J.F. Johnson and R.S. Porter, eds.), Vol.2, p.733, Plenum Press, N.Y. (1973); M.J.S. Dewar and R.M. Riddle *J. Amer. Chem. Soc.*, 97, 6658 (1975); M.J.S. Dewar and A.C. Griffin, *J. Amer. Chem. Soc.*, 97, 6662 (1975).
25. H.J. Deutscher, F. Kuschel, H. Schubert and D. Demus, Deutsche Demokratische Republik, Patent, D.O.S. Nr 24 29 093 (1975).
26. R. Eidenschink, D. Erdmann, J. Krause and L. Pohl, *Angew Chem.*, 89, 103 (1977).
27. D. Coates, G.W. Gray and D. McDonnell, unpublished results.
28. C. Weygand and R. Gabler, *Z. phys. Chem.*, B46, 270 (1940).
29. H. Schubert, R. Dehne and V. Uhlig, *Z. Chem.*, 12, 219 (1972).
30. G.W. Gray in *"Liquid Crystals and Plastic Crystals"*, (G.W. Gray and P.A. Winsor, eds.), Vol. 1, Chap. 4.1, Ellis Horwood, Chichester, England, (1974).
31. W.R. Young, I. Haller and D.C. Green, *I.B.M. Res. Rep.*, RC3827 (1972); *J. Org. Chem.*, 37, 3707 (1972); W.R. Young and D.C. Green, *I.B.M. Res. Rep.*, RC4121 (1972); *Mol. Cryst. Liq. Cryst.*, 26, 7, (1974).
32. A.C. Griffin, D.L. Wertz and A.C. Griffin, Jr., *Mol. Cryst. Liq. Cryst.*, 44, 267, (1978).
33. E.M. Barrall II and J.F. Johnson in *"Liquid Crystals and Plastic Crystals"* (G.W. Gray and P.A. Winsor, eds.), Vol.2, Chap.10, Ellis Horwood, Chichester, England, (1974).
34. I. Teucher, C.M. Paleos, and M.M. Labes, *Mol. Cryst. Liq. Cryst.*, 11, 187 (1970); H. Hirata, S.M. Waxman, I. Teucher and M.M. Labes, *Mol. Cryst. Liq. Cryst.*, 20, 343 (1973).
35. J. van der Veen and T.C.J.M. Hegge, *Angew. Chem.*, 86, 378 (1974).
36. D.J. Byron, G.W. Gray and B.M. Worrall, *J. Chem. Soc.*, 3706 (1965).
37. G.W. Gray, J.B. Hartley, A. Ibbotson and F. Marson, *J. Chem. Soc.*, 4359 (1955).
38. G.W. Gray and A. Mosley, *J. Chem. Soc. Perkin II*, 97, (1976).
39. G.W. Gray and D.G. McDonnell, unpublished results.
40. W.H. de Jeu, J. van der Veen and W.J.A. Goossens, *Solid State Commun.*, 12, 405 (1973).
41. W. Maier and A. Saupe, *Z. Naturforsch.*, 13a, 564 (1958); 14, 822 (1959); 15a, 287 (1960).
42. H. Stenschke, *Solid State Commun.*, 10, 653 (1972).
43. G.W. Gray and K.J. Harrison, *Symp. Faraday Soc.*, 5, 54 (1971).
44. G.W. Gray and K.J. Harrison, *Mol. Cryst. Liq. Cryst.*, 13, 37 (1971).
45. M.E. Neubert, L.T. Carlino, R. D'Sidocky and D.L. Fishel in *"Liquid Crystals and Ordered Fluids"* (J.F. Johnson and R.S. Porter, eds.), Vol. 2, p.293, Plenum Press, N.Y., (1973).
46. Y.Y. Hsu and D. Dolphin in *"Liquid Crystals and Ordered Fluids"* (J.F. Johnson and R.S. Porter, eds.), Vol.2, p.461, Plenum Press, N.Y., (1973).
47. J.E. Goldmacher, German Patent, 2,026,280 (1970).
48. G.W. Gray, *Mol. Cryst. Liq. Cryst.*, 7, 127 (1969).
49. M. Leclerq, J. Billard and J. Jacques, *Mol. Cryst. Liq. Cryst.*, 8, 367 (1969).
50. J.W. Goodby, G.W. Gray and D.G. McDonnell, *Mol. Cryst. Liq. Cryst.*, 34, 183 (1976).
51. R.D. Ennulat, *Mol. Cryst. Liq. Cryst.*, 8, 247 (1969); W. Elser, J.L.W. Pohlmann and P.R. Boyd, *Mol. Cryst. Liq. Cryst.*, 15, 175 (1971).
52. C. Destrade, H. Gasparoux and F. Guillon, *Mol. Cryst. Liq. Cryst.*, 40, 163 (1977).
53. D. Coates and G.W. Gray, *J. Phys. (Paris)*, 36, 365 (1975).

Chapter 2

INTERMOLECULAR FORCES

A. J. STONE

University Chemical Laboratory, Lensfield Road, Cambridge, CB2 1EW, England.

Introduction

The purpose of this Chapter is to survey the theory of intermolecular forces at a fairly elementary level, with emphasis on the orientational dependence of the potential, since this is the feature of particular relevance to studies of liquid crystals.

We begin, however, by looking at a typical intermolecular potential which is shown in figure 1 as a function of the distance r between the centres of the molecules. (By 'centre' we mean the centre of symmetry, if any, or the centre of mass, or any suitable point within the molecule. The choice is often arbitrary; e.g. for HD it might be the electrical centre, midway between the nuclei, or the centre of mass). Such a potential usually has a minimum, whose depth is conventionally denoted by ε at a value of r denoted by r_0. At larger values of r the potential decays to zero, while for smaller values of r it rises through zero at $r = \sigma$ and then rapidly to very high values. The depth of the well varies considerably, but for small closed-shell molecules it is typically of the order of a few kJ/mol, corresponding to a temperature of a few hundred kelvin, in accordance with the observed fact that solids liquefy

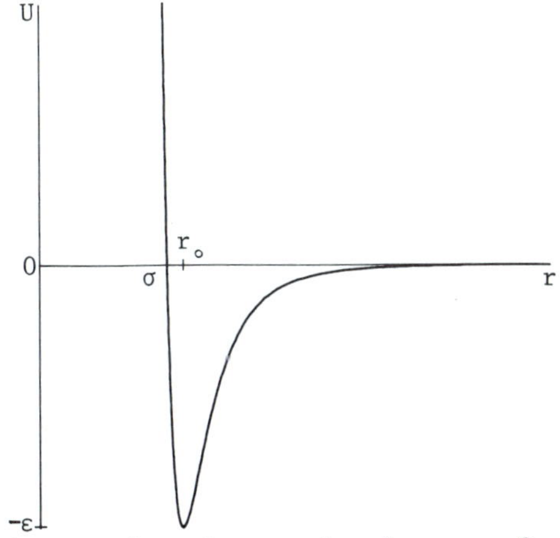

Fig. 1 *The energy U of interaction of two molecules as a function of the distance r between them.*

and vapourize at such temperatures.

The repulsion at short range is a consequence of the overlap of the electron clouds (it can be thought of as the energy needed to redistribute the electrons to avoid a breach of the Pauli exclusion principle, which forbids two electrons to be simultaneously in the same place with the same spin) and it therefore varies with distance roughly like $\exp(-r/\rho)$, where ρ is a distance of the order of σ. This energy of interaction is not pairwise additive: that is, the energy of interaction of three molecules A, B and C cannot be written as the sum of three pair interactions U_{AB}, U_{BC} and U_{AC}, because the presence of C modifies the interaction between A and B. These points are summarized in Table 1.

Table 1. *Contributions to the interaction energy of two molecules. L_A and L_B are the ranks of the lowest non-vanishing multipoles of the two molecules: 0 for a charge (monopole) 1 for a dipole, 2 for a quadrupole, etc.*

	Functional Form	Sign	Pairwise Additive?
Short range			
Overlap	$\exp(-r/\rho), \rho \approx \sigma$	± but ++ for small r	No
Long range			
Electrostatic	r^{-n}, $n>L_A+L_B+1$	±	Yes
Induction	r^{-n}, $n>\min(2L_A,2L_B)+4$	- (for ground states)	No
Dispersion	r^{-n}, $n>6$	- (for ground states)	Yes, approximately
Resonance	r^{-n}, $n>3$	± (degenerate states only)	No
Magnetic	r^{-n}, $n>3$	± weak	Yes

The interaction energy at long range can be broken down into a number of distinct contributions, with characteristics shown in Table 1. The electrostatic term is a straightforward interaction between the charge distributions of the two molecules; the induction and dispersion terms arise from the perturbation of the charge distribution of one molecule in the presence of the other, as we shall see in the next section. The resonance term arises only when at least one of the molecules is in a degenerate state, and will be ignored here. Likewise the magnetic term contributes only when there is electronic degeneracy, and is even then very weak; there is a further term arising from the magnetic interaction of any nuclear spins which may be present, but this is several orders of magnitude weaker still.

We have so far assumed that the relative orientations of the two molecules have been kept fixed while the intermolecular distance r is varied. For the simplest case, where the 'molecules' are both closed-shell atoms, there is no other coordinate needed to describe the system, and it is possible to obtain the intermolecular potential experimentally as a function of r in a fairly simple yet extremely accurate form [1]. In general, however, when the two molecules contain N_A and N_B atoms, there are $(3N_A + 3N_B - 6)$ degrees of freedom in all, and even if we disregard the dependence of the potential on the internal degrees of freedom $\{(3N_A-6+3N_B - 6)$ for non-linear molecules$\}$ we are left with six variables describing the relative position of the molecules. If the molecules are linear, or if, as in simple models of liquid crystals, they can be treated as if they were cylindrically symmetric, then the potential is independent of rotation of either molecule about its symmetry axis and the number of variables needed drops to four. This is still a substantial number, and part of the problem is to find a compact, efficient and readily comprehensible way of describing the intermolecular potential. We return to this problem later, but first review the perturbation-theory description of the long-range potential.

Long-range Forces

This is a brief summary of the application of perturbation theory to two interacting molecules at long range. Fuller accounts may be found, for example, in the articles by Buckingham [2] and the books by Margenau and Kestner [3] and Hirschfelder, Curtiss and Bird [4].

For the isolated, non-interacting molecules we have

$$H_A|m_A\rangle = W_m^A|m_A\rangle,$$

$$H_B|n_B\rangle = W_n^B|n_B\rangle. \tag{1}$$

H_A is the internal (vibronic) hamiltonian of molecule A, and $|m_A\rangle$ is one of the vibronic states of A. For the joint system, in the absence of any interactions, we have

$$(H_A + H_B)|m_An_B\rangle = (W_m^A + W_n^B)|m_An_B\rangle, \tag{2}$$

where $|m_A n_B\rangle \equiv |m_A\rangle|n_B\rangle$. Note that although the state $|m_A\rangle$ must be antisymmetric with respect to exchange of electrons within molecule A, and $|n_B\rangle$ with respect to exchange within molecule B, we take no account of exchange of electrons in A with electrons in B. This neglect of intermolecular electron exchange is valid, and introduces no errors, if the molecules are so far apart that their electron clouds do not overlap. It is the fact that electron exchange cannot be neglected at short or even intermediate range which makes the theory so difficult there [5].

The perturbation arising from the interaction of the two molecules is now

$$H' = \sum_{ab} e_a e_b / 4\pi\varepsilon_o r_{ab}, \tag{3}$$

where e_a and e_b are the charges of particles a and b, r_{ab} is the distance between them, and the sum is taken over all particles (electrons and nuclei) a in molecule A and b in molecule B. Then standard perturbation theory gives

$$W = W_A + W_B + U \tag{4}$$

with

$$U = \langle o_A o_B | H' | o_A o_B \rangle - \sum_{m_A n_B}{}' |\langle o_A o_B | H' | m_A n_B \rangle|^2 / (\Delta W_m{}^A + \Delta W_n{}^B) + \ldots, \tag{5}$$

where $|o_A o_B\rangle$ is the ground state of the combined system, the prime indicates that it is excluded from the sum and $\Delta W_m{}^A = W_m{}^A - W_o{}^A$.

This is conventionally separated into the contributions described in Table 1, namely the electrostatic energy:

$$U_{elect.} = \langle o_A o_B | H' | o_A o_B \rangle; \tag{6}$$

the induction energy (comprising two terms, describing respectively the distortion of the charge distribution of A by that of B, and *vice versa*):

$$U_{ind.} = - \sum_{m \neq o} |\langle o_A o_B | H' | m_A o_B \rangle|^2 / \Delta W_m{}^A - \sum_{n \neq o} |\langle o_A o_B | H' | o_A n_B \rangle|^2 / \Delta W_n{}^B \tag{7}$$

and the dispersion energy:

$$U_{disp.} = - \sum_{m \neq o} \sum_{n \neq o} |\langle o_A o_B | H' | m_A n_B \rangle|^2 / (\Delta W_m{}^A + \Delta W_n{}^B). \tag{8}$$

To proceed further we need a more helpful expression for H'. We can write it as

$$H' = \sum_a e_a \phi^B(\mathbf{r}_a), \tag{9}$$

where ϕ^B is the potential due to the charge distribution of B. Expanding this in a Taylor series about an origin at the centre of A yields

$$H' = \sum_a e_a \phi^B(A) - \sum_a e_a r_{a\alpha} F_\alpha{}^B(A) - \tfrac{1}{2}\sum_a e_a r_{a\alpha} r_{a\beta} F_{\alpha\beta}{}^B(A) - \cdots,$$

where $\phi^B(A)$ is now the potential at the centre of A, $F_\alpha{}^B(A) = -\nabla_\alpha^A \phi^B(A)$ is the electric field there, $F_{\alpha\beta}{}^B(A) = -\nabla_\alpha{}^A \nabla_\beta{}^A \phi^B(A)$ is the field gradient, and so on. Now we introduce the multipole moments of the charge distribution of A. The charge, or monopole moment, is

$$q^A = \sum_a e_a; \tag{10a}$$

the dipole moment is

$$\hat{\mu}_\alpha{}^A = \sum_a e_a r_{a\alpha}; \tag{10b}$$

the quadrupole moment is

$$\hat{\Theta}_{\alpha\beta}{}^A = \tfrac{1}{2}\sum_a 3e_a (r_{a\alpha} r_{a\beta} - r_a{}^2 \delta_{\alpha\beta}); \tag{10c}$$

and generally the nth moment (the 2^n-pole moment) is a tensor with n subscripts:

$$\hat{\xi}^A_{\alpha\beta\ldots\nu} = (-)^n (n!)^{-1} \sum_a e_a r_a^{2n+1}\, \partial/\partial r_{a\alpha} . \partial/\partial r_{a\beta} \ldots . \partial/\partial r_{a\nu}\, (1/r_a).$$

The definition (10c) ensures that $\Theta_{\alpha\alpha} = 0$, so that the quadrupole moment has zero scalar (spherically symmetric) part and describes only the anisotropy of the charge distribution. Higher moments are traceless with respect to any pair of suffixes, and fully symmetric with respect to permutation of suffixes. These properties imply that the 2^L-pole moment has 2L+1 independent components in general, though some of them may be zero if the molecule has any symmetry. Note that

$$\delta_{\alpha\beta} F_{\alpha\beta}{}^B(A) = F_{\alpha\alpha}{}^B(A) = -\nabla^2 \phi^B|_A = +\varepsilon_0{}^{-1} \rho^B(A),$$

where $\rho^B(A)$ is the charge density of B at the centre of A and is zero since the charge distributions do not overlap.

Using these multipole moments, we find

$$H' = q^A \phi^B(A) - \hat{\mu}_\alpha{}^A F_\alpha{}^B(A) - \tfrac{1}{3}\Theta_{\alpha\beta}{}^B\, F_{\alpha\beta}{}^B(A) - \cdots \quad . \tag{11}$$

Now the potential at A due to the charges of B is

$$\phi^B(A) = \sum_b e_b|r + \mathbf{r}_b|^{-1},$$

(where $\mathbf{r}_b$ is the position vector of particle b relative to the centre of B, and $\mathbf{r} = \mathbf{r}_B - \mathbf{r}_A$) which we can expand in a Taylor series:

$$\phi^B(A) = (4\pi\varepsilon_o)^{-1}\sum_b e_b\{r^{-1} + r_{b\alpha}\nabla_\alpha{}^B(r^{-1}) + \tfrac{1}{2}r_{b\alpha}r_{b\beta}\nabla_\alpha{}^B\nabla_\beta{}^B(r^{-1})+\ldots\},$$

$$= (4\pi\varepsilon_o)^{-1}\{q^B r^{-1} + \hat{\mu}_\alpha{}^B\nabla_\alpha{}^B(r^{-1}) + \tfrac{1}{3}\hat{\Theta}_{\alpha\beta}{}^B\nabla_\alpha{}^B\nabla_\beta{}^B(r^{-1})+\ldots\}. \quad (12)$$

Putting eqs. (11) and (12) together, and using the interaction tensors defined by

$$T = (4\pi\varepsilon_o)^{-1}r^{-1},$$

$$T_\alpha = (4\pi\varepsilon_o)^{-1}\nabla_\alpha{}^B = -(4\pi\varepsilon_o)^{-1}\nabla_\alpha{}^A r^{-1},$$

$$= (4\pi\varepsilon_o)^{-1}r_\alpha/r^3,$$

$$T_{\alpha\beta} = (4\pi\varepsilon_o)^{-1}\nabla_\alpha{}^B\nabla_\alpha{}^B r^{-1},$$

$$= (4\pi\varepsilon_o)^{-1}(3r_\alpha r_\beta - r^2\delta_{\alpha\beta})/r^5, \quad (13)$$

etc., we obtain

$$H' = q^A q^B T + (q^A\hat{\mu}_\alpha{}^B - \hat{\mu}_\alpha{}^A q^B)T_\alpha + (\tfrac{1}{3}q^A\hat{\Theta}_{\alpha\beta}{}^B - \hat{\mu}_\alpha{}^A\hat{\mu}_\beta{}^B + \tfrac{1}{3}\hat{\Theta}_{\alpha\beta}{}^A q^B)T_{\alpha\beta} + \ldots \quad .(14)$$

Eq. (14) is the multipole expansion of the perturbation, and it converges rapidly if the distance r between the molecules is much larger than the distance from the centre of either molecule to the edge of its charge distribution. The multipole moments of each molecule are referred to an origin at the centre of that molecule; if the origin is moved the individual terms in the expansion will change and convergence may be impaired, though the sum will be unaffected.

It is often convenient to express the multipole expansion in terms, not of cartesian multipole moment tensors, but of spherical tensor equivalents. The 2^L-pole moment has 2L+1 independent components, which are

$$Q_{L,m}{}^A = \sum_a e_a r_a{}^L C_{L,m}(\mathbf{r}_a), \quad m = -L,-L+1\ldots,L, \quad (15)$$

where $C_{L,m}(\mathbf{r}_a)$ is a modified spherical harmonic:

$$C_{L,m}(\theta,\phi) = \{4\pi/(2L+1)\}^{\frac{1}{2}}Y_{L,m}(\theta,\phi), \quad (16)$$

and the angles θ and ϕ implied in $C_{L,m}(\mathbf{r}_a)$ are the polar angles describing the direction of the vector $\mathbf{r}_a$. The modified spherical harmonics have the property that

$$C_{L,m}(0,0) = \delta_{m0}. \tag{17}$$

In spherical tensor form the perturbation becomes [6,7]

$$H' = (4\pi\varepsilon_o)^{-1} \sum_{LmL'm'M} (-)^L [L;L'] r^{-L-L'-1} \hat{Q}^A_{L,m} \hat{Q}^B_{L',m'} C_{L+L',M}(r) \begin{pmatrix} L & L' & L+L' \\ m & m' & M \end{pmatrix}, \tag{18}$$

where

$$[L;L'] = \{(2L+2L'+1)!/(2L)!(2L')!\}^{\frac{1}{2}},$$

and $\begin{pmatrix} L & L' & L+L' \\ m & m' & M \end{pmatrix}$ is a Wigner 3-j symbol which is related to the Clebsch-Gordan coefficient [8]. Now we may substitute either eqs. (13) or (18) in the first-order perturbation expression (6) for $U_{\text{elect.}}$. We obtain an expression involving just the expectation values of the multipole moments in the ground states of the two molecules:

$$\begin{aligned} U_{\text{elect.}} &= q^A q^B T + (q^A \mu_\alpha{}^B - \mu_\alpha{}^A q^B) T_\alpha \\ &+ (\tfrac{1}{3} q^A \Theta_{\alpha\beta}{}^B - \mu_\alpha{}^A \mu_\beta{}^B + \tfrac{1}{3} \Theta^A_{\alpha\beta} q^B) T_{\alpha\beta} + \ldots, \\ &= (4\pi\varepsilon_o)^{-1} \sum_{LmL'm'M} (-)^L [L;L'] r^{-L-L'-1} \\ &\times Q_{L,m} Q_{L',m'} C_{L+L',M}(\mathbf{r}) \begin{pmatrix} L & L' & L+L' \\ m & m' & M \end{pmatrix}. \end{aligned} \tag{19}$$

The multipole moments which appear here are referred to laboratory axes, and it is usually more convenient to refer them to axes fixed in the molecule. This is readily done:

$$Q_{L,m}{}^A = \sum_k \tilde{Q}_{L,k}{}^A D^L_{m,k}(\Omega_A)^*,$$

where the tilde denotes tensor elements evaluated in molecular axes, $\Omega_A \equiv (\alpha_A \beta_A \gamma_A)$ is a set of Euler angles describing the orientation of the molecule, and $D^L_{m,k}$ is a Wigner rotation matrix defined in Chapter 3 [8]. Then eq. (19) becomes

$$U_{\text{elect.}} = (4\pi\varepsilon_o)^{-1} \sum_{LL'kk'} (-)^{L+L'} [L;L'] r^{-L-L'-1} \tilde{Q}^A_{L,k} \tilde{Q}_{L',k'}{}^B S^{kk'}_{LL'L+L'}, \tag{20}$$

where the function

$$S^{kk'}_{LL'J} = (i)^{L-L'-J} \sum_{mm'M} D^L_{m,k}(\Omega_A)^* D^{L'}_{m',k'}(\Omega_B)^* C_{J,M}(\mathbf{r}) \begin{pmatrix} L & L' & J \\ m & m' & M \end{pmatrix} \tag{21}$$

contains all the orientational dependence of this part of the potential.

In the induction energy we obtain terms of the form

$$r^{-L_1-L_2-L_1'-L_2'-2} Q^A_{L_1,m_1} Q^A_{L_2,m_2} \sum_{n\neq o} \langle o_B|Q_{L_1',m_1'}|n_B\rangle\langle n_B|Q_{L_2',m_2'}|o_B\rangle/\Delta W^B_n. \quad (22)$$

Here the summation is a polarisability; it describes the magnitude of the $2^{L_2'}$-pole moment induced by the L_1'th gradient of the electrostatic potential (or the $2^{L_1'}$-pole moment induced by the L_2'th gradient of the potential). Thus we can understand (22) physically as describing the induction of a $2^{L_2'}$-pole moment in B by the L_1'th derivative at B of the potential due to the 2^{L_2}-pole moment of A, and the interaction of this induced moment with the 2^{L_1}-pole moment of A.

The dispersion energy involves only the states where both molecules are excited. It can be reduced to an expression involving an integral over products of polarizabilities at imaginary frequency and the leading term is proportional to r^{-6} [9]. In cartesian tensor form

$$U_{disp.} = -(\hbar/2\pi)T_{\alpha\beta}T_{\gamma\delta}\int_0^\infty d\nu\ \alpha^A_{\alpha\gamma}(i\nu)\alpha^B_{\beta\delta}(i\nu), \quad (23)$$

where

$$\alpha^A_{\alpha\gamma}(i\nu) = 2\sum_{n\neq o} [\Delta W^A_n/((\Delta W^A_n)^2+\hbar^2\nu^2)]\mathrm{Re}(\langle o_A|\hat{\mu}_\alpha|n_A\rangle\langle n_A|\hat{\mu}_\gamma|o_A\rangle). \quad (24)$$

Alternatively, the Unsöld approximation leads to an expression involving products of static polarizabilities, whose leading term was first obtained by London [10]:

$$U_{disp.} \approx -\tfrac{1}{4}\overline{\Delta W^A}\ \overline{\Delta W^B}/(\overline{\Delta W^A}+\overline{W^B})T_{\alpha\beta}T_{\gamma\delta}\alpha^A_{\alpha\gamma}\alpha^B_{\beta\delta} + \ldots ,$$

where $\overline{\Delta W^A}$ is some average of the excitation energies ΔW^A_m. For atoms, where $\alpha_{\alpha\gamma} = \alpha\ \delta_{\alpha\gamma}$, this reduces to the London formula

$$U_{disp} \approx \{3\alpha^A\alpha^B/2(4\pi\varepsilon_0)^2r^6\}\overline{\Delta W^A}\,\overline{\Delta W^B}|(\overline{\Delta W^A}+\overline{\Delta W^B})+\ldots . \quad (25)$$

The induction and dispersion contributions to the energy can both be put into a form like (20), as a sum of products of orientation dependent S-functions with coefficients depending on the properties of the molecules referred to molecular axes. For example, the contribution to the dispersion energy from the dipole-dipole polarizability takes the following form for linear (or cylindically symmetrical) molecules in the London approximation [6,7]:

$$U_{disp.} \approx -(15/2)(4\pi\varepsilon_o)^{-2} r^{-6}\, \overline{\Delta W^A}\, \overline{\Delta W^B} / (\overline{\Delta W^A} + \overline{\Delta W^B})$$

$$\times \sum_{LL'J} \alpha_L{}^A \alpha_{L'}{}^B S^{00}_{LL'J}\; i^{-L-L'-J} \begin{pmatrix} 2 & 2 & J \\ 0 & 0 & 0 \end{pmatrix} \begin{Bmatrix} 1 & 1 & L \\ 1 & 1 & L' \\ 2 & 2 & J \end{Bmatrix}, \quad (26)$$

where the quantity in braces is a Wigner 9-j symbol, and

$$\alpha_o = -\sqrt{3}\,\bar{\alpha} = -\sqrt{\tfrac{1}{3}}(\alpha_{||} + 2\alpha_{\perp}),$$

$$\alpha_2 = (10/3)^{\frac{1}{2}}\,\Delta\alpha = (10/3)^{\frac{1}{2}}(\alpha_{||} - \alpha_{\perp}),$$

are the isotropic and anisotropic components of the polarizability in spherical tensor form. The indices L and L' take only the values of 0 and 2, and the index J is limited to even values between $|L-L'|$ and $|L+L'|$. Consequently there are contributions with radial dependence r^{-6} whose angular dependence is described by S_{000}, S_{022}, S_{202}, S_{220}, S_{222} and S_{224}. Of these, the S_{000} term is independent of orientation and is the familiar London expression (25). S_{022} is independent of the orientation of molecule 1, and S_{202} is independent of the orientation of molecule 2, leaving the terms S_{220}, S_{222} and S_{224} which depend on the orientation of both molecules. These are the terms, proportional to the product $\alpha_2{}^A\alpha_2{}^B$ of the polarizability anisotropy, which provide the orienting potential in the Maier-Saupe theory of the nematic mesophase [10] discussed in Chapter 4.

Higher polarizabilities yield dispersion energy terms which vary with distance like r^{-8}, r^{-10}, etc. These may be important at short range - they contribute for example about one-sixth as much as the r^{-6} term to the energy of two argon atoms at the minimum-energy separation.

The application of such formulae to large molecules such as liquid crystals runs into several difficulties. The accurate formulae call for the polarizabilities over the full range of imaginary frequencies, and although they can be evaluated in principle the task is not an easy one. It is known {and indeed is apparent from the formula (24)} that $\alpha(i\nu)$ is a real function of ν, and decreases monotonically to zero as ν runs from zero to infinity. Various empirical formulae have been produced to avoid the difficulty of a complete calculation - see the discussion by Claverie [12].

A more fundamental difficulty is that the multipole expansion, on which this approach is based, diverges if the molecules are so close that their charge distributions can overlap at some orientation. For highly non-spherical molecules there are then large accessible regions of configuration space, including probably the region around the potential minimum, where the multipole expansion is invalid. Stiles [13] and Perram and Stiles [14] have suggested an alternative approach using ellipsoidal harmonics, which extends the region of convergence substantially, but the formulae become extremely cumbersome and the method is not sufficiently fully developed to be generally useful at present.

Short-range Forces

At short range, then, the multipole expansion must be abandoned, and it becomes necessary to return to the original interaction hamiltonian of eq. (3). Further difficulties arise, however, from the fact that the molecular charge distributions now overlap, so that proper account must be taken of the indistinguishability of the electrons. Indeed, although the total hamiltonian (H_A+H_B+H') is well-defined, there is now no natural way to define the unperturbed hamiltonian or the unperturbed wavefunctions, since the obvious choice for the former is (H_A+H_B) which is not symmetric with respect to electron permutations, and the obvious choice for the latter is the set of functions $|m\rangle_A|n_B\rangle$, which are not antisymmetric, and if antisymmetrized are not orthogonal, so that they cannot be eigenfunctions of a hermitian zeroth-order hamiltonian. Furthermore, they constitute an overcomplete set of functions. Indeed, the very notion of the perturbation expansion becomes rather shaky, since the 'order' of a particular term cannot be defined unambiguously [5]. Some progress has been made with perturbation methods, but they cannot be used for large molecules except at a very low level of accuracy.

The alternative approach is to use a variational method, and to compare the energy obtained from a 'supermolecule' calculation on the entire interacting system with the energies of the isolated molecules. This too has its difficulties. The intermolecular potential (of the order of 10^{-3} hartree) is obtained by subtracting two quantities of the order of several hundred hartree, for molecules of quite modest size, and this type of procedure leads to considerable loss of precision. Moreover, in calculations where the basis set is not very good, the presence of the basis functions of molecule B can allow a better description of the molecular orbitals of molecule A, so that the energy of the combined system is decreased spuriously [15]. This is the so-called 'ghost orbital effect'. It is therefore necessary to recalculate the energy of the 'isolated' molecules for each orientation, each in the presence of the basis functions (but not of course the electrons and nuclei) of the other. This not only increases the labour considerably, but has an adverse effect on precision, since the baseline for the intermolecular energy, namely the energy of the individual molecules, is not a constant but itself depends on the configuration.

In view of all these difficulties, it is not surprising that few calculations of any quality have been done to find the interaction potential between pairs of molecules, and accurate calculations for liquid crystal molecules seem totally inaccessible at present.

Another approach is due to Gordon and Kim [16] who take the energy to be the integral of an energy density which is a local function of electron density. This is valid for a uniform electron gas, for which the functional form of the energy density in terms of electron density is known. The electron density for a pair of overlapping closed-shell molecules is assumed to be the sum of the electron densities for the individual molecules and since the energy expression is non-linear this leads to an interaction energy. Unfortunately this method, although very successful for atoms or very small molecules, seems unlikely at present to be easily applicable to larger molecules.

Phenomenological Descriptions of the Intermolecular Potential

Ab initio theoretical methods therefore provide a good understanding of the physical processes by which molecules interact, but have severe limitations as a source of accurate numerical potentials. Consequently many workers have used phenomenological descriptions, either as a straightforward model unrelated to any particular pair of molecules, or as a basis for describing, by means of adjustable parameters fitted to experimental data, the actual potential between two molecules.

One popular form is the atom-atom potential,

$$U = \sum_{ab} \varepsilon_{ab} f(r_{ab}/\sigma_{ab}), \qquad (27)$$

where ε_{ab} is an energy scaling parameter and σ_{ab} a distance scaling parameter for the interaction between atoms a on molecule A and b on molecule B. The function f is often taken to be a Lennard-Jones potential

$$f = 4\{(r/\sigma)^{-12} - (r/\sigma)^{-6}\} \qquad (28)$$

in which case the attractive dispersion force is also included in a crude way. For small molecules this provides a reasonable description, though it may be necessary to extend the sum to additional force centres to provide a more accurate description. Raich and Gillis [17] found it necessary, in describing the interaction between nitrogen atoms, to take three force centres on each molecule, two at the atoms and one at the centre of the bond. For larger molecules, on the other hand, it may be necessary to lump several atoms into a single force centre to obtain a manageable description. Evans and Watts [18], in describing the interaction between two benzene molecules, used six centres on each molecule not only for the repulsive part of the potential but also for the long-range part.

Another approach is due to Kihara [19]. Here the potential is expressed in the form

$$U = \varepsilon f(s/\sigma), \tag{29}$$

where s is the distance, not between molecular centres, but between a pair of convex hard cores representing the two molecules. The difficulty presented by this model is that the calculation of the distance s is difficult and time-consuming; also ε is wrongly taken to be independent of orientation.

Berne and Pechukas [20] have proposed a 'gaussian overlap model' in which the overlap between two ellipsoidal charge distributions is first calculated. If the charge has a prolate ellipsoidal gaussian distribution, given in local axes by

$$\rho(\mathbf{r}) = \exp\{-z^2/\sigma_{\parallel}{}^2-(x^2+y^2)/\sigma_{\perp}{}^2\}, \tag{30}$$

with $\sigma_{\parallel} > \sigma_{\perp}$, then the overlap between two such distributions is of the form

$$S(\Omega,r) = \varepsilon(\Omega)\ \exp\{-r^2/\sigma(\Omega)^2\}. \tag{31}$$

Here r is the distance between the centres of the charge distributions, and $\varepsilon(\Omega)$ and $\sigma(\Omega)$ are functions of the relative orientation of the two molecules. Explicitly,

$$\varepsilon(\Omega) = \varepsilon_0[1-\chi^2(\hat{\mathbf{z}}.\hat{z}_2)^2]^{-\frac{1}{2}}, \tag{32}$$

$$\sigma(\Omega)=\sigma_0\left[1-\tfrac{1}{2}\chi[(\hat{\mathbf{r}}.\hat{\mathbf{z}}_1+\mathbf{r}.\hat{z}_2)^2\{1+\chi(\hat{\mathbf{z}}_1.\hat{\mathbf{z}}_2)\}^{-1}+\{(\hat{\mathbf{r}}.\hat{\mathbf{z}}_1-\hat{\mathbf{r}}.\hat{\mathbf{z}}_2)^2\{1-\chi(\hat{\mathbf{z}}_1.\hat{\mathbf{z}}_2)\}^{-1}]\right]^{-\frac{1}{2}} \tag{33}$$

where $\hat{\mathbf{z}}_1$ and $\hat{\mathbf{z}}_2$ are unit vectors along the axes of the molecules, $\hat{\mathbf{r}}=\mathbf{r}/r$ is a unit vector in the direction of $\mathbf{r}$, and

$$\chi = (\sigma_{\parallel}{}^2-\sigma_{\perp}{}^2)/(\sigma_{\parallel}{}^2+\sigma_{\perp}{}^2).$$

Now Berne and Pechukas [20] suggest that the functional forms $\varepsilon(\Omega)$ and $\sigma(\Omega)$ may give a reasonable description of the well-depth and the van der Waals 'diameter' as functions of orientation so that the potential could be written in the form

$$U = \varepsilon(\Omega)f\{r/\sigma(\Omega)\}. \tag{34}$$

Kushick and Berne [21] have used such a potential in a molecular dynamics simulation of a liquid crystal for both two and three-dimensions. This and the Monte Carlo technique are described in Chapter 9.

Potential functions of the general form of (34) are known as Corner potentials, after their first proposer [22]. Walmsley [23] has given a recent account of such potentials, and it seems likely that a Corner description of the short-range forces, together with a multipole expansion of the long-range terms, will provide a suitable framework for the development of reasonably reliable potential functions for large molecules. There is however a long way to go before such potential functions make the transition from idealized models to representations, however crude, of a real potential. One unsatisfactory feature of the Corner potential (34), for example, is that the potential as a function of r is simply scaled according to $\sigma(\Omega)$, so that where $\sigma(\Omega)$ is large the potential is softer as we see from figure 2. Another way of expressing this deficiency is that the force, $-\partial U/\partial r$, is inversely proportional to $\sigma(\Omega)$:

$$-\partial U/\partial r = -(\varepsilon/\sigma)f'(r/\sigma). \tag{35}$$

A potential function which might be more suitable, and which at least does not suffer from this particular deficiency, would be one of the form

$$U = \varepsilon(\Omega)f\{r-\sigma(\Omega)\}, \tag{36}$$

where the potential is shifted rather than scaled.

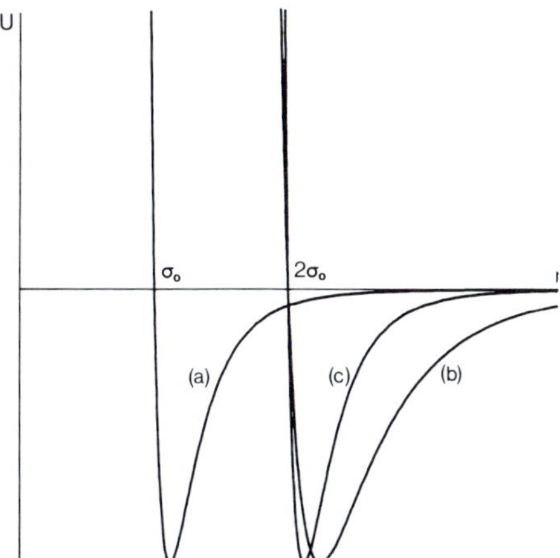

Fig. 2 *Scaled vs. shifted potential. (a) shows some function $U=f(r)$ with $\sigma=\sigma_0$; (b) shows the function $f(r/2)$ obtained by scaling to $\sigma=2\sigma_0$; (c) shows the function $f(r-\sigma_0)$ obtained by shifting to $\sigma=2\sigma_0$.*

Description of Orientation Dependent Functions

Various of the empirical potential functions are expressed in terms of other functions such as $\sigma(\Omega)$ and $\varepsilon(\Omega)$ of the relative orientation Ω of the two molecules. Several coordinate systems and sets of expansion functions have been used to describe such functions. Most of them were proposed initially for expanding the intermolecular potential itself, in which case the coefficients in the expansion are still functions of r.

Pople [24] described the potential between two linear molecules in the form

$$U = \sum_{L_1 L_2 m} u_{L_1 L_2 : m}(r) C_{L_1,m}(\theta_1 \phi_1) C_{L_2,m}(\theta_2 \phi_2), \quad (37)$$

where the angles θ_1, ϕ_1, θ_2 and ϕ_2 are defined in figure 3. For (37) to be real, the condition $(u_{L_1 L_2 : m})^* = u_{L_1 L_2 : -m}$ must be imposed. If molecule 1 is centrosymmetric as well as linear then $u_{L_1 L_2 : m} = 0$ for L_1 odd; similarly for molecule 2. This function contains five variables but is independent of $(\phi_1 - \phi_2)$; and it is manifestly invariant with respect to a change of laboratory axes, since only $(\phi_1 - \phi_2)$ depends on an arbitrary definition of axes.

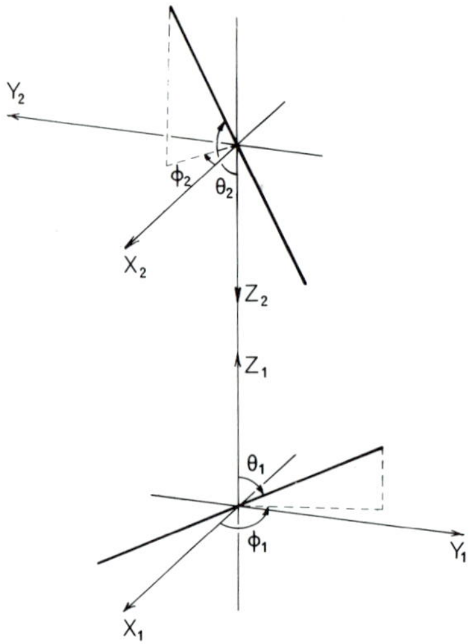

Fig. 3 *Definition of angular variables for two linear molecules (Pople [24]).*

Steele [25] generalised this type of expansion to deal with non-linear molecules. Here the orientation is described by Euler angles, β and α describe the direction of an arbitrary axis within the molecule (usually the highest-order symmetry axis) like θ and ϕ respectively in figure 3, and γ describes the angle of rotation of the molecule about that axis as in figure 4. Note that Steele took the Z_2 axis parallel to Z_1, whereas Pople took Z_2 antiparallel to Z_1; Steele's definition leads to an asymmetry between the two molecules but is perhaps more natural nevertheless. In terms of these coordinates, the energy becomes

$$U = \sum_{L_1 L_2 m k_1 k_2} u_{L_1 L_2 : k_1 k_2 m} D^{L_1}_{m,k_1}(\Omega_1) D^{L_2}_{-m,k_2}(\Omega_2). \quad (38)$$

Here there are seven variables: $\Omega_1 \equiv (\alpha_1, \beta_1, \gamma_1)$, $\Omega_2 \equiv (\alpha_2, \beta_2, \gamma_2)$ and r, but the energy is independent of $(\alpha_1 + \alpha_2)$. Again the variables are all internal and the expression (38) is manifestly invariant with respect to change of laboratory axes.

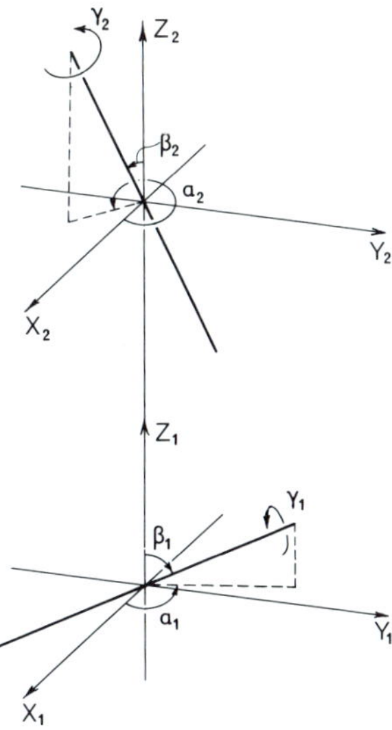

Fig. 4 *Definition of angular variables for two non-linear molecules (Steele [25]).*

It is not always convenient, however, to refer the angles to a local frame of reference. It may well be more natural to use laboratory axes, defined perhaps in relation to an external field or to the director in a liquid crystal. In this case the angles can be defined as in figure 5, and the energy written as

$$U = \sum u(LL'J,kk',mm'M;\ r) D^{L}_{m,k}(\Omega_1) D_{m',k'}(\Omega_2) C_{J,M}(\theta,\phi). \qquad (39)$$

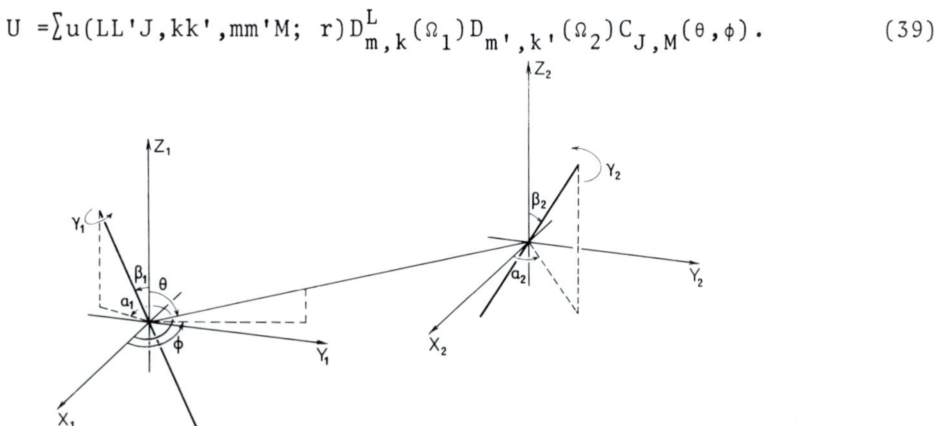

Fig. 5 *General definition of angular variables for two non-linear molecules.*

Here the coordinates do refer explicitly to an external frame of reference, and (39) is not as it stands invariant under a change of frame. However it is not difficult to show [8,26] that for invariance the terms in (39) must appear only in expressions of the form which we have already encountered in eq. (21), namely

$$S^{kk'}_{LL'J} = (i)^{L-L'-J} \sum_{mm'M} \begin{pmatrix} L & L' & J \\ m & m' & M \end{pmatrix} D^{L}_{m,k}(\Omega_1) D^{L'}_{m',k'}(\Omega_2) C_{J,M}(\theta,\phi) \qquad (40)$$

and then the energy can be expressed as

$$U = \sum u^{kk'}_{LL'J}(r) S^{kk'}_{LL'J}(\Omega), \tag{41}$$

where $\Omega \equiv (\Omega_1, \Omega_2, \theta, \phi)$. The phase factor $(i)^{L-L'-J}$ in eq. (40) makes $S^{kk'}_{LL'J}$ invariant under exchange of the roles of the two molecules, and makes $S^{00}_{LL'J}$ real.

This type of expression was first used, with a different phase and normalisation, in nuclear theory by Biedenharn and Rose [27] and for the distribution function by Blum and Toruella [26]; its application to intermolecular forces has been described by Tough [7] and by Stone and Tough [6]. It contains 9 variables, so three are redundant; however this provides the flexibility to choose the coordinate system in different ways, using either an arbitrary laboratory axis system, or the system used by Steele ($\theta=\phi=0$) or other possibilities. For example, an axis system referred to the axes of molecule 1 would be obtained by setting $\alpha_1=\beta_1=\gamma_1=0$.

The S-function expansion also has the minor virtue that the various contributions to the energy tend to appear as single terms; thus the energy of interaction of two linear dipoles, which requires several terms in a Pople or Steele expansion, is represented by the single term

$$U(\text{dipole-dipole}) = \sqrt{30}\,\mu_1\mu_2(4\pi\varepsilon_0 r^3)^{-1} S^{00}_{112}. \tag{42}$$

A further and more significant advantage is that it is very easy to determine the restrictions arising from any symmetry which may be present. The Wigner rotation matrices or symmetric top functions $D^L_{m,k}$ behave under molecular symmetry operations like sets of spherical harmonics $Y_{L,k}$, and only those combinations of the $D^L_{m,k}$ can appear which are symmetric under symmetry operations of the molecule; these properties are described in greater detail in Chapter 3. For example, rotation of the molecule through $2\pi/n$ about the molecular z axis replaces $D^L_{m,k}(\alpha,\beta,\gamma)$ by $D^L_{m,k}(\alpha,\beta,\gamma-2\pi/n) = \exp(2\pi ik/n) D^L_{m,k}(\alpha,\beta,\gamma)$. If the molecule has a C_n symmetry axis, such a rotation must leave the energy unchanged, so terms involving $D^L_{m,k}$ can only appear if $k = 0 \pmod n$. It follows that the only S-functions $S^{kk'}_{LL'J}$ which can appear are those for which $k = 0 \pmod n$. In particular, we see that for linear molecules, $k = k' = 0$, so that we can use the simpler functions

$$S_{LL'J} \equiv S^{00}_{LL'J} = (i)^{L-L'-J} \sum_{mm'M} \begin{pmatrix} L & L' & J \\ m & m' & M \end{pmatrix} C_{L,m}(\beta_1\alpha_1) C_{L',m'}(\beta_2\alpha_2) C_{J,M}(\theta,\phi). \tag{43}$$

The restrictions implied by the presence of other symmetry properties are given in Table 2.

Table 2. *Restrictions imposed by symmetry on the expansion (41) for the energy.*

	Symmetry Property	Consequence
A. of the system as a whole		
1.	Molecules identical	$u^{kk'}_{LL'J} = u^{k'k}_{L,LJ}$
2.	Both molecules linear	$L+L'+J$ even
3.	Both molecules centrosymmetric	L, L', J all even
4.	Time reversal	$u^{kk'}_{LL'J} = (-)^{k+k'} u^{-k-k'}_{LL'J}$
B. of molecule 1 (similar rules hold for molecule 2)		
1.	Centrosymmetric	L even
2.	$C_n^{(z)}$ axis	$k = 0 \pmod{n}$
3.	σ_h reflection	$k + L$ even
4.	C_2 rotation*	$u^{kk'}_{LL'J} = (-)^L e^{2\chi ik} u^{-kk'}_{LL'J}$
	(a) $C_2^{(x)}$	$u^{kk'}_{LL'J} = (-)^L u^{-kk'}_{LL'J}$
	(b) $C_2^{(y)}$	$u^{kk}_{LL'J} = (-)^{L+k} u^{-kk'}_{LL'J}$
5.	σ_v reflection*	$u^{kk'}_{LL'J} = (-)^k e^{2\chi ik} u^{-kk'}_{LL'J}$
	(a) σ_v^{xz}	$u^{kk'}_{LL'J} = (-)^k {}^{-kk'}_{LL'}$
	(b) σ_v^{yz}	$u^{kk'}_{LL'J} = u^{-kk'}_{LL'J}$

*The C_2 rotation considered is about an axis which lies in the xz plane and makes an angle χ with the x axis. The σ_v reflection plane contains the z axis and makes an angle χ with the x axis.

The equation defining the S-function, eq. (40), is a triple sum over products of quantities which, while available in tabulated form [8], are nevertheless rather clumsy to handle explicitly. Fortunately direct evaluation of (40) is not needed, as any $S^{00}_{LL'J}$ can be obtained in terms of other functions with smaller values of $L + L' + J$ by means of a recurrence relation [28], and ultimately in terms of the scalar products $\hat{\mathbf{z}}_1.\hat{\mathbf{r}}$, $\hat{\mathbf{z}}_2.\mathbf{r}$ and $\hat{\mathbf{z}}_1.\hat{\mathbf{z}}_2$, where $\hat{z}_1$ and $\hat{\mathbf{z}}_2$ are unit vectors along the molecular axes and $\hat{\mathbf{r}}$ is a unit vector in the direction of $\mathbf{r}$. Some of the first few members of the sequence are given in Table 3, and a fuller table has been given by Stone [28]. Consequently evaluation of a potential given in S-function form is simple and fast. The functions $S^{kk'}_{LL'J}$ required for non-linear molecules are also obtained easily, in terms of the $S^{00}_{LL'J}$, using a second recurrence relation [28].

If now we return to the gaussian overlap model, we can see that the functions $\varepsilon(\Omega)$ given in eqs. (32) and (33) are expressed in terms of the same scalar products that appear in the expressions for the S-functions in Table 3. Eqs. (32) and (33) can in fact be readily rewritten in S-function form:

$$\varepsilon(\Omega) = 3^{\frac{1}{2}}\varepsilon_o \; (3-\chi^2-\sqrt{20}\;\chi^2 S_{220})^{-\frac{1}{2}} \tag{44}$$

$$\sigma(\Omega) = 3\sigma_o[\varepsilon_o/\varepsilon(\Omega)]\{9-6\chi-\chi^2+\sqrt{280}\;\chi^2 S_{222}, -\sqrt{20}[\chi^2 S_{220}+(3\chi-2\chi^2)(S_{202}+S_{022})]\}^{-\frac{1}{2}}. \tag{45}$$

It is clear from this that there is scope for modifying these functions to give a better fit to experimental data without compromising the efficiency of evaluation which is one of the virtues of the gaussian overlap model.

Vector functions of the orientation of a pair of molecules can be similarly expanded using vector functions of direction called vector spherical harmonics. These are defined in terms of ordinary spherical harmonics $C_{\lambda\mu}(\theta,\phi)$ and a set of unit vectors $\mathbf{e}_\nu$:

$$\mathbf{C}^{M}_{L\lambda 1}(\theta,\phi) = \sum_{\mu\nu} C(\lambda 1 L;\mu\nu) C_{\lambda,\mu}(\theta,\phi)\mathbf{e}_\nu, \tag{46}$$

where $C(\lambda 1 L;\mu\nu)$ is a Clebsch-Gordan coefficient and

$$\mathbf{e}_o = \mathbf{e}_z, \qquad \mathbf{e}_{\pm 1} = \pm\sqrt{\tfrac{1}{2}}(e_x \pm i e_y). \tag{47}$$

The **V** functions which form a basis for the expansion of any vector function of the relative orientation of two molecules are defined by

$$\mathbf{V}^{kk'}_{LL'J;\lambda} = (i)^{L_1-L_2-\lambda+1} \sum_{mm'M} D^{L*}_{m,k}(\Omega_1)\; D^{L'*}_{m',k'}(\Omega_2)\; \mathbf{C}^{M}_{L\lambda 1}(\theta,\phi)\begin{pmatrix} L & L' & J \\ m & m' & M \end{pmatrix}. \tag{48}$$

The potential expressed in terms of S-functions can be readily differentiated [28] to yield the resulting forces and torques on the two molecules. For example the force $F_1^{(2)}$ and torque $G_1^{(2)}$ on molecule 1 due to its dipole-dipole interaction with molecule 2 are found to be

Table 3. *S functions $S^{00}_{LL'J}$ for small values of L, L' and J (further functions are tabulated by Stone[28]*

S_{000}	$= 1$
$\sqrt{3}\ S_{110}$	$= -\hat{\mathbf{z}}_1.\hat{\mathbf{z}}_2$
$\sqrt{3}\ S_{101}$	$= -\hat{\mathbf{z}}_1.\hat{\mathbf{r}}$
$\sqrt{3}\ S_{011}$	$= +\hat{\mathbf{z}}_2.\hat{\mathbf{r}}$
$\sqrt{30}\ S_{112}$	$= \hat{\mathbf{z}}_1.\hat{\mathbf{z}}_2 -3(\hat{\mathbf{z}}_1.\mathbf{r})(\hat{\mathbf{z}}_2.\mathrm{r})$
$2\sqrt{5}\ S_{220}$	$= 3(\hat{\mathbf{z}}_1.\hat{\mathbf{z}}_2)^2-1$
$\sqrt{70}\ S_{222}$	$= 2-3(\hat{\mathbf{z}}_1.\hat{\mathbf{r}})^2-3(\hat{\mathbf{z}}_2.\hat{\mathbf{r}})^2-3(\hat{\mathbf{z}}_1.\hat{\mathbf{z}}_2)^2$
	$+ 9(\hat{\mathbf{z}}_1.\hat{\mathbf{r}})(\hat{\mathbf{z}}_2.\hat{\mathbf{r}})(\hat{\mathbf{z}}_1.\hat{\mathbf{z}}_2)$

$$\mathbf{F}_1^{(2)} = -(15/7)^{\frac{1}{2}}(\mu_1\mu_2/4\pi\varepsilon_o r^4)\mathbf{V}^{00}_{112;3}, \tag{49}$$

$$\mathbf{G}_1^{(2)} = -(3\mu_1\mu_2/4\pi\varepsilon_o r^3)(\mathbf{V}^{00}_{111;2}+5^{\frac{1}{2}}\mathbf{V}^{00}_{112;2}). \tag{50}$$

The **V** functions, like the S-functions, are readily expressed in terms of the unit vectors $\hat{\mathbf{z}}_1,\hat{\mathbf{z}}_2$ and $\hat{\mathbf{r}}$, so that these vector quantities can be conveniently and rapidly calculated for the purpose of molecular dynamics calculations. They have been tabulated by Stone [28]. Consequently the need to calculate forces and torques for a molecular dynamics calculation does not restrict us to oversimplified models; we can look for realistic potentials and still carry out such calculations with reasonable efficiency.

References

1. J.A. Barker, *"Rare Gas Solids"*, M.L. Klein and J.A. Venables, eds, Academic Press (1975).
2. A.D. Buckingham, *Adv. Chem. Phys.* **12**, 107 (1967); A.D. Buckingham, *Intermolecular Interactions*, B. Pullman, ed., Wiley (1978).
3. H. Margenau and N.R. Kestner, *"Theory of Intermolecular Forces"*, Pergamon, (1969)
4. J.O. Hirschfelder, C.F. Curtiss and R.B. Bird, *"Molecular Theory of Gases and Liquids"*, Wiley, (1954).
5. P.R. Certain and L.W. Bruch, *MTP International Review of Science, Phys. Chem. Series I*, **1**, 113 (1972).

6. A.J. Stone and R.J.A. Tough, *Mol. Phys.* submitted for publication.
7. R.J.A. Tough, Ph.D. Thesis, University of Cambridge, (1977).
8. D.M. Brink and G.R. Satchler, *"Angular Momentum"* Oxford University Press, (1968).
9. A. Dalgarno, *Adv. Chem. Phys.* **12**, 143 (1967); A. Dalgarno and W.D. Davison, *Adv. Atomic Molec. Phys.* **2**, 1 (1965); J. Mahanty and B.W. Ninham, *"Dispersion Forces"* Academic Press,(1976).
10. F. London, *Z. Physik,* **63**, 245 (1930). *Z. Physik. Chem. B.* **11**, 221 (1930).
11. W. Maier and A. Saupe, *Z. Naturforsch,* **13a**, 564 (1958). *ibid* **14a**, 882 (1959). *ibid* **15a**, 287 (1960).
12. P. Claverie, *"Intermolecular Interactions"*, P. Pullman, ed., Wiley, (1978).
13. P.J. Stiles, *Chem. Phys. Letters,* **30**, 126 (1975).
14. J.W. Perram and P.J. Stiles, *Proc. Roy. Soc.,* **A349**, 125 (1976).
15. B. Liu and A.D. McLean, *J. Chem. Phys.,* **59**, 4557 (1973); N.S. Ostlund and D.L. Merrifield, *Chem. Phys. Letters,* **39**, 612 (1976).
16. R.G. Gordon and V.S. Kim, *J. Chem. Phys.,* **56**, 3122 (1972).
17. J.C. Raich and N.S. Gillis, *J. Chem. Phys.* **66**, 846 (1977).
18. D.J. Evans and R.O. Watts, *Molec. Phys.* **31**, 83; **32**, 93 (1976).
19. T. Kihara, *Rev. Mod. Phys.* **25**, 831 (1953); T. Kihara, K. Sakai and A. Koide, *Chem. Phys. Letters,* **47**, 416 (1971).
20. B.J. Berne and P. Pechukas, *J. Chem. Phys.* **56**, 4213, (1972).
21. J. Kushick and B.J. Berne, *J. Chem. Phys.* **64**, 1363 (1976).
22. J. Corner, *Proc. Roy. Soc.* **A192**, 275 (1948).
23. S.J. Walmsley, *Chem. Phys. Letters,* **49**, 390 (1977).
24. J.A. Pople, *Proc. Roy. Soc.* **A221**, 498 (1954).
25. W.A. Steele, *J. Chem. Phys.* **39**, 3197 (1963).
26. L. Blum and A.J. Torruella, *J. Chem. Phys.* **51**, 303 (1972).
27. L.C. Biedenharn and M.E. Rose, *Rev. Mod. Phys.* **25**, 729 (1953).
28. A.J. Stone, *Mol. Phys.* **36**, 241 (1978).

Chapter 3

DISTRIBUTION FUNCTIONS AND ORDER PARAMETERS

CLAUDIO ZANNONI

Department of Chemistry, The University, Risorgimento 4, 40136 Bologna, Italy

Introduction

The most fundamental characteristic of liquid crystals, at least from a microscopic point of view, is the presence of long-range orientational order while positional order is limited or absent altogether [1]. In this Chapter we shall discuss in some detail the way to describe this ordering. We shall also introduce distribution functions and other concepts from statistical mechanics which can help us in calculating order parameters as well as other relevant averages. In the following two sections we recall some important definitions and concepts from statistical mechanics and, where necessary, provide the generalizations needed to treat anisotropic fluids. We discuss in particular the singlet and pair distributions and give equations for the relevant thermodynamic observables in terms of these distributions. In the third section we shall introduce order parameters by expanding the singlet distribution in a complete basis set. The effect of molecular and mesophase symmetry on the orientational order parameters will be examined. In the final section a general expansion of the pair distribution will be given. The coefficients in this expansion will be interpreted as correlations and their importance will be discussed.

Distributions

We start by recalling some basic definitions from statistical mechanics and by describing the notation involved. We consider a macroscopic system of N classical rigid particles with position (e.g. centre of mass) specified by a vector $\mathbf{r}$ and orientation defined by the three Euler angles $(\alpha\beta\gamma) \equiv \Omega$. The equilibrium configurational partition function for such a system can be written as [2]

$$Q_N = (1/N!)\int\{d\mathbf{X}^N\}\exp[-\beta U(\{\mathbf{X}^N\})], \qquad (1)$$
$$\equiv Z_N/N!$$

where $U(\{\mathbf{X}^N\})$ is the potential energy of the N particles and, as usual, the β in the Boltzmann factor is $\beta \equiv (kT)^{-1}$ where k is the Boltzmann constant and T is the absolute temperature. For economy of notation we have used $\mathbf{X}$ to indicate the six variables $(\mathbf{r},\Omega)$ and the curly brackets to denote collectively N variables. Thus $\{X^N\} \equiv (X_1,X_2,\ldots,X_N)$ and, similarly, $\{dX^N\} \equiv d\mathbf{X}_1 \cdot d\mathbf{X}_2 \cdots d\mathbf{X}_N$. Each volume element $d\mathbf{X}$ is equivalent to $d\mathbf{r}d\Omega$, where $d\mathbf{r} \equiv dr_x dr_y dr_z$ and $d\Omega \equiv d\alpha \sin\beta d\beta d\gamma$. To complete the remarks about notation we mention that in the absence of possible ambiguities we shall use only one integration sign to indicate the, possibly multiple, integration over all the variables whose volume elements appear. Integration is extended to the sample volume V for positions and to the usual domains $0 \leqslant \alpha \leqslant 2\pi$, $0 \leqslant \beta \leqslant \pi$ and $0 \leqslant \gamma \leqslant 2\pi$ for angles [3].

The probability of finding n particles out of the given N in the range $X_1+dX_1, X_2+dX_2, \ldots, X_n+dX_n$, can be written as the n-particle distribution

$$P^{(n)}(\{X^n\}) = (N!/(N-n)!Z_N)\int\{d\mathbf{X}^N_{n+1}\}\exp[-\beta U(\{X^N\})], \tag{2}$$

where, with a slight extension of the previous convention, we have introduced $\{d\mathbf{X}^N_m\} \equiv dX_m d\mathbf{X}_{m+1} \cdots d\mathbf{X}_N$. Notice that $P^{(n)}$ is not normalized to one but to the number of n-plets that can be formed by choosing n variables out of N i.e.

$$\int\{d\mathbf{X}^n\}P^{(n)}(\{\mathbf{X}^n\}) = N!/(N-n)! \ . \tag{3}$$

Here we shall not need to concern ourselves very much with the general n-particle distribution but mainly with the one- and two-particle distributions, or, as they are often called, the singlet and the pair distributions. These can be used to define the canonical ensemble average $\langle A \rangle$ of any property depending on position and orientation of one or two particles respectively. Thus

$$\langle A(\mathbf{X}_1)\rangle = (1/N)\int dX_1 A(\mathbf{X}_1)P^{(1)}(\mathbf{X}_1) \tag{4}$$

and

$$\langle A(X_1,X_2)\rangle = [1/N(N-1)] \int d\mathbf{X}_1 dX_2 A(\mathbf{X}_1,X_2)P^{(2)}(X_1,X_2). \tag{5}$$

The singlet distribution $P^{(1)}$ which gives the probability of finding a molecule at a particular position and orientation is

$$P^{(1)}(\mathbf{r}_i,\Omega_1) = (N/Z_N)\int\{d\mathbf{r}^N_2\}\{d\Omega^N_2\}\exp[-\beta U(\{r^N,\Omega^N\})]. \tag{6}$$

Similarly the pair distribution function, giving the probability of finding simultaneously a particle in a volume $d\mathbf{r}_1 d\Omega_1$ centred at $\mathbf{r}_1,\Omega_1$ and a second one in a volume $d\mathbf{r}_2 d\Omega_2$ centred at $\mathbf{r}_2,\Omega_2$ is

$$P^{(2)}(r_1,\Omega_1;r_2,\Omega_2) = [N(N-1)/Z_N]\int\{dr^N_3\}\{d\Omega^N_3\}\exp[-\beta U(\{r^N,\Omega^N\})]. \tag{7}$$

Inspection of eqs. (4) and (5) shows that formal alternative definitions of $P^{(1)}$ and $P^{(2)}$ are [2]

$$P^{(1)}(\mathbf{r}_1,\Omega_1) = N\langle\delta(\mathbf{r}_1-\mathbf{r}_1')\delta(\Omega_1-\Omega_1')\rangle \tag{8}$$

and

$$P^{(2)}(\mathbf{r}_1,\Omega_1;\mathbf{r}_2,\Omega_2) = N(N-1)\langle\delta(\mathbf{r}_1-\mathbf{r}_1')\delta(\Omega_1-\Omega_1')\delta(\mathbf{r}_2-\mathbf{r}_2')\delta(\Omega_2-\Omega_2')\rangle. \tag{9}$$

Similarly we have in general

$$P^{(n)}(\mathbf{X}_1,\ldots,\mathbf{X}_n) = [N!/(N-n)!]\langle\delta(\mathbf{X}_1-\mathbf{X}_1')\ldots(\mathbf{X}_n-\mathbf{X}_n')\rangle, \tag{10}$$

where $\delta(\mathbf{a}-\mathbf{b})$ is a Dirac delta function and the integration implied in the ensemble average $\langle\cdots\rangle$ is over the primed variables. Eqs. (8-10) can be rewritten in a convenient way using the Fourier integral representation of the spatial delta function

$$\delta(\mathbf{r}) = (2\pi)^{-3}\int d\mathbf{k}\ \exp(i\mathbf{k}.\mathbf{r}) \tag{11}$$

and the representation of the angular delta function

$$\delta(\Omega-\Omega') = \sum[(2L+1)/8\pi^2]D^L_{m,n}(\Omega)D^{L*}_{m,n}(\Omega'), \tag{12}$$

where the functions $D^L_{m,n}(\Omega)$ are the Wigner rotation matrices[3]; we define them and give their important properties in the Appendix. The sum in eq. (12) runs over L, m and n; we shall normally take the summations on the right hand side of an equation to run over all indices not appearing on the left hand side. Replacing the delta functions in eqs. (8-10) gives some general expansions for the probability distributions. As an example

$$P^{(1)}(\mathbf{r}_1,\Omega_1) = N\int d\mathbf{k}\sum c_{Lmm}(\mathbf{k})\exp(i\mathbf{k}.\mathbf{r}_1)D^L_{m,n}(\Omega_1); \tag{13}$$

the coefficients

$$c_{Lmn}(\mathbf{k}) = [(2L+1)/64\pi^5]\langle\exp(-i\mathbf{k}.\mathbf{r}')D^{L*}_{m,n}(\Omega')\rangle, \tag{14}$$

will be identified later with the order parameters for the system.

It is possible to give a limiting expression for the n-particle distribution valid for very low densities. In fact, when the distance between particles is very large and their reciprocal influence negligible, the probabilities of finding them in their respective volume elements become statistically independent. Therefore, in the limit that $|\mathbf{r}_i-\mathbf{r}_j|\to\infty$ for all pairs of particles i and j, the n-particle joint distribution must yield simply the product of n single particle distributions

$$P^{(n)}(\mathbf{X}_1,\mathbf{X}_2,\ldots,\mathbf{X}_n) = [N!/(N-n)!N^n]P^{(1)}(\mathbf{X}_1)P^{(1)}(\mathbf{X}_2)\ldots P^{(1)}(\mathbf{X}_n). \tag{15}$$

We can use this property to introduce reduced n-particle distributions or correlations, which tend to unity as the inter-particle distances tend to infinity or, in practice, when these distances become many orders of magnitude larger than typical intermolecular distances. Thus we define

$$g^{(n)}(\{\mathbf{X}^n\}) \equiv P^{(n)}(\{\mathbf{X}^n\}) / \prod_{i=1}^{n} P^{(1)}(\mathbf{X}_i). \tag{16}$$

The most important of this family of $g^{(n)}$ is the pair correlation function

$$g^{(2)}(\mathbf{X}_1,\mathbf{X}_2) \equiv P^{(2)}(\mathbf{X}_1,\mathbf{X}_2)/P^{(1)}(\mathbf{X}_1)P^{(1)}(\mathbf{X}_2). \tag{17}$$

For a uniform system the physical properties are invariant under translation and the interaction energy $U(\{X^N\})$ depends only on relative distances. Therefore for an ordinary isotropic fluid or for a nematic, but not for a smectic, we can write

$$P^{(1)}(\mathbf{r}_1,\Omega_1) = \rho f(\Omega_1) \tag{18}$$

and

$$P^{(2)}(\mathbf{r}_1,\Omega_1;\ \mathbf{r}_2,\Omega_2) = \rho^2 f(\Omega_1) f(\Omega_2) g^{(2)}(\mathbf{r}_{12},\Omega_1,\Omega_2), \tag{19}$$

where $\rho \equiv N/V$ is the number density and $f(\Omega_1)$ is a purely orientational singlet distribution normalized to unity:

$$f(\Omega_1) = (V/Z_N)\int\{d\mathbf{X}_2^N\}\exp[-\beta U(\{\mathbf{X}^N\})], \tag{20}$$

with

$$\int d\Omega_1 f(\Omega_1) = 1.$$

For an isotropic molecular fluid we have simply

$$P^{(1)}(\mathbf{r}_1,\Omega_1) = \rho/8\pi^2$$

and

$$P^{(2)}(\mathbf{r}_1,\Omega_1;\ \mathbf{r}_2,\Omega_2) = (\rho/8\pi^2)^2 g^{(2)}(\mathbf{r}_{12},\Omega_1,\Omega_2). \tag{21}$$

If, moreover, the constituent particles are spherical (e.g. atoms) the pair correlation function $g^{(2)}(\mathbf{r}_{12},\Omega_1,\Omega_2)$ depends on inter-particle distance alone and $g(r_{12})$ is called the radial distribution function.

It is helpful to gain greater physical understanding of the pair distribution by considering some limiting situations. For very dilute fluid systems, the density is so low that configurations with three or more particles interacting simultaneously are extremely rare and can be neglected. Therefore, in this limit,

$$g^{(2)}(\mathbf{X}_1,\mathbf{X}_2) = \exp[-\beta U(\mathbf{X}_1,\mathbf{X}_2)],\quad \rho\to 0. \tag{22}$$

In accord with the definition, eq. (17), we have for N>>1

$$\lim_{r_{12}\to\infty} g^{(2)}(\mathbf{X}_1,\mathbf{X}_2) = 1, \tag{23}$$

corresponding to no correlation at all. If the particles have a hard impenetrable core, there is clearly a vanishing probability of finding a second molecule nearer than a minimum approach distance σ from the first one. Then

$$g^{(2)}(\mathbf{X}_1,\mathbf{X}_2) = 0, \quad r_{12}<\sigma(\Omega_1,\Omega_2). \tag{24}$$

Another limiting situation where we can sketch the behaviour of $g^{(2)}$ is that of an ideal solid. Imagine this solid composed of spherical particles exactly positioned at the lattice sites. Every particle will have z_1 nearest neighbours at a distance r_1, z_2 at a distance r_2 and so on. Therefore

$$\begin{aligned} g(r) &= (1/4\pi r^2\rho)\sum_i z\ \delta(r-r_i), \\ &= (1/4\pi r^2\rho)\lim_{\varepsilon\to o}\sum_i z_i(4\pi\varepsilon)^{-\frac{1}{2}}\exp\{-(r-r_i)^2/4\varepsilon\}, \end{aligned} \tag{25}$$

where the second equality follows from simply writing the delta function as a gaussian of vanishing width. We see that for an idealized solid g(r) consists of a series of peaks corresponding to the various shells of neighbours. In a real crystal the positions will not be defined with absolute certainty due to the possibility of thermal oscillations etc. and the peaks will obviously be smeared out; e.g. they could be more or less sharp gaussians instead of delta functions. In a liquid the peaks will be even more diffuse and we could have something like the behaviour sketched in figure 1. The successive peaks in g(r) can still be associated

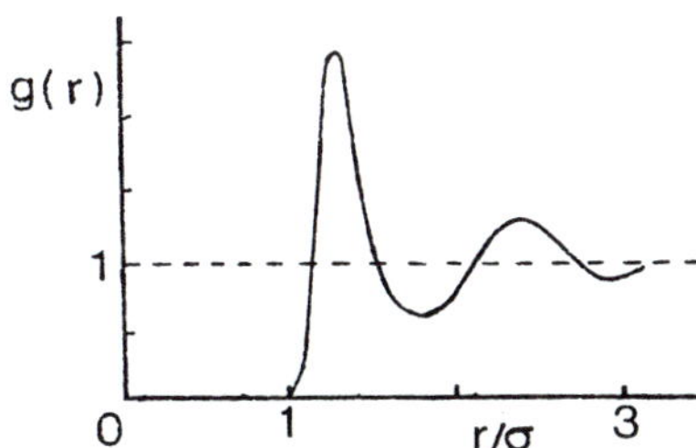

Fig. 1 *The possible behaviour of the radial distribution function in a liquid; σ is the distance of closest approach for two particles.*

with the presence of average shells of neighbours and give an indication of short-range order (structure) in a fluid. Notice the fundamental difference between g(r) in a fluid and in the idealized solid. In the fluid g(r) decays to one in the absence of correlations while in the solid g(r) keeps oscillating even when the separation r is extremely large. In other words in a solid we have long-range positional order, which vanishes at the melting transition.

Thermodynamic Properties

The configurational partition function and the distributions introduced in the previous section provide a convenient link between microscopic or molecular properties and thermodynamic observables. In general we have for the configurational contributions to the Helmholtz free energy A, entropy S, pressure P and internal energy U the expressions [2,4],

$$A = -kT \ln Q_N, \tag{26}$$

$$S = -(\partial A/\partial T)_V, \tag{27}$$

$$P = -(\partial A/\partial V)_T, \tag{28}$$

$$\begin{aligned} U &= A + TS, \\ &= -T^2(\partial(A/T)/\partial T)_V, \\ &= kT^2(\partial(\ln Q_N)/\partial T)_V \cdot \end{aligned} \tag{29}$$

Other important observables are the specific heat at constant volume C_V and the isothermal compressibility κ_T,

$$\begin{aligned} C_V &= (\partial U/\partial T)_V = T(\partial S/\partial T)_V, \\ &= -T(\partial^2 A/\partial T^2)_V, \end{aligned} \tag{30}$$

$$\kappa_T = -(1/V)(\partial V/\partial P)_T. \tag{31}$$

Since C_V is positive eq. (30) implies that the free energy at constant volume is a downward concave function of temperature.

We recall also the Ehrenfest scheme of classifying phase transitions [4,5], which, although not of general applicability, sometimes gives a useful practical framework. According to the scheme a transition is classified as being of nth order if the first (n-1) derivatives of the free energy are continuous across the phase transition while the nth shows a discontinuity. Thus in a first order transition, e.g. melting or condensation away from the critical point, there is a finite jump in the first derivative of the free energy, the entropy, and a latent heat $T\Delta S$. In a higher order transition, however, the entropy change is continuous across the transition. This idealized behaviour is sketched in figure 2. The nematic-isotropic phase transition is a first order transition with a small latent heat. For this reason it is often called a weak first order phase transition. The prediction and the location of this phase transition is central to any statistical theory of liquid crystals. A cautionary remark is thus perhaps not inappropriate here. If we look back at eq. (1), we see that the partition function Q_N is a continuous and infinitely differentiable function of inverse temperature

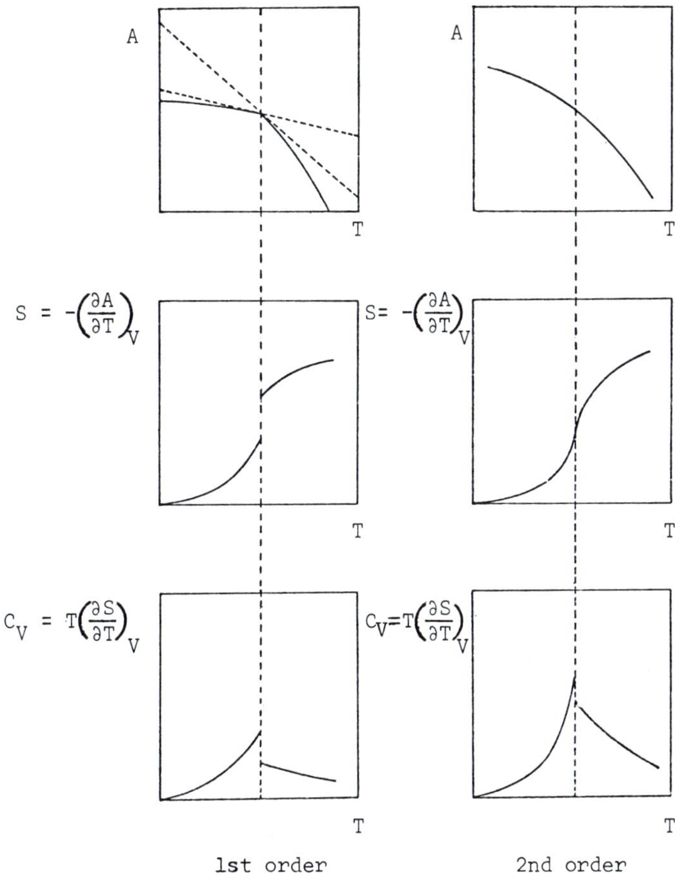

Fig. 2 *A sketch of the behaviour of the free energy, A, and of its temperature derivatives at a first order and second order phase transition, according to the Ehrenfest scheme.*

for any finite N. This implies that the free energy and its derivatives are also continuous and differentiable. Strictly, therefore, we cannot have a true phase transition for a system of finite size [2a]. We have to go to the so-called thermodynamic limit, that is calculate the free energy per particle and take the limit

$$a(\rho, T) = \lim_{\substack{N\to\infty \\ V\to\infty}} A(V,T)/N \ , \ \rho = \lim_{\substack{N\to\infty \\ V\to\infty}} N/V = \text{constant.}$$

This limiting free energy can have the discontinuities which correspond to phase transitions. Another important reason for going to the thermodynamic limit, by letting N tend to infinity, while keeping the density N/V constant, is that only in this limit do the various ensembles used in statistical thermodynamics become exactly equivalent [2].

Let us now write down explicit expressions for energy, pressure and specific heat as ensemble averages using the distributions we have defined previously.

Energy

Assume that the total potential energy can be decomposed as a sum of pairwise interactions;

$$U(\{X^N\}) = \sum_{1\leqslant i<j\leqslant N} U(X_i,X_j). \tag{32}$$

Then the total average potential energy is

$$\begin{aligned} \langle U\rangle = \langle U(\{\mathbf{X}^N\})\rangle &= \tfrac{1}{2}\int d\mathbf{X}_1 d\mathbf{X}_2 U(\mathbf{X}_1,\mathbf{X}_2)P^{(2)}(X_1,X_2), \\ &\equiv \tfrac{1}{2}\rho^2 \int d\mathbf{X}_1 d\mathbf{X}_2 U(X_1,\mathbf{X}_2)G(\mathbf{X}_1,X_2), \end{aligned} \tag{33}$$

for uniform fluids $U(\mathbf{X}_1,\mathbf{X}_2) = U(\mathbf{r}_{12},\Omega_1,\Omega_2)$ and we find

$$\begin{aligned} \langle U\rangle &= \tfrac{1}{2}V\rho^2\int d\mathbf{r}_{12}d\Omega_1\; d\Omega_2 U(\mathbf{r}_{12},\Omega_1,\Omega_2)f(\Omega_1)f(\Omega_2)g^{(2)}(\mathbf{r}_{12},\Omega_1,\Omega_2), \\ &\equiv \tfrac{1}{2}V\rho^2\int d\mathbf{r}_{12}d\Omega_1\; d\Omega_2 U(\mathbf{r}_{12},\Omega_1,\Omega_2)G(\mathbf{r}_{12},\Omega_1,\Omega_2). \end{aligned} \tag{34}$$

Thus we see that for an anisotropic fluid the key quantity to be calculated is $P^{(2)}(\mathbf{r}_{12},\Omega_1,\Omega_2) \equiv \rho^2 G(\mathbf{r}_{12},\Omega_1,\Omega_2)$ and not $g^{(2)}$ as in an ordinary isotropic fluid. The reason is, of course, that in an anisotropic fluid the orientational distribution $f(\Omega)$ is not a constant so that $g^{(2)}$ does not have a simple normalization. We have therefore introduced for convenience the reduced distribution G, which is normalized according to

$$\int d\mathbf{r}_{12}d\Omega_1\; d\Omega_2 G(\mathbf{r}_{12},\Omega_1,\Omega_2) = V(1-N^{-1}) \tag{35}$$

and has simple properties,

$$G(\mathbf{r}_{12},\Omega_1,\Omega_2) = f(\Omega_1)f(\Omega_2),\quad r_{12}\to\infty \tag{36}$$

and

$$G(\mathbf{r}_{12},\Omega_1,\Omega_2) = f(\Omega_1)f(\Omega_2)\exp\{-\beta U(\mathbf{r}_{12},\Omega_1,\Omega_2)\},\quad \rho\to 0. \tag{37}$$

Pressure

An equation linking the pressure to the other thermodynamic variables, the so-called equation of state, can also be given an ensemble average representation by combining eqs. (26) and (28). The volume dependence of the configurational integral Q_N can be made explicit by transforming to dimensionless coordinates: $\mathbf{s}_i = \mathbf{r}_i/V^{1/3}$. We then find the virial equation

$$P = \rho kT - (1/3V)\sum \mathbf{r}_i . \partial U(\{\mathbf{X}^N\})/\partial \mathbf{r}_i\rangle. \tag{38}$$

For a pair additive potential eq. (38) can be reduced to

$$P = \rho kT-(1/6)\int d\mathbf{X}_1\ d\mathbf{X}_2 P^{(2)}(X_1,X_2)\mathbf{r}_{12}\cdot\partial U(X_1,X_2)/\partial\mathbf{r}_{12}. \tag{39}$$

This can be simplified further, for a uniform fluid, to

$$P = \rho kT-(\rho^2/6)\int d\mathbf{r}_{12}d\Omega_1 d\Omega_2 G(\mathbf{r}_{12},\Omega_1,\Omega_2)\mathbf{r}_{12}\cdot\partial U(\mathbf{r}_{12},\Omega_1,\Omega_2)/\partial\mathbf{r}_{12}, \tag{40}$$

which reduces, for isotropic fluids, to

$$P = \rho kT-(\rho^2/384\pi^4)\int d\ _{12}d\Omega_1 d\Omega_2 g^{(2)}(\mathbf{r}_{12},\Omega_1,\Omega_2) \times r_{12}\cdot\partial U(\mathbf{r}_{12},\Omega_1,\Omega_2)/\partial\mathbf{r}_{12}. \tag{41}$$

For spherically symmetric particles and isotropic intermolecular interactions we recover the familiar expression

$$P = \rho kT-(2\pi\rho^2/3)\int dr_{12}g(r_{12})r_{12}^3\partial U(r_{12})/\partial r_{12}. \tag{42}$$

Eqs. (38-42) hold, of course, for any potential that is a continuous differentiable function of positions. However, their validity can be shown to extend also to the important hard-core potentials. Consider, as a simple example, the hard-sphere potential

$$U(r_{12}) = \begin{cases} 0 & \text{when } r_{12} \geq \sigma \\ \infty & \text{when } r_{12} < \sigma . \end{cases}$$

Eq. (42) can be rewritten as [2b]

$$P/kT = \rho + (2\pi\rho^2/3)\int_0^\infty dr\ r^3 y(r)dH(r-\sigma)/dr, \tag{43}$$

where $y(r) \equiv g(r)\exp\{\beta U(r)\}$ is assumed to be an everywhere continuous function of r and $H(r-\sigma) \equiv \exp\{-\beta U(r)\}$. Integration can now be easily performed noticing that $H(r-\sigma)$ is a unit step function. The derivative of a unit step function is a delta function so we have at once

$$P/kT = \rho + (2\pi\rho^2\sigma^3/3)y(\sigma), \tag{44}$$

where $y(\sigma)$ is $\lim_{\varepsilon\to 0^+} g(\sigma+\varepsilon)$.

This equation is particularly important in computer simulations, where the pair distribution at the contact point $y(\sigma)$ can be calculated numerically [6,7]. Eq. (38) can be generalized to systems composed of hard anisotropic particles [7], at least in principle.

Specific Heat

Some manipulation of eq. (30) shows that the specific heat can be written as

$$C_V = (\langle U^2\rangle - \langle U\rangle^2)/kT^2, \tag{45}$$

$$= \tfrac{1}{2}\rho^2 \int dX_1 dX_2 U(X_1,X_2)\partial G(X_1,X_2)/\partial T, \tag{46}$$

for temperature independent intermolecular potentials. Eq. (45) shows that C_V is a positive quantity determined essentially by the fluctuations in the internal energy.

Compressibility

Consider a macroscopic subvolume V_A of the total volume V and the mean square fluctuation in the number of particles N_A occupying it. Clearly

$$\langle N_A\rangle = \int_{V_A} d\mathbf{X}_1 P^{(1)}(\mathbf{X}_1) \tag{47}$$

and

$$\langle N_A(N_A-1)\rangle = \int_{V_A} d\mathbf{X}_1 dX_2 P^{(2)}(\mathbf{X}_1,X_2). \tag{48}$$

Therefore

$$\langle N_A^2\rangle - \langle N_A\rangle^2 = \langle N_A\rangle + \int_{V_A} dX_1 dX_2 [P^{(2)}(\mathbf{X}_1,X_2) - P^{(1)}(\mathbf{X}_1)P^{(1)}(X_2)] \tag{49}$$

and, remembering [2] that κ_T can also be written as $\kappa_T = (V/kT)\langle \Delta N^2\rangle/\langle N\rangle^2$,

$$\kappa_T = (1/kT)\{(1/\rho) + (1/\rho^2 V_A)\iint_{V_A} dX_1 d\mathbf{X}_2 [P^{(2)}(\mathbf{X}_1,X_2) - P^{(1)}(X_1)P^{(1)}(X_2)]\}. \tag{50}$$

For a uniform fluid this reduces to

$$\kappa_T = (1/kT)\{(1/\rho) + \int_{V_A} d\mathbf{r} d\Omega_1 d\Omega_2 [G(\mathbf{r},\Omega_1,\Omega_2) - f(\Omega_1)f(\Omega_2)]\} \tag{51}$$

and for an isotropic fluid to the equation given by Steele [8],

$$\kappa_T = (1/kT)\{(1/\rho) + (1/64\pi^2)\int d\mathbf{r} d\Omega_1 d\Omega_2 [g^{(2)}(\mathbf{r},\Omega_1,\Omega_2) - 1]\}. \tag{52}$$

Near a phase transition, where the compressibility diverges it follows from eqs. (50-52) that the area under the correlation G must also diverge. In other words, the range of the correlations, the correlation length, has to diverge.

Order Parameters

When studying phase transformations it is often useful to define an order parameter, that is a quantity which changes value on going from one phase to the other and can therefore be used to monitor the transition. In many instances there is a simple practical choice for this order parameter, such as the magnetization in a magnetic system [5], or the anisotropy in some tensor property for a nematic-isotropic transition [1,9]. From a molecular point of view, however, we should describe the passage from one phase to another in terms of the modifications that this produces in the distribution functions. It is natural, therefore, to introduce in a general way order parameters as expansion coefficients of, say, the singlet distribution in a suitable basis set. Thus, if the distribution depends on positions and orientations, we will have positional, orientational and mixed positional-orientational order parameters [1b]. These parameters will, of course, alter with the variable producing the phase change, and, if properly chosen, will vanish when the transition to the more symmetric phase takes place. The most characteristic and, perhaps, the most important order parameters for liquid crystal phases are the orientational ones. These define the distribution function $f(\Omega)$,

$$f(\Omega) = (1/N)\int d\mathbf{r} P^{(1)}(\mathbf{r},\Omega), \tag{53}$$

which, of course, for a translationally invariant fluid such as a nematic, reduces to

$$f(\Omega) = P^{(1)}(\Omega)/\rho.$$

The ensemble average of any single-particle orientational function $A(\Omega)$ can be written in terms of $f(\Omega)$ as

$$\bar{A} = \int d\Omega f(\Omega) A(\Omega). \tag{54}$$

We use an upper bar instead of the usual angular brackets to indicate this purely orientational average. Any well behaved function of the three Euler angles $f(\Omega)$ can be expanded in a Wigner series [3], that is in a basis of Wigner rotation matrices;

$$f(\Omega) = \sum f_{Lmn} D^L_{m,n}(\Omega). \tag{55}$$

Multiplying both sides of eq. (55) by D^{L*}_{mn} and integrating over the angles we find

$$f_{Lmn} = [(2L+1)/8\pi^2]\overline{D^{L*}_{m,n}}, \tag{56}$$

because of the orthogonality of the Wigner rotation matrices and the definition in eq. (54). The averages $\overline{D^L_{m,n}}$, which completely define $f(\Omega)$, are just the orientational order parameters. We give the explicit expressions for the rotation matrices of rank $L = 0,1,2,4$ in the Appendix. There can be up to $(2L+1)^2$ order parameters of rank L although this number can be drastically reduced by exploiting the symmetry properties of the mesophase and of its constituent particles. For example in a uniaxial phase the singlet distribution must be invariant under rotation about the director i.e. the axis of symmetry. If this is chosen to be z, it follows that m must be zero in the average $\overline{D^L_{m,n}}$. In addition if a uniaxial mesophase has a symmetry plane perpendicular to the director

($D_{\infty h}$ symmetry) then only terms with even L can appear in eq. (55). To simplify further one has to investigate the symmetry of the molecules forming the mesophase. Let us consider the simplest case of all, which arises when the molecules are cylindrically symmetric. In this limit rotation about the molecular symmetry axis should not modify the distribution $f(\Omega)$, which implies $n = 0$ in eq. (55). In other words for a cylindrically symmetric mesophase composed of cylindrically symmetric particles the singlet orientational distribution has to depend only on the angle β between the director and the molecular symmetry axis. Accordingly we have $f(\beta) = f(\Omega)/4\pi^2$, where

$$f(\beta) = \sum f_L D^L_{0,0}(\beta), \quad L \text{ even}, \tag{57}$$

and

$$\int d\beta \sin\beta \; f(\beta) = 1.$$

In this important but, strictly, unrealistic case the expansion coefficients are $f_L = [(2L+1)/2]\, \bar{P}_L$ and the order parameters are just averages of the even Legendre polynomials. Obviously a knowledge of $f(\Omega)$ implies that all the order parameters can be calculated. *Vice versa* a knowledge of $\bar{P}_2$, $\bar{P}_4$ etc. gives us progressively more information on the singlet distribution function and allows a test of the various theoretical models put forward for nematics or other liquid crystals.

We cannot say very much about the rate of convergence of expansion (57). However, since the number of zeros in the Legendre polynomials increases with their rank we expect the $\bar{P}_L$ to decrease if the distribution $f(\Omega)$ is peaked at $\beta=0$ and monotonically non-increasing. If this is the case we expect the expansion coefficients f_L to decrease with L at low order. As the order increases, the convergence will get worse as all the $\bar{P}_L$ tend to their limit $\bar{P}_L = 1$ (L even) for complete order. We do not expect convergence to be very rapid in the region of intermediate order: $\bar{P}_2 \sim 0.4 \div 0.6$. Thus, unless we have a technique able to yield directly $f(\beta)$ [10] or, in general, $f(\Omega)$ the determination of order parameters of rank as high as possible is a problem of great importance. We do not want to discuss the problems connected with the practical measurement of the order parameters here, since these are treated in detail in other Chapters. However, we think it worthwhile to examine briefly how the order parameters $\overline{D^L_{m,n}}$ can be related to measureable quantities.

Let us consider, as an example, the determination of a second rank tensor property during a time sufficiently long to allow observation of statically averaged quantities. If $F^{(L,m)}$ are the irreducible spherical components of the tensor **F**, as defined in the Appendix and if **F** is symmetric, then its only non-zero components will be those of rank L=0 and L=2. Let us consider the anisotropic, L=2, component. The tensor **F** is measured in a laboratory frame and its components can be related to the molecule fixed components $F'^{(2,m)}$ by

$$F^{(2,n)} = \sum D^{2*}_{n,m}(\Omega) F'^{(2,m)}.$$

Taking an ensemble average we have, for a uniaxial mesophase,

$$\overline{F^{(2,n)}} = \overline{F^{(2,0)}}\delta_{n0},$$

$$= \sum \overline{D^{2*}_{0,m}} F'^{(2,m)}, \tag{58}$$

where we have taken the laboratory z axis to be parallel to the director. If the molecule has cylindrical symmetry, in the sense that $\overline{D^2_{0,m}} = \overline{D^2_{0,0}}\delta_{m0}$, then we have

$$\bar{P}_2 \equiv \overline{D^2_{0,0}}$$

$$= \overline{F^{(2,0)}} / F'^{(2,0)}. \tag{59}$$

Thus, the orientational order parameter $\bar{P}_2$ can be obtained in practice from the measured anisotropy in $\bar{\mathbf{F}}$.

A similar analysis can obviously be given for a tensor of higher rank. However, while it is relatively easy to determine $\bar{P}_2$ from the anisotropy of some second rank tensor **F** (e.g. the dipolar coupling between two nuclei) it is much more difficult to find a tensor quantity of higher rank which can be measured directly. One is then forced to resort to somewhat indirect methods. For example the fourth rank order parameter $\bar{P}_4$ can be extracted from measurements of the mean square value of a second rank quantity [11,12]. In the limit of cylindrical symmetry, we have, in fact,

$$\overline{(F^{(2,0)})^2} = \{(1/5)+(2/7)\bar{P}_2 + (18/35)\bar{P}_4\}(F'^{(2,0)})^2. \tag{60}$$

In the Raman scattering technique described by Pershan [11], in Chapter 17 the molecular quantity **F'** is the differential polarizability tensor for a certain localized Raman mode.

Information about all the parameters is contained, in principle, in the single molecule coherent contribution $(d\sigma/d\Omega)^{coh}_s$ to the cross section for neutrons scattered by a monodomain nematic [13,14]. In fact, assuming that the nematogenic molecules are rigid, we have

$$(d\sigma/d\Omega)^{coh}_s = \sum_{i,j} a_i a_j \overline{\exp(i\mathbf{Q}.\mathbf{r}_{ij})}, \tag{61}$$

where Q is the scattering vector, $\mathbf{r}_{ij}$ the vector joining nuclei i and j, which have bound coherent scattering lengths a_i and a_j respectively.

Rayleigh expansion of the exponential gives

$$\overline{\exp(i\mathbf{Q}.\mathbf{r}_{ij})} = \sum (i)^L (2L+1) j_L(Qr_{ij})\ \overline{D^L_{0,0}(Q-r_{ij})}, \tag{62}$$

where we use the notation $D^L_{m,n}(B-A)$ to indicate a rotation from A to B; e.g. here, from $\mathbf{r}_{ij}$ to $\mathbf{Q}$. $j_L(Qr_{ij})$ is a spherical Bessel function of rank L. Using the properties of the Wigner rotation matrices we can write

$$\begin{aligned}\overline{D^L_{0,0}(Q-r_{ij})} &= \sum \overline{D^L_{m,0}(r_{ij}-L)} D^{L^*}_{m,0}(Q-L)\,,\\ &= \sum \overline{D^L_{m,n}(M-L)} D^L_{n,0}(r_{ij}-M) D^{L^*}_{m,0}(Q-L)\,,\end{aligned} \tag{63}$$

where $D^L_{m,n}(M-L)$ indicates a rotation from a laboratory to a molecule fixed frame. For a uniaxial mesophase and cylindrically symmetric molecules we have $\overline{D^L_{m,n}} = \bar{P}_L\ \delta_{m0}\ \delta_{n0}$ if the laboratory z axis is along the director and the molecular z axis is parallel to the axis of cylindrical symmetry. Substitution in eq. (62) gives eventually the desired expression for the cross section in terms of the order parameters

$$(d\sigma/d\Omega)^{coh}_s = \sum \{a_i a_j (i)^L (2L+1) j_L(Qr_{ij}) D^L_{0,0}(Q-L) D^L_{0,0}(r_{ij}-M\} \bar{P}_L. \tag{64}$$

Molecular Symmetry and Mesophase Symmetry

As we have seen in the previous section the general expansion of the singlet orientational distribution $f(\Omega)$ can be simplified when the mesophase and its constituent molecules are both cylindrically symmetric. In general, however, the exploitation of the symmetry of the molecule and the phase is not quite so intuitive. Thus we wish to introduce a formal procedure for determining the independent order parameters. Admittedly, we have not yet defined what we mean by the symmetry of a phase. This problem was recently discussed by Goshen *et al.* [15]. They noticed that Landau's original suggestion [2c] of defining the symmetry of a fluid as that of the singlet distribution $P^{(1)}(\mathbf{r})$ and that of nematics by the symmetry of $P^{(2)}(\mathbf{r}_1,\mathbf{r}_2)$ is inadequate,but they did not propose an alternative definition. However, following our approach it seems clear that Landau's definition only needs generalization to the case of anisotropic particles. We shall therefore define the symmetry group of a phase as the group of transformations of the laboratory system that leave the singlet distribution $P^{(1)}(\mathbf{r},\Omega)$ as well as the higher ones invariant. Similarly, we can define an effective symmetry for the molecule in terms of the group of molecular transformations leaving the singlet distribution unchanged.

The same considerations hold, of course, for the purely orientational distribution $f(\Omega)$ which is of primary concern to us. In group theoretical language [16] we would say that $f(\Omega)$ belongs to the totally symmetric representation of the group of the molecule and of the mesophase. Therefore one way of applying symmetry is to project the distribution onto the totally symmetric representation of these groups [8,16-18]. We shall actually use this method in a later section to simplify the pair distribution. Another way, strictly related to the previous one, which we find convenient for the simplification of order parameters depends on their definition;

$$\overline{D^L_{m,n}} = \int d\Omega f(\Omega) D^L_{m,n}(\Omega). \tag{65}$$

Now a symmetry transformation O does not alter $\int d\Omega$ and also leaves $f(\Omega)$ invariant. It then follows that the action of O on the order parameters depends solely on its effect on the Wigner rotation matrices. Since these are just a representation of the rotation $D(\alpha\beta\gamma)$ in an angular momentum basis $|Lm\rangle$;

$$D^L_{m,n}(\alpha\beta\gamma) \equiv \langle Lm|D(\alpha\beta\gamma)|Ln\rangle,$$

it follows that to investigate the effect of a molecular symmetry operation O_M we need to determine the matrix elements

$$\langle Lm|O_M D(\Omega)|Ln\rangle.$$

Similarly, the application of a laboratory transformation such as a symmetry operation of the mesophase O_L yields

$$\langle Lm|O_L D(\Omega)|Ln\rangle.$$

Every symmetry operation for the molecule or for the phase gives a relation that the order parameters have to satisfy, that is

$$\overline{D^L_{m,n}} = \overline{\langle Lm|O_M D(\Omega)|Ln\rangle}, \tag{66}$$

for the molecule and

$$\overline{D^L_{m,n}} = \overline{\langle Lm|O_L}\,\overline{D(\Omega)|Ln\rangle}, \tag{67}$$

for the phase. Any point operation can be written as a certain combination of rotations and inversion. Let us consider therefore each of these at a time. Thus suppose that the symmetry operation of the mesophase O_L is a rotation; then it will transform the original laboratory frame L into another frame L' and give

$$\begin{aligned}\langle Lm|O_L D(M\text{-}L)|Ln\rangle &= \langle Lm|D(L\text{-}L')D(M\text{-}L)|Ln\rangle,\\ &= \sum_q D^L_{m,q}(L\text{-}L')D^L_{q,n}(M\text{-}L),\end{aligned} \tag{68}$$

where for greater clarity we have written the rotation from the laboratory to the molecular frame as D(M-L) instead of D(Ω). Quite analogously, if the molecule has a rotation O_M amongst its symmetry operations, we find

$$\langle Lm|O_M D(M\text{-}L)|Ln\rangle = \langle Lm|D(M\text{-}L)D(M'\text{-}M)|Ln\rangle,$$
$$= \sum_q D^L_{m,q}(M\text{-}L)D^L_{q,n}(M'\text{-}M). \quad (69)$$

Notice that successive rotations D are applied in reverse order, that is from left to right (cf. the Appendix, eq. A8). If a symmetry operation O contains the inversion operation, I, it is clear that to investigate its effect we also need the matrix elements $\langle Lm|I|Ln\rangle$. These are obtained by remembering [16a] that $I|Lm\rangle = (-)^L|Lm\rangle$. Therefore [17]

$$\langle Lm|I|Ln\rangle = (-)^L\delta_{mn}. \quad (70)$$

The inversion and rotation operations commute:

$$\langle Lm|ID|Ln\rangle = \sum_q \langle Lm|I|Lq\rangle\langle Lq|D|Ln\rangle,$$
$$= (-)^L D^L_{m,n},$$
$$= \langle Lm|DI|Ln\rangle.$$

Therefore, for our purposes, the inversion has the same effect in either a laboratory or a molecular frame.

Let us illustrate the previous arguments with a simple example. Suppose that the mesophase has a plane of symmetry perpendicular to the z axis, that is $O_L = \sigma_h = \sigma(xy)$. Since we can write [16] $\sigma(xy) = ID(\pi 00)$ we find

$$\langle Lm|\sigma(xy)_L D|Ln\rangle = \sum \langle Lm|I|Lq\rangle\langle Lq|D(\pi 00)|Lp\rangle\langle Lp|D|Ln\rangle,$$
$$= (-)^L\delta_{mq}\exp(-iq\pi)\delta_{qp}D^L_{p,n},$$
$$= (-)^{L+m}D^L_{m,n}.$$

A symmetry plane σ(xy) in the molecule would give instead

$$\langle Lm|\sigma(xy)_M D|Ln\rangle = (-)^{L+n}D^L_{m,n}.$$

Proceeding in a similar way we obtain the results reported in Table 1 for the effect of various molecular symmetry operations on the Wigner functions. The effect of mesophase symmetries can also be easily obtained from Table 1 remembering that the order of operations is changed so that for example, the first subscript, m, in $D^L_{m,n}$ is changed instead of n.

Table 1 *Effect of various molecular symmetry operations O_M on the Wigner rotation matrices. The notation for the symmetry operators [16e] is as follows. I stands for the inversion; $\sigma(xy)$ for a symmetry plane perpendicular to z; σ_ϕ for a plane making an angle ϕ with (zx). A rotation of $2\pi/k$ about z is written as $C_k(z)$ and a continuous rotation of γ about z as $C_\gamma(z)$. A rotation of π about an axis perpendicular to z and making an angle ϕ with x is indicated by $C_2(\phi)$. S_k stands for a k-fold rotoreflection axis.*

Operator O_M	$\langle Lm\|O_M D\|Ln\rangle$
$I = S_2$	$(-)^L D^L_{m,n}$
$\sigma(xy)$	$(-)^{L+n} D^L_{m,n}$
$\sigma(xz)$	$(-)^n D^L_{m,-n}$
$\sigma(yz)$	$D^L_{m,-n}$
σ_ϕ	$(-)^n \exp(-i2n\phi) D^L_{m,-n}$
$C_2(z)$	$(-)^n D^L_{m,n}$
$C_k(z)$	$\exp(-in2\pi/k) D^L_{m,n}$
$C_\gamma(z)$	$\exp(-in\gamma) D^L_{m,n}$
$C_2(x)$	$(-)^L D^L_{m,-n}$
$C_k(x)$	$(-)^{L-n} \exp(-in2\pi/k) D^L_{m,-n}$
$C_\gamma(x)$	$(-)^{L-n} \exp(-in\gamma) D^L_{m,-n}$
$C_2(y)$	$(-)^{L-n} D^L_{m,-n}$
$C_\beta(y)$	$\sum_q D^L_{m,q} d^L_{q,n}(\beta)$
$C_2(\phi)$	$(-)^L \exp(-i2n\phi) D^L_{m,-n}$
S_k	$(-)^{L+n} \exp(-in2\pi/ D^L_{m,-n}$

Having studied the effect of symmetry transformations on the Wigner rotation matrices it remains to substitute eqs.(66) and (67) to obtain the corresponding symmetries of the order parameters. Thus, in the example we have just seen, a $\sigma(xy)$ plane in the laboratory gives

$$\overline{D^L_{m,n}} = (-)^{L+m} \overline{D^L_{m,n}},$$

while a $\sigma(xy)$ operation in the molecule yields

$$\overline{D^L_{m,n}} = (-)^{L+n}\overline{D^L_{m,n}}.$$

In general an additional symmetry for the order parameters follows immediately from the reality of $f(\Omega)$ and the relation $D^{L*}_{m,n} = (1)^{m-n}D^L_{-m,-n}$ for the Wigner functions. Thus

$$\overline{D^{L*}_{m,n}} = (-)^{m-n}\overline{D^{L*}_{-m,-n}}. \tag{71}$$

The symmetry relations just determined allow us to find the relevant order parameters for molecules and phases of various symmetry. As an example we give in Table 2 the surviving order parameters of second and fourth rank for a uniaxial phase and various molecular symmetries. Finally, it is perhaps worth mentioning that the relations we have found only tell us about the order parameters that can be different from zero. Thus they indicate which order parameters we can try to measure but, of course, do not say anything about their magnitudes. Indeed it may well turn out that molecules of low symmetry exhibit order parameters consistent, within experimental error, with some higher symmetry.

Ordering Matrix

An alternative definition for order parameters in uniaxial phases can be obtained by expanding the singlet orientational distribution $f(\Omega)$ in terms of the direction cosines $\ell_\alpha(\alpha = x,y,z)$ of the director with respect to a molecule fixed frame [19]. Thus

$$f(\Omega) = (1/8\pi^2)\ \{1+5\sum_{\alpha,\beta} S_{\alpha\beta}\ell_\alpha\ell_\beta + 9\sum_{\alpha,\beta,\gamma,\delta} S_{\alpha\beta\gamma\delta}\ \ell_\alpha\ell_\beta\ell_\gamma\ell_\delta+\ldots\}, \tag{72}$$

where

$$S_{\alpha\beta} = (3\overline{\ell_\alpha\ell_\beta}-\delta_{\alpha\beta})/2 \tag{73}$$

is called the Saupe ordering matrix and

$$\begin{aligned} S_{\alpha\beta\gamma\delta} = {} & [35\overline{\ell_\alpha\ell_\beta\ell_\gamma\ell_\delta} - 5(\overline{\ell_\alpha\ell_\beta}\delta_{\gamma\delta}+\overline{\ell_\alpha\ell_\gamma}\delta_{\beta\delta}+\overline{\ell_\alpha\ell_\delta}\delta_{\beta\gamma}+\overline{\ell_\beta\ell_\gamma}\delta_{\alpha\delta} \\ & + \overline{\ell_\beta\ell_\delta}\delta_{\alpha\gamma}+\overline{\ell_\gamma\ell_\delta}\delta_{\alpha\beta}) + (\delta_{\alpha\beta}\delta_{\gamma\delta}+\delta_{\alpha\gamma}\delta_{\beta\delta}+\delta_{\alpha\delta}\delta_{\beta\gamma})]/8. \end{aligned} \tag{74}$$

In general, expansions in Wigner matrices are more convenient for theoretical manipulations in view of their simple transformation properties. However the ordering matrix is used almost universally, as for example in nuclear magnetic resonance, so we give in Table 3 the relation between $S_{\alpha\beta}$ and $\overline{D^2_{0,n}}$. Notice that from eq. (73) it follows at once that the second rank S matrix is traceless

$$\sum_\alpha S_{\alpha\alpha} = 0.$$

Table 2. *Independent orientational order parameters of second and fourth rank for uniaxial phases and molecules of various symmetry. n_2 and n_4 indicate the number of independent real quantities to be determined. Schönflies notation is used for the point group symbols.*

Molecular point group	n_2	$\overline{D^2_{0,n}}$	n_4	$\overline{D^4_{0,n}}$
C_1,C_i	5	$\overline{D^2_{0,0}},\overline{D^2_{0,1}},\overline{D^2_{0,2}}$	9	$\overline{D^4_{0,0}},\overline{D^4_{0,1}},\overline{D^4_{0,2}},\overline{D^4_{0,3}},\overline{D^4_{0,4}}$
C_s,C_2,C_{2h}	3	$\overline{D^2_{0,0}},\overline{D^2_{0,2}}$	5	$\overline{D^4_{0,0}},\overline{D^4_{0,2}},\overline{D^4_{0,4}}$
C_{2v},D_2,D_{2h}	2	$\overline{D^2_{0,0}},\overline{D^2_{0,2}}=\overline{D^{2*}_{0,2}}$	3	$\overline{D^4_{0,0}},\overline{D^4_{0,2}}=\overline{D^{4*}_{0,2}},\overline{D^4_{0,4}}=\overline{D^{4*}_{0,4}}$
C_3,S_6	1	$\overline{D^2_{0,0}}$	3	$\overline{D^4_{0,0}},\overline{D^4_{0,3}}$
C_4,C_{4h},S_4	1	$\overline{D^2_{0,0}}$	3	$\overline{D^4_{0,0}},\overline{D^4_{0,4}}$
C_{3v},D_3,D_{3d}	1	$\overline{D^2_{0,0}}$	2	$\overline{D^4_{0,0}},\overline{D^4_{0,3}}=\overline{D^{4*}_{0,3}}$
C_{4v},D_{2d},D_{4h},D_4	1	$\overline{D^2_{0,0}}$	2	$\overline{D^4_{0,0}},\overline{D^4_{0,4}}=\overline{D^{4*}_{0,4}}$
C_5,C_{5h},C_{5v}	1	$\overline{D^2_{0,0}}$	1	$\overline{D^4_{0,0}}$
D_{4d},D_5,D_{5h},D_{5d}	1	$\overline{D^2_{0,0}}$	1	$\overline{D^4_{0,0}}$
C_{3h},C_6,C_{6h},C_{6v}	1	$\overline{D^2_{0,0}}$	1	$\overline{D^4_{0,0}}$
D_{3h},D_6,D_{6h},D_{6d}	1	$\overline{D^2_{0,0}}$	1	$\overline{D^4_{0,0}}$
$C_\infty,C_{\infty v},C_{\infty h},D_{\infty h}$	1	$\overline{D^2_{0,0}}$	1	$\overline{D^4_{0,0}}$
T,T_h,T_d,O,O_h	0		1	$\overline{D^4_{0,0}}$
Y,Y_h,KK_h	0		0	

Table 3 *The relation between Saupe's ordering matrix elements* $S_{\alpha\beta}$ *and the order parameters* $\overline{D^2_{0,n}}$

$$S_{zz} = \overline{D^2_{0,0}}$$

$$S_{xx}-S_{yy} = (6)^{\frac{1}{2}} \text{ Re } \overline{D^2_{0,2}} = (3/2)^{\frac{1}{2}} (\overline{D^2_{0,2}} + \overline{D^2_{0,-2}})$$

$$S_{xy} = -(3/2)^{\frac{1}{2}} \text{Im } \overline{D^2_{0,2}} = i\ (3/8)^{\frac{1}{2}} (\overline{D^2_{0,-2}} - \overline{D^2_{0,2}})$$

$$S_{xz} = -(3/2)^{\frac{1}{2}} \text{ Re } \overline{D^2_{0,1}} = (3/8)^{\frac{1}{2}} (\overline{D^2_{0,-1}} - \overline{D^2_{0,1}})$$

$$S_{yz} = (3/2)^{\frac{1}{2}} \text{ Im } \overline{D^2_{0,1}} = i\ (3/8)^{\frac{1}{2}} (\overline{D^2_{0,1}} + \overline{D^2_{0,-1}})$$

Smectics

In a system with some translational order the distribution functions depend on position as well as orientation. We consider here the singlet distribution $P^{(1)}(\mathbf{r},\Omega)$ for such systems and introduce order parameters for smectic phases. We shall assume, for simplicity, that the system is contained in a large cubic box of edge λ. Then the functions $\psi_k = \lambda^{-3/2} \exp(i\mathbf{k}.\mathbf{r})$, where the lattice vector $\mathbf{k}$ has components $k_\alpha = 0$, $\pm 2\pi/\lambda$, $\pm 4\pi/\lambda,\ldots,\pm n2\pi/\lambda$, $\alpha = x,y,z$, constitute a suitable orthonormal set and we can write

$$P^{(1)}(\mathbf{r},\Omega) = \sum P_{Lmn}(\mathbf{k})\exp(i\mathbf{k}.\mathbf{r})D^L_{m,n}(\Omega), \tag{75}$$

with

$$P_{Lmn}(\mathbf{k}) = \{(2L+1)/8\pi^2V\}\int_V d\mathbf{r}d\Omega\ P^{(1)}(\mathbf{r},\Omega)\exp(-i\mathbf{k}.\mathbf{r})D^{L*}_{m,n}(\Omega), \tag{76}$$

because of the orthogonality condition

$$(1/\lambda^3)\int_V d\mathbf{r}\ \exp(i\mathbf{k}.\mathbf{r})\exp(-i\mathbf{k}'.\mathbf{r}) = \delta(\mathbf{k}-\mathbf{k}'). \tag{77}$$

Eq. (75) is a very general expression which holds in principle for crystals, smectics and nematics as special cases. Thus for a monoatomic lattice, $L = m = n = 0$, we have the Fourier expansion [2c]

$$P^{(1)}(\mathbf{r},\Omega) = \sum P_{000}(\mathbf{k})\exp(i\mathbf{k}.\mathbf{r}), \tag{78}$$

where $\mathbf{k}$ is a vector in reciprocal space and the spatial integrations in eq. (76) can be limited to the repeat unit. For a uniform nematic or an isotropic liquid eq. (75) becomes

$$P^{(1)}(\mathbf{r},\Omega) = \sum P_{Lmn}(0)D^L_{m,n}(\Omega). \tag{79}$$

Here the container size is assumed to be large, in the sense that $\lambda \to \infty$, so that finite size effects are negligible and we can only have $\mathbf{k}=0$.

For a smectic phase with positional order in one dimension only (e.g. a smectic A or C phase) we can restrict the series (75) to

$$p^{(1)}(z,\Omega) = P_{Lmn}(k)\exp(ikz)D^L_{m,n}(\Omega), \tag{80}$$

where $k \equiv k_z$ and z is parallel to the layer normal. The expansion can be simplified by using the symmetry of the molecule and the mesophase. For example in a smectic A, with uniaxial symmetry about z,

$$p^{(1)}(z,\Omega) = \sum P_{L0n}(k)\exp(ikz)D^L_{0,n}(\Omega). \tag{81}$$

Now suppose that the phase also has a symmetry plane $\sigma(xy)$, perpendicular to the z direction; then

$$P_{L0n}(k) = (-)^L P_{L0n}(-k).$$

Consequently, for k=0, this implies that L has to be even. If L is even because of molecular symmetry then $P_{L0n}(k) = P_{L0n}(-k)$ and eq. (81) can be reduced to

$$p^{(1)}(z,\Omega) = \sum_{M=0}^{\infty} \sum (2-\delta_{M0})P_{L0n}(2\pi M/d)\cos(2\pi Mz/d)D^L_{0,n}(\Omega), \tag{82}$$

where d is the average interlayer spacing and the notation $|k| = (2\pi M/d)$ has been introduced. Of course, for cylindrically symmetric ($D_{\infty h}$) molecules, this reduces to the expression given by Wojtowicz [1b],

$$P(z,\cos\beta) = \sum_{M=0}^{\infty} P_{L;M}\cos(2\pi Mz/d)P_L(\cos\beta), \text{ L even}, \tag{83}$$

which we normalize so that

$$\int_0^d dz\int_0^\pi d\beta\sin\beta P(z,\cos\beta) = N. \tag{84}$$

Then

$$P_{L;M} = (2L+1)\langle\cos(2\pi Mz/d)P_L(\cos\beta)\rangle/2d, \tag{85}$$

$$P_{0;0} = N/2d, \tag{86}$$

$$P_{0;M} = \langle\cos(2\pi Mz/d)\rangle/2d, \tag{87}$$

$$P_{L;0} = (2L+1)\langle P_L\rangle/2d, \tag{88}$$

where

$$\langle A\rangle = \int_0^d dx\int_0^\pi d\beta\sin\beta P(z,\cos\beta)A \tag{89}$$

and the coefficients $P_{L;M}$ represent the order parameters for the system. In particular the terms $P_{0;M}$ and $P_{L;0}$ give the positional and orientational order respectively.

Orientational Order Parameters for More Complex Phases

A group theoretical classification of all the possible ordered phases has been given by Goshen *et al.* [15]. Here, however, we do not want to give all the possible order parameters for every phase; instead we begin by making some general remarks, and then say something about the smectic C phase. Firstly we point out that phases possessing translational order should be given a space group like classification [15]. Thus the symmetrization of the distribution functions and of the order parameters should be done by applying all the operations of the relevant space group. This can be rather involved. Notice, however, that the purely orientational order parameters, which define the distribution $f(\Omega)$, (cf. eq. 53) can be classified simply according to the point group for the system. This is, of course, because the purely orientational order parameters are invariant under translation; they correspond to $k=0$ in expansion (75). Thus to find out the orientational parameters for a phase of a given crystallographic point group we can use exactly the same arguments as before. Consider an ideal smectic C for example. This phase is a tilted version of the smectic A phase, where the director or major optic axis make an angle θ, known as the tilt angle, with the layer normal. The point group of this model is then C_{2h}, with Z parallel to the layer and Y parallel to the layer normal; here we use X,Y,Z for the laboratory frame. According to Table 1 the symmetry operations of C_{2h} give

$$\overline{D^L_{m,n}} = (-)^m \overline{D^L_{m,n}} = (-)^{L+m} \overline{D^L_{m,n}}; \tag{90}$$

thus L and m have to be even. For second rank the possible order parameters should then be

$$\overline{D^2_{0,n}} \quad ; \quad \overline{D_{2,n}} = (-)^n \overline{D^{2*}_{-2,n}}. \tag{91}$$

Notice that, even if the particles constituting the phase have cylindrical symmetry, we still have two independent order parameters, corresponding to the biaxiality of the phase. This is perhaps more transparent in cartesian coordinates. Let us define an ordering matrix **Q**

$$Q_{\alpha\beta} = (3\overline{\ell_\alpha \ell_\beta} - \delta_{\alpha\beta})/2 \tag{92}$$

formally identical to the Saupe ordering matrix. However, here it is the phase that is not cylindrically symmetric instead of the particles, and so we take ℓ_α to be the direction cosines for a molecule in the laboratory frame. Now, from the assumed symmetry of the smectic C it is clear that

$$\mathbf{Q} = \begin{pmatrix} Q_{XX} & Q_{XY} & 0 \\ Q_{XY} & Q_{YY} & 0 \\ 0 & 0 & Q_{ZZ} \end{pmatrix}. \tag{93}$$

This can be diagonalized by a rotation of θ about Z, provided $\tan 2\theta = 2Q_{XY}/(Q_{XX}-Q_{YY})$. The angle θ is just the tilt angle, which represents the orientation of the director in the (XY) plane. The diagonalized form of **Q** is

$$\mathbf{Q} = \begin{pmatrix} -\tfrac{1}{2}Q_{ZZ} + \xi & 0 & 0 \\ 0 & -\tfrac{1}{2}Q_{ZZ} - \xi & 0 \\ 0 & 0 & Q_{ZZ} \end{pmatrix},$$

where $\xi = \tfrac{1}{2}\{(Q_{XX}-Q_{YY})^2 + 4Q_{XY}^2\}^{\frac{1}{2}}$ is a biaxiality parameter.

Pair Distribution

In the previous section we have seen how symmetry can be used to limit the number of order parameters and hence the possible form of the singlet distribution function. Similar arguments can be applied to the pair distribution $G(X_1,X_2)$ or, rather, $G(r_{12},\Omega_1,\Omega_2)$ since we shall confine our attention to translationally invariant liquid crystals and so ignore the smectic phases. We use the methods developed by Steele [8], Jepsen and Friedman[18] and Blum and Torruella [17] for ordinary fluids. Let us start by noticing that $G(\mathbf{r}_{12},\Omega_1,\Omega_2)$ when written in a laboratory frame has to depend on the intermolecular separation r_{12} as well as the orientations Ω_1,Ω_2 of molecules 1 and 2, together with the orientation of the intermolecular vector Ω_r. Therefore we can write down immediately

$$G(r_{12},\Omega_1,\Omega_2) = \sum G^{m_1m_2m;n_1n_2}_{L_1L_2L}(r_{12})D^{L_1}_{m_1,n_1}(\Omega_1)D^{L_2}_{m_2,n_2}(\Omega_2)D^{L}_{m,o}(\Omega_r), \tag{94}$$

where this distribution should be invariant under the symmetry operations of the molecules and of the phase. Moreover permutation of identical molecules should leave G unaltered. We can therefore symmetry adapt G so that it transforms according to the totally symmetric representation of the group of transformations just mentioned. Projection onto the totally symmetric representation of a group can be achieved [8,16] by application of a projection operator P

$$P = (1/n_s)\sum_{s=1}^{n_s} O_s, \tag{95}$$

for a group of n_s transformations O_s. For a continuous group eq. (95) becomes

$$P = (1/\int ds)\int ds O_s \tag{96}$$

with an obvious change of notation.

It is clear that application of P to eq. (94) again requires investigating the effect of the operations O_s on the Wigner functions. Since we have already examined the effect of rotations and reflections the only new effect we have to study is that of permuting 1 and 2. This is rather easy, since, apart from exchanging subscripts, the permutation transforms $\mathbf{r}_{12}$ in $\mathbf{r}_{21} = -\mathbf{r}_{12}$ and then $D^L_{m,o}(r\text{-}L)$ into $(-)^L D^L_{m,o}(r\text{-}L)$. Thus, for identical molecules, we obtain the condition

$$G^{m_1 m_2 m; n_1 n_2}_{L_1 L_2 L}(r) = (-)^L G^{m_2 m_1 m; n_2 n_1}_{L_2 L_1 L}(r). \tag{97}$$

Consider a rotationally invariant fluid, such as a nematic above its clearing point or an unoriented nematic. In this case the phase symmetry is that of the full rotation group and the projector P is simply $(1/8\pi^2)\int d\Omega D(\Omega)$. Application of P gives

$$G(r_{12},\Omega_1,\Omega_2) = \sum G^{m_1 m_2 m; n_1 n_2}_{L_1 L_2 L} D^{L_1}_{q_1 n_1}(\Omega_1) D^{L_2}_{q_2,n_2}(\Omega_2) D^{L}_{q,o}(\Omega_r)$$

$$\times \int d\Omega D^{L_1}_{m_1,q_1} D^{L_2}_{m_2,q_2} D^{L}_{m,q}/8\pi^2. \tag{98}$$

Use of the result for the integral of three rotation matrices [20] (cf. eq. A6) yields the most general form for the rotationally invariant distribution G as

$$G(r_{12},\Omega_1,\Omega_2) = \sum G^{q_1 q_2; n_1 n_2}_{L_1 L_2 L}(r_{12}) D^{L_1}_{q_1,n_1}(\Omega_1) D^{L_2}_{q_2,n_2}(\Omega_2) D^{L}_{-q_1-q_2,o}(\Omega_r), \tag{99}$$

where

$$G^{q_1 q_2; -n_1 n_2}_{L_1 L_2 L}(r_{12}) = \{(-)^{m-q_1-q_2}/(2L+1)\} C(L_1 L_2 L; m_1 m_2) C(L_1 L_2 L; q_1 q_2) \delta_{m_1+m_2,-m}$$

$$\times\, G^{m_1 m_2 m; n_1 n_2}_{L_1 L_2 L}(r_{12}) \tag{100}$$

and C(abc;de) are Clebsch-Gordan coefficients [3]. Notice that the requirement of rotational invariance has severely limited the form of eq. (94); in fact the Clebsch-Gordan coefficients cause the ranks L_1, L_2, L to satisfy the triangular condition; $|L_1 - L_2| \leqslant L \leqslant L_1 + L_2$.

Let us now make the further assumption that the intermolecular vector is spherically distributed so that rotating $\mathbf{r}_{12}$ does not change G. A projection similar to the previous one gives $\delta_{L,0}\delta_{q_1,-q_2}$ in eq. (99). Since $C(L_1 L_2 0; q\,{-q}) = (-)^{L_1 - q}(2L_1+1)^{-\frac{1}{2}}\delta_{L_1 L_2}$, we find

$$G(r_{12},\Omega_1,\Omega_2) = \sum(-)^q G_L^{n_1 n_2}(r_{12}) D^L_{q,n_1}(\Omega_1) D^L_{-q,n_2}(\Omega_2)(2L+1)/64\pi^4, \quad (101)$$

where

$$G_L^{n_1 n_2}(r_{12}) = \sum(-)^{m_1} G_{LL0}^{m_1 -m_1 0;n_1 n_2}(8\pi^2/2L+1)^2 \quad (102)$$

and we have removed unnecessary subscripts. Eq. (101) can be written in a more transparent form by using the property

$$D^L_{q,n_1}(\Omega_1) = D^{L*}_{n_1,q}(-\Omega_1), = (-)^{n_1-q} D^L_{-n_1,-q}(-\Omega_1)$$

and the closure of the Wigner functions eq. (A8). Thus we find

$$G(r_{12},\Omega_1,\Omega_2) = \sum(-)^{n_1}\{(2L+1)/64\pi^4\} G_L^{n_1 n_2}(r_{12}) D^L_{-n_1,n_2}(\Omega_{12}), \quad (103)$$

where Ω_{12} is the angle of rotation from molecule 1 to molecule 2. This shows that for a rotationally invariant fluid the reduced pair distribution $G(r_{12},\Omega_1,\Omega_2)$, which does not depend on the orientation of the intermolecular vector must be a function of relative orientations only. The same conclusion obviously applies to $G(r_{12},\Omega_1,\Omega_2)$ for any fluid where the distribution of intermolecular vectors is spherically symmetric. For such a situation, if the molecules are cylindrically symmetric, $n_1=n_2=0$ and

$$G(\mathbf{r}_{12},\Omega_1,\Omega_2) = G(r_{12},\Omega_{12}), = \sum\{(2L+1)/64\pi^4\} G_L^{00}(r_{12}) D^L_{0,0}(\Omega_{12}). \quad (104)$$

This simple equation has some important features. First of all because it is fully rotationally invariant it can be used to describe the onset of orientational order at an isotropic-nematic transition in the absence of external fields. Remember in fact that when we have introduced order parameters from the singlet distribution we have, more or less, assumed implicitly the existence of an aligned mesophase with a uniform director. This, in turn, implies the existence of an external field to break the full rotational symmetry of the mesophase above the isotropic transition. While this is perfectly legitimate it would be more satisfactory to be able to discuss how long-range ordering is set up in a system for which the nematic phase had been formed in the absence of a field. It is physically intuitive that to do this we have to resort to the pair distribution since this can tell us about the extent of orientational correlations without having to resort to a laboratory principal axis system.

Let us interpret the coefficients $G_L^{00}(r_{12})$ in expansion (104). From the definition of $G(r_{12},\Omega_{12})$ and the orthogonality of the Legendre polynomials we find

$$G_L(r_{12}) = \int d\Omega_1 d\Omega_2 G(r_{12},\Omega_{12}) D^L_{0,0}(\Omega_{12}),$$
$$= \overline{P_L\{\cos\beta_{12}(r_{12})\}} G^{00}_0(r_{12}),$$
$$\equiv G^{00}_L(r_{12}) G^{00}_0(r_{12}). \quad (105)$$

Thus $G_L(r_{12})$ is an Lth rank angular correlation. For L=0,

$$G^{00}_0(r_{12}) = \int d\Omega_1 d\Omega_2 G(r_{12},\Omega_{12}),$$

is simply the centre of mass pair distribution. Any $G_L(r_{12})$ with $L\neq 0$, say $G_2(r_{12})$, expresses the correlation in orientation between a molecule at the origin and another, distance r_{12} from it; we have, of course $G_L(0) = 1$. For short interparticle separations *e.g.* when r_{12} is the average nearest neighbour distance, $G_2(r_{12})$ can be taken to define the short-range order. For large separations $G_2(r_{12})$ is a measure of the long-range order. In this limit the pair distribution $G(\mathbf{r}_{12},\Omega_2,\Omega_2)$ reduces to the product of two singlet orientational distribution functions $f(\Omega_1)f(\Omega_2)$ (cf. eq. 36) and, similarly, the symmetrized distribution $\mathcal{P}G(\mathbf{r}_{12},\Omega_1,\Omega_2)$ tends to the symmetrized product $\mathcal{P}\{f(\Omega_1)f(\Omega_2)\}$. For a rotationally invariant phase we have

$$P\{f(\Omega_1)f(\Omega_2)\} = \sum\{(2L+1)/64\pi^4\} \overline{D^L_{n_1,n_2}(\Omega_{12}) D^L_{n_1,n_2}(\Omega_{12})}, \quad (106)$$

with $n_1 = n_2 = 0$ for uniaxial molecules. Thus we find

$$G_L(r_{12}) = \sum_q \overline{D^{L*}_{q,0}(\Omega_1)}\ \overline{D^L_{q,0}(\Omega_2)}, \quad (107)$$

for large r_{12}. Now, if r_{12} is a separation much larger than any molecular distance but small compared to the coherence length of the director we obtain

$$G_L(r_{12}) = \overline{D^L_{0,0}(\Omega_1)}\ \overline{D^L_{0,0}(\Omega_2)}$$
$$= \overline{P}^2_L , \quad (108)$$

for large r_{12} when we assume local uniaxial symmetry. Thus we expect the behaviour of, say, $G_2(r_{12})$ to change dramatically at the nematic-isotropic transition. Above the clearing point in fact $G_2(r_{12})$ should decay to zero with distance while in the nematic phase it should tend to the square of the order parameter. This trend is apparent in our computer simulations [21] described in Chapter 9 of the Maier-Saupe-Lasher[22] lattice model, where we find that in the ordered phase $G_2(r_{12})$ decays to a plateau value of $\overline{P_2}^2$ after a relatively short distance corresponding to a few shells of neighbours.

Let us add a few cautionary words, however, on what we mean by the large distance limit in eq. (107). We have mentioned before that the separation r_{12} although large compared to molecular distances, should be small compared to the coherence length of the director. This is, of course, because in an unoriented nematic and for truly macroscopic separations r_{12} the director will have to arrange itself to preserve the spherical symmetry of the phase. Thus for distances greater than some coherence length the correlation will be apparently lost. In the same way there is no net macroscopic order in an unoriented nematic while there is a long-range order given by the plateau of $G_2(r_{12})$.

Another advantage of introducing the pair correlation coefficients $G_L(r_{12})$ is that they can be used to describe the pretransitional effects associated with the onset of ordering in the isotropic phase when approaching the nematic transition. In this connection it is useful to define a coefficient G_L by

$$G_L \equiv \int dr_{12} r_{12}^2 G_L\,(r_{12}),$$
$$= \int dr_{12} r_{12}^2 d\Omega_1 d\Omega_2 G(r_{12},\Omega_{12}) P_L\{\cos\beta_{12}(r_{12})\},$$

as the area under the correlation $r^2 G_L\,(r)$. They can be used to describe experiments in the pretransitional region such as the optical birefringence induced by a strong electric field (Kerr effect).

Nematics and Cholesterics

To examine briefly some differences between nematics and cholesterics, let us go back to the expansion (101) for a rotationally invariant phase. If the fluid is also centrosymmetric the pair distribution has to be invariant under inversion. Therefore we have for such a system that $(L_1 + L_2 + L)$ has to be even in eq. (101). The same is not true if the fluid is formed by optically active molecules. Thus for cholesterics we have that $(L_1 + L_2 + L)$ can be odd as well. The presence of these odd terms indicates that the probability of finding two molecules at a given orientation can depend also on the sign of the angle of the second molecule with respect to the first. In other words the presence of odd terms means that the characteristic twist of cholesterics can be obtained. This aspect is described in greater detail in Chapter 5.

Acknowledgement

It is a pleasure to thank Professor G.R. Luckhurst for many stimulating and valuable discussions. This work was supported in part by Laboratorio C.N.R. (Ozzano E.).

Appendix

Irreducible tensors and Wigner rotation matrices

A tensor of rank n is a quantity that transforms under rotation as the nth direct power of a vector. The 3^n dimensional representation of the rotation group realized in this way can be decomposed into a set of irreducible representations $D^{(L)}$ each of dimension (2L+1). If we describe the rotations in terms of Euler angles ($\alpha\beta\gamma$) (cf. Chapter 17) then the matrix elements of the irreducible representation in a basis where J^2, J_z (*i.e.* the angular momentum and its projection) are diagonal, are [3]

$$D^L_{m,n}(\alpha\beta\gamma) = \langle Lm|e^{-i\alpha J_z}e^{-i\beta J_y}e^{-i\gamma J_z}|Ln\rangle. \tag{A1}$$

The matrix elements $D^L_{m,n}(\alpha\beta\gamma)$ are called Wigner rotation matrices, Wigner functions or generalized spherical harmonics. Combinations of ordinary tensor components transforming according to the representation $D^{(L)}$ are called irreducible tensor components of rank L and denoted by $T^{(L,m)}$, *e.g.*

$$T^{(L,m)'} = \sum_n D^L_{n,m}(\Omega)T^{(L,n)}. \tag{A2}$$

Equation (A2) illustrates the main reason for the usefulness of irreducible tensors in problems involving rotations *i.e.* that their transformation properties are very simple. The set of (2L+1) components, $T^{(L)}$, is called an irreducible tensor of rank L. The Euler angles Ω in eq. (A2) determine the rotations which carry the original (unprimed) coordinate system into the rotated (primed) one. We follow the convention of Rose [3] for the Euler angles. From eq. (A1) it is apparent that we can express $D^L_{m,n}(\alpha\beta\gamma)$ as

$$D^L_{m,n}(\alpha\beta\gamma) = e^{-im\alpha}d^L_{m,n}(\beta)e^{-in\gamma}, \tag{A3}$$

where the real quantities,

$$d^L_{m,n}(\beta) \equiv \langle Lm|e^{-i\beta J_y}|Ln\rangle, \tag{A4}$$

are called reduced or small Wigner matrices.

The functions $D^L_{m,n}(\alpha\beta\gamma)$ constitute a complete orthogonal set spanning the space of the angles α,β,γ. Some of their properties, which we frequently use, are:

Orthogonality

$$\int d\alpha \sin\beta d\beta d\gamma \; D^{L*}_{m,n}(\alpha\beta\gamma) D^{L'}_{m',n'}(\alpha\beta\gamma) = \{8\pi^2/(2L+1)\}\delta_{m,m'}\delta_{n,n'}\delta_{L,L'} \; . \quad (A5)$$

Integral of three Wigner rotation matrices

$$\int d\alpha \sin\beta d\beta d\gamma \; D^{L''*}_{m'',n''}(\alpha\beta\gamma) D_{m',n'}(\alpha\beta\gamma) D^{L}_{m,n}(\alpha\beta\gamma)$$

$$= \{8\pi^2/(2L''+1)\}\delta_{m+m',m''}\delta_{n+m',m''} C(LL'L'',;mm')C(LL'L'';nn'), \quad (A6)$$

where C(abc;de) is a Clebsch-Gordan coefficient [3]. An equivalent expression can be obtained in terms of 3-j symbols [20] remembering that

$$C(abc;de) = (-)^{a-b-f}\delta_{d+e+f,0}(2c+1)^{\frac{1}{2}} \begin{pmatrix} a & b & c \\ d & e & f \end{pmatrix}. \quad (A7)$$

Closure

$$\sum_n D^{L}_{m,n}(\alpha_1\beta_1\gamma_1) D^{L}_{n,m'}(\alpha_2\beta_2\gamma_2) = D^{L}_{m,m'}(\alpha\beta\gamma), \quad (A8)$$

where $(\alpha\beta\gamma)$ is the resultant of $(\alpha_1\beta_1\gamma_1)$ then $(\alpha_2\beta_2\gamma_2)$.

Symmetry

$$D^{L*}_{m,n}(\alpha\beta\gamma) = (-)^{m-n}D^{L}_{-m,-n}(\alpha\beta\gamma) = D^{L}_{n,m}(-\gamma-\beta-\alpha). \quad (A9)$$

Products

$$D^{L'}_{m',n'}D^{L''}_{m'',n''} = \sum_{L=|L'-L'|}^{L'+L''} C(L'L''L;m'm'')C(L'L''L;n'n'')D^{L}_{m'+m'',n'+n''}. \quad (A10)$$

Special cases

$$D^{L}_{m,0}(\alpha\beta 0) = \{4\pi/(2L+1)\}^{\frac{1}{2}} Y^{*}_{L,m}(\beta\alpha), \quad (A11)$$

$$D^{L}_{0,0}(0\beta 0) = d^{L}_{0,0}(\beta) = P_L(\cos\beta),$$

where $Y_{L,m}$ is a spherical harmonic and P_L a Legendre polynomial.

We now give the explicit expressions for the Wigner rotation matrices (and implicitly the order parameters) of rank L = 0,1,2,4. From eq.(A3) we see that what we really need are expressions for the small matrices $d^{L}_{m,n}(\beta)$. Some partial tabulations can be found in the literature [20]. In general however the $d^{L}_{m,n}$ can be obtained from the Wigner formula [3]

$$d^{L}_{m,n}(\beta) = \sum_{\chi} C^{Lmn}_{qp} \cos^{q}(\beta/2)\sin^{p}(\beta/2), \tag{A12}$$

where

$$q \equiv 2L + m-n -2\chi,$$

$$p \equiv n-m + 2\chi,$$

$$C^{Lmn}_{qp} \equiv (-)^{\chi} \frac{\{(L+m)!(L-m)!(L+n)!(L-n)!\}^{\frac{1}{2}}}{(L-n-\chi)!(L+m-\chi)!(\chi+n-m)!\chi!}$$

and the sum is over the values of the integer χ for which the factorial arguments are greater or equal to zero. The number of $d^{L}_{m,n}$ of a given rank L to be calculated with this rather unwieldy formula is, in principle $(2L+1)^{2}$. This very high number of $d^{L}_{m,n}$ for even moderately high L can be drastically reduced if symmetry relations are taken into account

$$d^{L}_{m,n}(\beta) = (-)^{m-n} d^{L}_{-m,-n}(\beta), \tag{A13}$$

$$= (-)^{m-n} d^{L}_{n,m}(\beta). \tag{A14}$$

Another helpful relation apparently never used in this context is

$$d^{L}_{m,n}(\beta) = (-)^{L+m} d^{L}_{m,-n}(\pi-\beta). \tag{A15}$$

We shall find this rather advantageous since it gives

$$d_{m,-n}(\beta) = (-)^{L+m} \sum C^{Lmn}_{qp} \sin^{q}(\beta/2)\cos^{p}(\beta/2), \tag{A16}$$

because $\cos(\pi/2-\beta/2) = \sin(\beta/2)$ and $\sin(\pi/2-\beta/2)=\cos(\beta/2)$. Eq. (A16) shows the important fact that if the $d^{L}_{m,n}(\beta)$ are written as a combination of $\cos(\beta/2)$ and $\sin(\beta/2)$ powers then $d^{L}_{m,-n}$ is obtained by simply exchanging $\sin(\beta/2)$ and $\cos(\beta/2)$ and multiplying the resulting expression by $(-)^{L+m}$. As an example, from

$$d^{4}_{1,1} = c^{2}(c^{6}-15c^{4}s^{2}+30c^{2}s^{4}-10s^{6}),$$

we have at once

$$d^{4}_{1,-1} = -s^{2}(s^{6}-15s^{4}c^{2}+30s^{2}c^{4}-10c^{6}),$$

where $c \equiv \cos(\beta/2)$ and $s \equiv \sin(\beta/2)$. Applications of (A13-A15) allow the number of $d^{L}_{m,n}$ to be calculated to be reduced from $(2L+1)^{2}$ to $(L+1)(L+2)/2$; thus for L=4 only 15 elements are needed instead of 81.

In Table A1 we give explicit expressions for the most important cases L=0,1,2,4.

Table A1 *Explicit expressions for the small Wigner matrices* $d^L_{m,n}(\beta)$; *L=0, 1, 2, 4. Here c ≡ cos(β/2) and s ≡ sin(β/2).*

L = 0

$$d^0_{0,0} = 1$$

L = 1

$$d^1_{1,1} = d^1_{-1,-1} = c^2$$

$$d^1_{1,0} = d^1_{-1,0} = -d^1_{0,1} = d^1_{0,-1} = -\sqrt{2}sc$$

$$d^1_{1,-1} = d^1_{-1,1} = s^2$$

$$d^1_{0,0} = -1 + 2c^2$$

L = 2

$$d^2_{2,2} = d^2_{-2,-2} = c^4$$

$$d^2_{2,1} = -d^2_{-2,-1} = -d^2_{1,2} = d^2_{-1,-2} = -2c^3s$$

$$d^2_{2,0} = d^2_{-2,0} = d^2_{0,2} = d^2_{0,-2} = \sqrt{6}\ c^2 - \sqrt{6}\ c^4$$

$$d^2_{2,-1} = -d^2_{-2,1} = -d^2_{-1,2} = d^2_{1,-2} = -2s^3c$$

$$d^2_{2,-2} = d^2_{-2,2} = s^4$$

$$d^2_{1,1} = d^2_{-1,-1} = -3c^2+4c^4$$

$$d^2_{1,0} = -d^2_{-1,0} = -d^2_{0,1} = d^2_{0,-1} = \sqrt{6}\ cs - 2\sqrt{6}\ c^3s$$

$$d^2_{1,-1} = d^2_{-1,1} = 3s^2-4s^4$$

$$d^2_{0,0} = 1 - 6c^2+6c^4$$

L = 4

$$d^4_{4,4} = d^4_{-4,-4} = c^8$$

(Table A1 continued)

$$d^4_{4,3} = -d^4_{-4,-3} = -d^4_{3,4} = d^4_{-3,-4} = -2\sqrt{2}\ c^7 s$$

$$d^4_{4,2} = d^4_{-4,-2} = d^4_{2,4} = d^4_{-2,-4} = 2\sqrt{7}\ c^6 s^2$$

$$d^4_{4,1} = -d_{-4,-1} = -d^4_{1,4} = d^4_{-1,-4} = -2\sqrt{14}\ c^5 s^3$$

$$d^4_{4,0} = d^4_{-4,0} = d^4_{0,4} = d^4_{0,-4} = \sqrt{70}\ c^4 s^4$$

$$d^4_{4,-1} = -d^4_{-4,1} = -d^4_{-1,4} = d^4_{1,-4} = -\ 2\sqrt{14}\ c^3 s^5$$

$$d^4_{4,-2} = d^4_{-4,2} = d^4_{-2,4} = d^4_{2,-4} = 2\sqrt{7}\ c^2 s^6$$

$$d^4_{4,-3} = -d^4_{-4,3} = -d^4_{3,-4} = d^4_{-3,4} = -2\sqrt{2}\ c\ s^7$$

$$d^4_{4,-4} = d^4_{-4,4} = s^8$$

$$d^4_{3,3} = d^4_{-3,-3} = c^6(c^2-7\ s^2)$$

$$d^4_{3,2} = -d^4_{-3,-2} = -d^4_{2,3} = d^4_{-2,-3} = -\sqrt{14}\ c^5 s(c^2-3s^2)$$

$$d^4_{3,1} = d^4_{-3,-1} = d^4_{1,3} = d^4_{-1,-3} = \sqrt{7}\ c^4 s^2(3c^2-5s^2)$$

$$d^4_{3,0} = -d^4_{-3,0} = -d^4_{0,3} = d^4_{0,-3} = -\ 2\sqrt{14}\ c^3 s^3(c^2-s^2)$$

$$d^4_{3,-1} = d^4_{-3,1} = d^4_{-1,3} = d^4_{1,-3} = \sqrt{7}\ c^2 s^4(5c^2-3s^2)$$

$$d^4_{3,-2} = -d^4_{-3,2} = -d^4_{-2,3} = d^4_{2,-3} = -\sqrt{14}\ c\ s^5(3c^2-s^2)$$

$$d^4_{3,-3} = d^4_{-3,3} = s^6(7\ c^2-s^2)$$

$$d^4_{2,2} = d^4_{-2,-2} = c^4(c^4-12c^2s^2+15s^4)$$

$$d^4_{2,1} = -d^4_{-2,-1} = -d^4_{1,2} = d^4_{-1,-2} = -\sqrt{2}\ c^3 s(3\ c^4-15\ c^2s^2+\ 10\ s^4)$$

$$d^4_{2,0} = d^4_{-2,0} = d^4_{0,2} = d^4_{0,-2} = \sqrt{10}\ c^2 s^2(3\ c^4-8c^2s^2+\ 3\ s^4)$$

$$d^4_{2,-1} = -d^4_{-2,1} = -d^4_{-1,2} = d^4_{1,-2} = -\sqrt{2}c\ s^3(10\ c^4-15\ c^2s^2+3\ s^4)$$

$$d^4_{2,-2} = d^4_{-2,2} = s^4(15\ c^4-\ 12\ c^2s^2+\ s^4)$$

$$d^4_{1,1} = d^4_{-1,-1} = c^2(c^6-15\ c^4s^2\ +\ 30\ c^2s^4-\ 10\ s^6)$$

$$d^4_{1,0} = -d^4_{-1,0} = -d^4_{0,1} = d^4_{0,-1} = -2\sqrt{5}\ cs(c^6-6c^4s^2+6\ c^2s^4-s^6)$$

$$d^4_{1,-1} = d^4_{-1,1} = s^2(10\ c^6-30\ c^4s^2\ +\ 15\ c^2s^4\ -\ s^6)$$

$$d^4_{0,0} = c^8\ -\ 16\ c^6s^2\ +\ 36\ c^4s^4\ -\ 16\ c^2s^6\ +\ s^8$$

References

1a. see *e.g.* P.G. de Gennes, *The Physics of Liquid Crystals,* Oxford U.P., (1974).
1b. E.B. Priestley, P.J. Wojtowicz and P. Sheng (Eds), *Introduction to Liquid Crystals,* Plenum Press, (1975).
1c. G.W. Gray and P. A. Winsor (Eds), *Liquid Crystals and Plastic Crystals,* Ellis Horwood, (1974).
2a. R. Balescu, *Equilibrium and Non-equilibrium Statistical Mechanics,* Wiley, (1975).
2b. J.P. Hansen and I.R. McDonald, *Theory of Simple Fluids,* Academic Press, (1976).
2c. L. Landau and E.M. Lifshitz, *Statistical Physics,* 2nd Edn, Pergamon Press, (1969).
3. M.E. Rose, *Elementary Theory of Angular Momentum,* Wiley, (1957).
4. A.B. Pippard, *Elements of Classical Thermodynamics,* Cambridge U.P., (1966).
5. H.E. Stanley, *Introduction to Phase Transitions and Critical Phenomena,* Oxford U.P., (1971).
6. N. Metropolis, A.W. Rosenbluth, M.N. Rosenbluth, A.H. Teller and E. Teller, *J. Chem. Phys.* **21**, 1087 (1953).
7. J. Viellard-Baron, *J. Chem. Phys.* **56**, 4729 (1972); *Mol. Phys.* **28**, 809 (1974).
8. W.A. Steele, *J. Chem. Phys.* **39**, 3197 (1963).
9. A. Saupe, *Angew Chem. Int. Ed.* **7**, 97 (1968).
10. G.R. Luckhurst, Chapter 7 of Ref. 1c.
11. E.D. Priestley and P.S. Pershan, *Mol. Cryst. Liq. Cryst.* **23**, 369 (1973); S. Jen, N.A. Clark, P.S. Pershan and E.B. Priestley, *Phys. Rev. Letts,* 26, 1552 (1973).
12. G.R. Luckhurst and A. Sanson, *Mol. Phys.* **24**, 1297 (1972); 28, 809 (1974); G.R. Luckhurst, M. Setaka and C. Zannoni, *Mol. Phys.* 28, 49 (1974); G.R. Luckhurst and R.N. Yeates, *J. Chem. Soc. Faraday II,* 72, 996 (1976).
13. P.G. de Gennes, *C.R. Acad. Sci. Paris, Sêr* B274, 142 (1972).
14. R. Pynn, *J. Phys. Chem. Solids,* 34, 735 (1973); *Acta Cryst.* **A31**, 323 (1975); M. Kohli, K. Otnes, R. Pynn and T. Riste, *Z. Physik* **B24**, 147 (1976).
15. S. Goshen, D. Mukamel and S. Shrikman, *Mol. Cryst. Liq. Cryst.* **31**, 171 (1975).
16a. L. Landau and E.M. Lifshitz, *Quantum Mechanics,* Pergamon Press, (1966).
16b. M. Lax, *Symmetry Principles in Solid State and Molecular Physics,* Wiley, (1974).
16c. P.W. Atkins, M.S. Child and C.S.G. Phillips, *Tables for Group Theory,* Oxford U.P., (1970).
17. L. Blum and A.J. Torruella, *J. Chem. Phys.* **56**, 303 (1972).
18. D.W. Jepsen and H.L. Friedman, *J. Chem. Phys.* **38**, 846 (1963).
19. A.D. Buckingham, *Disc. Faraday Soc.,* 43, 205 (1967).
20. D.M. Brink and G.R. Satchler, *Angular Momentum,* Clarendon Press, (1968).
21. C. Zannoni, to be published.
22. G. Lasher, *Phys. Rev.* A5, 1350 (1972); P.A. Lebwohl and G. Lasher, *Phys. Rev.* **A6**, 426 (1972).

Chapter 4

MOLECULAR FIELD THEORIES OF NEMATICS

G. R. LUCKHURST

Department of Chemistry, The University, Southampton, SO9 5NH, England.

Introduction

A liquid crystal differs in many ways from the isotropic phase from which it is formed but the difference of prime importance is the extent of long-range orientational order. It is this difference which any molecular theory of liquid crystals should first explain. The development of such theories must begin with the formulation of the intermolecular potential and the identification of those features which are responsible for the transition from the liquid to the liquid crystal. Many years ago Born [1] realised that it is the anisotropic component of the pair potential which causes the order-disorder transition in nematics. At that time our knowledge of anisotropic potentials was extremely limited although something was known about the electrostatic interaction between a pair of electric dipoles. Born based his theory of nematics on this interaction but we now know that an electric dipole is not essential for the formation of a nematic, as we can see from the existence of such a phase for p-quinquephenyl. It was not until some years later that London [2] developed his theory of anisotropic dispersion forces which were later to be employed by Maier and Saupe [3] in one of the more successful theories of nematics. London dispersion forces, like the dipolar interaction, are long range. However the short-range contribution to the pair potential may also be anisotropic and these repulsive forces, which reflect the molecular shape, have also been proposed as those responsible for the structure of the liquid crystal phase.

The identification of the essential features of the pair potential is, of course, just the first problem for we must then tackle the statistical problem of developing expressions for the equilibrium properties of the system. As de Gennes [4] has written "The statistical mechanics of liquids is difficult; the statistical mechanics of nematics is still worse". It is easy to appreciate the reason for such a view since the pair potential appropriate for a nematic is clearly far more complex than that for a simple liquid. However as we shall see this view is too pessimistic, for a molecular theory of nematics can be relatively straightforward and indeed more successful than that for normal liquids. This seemingly paradoxical situation obtains provided we are not concerned with the complete behaviour of a nematic but restrict our attention to the essential differences between a nematic and the corresponding isotropic phase.

In this Chapter we concentrate on theories of nematics based on the long-range part of the anisotropic intermolecular potential; those founded on the short-range components are to be found in Chapters 7 and 8. The statistical mechanics of systems interacting via long-range anisotropic forces are conveniently handled with the aid of the molecular field approximation [5] and this is outlined in the following section. We shall then use this approximation together with a particularly simple pair potential to derive the principle results of the Maier-Saupe theory. The predictions of the theory are then compared with the results of computer simulation studies and with those of experiments on real nematogens. Having thus established the range of validity of the theory the form of the pair potential is extended to allow for a more realistic interaction between cylindrically symmetric molecules as well as to include deviations from molecular cylindrical symmetry.

The Molecular Field Approximation

Evaluation of the equilibrium configurational partition function

$$Q_N = (1/N!)\int\{d\mathbf{X}^N\}\exp[-\beta U(\{\mathbf{X}^N\})], \tag{1}$$

from which we obtain the thermodynamic properties, exemplifies the problem highlighted by de Gennes. Here we adopt the notation introduced in Chapter 3 so that $\{\mathbf{X}^N\}$ denotes both the spatial ($\mathbf{r}$) and orientational (Ω) coordinates of the N classical, rigid particles in the ensemble. The extreme difficulty in evaluating this multi-dimensional integral stems from the complicated dependence of the potential energy on the coordinates of the particles in a system with a density as high as that in a nematic. This difficulty does not occur in the gas phase where the density is so low that the particles behave independently of each other and the total partition function can be written as the product of N single particle partition functions each of which is readily evaluated.

The molecular or mean field approximation also reduces the total partition function to a product of single particle partition functions by writi the energy $U(\{\mathbf{X}^N\})$ as the sum of N energies each of which depends on the coordinates of an individual particle. In other words each molecule is viewed as moving in a field generated via its interaction with all the other molecules. This approximation will only be valid when the strength of the field experienced by a particular molecule is insensitive to the coordinates of every molecule except that being considered. We might anticipate therefore that the mean field approximation will be good for liquid crystals because of the inherent long-range order. There are various routes to the potential or pseudo-potential used to represent the interaction of a molecule with the mean field. One procedure involves the more or less intuitive averaging of the anisotropic pair potential over the coordinates of one particle [6] while a more rigorous approach starts with the partition function which is then factored into positional and orientational contributions [7]. An alternative derivation is based on a solution of the Kirkwood integral equations for the spatial and orientational distribution functions [8]. Here we shall adopt yet another approach, advocated by Woo and his colleagues [9] which develops the Bogoliubov-Born-Green-Kirkwood-Yvon hierarchy of equations for the distribution functions [10]. This approach is attractive, not only because it is simple, but also because the nature of the mean field approximation is particularly clear. In addition the method of extending the approximation to higher order is also apparent.

Our starting point is the singlet probability density $P^{(1)}(\mathbf{X}_1)$ which, as we saw in Chapter 3, is given by

$$P^{(1)}(\mathbf{X}_1) = (N/Z_N)\int\{d\mathbf{X}_2^N\}\exp[-\beta U(\{\mathbf{X}^N\})], \tag{2}$$

where $Z_N/N!$ is defined to be Q_N. To proceed we make the usual approximation [10] and replace the total potential energy by a sum of pair potentials

$$U(\{\mathbf{X}^N\}) = \sum_{i<j} U(\mathbf{X}_i,\mathbf{X}_j). \tag{3}$$

This approximation may not be valid if $U(\mathbf{X}_i,\mathbf{X}_j)$ is the potential for two isolated molecules but does seem to work if $U(\mathbf{X}_i,\mathbf{X}_j)$ is replaced by an effective potential. Differentiation of the singlet distribution with respect to the coordinates of particle 1 yields the apparently simpler result

$$\nabla_1 P^{(1)}(\mathbf{X}_1) = -\beta\int d\mathbf{X}_2 \nabla_1 U(\mathbf{X}_1,\mathbf{X}_2)P^{(2)}(\mathbf{X}_1,\mathbf{X}_2). \tag{4}$$

The simplification is, of course, an illusion because the pair distribution is given by another multi-dimensional integral

$$P^{(2)}(\mathbf{X}_1,\mathbf{X}_2) = \{N(N-1)/Z_N\}\int\{d\mathbf{X}_3^N\}\exp[-\beta U(\{\mathbf{X}^N\})]. \tag{5}$$

Equation (4) is, in fact, just the first in a hierarchy of equations each of which links the gradient of one distribution function to an integral involving the next distribution; in general

$$\nabla_1 P^{(n)}(\{\mathbf{X}_n^N\}) = -\beta \sum_{j=2}^{n} \nabla_1 U(\mathbf{X}_1,\mathbf{X}_j) P^{(n)}(\{\mathbf{X}_n^N\})$$

$$-\beta \int d\mathbf{X}_{n+1} \nabla_1 U(\mathbf{X}_1,\mathbf{X}_{n+1}) P^{(n+1)}(\{\mathbf{X}_{n+1}^N\}). \tag{6}$$

There are then an infinite set of coupled integro-differential equations and to solve them it is necessary to break the coupling by introducing a closure approximation which relates a higher distribution function to its lower counterparts.

In earlier theories of simple liquids it was usual to relate the triplet distribution to the pair distribution as in the Kirkwood superposition approximation [10]. This is the first point at which the coupling can sensibly be broken because the singlet distribution in a simple fluid is just a constant related to the number density ρ. In contrast the singlet distribution is not a constant for a nematic but is related to the important orientational singlet distribution $f(\Omega_1)$ by

$$P^{(1)}(\mathbf{r}_1,\Omega_1) = \rho f(\Omega_1), \tag{7}$$

where Ω_1 denotes the Euler angles defining the molecular orientation. We shall therefore effect the closure by writing the pair distribution function in terms of the orientational singlet distribution. We begin with the definition of the pair correlation function $g(\mathbf{r}_{12},\Omega_1,\Omega_2)$ for a homogeneous phase

$$P^{(2)}(\mathbf{r}_1,\Omega_1;\mathbf{r}_2,\Omega_2) = \rho^2 f(\Omega_1) f(\Omega_2) g(\mathbf{r}_{12},\Omega_1,\Omega_2), \tag{8}$$

where $\mathbf{r}_{12}$ denotes the vector connecting particles 1 and 2. The integro-differential equation for the singlet distribution function can now be written as

$$\nabla_{\Omega_1} \ln f(\Omega_1) = -\beta\rho \int d\mathbf{r}_2 d\Omega_2 \nabla_{\Omega_1} U(\mathbf{r}_{12},\Omega_1,\Omega_2) f(\Omega_2) g(\mathbf{r}_{12},\Omega_1,\Omega_2), \tag{9}$$

where ∇_{Ω_1} denotes differentiation with respect to the angular coordinates. This result is still exact although we are no further forward because the pair correlation function is dependent on the Euler angles Ω_1 and Ω_2. To progress we should clearly ignore this dependence on the orientations of the two particles although it is not necessary to neglect the dependence on the intermolecular vector $\mathbf{r}_{12}$. This is the molecular field approximation which reduces eq. (8) for the pair distribution to

$$P^{(2)}(\mathbf{r}_1,\Omega_2;\mathbf{r}_2,\Omega_2) = \rho^2 f(\Omega_1) f(\Omega_2) g(\mathbf{r}_{12}), \tag{10}$$

that is the probability density for finding two molecules with particular orientations is independent of their separation. In other words we ignore short-range orientational ordering in excess of that imposed by the existence of the long-range ordering present in the mesophase. Computer simulation experiments provide us with a direct test of the molecular field approximation and we return to this point shortly. The molecular field approximation allows eq. (9) to be integrated and the resulting orientational singlet distribution function is

$$f(\Omega_1) = Z^{-1}\exp\{-\beta\rho\int d\mathbf{r}_2 d\Omega_2 U(\mathbf{r}_{12},\Omega_1,\Omega_2)f(\Omega_2)g(\mathbf{r}_{12})\}, \qquad (11)$$

where the constant of integration Z can be obtained from the normalisation requirement

$$\int d\Omega_1 f(\Omega_1) = 1. \qquad (12)$$

The Boltzmann factor in eq.(11) for the singlet distribution function may be identified with the pseudo-potential $U(\Omega_1)$ for molecule 1 and if this is placed at the origin of the coordinate system

$$U(\Omega_1) = \rho\int d\mathbf{r}_{12} d\Omega_2 U(\mathbf{r}_{12},\Omega_1,\Omega_2)f(\Omega_2)g(\mathbf{r}_{12}); \qquad (13)$$

with this identification the constant Z becomes the single particle orientational partition function. To conclude, the pseudo-potential and hence the orientational singlet distribution function is obtained from the pair potential, in the mean field limit, by taking a set of averages - over all orientations of the intermolecular vector, all intermolecular separations and all orientations of the second molecule. We shall see how to take these averages in the next section.

In this derivation we have made, implicitly, an assumption about the symmetry of the mesophase and we must now make this explicit. In an unperturbed mesophase the director adopts all orientations with respect to the laboratory frame with equal probability; that is, it has the symmetry of the full rotation group. The orientational singlet distribution function and pseudo-potential must conform to the same symmetry and so they will both adopt constant values of no particular relevance to the mesophase. This difficulty may be overcome in two ways. In one we would restrict our attention to a region of the mesophase which was small in comparison with the coherence volume. The director would be uniform throughout this region which would then possess the symmetry of the phase. Alternatively we could apply an external field to the sample and so break its full rotational symmetry [11]. Of course the symmetry of the external field must be the same as that of the phase under investigation; thus for nematics we would have to add a term of the form

$$U_{ext} = \sum_m F^{(2,m)} \sum_{i=1}^{N} Y_{2,m}(\omega_i) \tag{14}$$

to the total potential energy, where ω_i denotes the orientation of molecule i with respect to the applied field. The calculation would then be carried through as before and as we are interested in the unperturbed phase, we would take the limit as the coupling parameters $F^{(2,m)}$ go to zero. If we assume that this procedure does not introduce any mathematical difficulties then we may take the limit at once and remove the external field contribution from the potential energy. However, we must remember that the system now has to conform to the broken symmetry which for a nematic is $D_{\infty h}$. With either device the pseudo-potential and singlet distribution need no longer be scalars but adopt the symmetry of the ordered phase.

Having determined the orientational singlet distribution function we can then calculate the average of any single particle orientational function. The most important of these are the orientational order parameters which, in general, are given by the Wigner rotation matrices

$$\overline{D^L_{m,n}} = \int d\Omega_1 D^L_{m,n}(\Omega_1) f(\Omega_1). \tag{15}$$

The large number of order parameters is restricted by the symmetry of both the mesophase and the constituent molecules as we saw in Chapter 3. The more important order parameters are considered in subsequent sections.

The configurational contribution to the internal energy can be calculated from the pair distribution function because of the assumed pairwise additivity

$$U = (\rho^2 V/2) \int d\mathbf{r}_{12} d\Omega_1 d\Omega_2 U(\mathbf{r}_{12}, \Omega_1, \Omega_2) f(\Omega_1) f(\Omega_2) g(\mathbf{r}_{12}), \tag{16}$$

where we have again placed molecule 1 at the origin and now integrated over the volume element $d\mathbf{r}_1$. This contribution may be separated into scalar and anisotropic parts because the pair potential can also be so divided (cf. Chapter 2); we shall return to the significance of this separation shortly.

To determine the order and position of the nematic-isotropic phase transition we need the free energy and so we must now see how to determine the configurational contribution to the entropy. This is related to the N-body distribution function by [9]

$$S = -(k/N!) \int \{d\mathbf{X}^N\} P^{(N)}(\{\mathbf{X}^N\}) \ln[P^{(N)}(\{\mathbf{X}^N\})/N!]. \tag{17}$$

In the molecular field limit we may write $P^{(N)}(\{\mathbf{X}^N\})$ as

$$P^{(N)}(\{\mathbf{X}^N\}) = \rho^N \prod_{i=1}^{N} f(\Omega_i) g^{(N)}(\{\mathbf{r}^N\}) \tag{18}$$

and so

$$S = -Nk\int d\Omega_1 f(\Omega_1)\ln f(\Omega_1) - Nk\ln\rho + k\ln N!$$

$$-(k\rho^N/N!)\int\{d\mathbf{r}^N\}g^{(N)}(\{\mathbf{r}^N\})\ln g^{(N)}(\{\mathbf{r}^N\}). \quad (19)$$

Calculation of the Helmholtz free energy for the ordered and disordered phases appears to demand a knowledge not only of the pair correlation function $g^{(2)}(\mathbf{r}_{12})$ but also the N-body correlation function $g^{(N)}(\{\mathbf{r}^N\})$ for both phases. Their evaluation is an extremely difficult task even for simple liquids although some attempts have been made to obtain $g^{(2)}(\mathbf{r}_{12})$ for nematics from the integro-differential equation for the two particle distribution function [9]. However we shall adopt an alternative procedure and assume that the spatial distributions are the same in both the isotropic and nematic phases. This enables us to determine the difference in behaviour between the two phases which should, at this stage, be the major aim of any molecular theory of nematics.

We conclude this section with a caveat concerning the calculation of the pseudo-potential from the pair potential and its use to determine the orientational singlet distribution function. The Helmholtz free energy, A, evaluated from the singlet distribution is, as we shall see, a function of the various order parameters $\overline{D^L_{m,n}}$. For the pseudo-potential to be correct, within the limitations of the mean field approximation, the free energy must be a minimum with respect to variations in the order parameters [4, 12]. That is the additional constraints

$$(\partial A/\partial\overline{D^L_{m,n}}\,\partial\overline{D^{L'}_{m',n'}}) = 0 \quad (20)$$

must be satisfied to ensure equilibrium and this requirement is not necessarily guaranteed by the particular averaging procedures used to obtain the pseudo-potential.

The Maier-Saupe Theory of Nematics

A derivation

In their derivation Maier and Saupe assumed that anisotropic dispersion forces were responsible for the existence of the nematic phase [3]. However, it is now known, as we shall see in the following section, that this assumption is not essential; all that is needed to obtain their results is an anisotropic potential with a particular dependence on the molecular orientations [6]. In this section we consider a simple pair potential which is consistent with the results of the Maier-Saupe theory. We shall also assume, like Maier and Saupe, that the constituent molecules are both rigid and possess $D_{\infty h}$ symmetry even though no nematogen conforms to this ideal.

The pair potential can, in general, be separated into scalar $U_o(r_{12})$ and anisotropic $U_a(\mathbf{r}_{12},\Omega_1,\Omega_2)$ components; we shall not be concerned with the scalar component; indeed it is important to realise that its specification is not an essential feature of the theory. We write the anisotropic term in the simple form

$$U_a(r_{12},\beta_{12}) = -u(r_{12})P_2(\cos\beta_{12}), \tag{21}$$

where $P_2(\cos\beta)$ is the second Legendre polynomial; $u(r_{12})$ is defined to be positive so that the molecules tend to align with their symmetry axes parallel. The potential is a function of the intermolecular separation r_{12} but not the orientation of the intermolecular vector; it also depends on the angle β_{12} between the two molecular symmetry axes alone. The mean field approximation is now implemented by taking the three averages, defined by eq. (13), of the pair potential. In principle the total pair potential should be averaged but in practice only the anisotropic component need be considered because the scalar part contributes a constant to the pseudo-potential which then cancels on evaluating the singlet distribution function. In addition our assumed equality of the radial distribution function in both phases means that the constant contribution to $U(\Omega_1)$ will not influence the difference in properties of the two phases. We proceed therefore with just the anisotropic component, $U_a(r_{12},\beta_{12})$, of the pair potential.

The averages to be taken are over the coordinates of the second molecule and so it is convenient to separate the coordinates of the two particles in the expression for the pair potential. This we can do with the aid of the spherical harmonic addition theorem [13] which allows us to write eq. (21) as

$$U_a(r_{12},\omega_1,\omega_2) = -u(r_{12})(4\pi/5)\sum_n Y_{2,n}^*(\omega_1)Y_{2,n}(\omega_2), \tag{22}$$

where ω denotes the spherical polar coordinates of the molecular symmetry axis in the laboratory frame.

The pseudo-potential in eq. (13) involves the pair correlation function $g^{(2)}(\mathbf{r}_{12})$ which for a normal fluid depends only on the intermolecular separation r_{12}. However in a uniaxial phase such as a nematic $g^{(2)}(\mathbf{r}_{12})$ is expected to depend on the orientation ω_{12} of the intermolecular vector as well as its length. The extent of this dependence is not known and in any case, we have assumed that the distribution of the molecular centres is the same in both phases; indeed in an alternative derivation this assumption follows naturally from the expansion of the partition function [7]. Consequently we take the pair correlation function to depend on the separation r_{12} alone in both the disordered and the ordered phase; we introduce, therefore, the radial distribution function $g(r_{12})$ which is given by

$$g(r_{12}) = g^{(2)}(r_{12})/4\pi. \tag{23}$$

The orientational pseudo-potential may now be written as

$$U(\beta_1) = -\sum_n \rho\int(d\omega_{12}/4\pi)\int dr_{12}4\pi r_{12}^2 u(r_{12})g(r_{12})$$

$$\times\int d\alpha_2 \sin\beta_2 d\beta_2 d\gamma_2 (4\pi/5)Y_{2,n}(\omega_1)Y_{2,n}^*(\omega_2)f(\beta_2)/4\pi^2; \tag{24}$$

here we have used the reduced singlet distribution because, for cylindrically symmetric particles $f(\Omega)$ depends only on the angle between the director and molecular symmetry axis. Evaluation of the first integral in eq. (24) gives unity while we are unable to determine the second integral because the radial distribution function is not known and so we define the average

$$\varepsilon = \rho\int dr_{12}4\pi r_{12}^2 u(r_{12})g(r_{12}). \tag{25}$$

The third integral, over $d\Omega_2$, is readily evaluated because of the definition in eq. (15) and gives $\bar{P}_2 P_2(\cos\beta_1)\delta_{0n}$ where the relations between the spherical harmonics and Legendre polynomials, given in the Appendix of Chapter 3, have been used. The pseudo-potential obtained from this simple anisotropic pair potential is then

$$U(\beta) = -\varepsilon\bar{P}_2 P_2(\cos\beta), \tag{26}$$

where we have omitted the subscript labelling a particular molecule. The orientational singlet distribution function is

$$f(\Omega) = Z^{-1}\exp\{\beta\varepsilon\bar{P}_2 P_2(\cos\beta)\}, \tag{27}$$

where

$$Z = \int d\alpha \sin\beta d\beta d\gamma \exp\{\beta\varepsilon\bar{P}_2 P_2(\cos\beta)\}; \tag{28}$$

n.b. it is important not to confuse the angle β with the Boltzmann factor $(kT)^{-1}$. The orientational contribution to the internal energy follows immediately from eqs. (16) and (27); it is

$$U = -N\varepsilon\bar{P}_2^2/2. \tag{29}$$

The entropy resulting from the orientational order is found, from eq. (19), to be

$$S = -(N\varepsilon\bar{P}_2^2/T) + Nk\ln\tilde{Z} + Nk\ln 8\pi^2, \tag{30}$$

where the orientational partition function $\tilde{Z}$ is defined by

$$\tilde{Z} = \int d\beta \sin\beta \exp\{\beta\varepsilon\bar{P}_2 P_2(\cos\beta)\}. \tag{31}$$

The orientational contribution to the Helmholtz free energy is then

$$A = (N\varepsilon\bar{P}_2^2/2) - NkT\ln\tilde{Z}, \tag{32}$$

where we have omitted the term $NkT\ln 8\pi^2$ which is the contribution to the free energy in the isotropic phase from the orientational degrees of freedom. As a consequence the free energy A will vanish in the limit of complete disorder. Finally we note that the derivative of this expression for A with respect to the order parameter $\bar{P}_2$ is zero and corresponds to a minimum in the free energy; the constraint imposed by eq. (20) is therefore satisfied.

Predictions

We now calculate a variety of properties of the Maier-Saupe model but defer comparison with experiment. Let us begin with the orientational Helmholtz free energy which requires the second rank order parameter $\bar{P}_2$ for its evaluation. The order parameter is obtained, as a function of the reduced variable kT/ε, by solving the so-called consistency condition [12, 14]

$$\bar{P}_2 = \tilde{Z}^{-1}\int d\beta \sin\beta P_2(\cos\beta) \exp\{\beta\varepsilon\bar{P}_2 P_2(\cos\beta)\}; \tag{33}$$

the results of such calculations are shown in figure 1. For all values

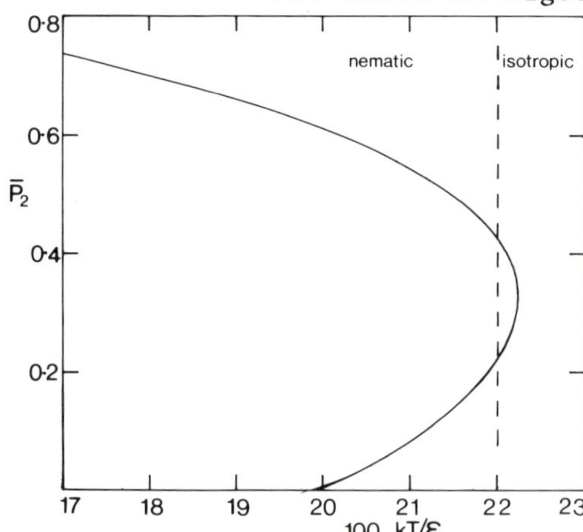

Fig. 1 *The temperature dependence of the order parameter $\bar{P}_2$ determined from the consistency condition, eq. (33).*

of kT/ε the solution $\bar{P}_2 = 0$, corresponding to the isotropic phase, is obtained while below kT/ε of 0.222 a second solution with $\bar{P}_2 \neq 0$ appears; this is the nematic mesophase. Within the temperature range kT/ε equals 0.195 to 0.222 there is a third solution with a lower value of $\bar{P}_2$.

The stability of the phases corresponding to the various solutions is determined by the orientational free energy evaluated from eq. (32) with the appropriate value of the order parameter. It is found that above kT/ε of 0.2203 the disordered phase is stable while below this value the phase with highest order is stable. The phase with intermediate order is always metastable. Thus the point $kT_{NI}/\varepsilon \doteq 0.2203$ corresponds to the nematic-isotropic transition provided the volume is constant. Of course, for a system at constant pressure we require the Gibbs free energy G to locate the transition; however we need not be concerned with the additional complications which the use of G entails [12]. At the transition the gradient $(\partial A/\partial T)_V$ is discontinuous and the entropy change, $\Delta S^{(NI)}/Nk$, is 0.417. The nematic-isotropic transition is, therefore, predicted to be first order. The order parameters $\bar{P}_2$ and $\bar{P}_4$ change discontinuously from zero at the transition to 0.429 and 0.120, respectively; their temperature dependence is shown in figure 2. There are

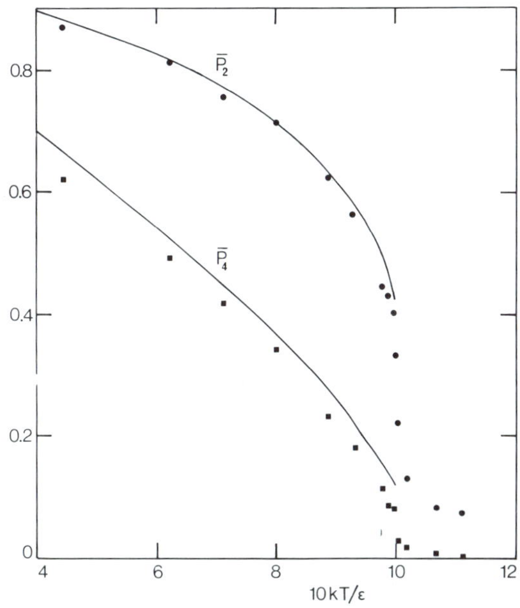

Fig. 2 *The dependence of the order parameters, $\bar{P}_2$ and $\bar{P}_4$, on the reduced temperature T/T_{NI}. The solid lines are obtained from the Maier-Saupe theory and the points are taken from a Monte Carlo simulation based on the same anisotropic pair potential [18].*

other properties which we might calculate but these are sufficient and we must now decide how to test such predictions of the Maier-Saupe theory.

Comparison with experiment

It is important to realise that there are two major elements in the Maier-Saupe theory. The first is the form adopted for the intermolecular potential while the second is the use of the mean field approximation. It would appear therefore that comparison with the properties of real nematogens would not be particularly illuminating because we would be unable to decide whether any disagreement stems from the failure of the mean field approximation, an incorrect form of the pair potential or both. However the advent of the computer simulation techniques described in Chapter 9 allows us to test the molecular field approximation alone because the precise form of the pair potential may be specified in the calculation.

Computer experiments There have been several simulations of particular relevance for such a comparison [15 - 18]. They were for a model system of particles restricted to a cubic lattice and interacting with the same purely anisotropic potential as that employed in our development of the Maier-Saupe theory, i.e. eq. (21) but with the interaction restricted to nearest neighbours. These calculations all demonstrate the existence of orientationally ordered and disordered phases separated by a first order transition and so confirm an important prediction of the Maier-Saupe theory. One of the Monte Carlo calculations [18] has been used to determine the second rank correlation coefficient $G_2(r_{12})$ (defined in Chapter 3 as the average $\overline{P_2\{\cos\beta_{12}(r_{12})\}}$) which is a measure of the angular correlation between pairs of molecules. According to the mean field approximation $G_2(r_{12})$ should be independent of the separation r_{12} and equal to $\bar{P}_2^2$. For separations greater than three times the lattice spacing, a, this prediction is found to be correct, as we can see from figure 3 of Chapter 9. However for smaller separations $G_2(r_{12})$ increases as r_{12} decreases in both the nematic and isotropic phases, although the magnitude of this effect decreases as the orientational order increases. The calculations reveal therefore that the mean field approximation is valid for relatively large separations but fails for short-range separations. What we must do now is investigate the effect of this partial failure of the approximation on the other properties of the nematogen. The Monte Carlo simulations also gave values for the nematic-isotropic transition temperature, the entropy of transition and the order parameters $\bar{P}_2$ and $\bar{P}_4$ at the transition. These results are given in detail in the table of Chapter 9 but, for convenience, some of the numbers are compared in Table 1 with the Maier-Saupe predictions. The Monte-Carlo calculations are performed as a function of u(a), that is the coefficient $u(r_{12})$ in eq. (21) evaluated at the lattice spacing. We can relate u(a) to the parameter ε in the pseudo-potential by recalling its definition eq. (25) and the form of the radial distribution function for a lattice {eq. (25) of Chapter 3}; this gives ε as 6u(a) for a cubic lattice. We see, from the results in Table 1, that the use

Table 1. *A comparison of the predictions of the Maier-Saupe theory with the results of a computer simulation experiment.*

	kT_{NI}/ε	$\Delta S^{(NI)}/Nk$	$\bar{P}_2^{(NI)}$	$\bar{P}_4^{(NI)}$
Maier-Saupe	0.2203	0.417	0.429	0.120
Monte-Carlo [17]	0.186±0.001	0.089±0.02	0.333±0.009	0.08±0.01*

*taken from reference [18]

of the mean field theory results in an overestimate of 17% in the transition temperature and over-estimates in $\bar{P}_2^{(NI)}$ and $\bar{P}_4^{(NI)}$ of about 25% and 40% respectively. Although these errors are large, that in the entropy of transition is enormous, indeed the entropy of transition predicted by the Maier-Saupe theory is over four times that obtained from the simulation experiment. Finally the temperature dependence of the order parameters $\bar{P}_2$ and $\bar{P}_4$ obtained from this experiment is compared with the predictions of the Maier-Saupe theory in figure 2; a reduced temperature T/T_{NI} scale has been used as we would for a real experiment. The agreement for the second rank order parameter is good at low reduced temperatures when the system is highly ordered, as we might have anticipated. However as T/T_{NI} increases the results of the simulation become increasingly smaller than the Maier-Saupe prediction. At the nematic-isotropic transition the discrepancy is approximately 0.09 although it should be remembered that the Monte Carlo calculation of properties near the phase transition is particularly difficult. The values of the fourth rank order parameter $\bar{P}_4$ obtained from the simulation are consistently about 0.05 lower than those predicted by the Maier-Saupe theory. Finally it is important to note that these comparisons have been made over an extremely wide range of reduced temperatures. Although this is possible in a simulation experiment it is not for real nematogens where the nematic phase invariably freezes at or about a reduced temperature of 0.9.

These comparisons show quite clearly that the quantitative predictions of the Maier-Saupe theory are invalidated by the use of the mean field approximation. Consequently we must not expect to be able to employ any mean field theory to interpret the behaviour of real nematics quantitatively although such theories are valuable in providing a qualitative understanding. However we might now be able to make such comparisons more meaningful by allowing for the limitations of the mean field approximation, based on the experience gained from computer simulation experiments. Of course it might be argued that a lattice is not a realistic model for a nematic, which is spatially disordered, and so might invalidate any allowance for the effects of the mean field approximation which we would make. However Monte Carlo calculations using the same anisotropic potential but with a scalar component given by a Lennard-Jones 6-12 potential allow us to remove the constraint imposed by the lattice [19]. This simulation study reveals the same discrepancies with the Maier-Saupe theory as those found for the lattice model.

Real experiments We can now compare the predictions of the Maier-Saupe theory with the behaviour of real nematogens and here our objective is to test the appropriateness of the assumed anisotropic pair potential bearing in mind the limitations imposed by the mean field approximation.

An important prediction of the theory is the nematic-isotropic transition temperature but to test this we must know ε, which is determined by the nature of the anisotropic intermolecular forces as well as the radial distribution function. If, as in the original theory, dispersion forces are taken to make the sole contribution to $U_a(r_{12},\beta_{12})$ then

$$\varepsilon \propto \alpha_2^2\overline{\Delta W}/\langle r_{12}^6\rangle , \tag{34}$$

where α_2 is a component of the anisotropic polarizability tensor and $\overline{\Delta W}$ is some average of the electronic excitation energies, quantities which were introduced in Chapter 2. It is difficult to calculate or measure these properties for molecules as complex as those that yield liquid crystals. None-the-less two quite different approaches have been used and while one estimates that kT_{NI}/ε is about 37 [20] the other gives a value of 2 [21]; despite the discrepancy both calculations are agreed that the Maier-Saupe value of 0.2203 is too small. That is, the anisotropy in the molecular polarizability is far too weak to account for the observed nematic-isotropic transitions; although allowance for the faults of the mean field approximation reduces the discrepancy slightly. However this test assumes that dispersion forces make the sole contribution to the anisotropic pair potential but, as we shall see, the same form of the pseudo-potential may be obtained without this assumption and then ε contains contributions from all anisotropic intermolecular forces. As a consequence we are not able to test the prediction of T_{NI} because our knowledge of the pair potential for nematogenic molecules is insufficient to obtain reliable estimates of ε.

The entropy of transition $\Delta S^{(NI)}$ is a difficult quantity to measure accurately as we can see from the wide range of results reported in Chapter 10 for 4,4'-dimethoxyazoxybenzene (PAA) and 4-methoxybenzylidene-

4'-n-butylanaline (MBBA). However an average value of $\Delta S^{(NI)}/Nk$ would appear to be 0.19 for PAA and 0.10 for MBBA. These results are not directly comparable with the Maier-Saupe value of 0.417 because this ignores the volume change at the transition. Given this change together with the coefficients of thermal expansion and isothermal compressibility of the isotropic phase at T_{NI}, it is possible to calculate the necessary corrections [22]. These show that the Maier-Saupe theory predicts an entropy of transition which is about 2 - 3 times too large [22]; this discrepancy is just what the failure of the mean field approximation would lead us to expect.

The second rank order parameter $\bar{P}_2$ has been measured by a variety of techniques and for many nematogens, of these we shall only consider PAA since its constituent molecules are the least flexible and so most likely to conform to the hypothetical particles of the Maier-Saupe theory. In addition the order parameter has been measured as a function of temperature at constant volume for PAA [23]; such measurements are of particular importance because they allow us to test the theory without the need to specify the volume dependence of the coefficient ε. The order parameter was determined using NMR spectroscopy and, as we shall see in Chapter 15, this technique is particularly reliable and is able to define $\bar{P}_2$ uniquely. The experimental results for PAA are shown in figure 3 as a function of temperature together with the curve (b) predicted

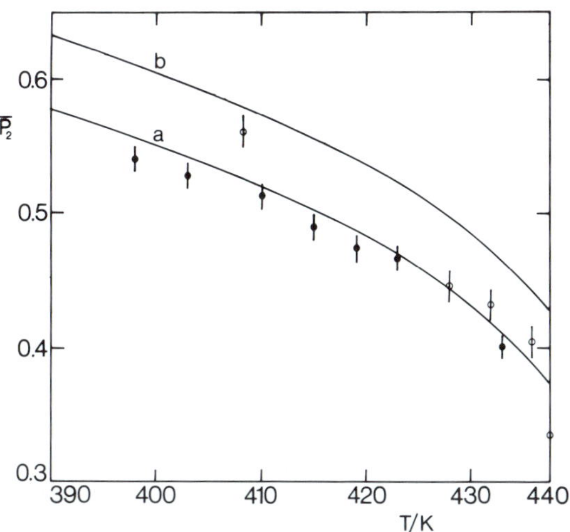

Fig. 3 *The temperature dependence of the order parameter $\bar{P}_2$ observed for PAA at constant volume. Curve (b) is predicted by the Maier-Saupe theory and curve (a) by the Humphries-James-Luckhurst theory. Computer simulation results are shown as open circles.*

by the Maier-Saupe theory. The observed values of $\bar{P}_2$ are slightly lower than theory predicts and this difference is essentially constant at 0.06. This discrepancy is again close to what we would expect from the failure of the mean field approximation and so should not necessarily be attributed to an inaccuracy in the anisotropic pair potential. Indeed the difference between theory and experiment exhibited by PAA in figure 3 is similar to that between theory and the results of the computer experiment, as we can see from the computer values included in the figure.

Measurements of the fourth rank order parameter $\bar{P}_4$ presents far greater difficulties than $\bar{P}_2$ because of the dearth of interactions of fourth rank. However, a variety of novel experiments have been devised to determine $\bar{P}_4$ and those using neutron scattering, Raman light scattering and electron resonance spectroscopy are described in Chapters 13, 17 and 19 respectively. Such experiments are at an early stage of development and a clear consensus on the magnitude of $\bar{P}_4$ has yet to emerge. For example, the results obtained by light scattering and fluorescence depolarization studies [24] tend to be lower than those determined by electron resonance spectroscopy [25]. Indeed negative values of $\bar{P}_4$ have even been observed and these are impossible to explain using the Maier-Saupe theory [26] or indeed one based on short-range anisotropic forces. We shall not deal, therefore, with the results obtained by light scattering experiments but turn instead to those determined by a coherent neutron scattering study of perdeuteriated PAA [27]. The experiment yields the ratio $\bar{P}_4/\bar{P}_2$ more accurately than the individual order parameters and the observed values of this ratio are shown as a function of reduced temperature in figure 16 of Chapter 13. The ratio at the nematic-isotropic transition is approximately 0.15 and if we combine this with the value for $\bar{P}_2^{(NI)}$ of 0.36, estimated from figure 2, we obtain $\bar{P}_4^{(NI)}$ equal to 0.05. This is considerably lower than the Maier-Saupe value of 0.120 and slightly lower than the computer simulation result of 0.08. However we must remember that the error is high for both experiments and so it is reasonable to conclude that the observed value for $\bar{P}_4^{(NI)}$ is consistent with the simple anisotropic potential.

It is difficult to compare the results for $\bar{P}_4$ at other temperatures because the measurements were made at constant pressure and not constant volume [27]. Indeed the vast majority of measurement of $\bar{P}_2$ were also made at constant pressure and so comparison with theory demanded a knowledge of the volume dependence of ε which results from the undoubted change in the radial distribution function with volume. Since this dependence is not known, an empirical approach was adopted and the coefficient ε was written as

$$\varepsilon=\varepsilon_0 V^{-\gamma}. \tag{35}$$

In the Maier-Saupe theory the distance dependence of $u(r_{12})$ is r_{12}^{-6} for London dispersion forces and so γ was set equal to 2 [3]. However in extensions of this theory no assumptions were made about the dependence of the anisotropic pair potential on the intermolecular separation and so γ was taken to be an adjustable parameter [6, 22]. Although the change in the molar volume throughout the nematic range is usually no more than a few percent the value adopted for γ does have a pronounced effect on the temperature dependence of $\bar{P}_2$ [6]. As a consequence it was possible to show that the best value of γ for PAA was 4, by comparing the calculated and observed temperature dependence of $\bar{P}_2$ [6]. Subsequently a more direct method of determining γ became available through measurements, by McColl and Shih [23], of the order parameter $\bar{P}_2$ as a function of both temperature and pressure. From their results they were able to determine the gradient

$$\Gamma = -(V/T)(\partial T/\partial V)_{\bar{P}_2} \tag{36}$$

and found that for PAA Γ was approximately constant and equal to 4.0±0.1. Since the gradient is evaluated for a constant order parameter it follows from eq. (33) that $\beta\varepsilon_0 V^{-\gamma}$ must also be constant and so Γ is predicted, from the modified pseudo-potential, to be a constant and equal to γ. At first, this appears to represent a minor success for the extension of the pseudo-potential for not only was Γ predicted to be a constant but the value obtained was identical to that determined by fitting the temperature dependence of the order parameter at constant pressure to theory.

This success was eventually found to be an illusion, for Cotter [28] has argued that for the Maier-Saupe theory to exhibit statistical-mechanical consistency γ cannot be treated as an adjustable parameter but must be assigned the value 1. Her argument rests on a statistical expression for the thermodynamic activity, z, which is related to the more familiar chemical potential [10]. The activity is important here because its temperature dependence is related to the variation of the internal energy with volume

$$(\partial \ln z/\partial \beta)_V = U/N - \rho^2(\partial U/\partial V)_\beta . \tag{37}$$

The statistical-mechanical expression for z was derived originally for spherical particles [10, 29] but is readily extended to molecules of arbitrary shape. The potential energy of an ensemble of N molecules can be separated into the energy of N-1 particles, $U(\{\mathbf{X}_2^N\})$, and the energy of particle 1 resulting from its interaction with these N-1 molecules

$$U(\{\mathbf{X}^N\}) = U(\{\mathbf{X}_2^N\}) + \Psi(\{\mathbf{X}_2^N\}, X_1). \tag{38}$$

The configurational integral, defined implicitly by eq. (1), can now be written as

$$Z_N = \int\{d\mathbf{X}_2^N\}d\mathbf{X}_1 \exp[-\beta U(\{\mathbf{X}_2^N\})]\exp[-\beta\Psi(\{\mathbf{X}_2^N\},\mathbf{X}_1)],$$

$$= Z_{N-1}\int d\mathbf{X}_1 \langle\exp[-\beta\Psi(\{\mathbf{X}_2^N\},\mathbf{X}_1)]\rangle, \tag{39}$$

where the angular brackets denote an average over the ensemble of N-1 particles. For a homogeneous system this average must be independent of the position $\mathbf{r}_1$ of particle 1; we may, therefore, integrate over this coordinate and so obtain

$$Z_N = VZ_{N-1}\int d\Omega_1 \langle\exp[-\beta\Psi(\{\mathbf{X}_2^N\},\Omega_1)]\rangle. \tag{40}$$

The thermodynamic activity is related to the configurational integrals for N and N-1 particles by [10, 29]

$$z/\rho = 8\pi^2 VZ_{N-1}/Z_N, \tag{41}$$

so that z/ρ tends to one as ρ goes to zero; thus the activity is

$$z/\rho = 8\pi^2/\int d\Omega_1 \langle\exp[-\beta\Psi(\{\mathbf{X}_2^N\},\Omega_1)]\rangle. \tag{42}$$

To use this result in eq. (37) and so test the consistency of the Maier-Saupe theory we must evaluate the average in eq. (42) with the aid of the molecular field approximation. This we can do by recognising that the singlet distribution function in eq. (2) is proportional to the average $\langle\exp[-\beta\Psi(\{\mathbf{X}_2^N\},\Omega_1)]\rangle$ and since the orientational distribution $f(\Omega_1)$ is normalised it must, therefore, be given by

$$f(\Omega_1) = \langle\exp[-\beta\Psi(\{\mathbf{X}_2^N\},\Omega_1)]\rangle/\int d\Omega_1 \langle\exp[-\beta\Psi(\{\mathbf{X}_2^N\},\Omega_1)]\rangle. \tag{43}$$

Comparison of this exact result with that obtained from the mean field approximation contained in eqs. (11) and (13), yields

$$\langle\exp[-\beta\Psi(\{\mathbf{X}_2^N\},\Omega_1)]\rangle = \exp\{-\beta U(\Omega_1)\}. \tag{44}$$

As we have seen the pseudo-potential is the sum of scalar and orientational components, z may consequently be written as a product of orientational and scalar activities. Eq. (37) must hold for both thermodynamic activities and so using the Maier-Saupe pseudo-potential, given in eq. (26), together with the orientational internal energy from eq. (29) we find that

$$\varepsilon = -V(\partial\varepsilon/\partial V)_T. \tag{45}$$

Consequently γ in eq. (35) can only take the value 1 if the theory is to be statistically consistent [28]. Hence in this form the theory is unable to account for the temperature dependence of the order parameter $\bar{P}_2$ at constant pressure or for the value of Γ (equal to 4) found for PAA [23].

It is important to realise, however, that this particular failure of the Maier-Saupe theory stems from the use of the mean field approximation and not necessarily from the form of the intermolecular potential. We can illustrate this point by considering the results of the computer simulation experiments with a lattice of particles. These calculations show that the order parameter $\bar{P}_2$ is constant for a given value of $\beta u(a)$ where a is the lattice spacing and $u(r_{12})$ gives the distance dependence of the pair potential in eq. (21). Thus to calculate Γ we need to know how u(a) depends on the molar volume. For a cubic lattice the spacing a is related to the volume by

$$a = a_o V^{1/3} \tag{46}$$

and so for a homogeneous potential with

$$u(r_{12}) = u_o r_{12}^{-n} \tag{47}$$

the volume dependence of u(a) is particularly simple. The value of Γ is readily evaluated to be n/3 and not equal 1 as imposed by the mean field approximation. The result obtained by McColl and Shih [23] would require that n is equal to 12; this would be a particularly steep potential, reminiscent of the repulsive term in the Lennard-Jones potential. However, as we know virtually nothing about the anisotropic pair potential for nematogenic molecules and because the computer simulations were for an over simplified model it would be unwise to take this analysis further.

Comparison of the Maier-Saupe theory of nematics with experiment has clearly demonstrated its limitations but the experience gained from computer simulations suggests that the flaws in the theory stem from the use of the molecular field approximation. Consequently there is no evidence to suggest that the long-range anisotropic pair potential employed in the theory is seriously in error and, by implication, that the short-range anisotropic forces are important in understanding the fundamental properties of nematogens. This presents us with a problem because such anisotropic repulsive forces undoubtedly exist, and more significantly, the short-range part of the scalar potential is now known to be responsible for the structure of normal liquids [30]; the long-range attractive component serves simply to maintain the high density.

One way of overcoming this difficulty is to invoke the existence of clusters of molecules with high orientational order. The notion is certainly not new and was introduced in the Maier-Saupe theory itself [3]; it has since been revised [31] and also used to rationalise more recent experimental and theoretical results [32]. The basic idea is that both long- and short-range anisotropic forces determine the molecular organisation in a nematic with the repulsive forces being responsible for the structure of the highly ordered clusters. The cluster is presumed to be less anisometric than a single molecule and as a result the short-range anisotropic forces will become less important. The other essential feature of this cluster model is that the clusters are not destroyed at the transition from the nematic to the isotropic phase. Thus the pair potential employed in the Maier-Saupe theory is to be thought of as that between clusters of molecules rather than single molecules.

The existence of such clusters has been questioned [21] possibly because it was thought that they must be rather like the aggregates of molecules originally envisaged in the swarm theory of nematics [4]. However such an inadequate description may be avoided by using the language of the distribution functions introduced in Chapter 3. Then the word cluster would be a shorthand for the statement that the pair distribution function $G(\mathbf{r}_{12},\Omega_1,\Omega_2)$ for parallel alignment is large and essentially constant over distances comparable to several molecular dimensions. In addition our model would require that the short-range part of $G(\mathbf{r}_{12},\Omega_1,\Omega_2)$ does not change at the isotropic-nematic transition. Consequently the variation in the orientational properties of the nematogen at the phase transition must be caused by that part of the pair potential for which the intermolecular separation is greater than the distance over which $G(\mathbf{r}_{12},\Omega_1,\Omega_2)$ is constant.

The success of the Maier-Saupe theory provides some evidence for the validity of this model but the availability of more direct evidence for the existence of clusters would be particularly convincing. X-ray and neutron scattering techniques are the most powerful probes of the distribution functions (cf. Chapter 13) and two experiments of particular relevance have been reported. In an X-ray study of 4-n-pentyl and 4-n-heptyl-4'-cyanobiphenyl [33] it was found that the intermolecular components at low $\mathbf{Q}$ are essentially the same in both the nematic and the isotropic phase. In other words the short-range molecular organisation is the same in the two phases. The form of the diffraction pattern [33] further indicates that there is a strong correlation between the axes of neighbouring molecules, in support of the notion that the repulsive forces determine the structure of a cluster. The other experiment involved coherent neutron scattering from perdeuteriated PAA [34]; this also indicated high local order in the nematic and isotropic phases, with the molecular planes stacked parallel. Additional but less direct evidence is available from a study of the Poley absorption for MBBA; it was found that the frequency of this far infra-red absorption did not vary appreciably through the nematic-isotropic transition [35]. Since the absorption is determined by the short-range intermolecular forces this observation implies that the short-range structure does not change significantly at the transition.

Conclusion

We complete this section by recalling that the Maier-Saupe theory of nematics contains two major assumptions. The first of these concerns the nature of the anisotropic pair potential, which is taken to be continuous and long range. The second assumption relates to the validity of the molecular field approximation which is employed to calculate the properties of an ensemble of particles interacting via such a potential. Comparison of these properties with the results of computer simulation experiments based on the same anisotropic pair potential allows us to study those limitations in the Maier-Saupe theory which stem from the use of the mean field approximation. Given this understanding of the theory it can be seen that the discrepancies revealed by comparison with the behaviour of real nematogens might also be attributed to the failure of the mean field approximation. Consequently there would not appear to be any evidence to support the view that the long-range anisotropic pair potential is fundamentally incorrect. However, there are indications that the detailed form of the pair potential is not quite right; we shall, therefore, refer briefly to this evidence and discuss possible modifications of the anisotropic pair potential in the remaining two sections.

Cylindrically Symmetric Molecules - the Complete Pair Potential

The Maier-Saupe theory predicts that, at constant volume, the order parameters $\bar{P}_L$ should be a universal function of the reduced temperature T/T_{NI}. Indeed this prediction is rigorously correct provided the anisotropic pair potentials for all nematogens differ only by a change in a single parameter; for example, in eqs. (21) and (47) this would be u_o. It is difficult to test this prediction precisely because the temperature dependence of $\bar{P}_2$, at constant volume, has only been determined for PAA. However a limited set of measurements suggest that the reduced volume V/V_{NI} is a universal function of T/T_{NI} [36]. Consequently it should be sufficient to test the Maier-Saupe prediction by using the wealth of order parameters measured at constant pressure. A selection of such results is shown in figure 4 for five different nematogens; the temperature dependence predicted by the Maier-Saupe theory, with the inverse volume dependence demanded by statistical consistency, is shown as the solid line. The agreement between theory and experiment is good but it is immediately obvious that $\bar{P}_2$ is not a universal function of the reduced temperature. It is reasonable to infer therefore that the anisotropic pair potential is more complicated than that given by eq. (21) and

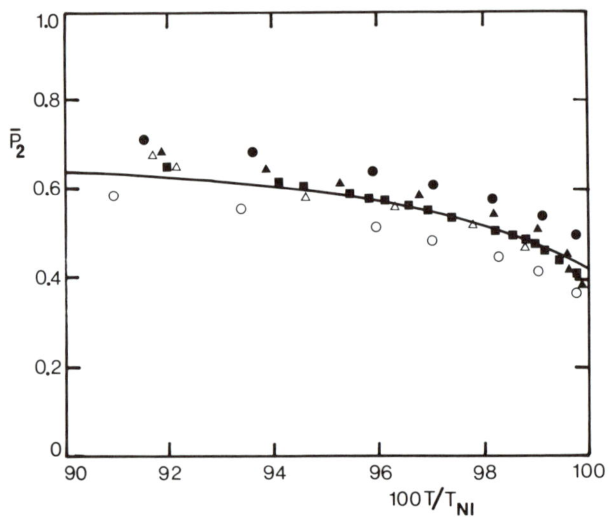

Fig. 4 *The temperature dependence of $\bar{P}_2$, at constant pressure, for 4,4'-dimethoxyazoxybenzene (○), 4,4'-diethoxyazoxybenzene (●), anisaldazine (■), 2,4-nonadienoic acid (▲) and 2,4-undecadienoic acid (△). The curve is predicted by the Maier-Saupe theory.*

employed in the Maier-Saupe theory. There are a variety of ways in which the pair potential can be extended and in this section we shall employ a perfectly general intermolecular potential for cylindrically symmetric particles. Of course no nematogenic molecule has such a symmetry and in the following section we consider modifications to the pair potential which result from deviations from cylindrical symmetry.

The pair potential for molecules of $D_{\infty h}$ symmetry can be expanded, quite generally, by using a product basis of modified spherical harmonics $C_{L,m}(\beta\alpha)$ [37, 38]. As we saw in Chapter 2 the expansion may be written in the form

$$U(r_{12},\Omega) = \sum u_{LL'J}(r_{12})S_{LL'J}(\Omega), \tag{48}$$

where

$$S_{LL'J}(\Omega) = (i)^{L-L'-J} \sum \begin{pmatrix} L & L' & J \\ m & m' & M \end{pmatrix} C_{L,m}(\beta_1\alpha_1)C_{L',m'}(\beta_2\alpha_2)C_{J,M}(\theta,\phi). \tag{49}$$

The symbol Ω denotes, collectively, the orientations of the molecular symmetry axes $(\alpha_1\beta_1, \alpha_2\beta_2)$ and intermolecular vector $(\theta\phi)$ in a laboratory frame; these angles are defined in figure 5 of Chapter 2. The unknown functional dependence of the expansion coefficients $u_{LL'J}$ on the intermolecular separation r_{12} and the large number of such coefficients, even in a truncated potential, makes it impossible to employ this general expansion in a computer simulation experiment. We are forced, therefore, to investigate the influence of the various terms in the pair potential on the properties of the nematic phase with the aid of the molecular field approximation. We recognise, of course, that the predictions of such calculations cannot be accurate although admittedly when the theories were first developed [6, 39] their objective was to remove the discrepancies between the Maier-Saupe theory and experiment. We now believe that the importance of these calculations rests on their ability to provide an understanding of the differences in the properties exhibited by various nematogens. It is in this spirit that we outline a mean field theory of nematics based on the general pair potential given in eqs. (48) and (49).

The orientational pseudo-potential is obtained by taking the same three averages as in the Maier-Saupe theory - over the orientations of the intermolecular vector, over the orientations of molecule 2 and over the intermolecular separation. The angular dependence of the pair potential is contained in the S-functions and so we begin with one of these. The dependence of $S_{LL'J}(\Omega)$ on the orientation of the intermolecular vector is contained in $C_{J,M}(\theta,\phi)$ and averaging this gives simply $\delta_{J0}\delta_{M0}$. The 3j coefficient in the S-function reduces to $\begin{pmatrix} L & L' & 0 \\ m & m' & 0 \end{pmatrix}$ and this is given by

$$\begin{pmatrix} L & L' & 0 \\ m & m' & 0 \end{pmatrix} = (-)^{L-m}\delta_{LL'}\delta_{m-m'}(2L+1)^{-\frac{1}{2}} . \tag{50}$$

The non-zero rotationally averaged S-functions are therefore

$$S_{LL0}(\Omega) = \sum(-)^{L-m}C_{L,m}(\beta_1\alpha_1)C_{L,-m}(\beta_2\alpha_2)\delta_{LL'}(2L+1)^{-\frac{1}{2}} . \tag{51}$$

We now average these remaining functions over the orientations of particle 2 using the reduced singlet distribution function $f(\beta_2)/4\pi^2$ and with the laboratory z axis parallel to the director. The requisite integral gives just the Lth rank order parameter

$$\int d\alpha_2 \sin\beta_2 d\beta_2 d\gamma_2 \; C_{L,-m}(\beta_2\alpha_2)f(\beta_2)/4\pi^2 = \bar{P}_L\delta_{om} \tag{52}$$

which is restricted to even values of L because of the $D_{\infty h}$ symmetry of the nematic mesophase. The distance dependence of the pair potential is contained in the expansion coefficients $u_{LL'J}(r_{12})$ and we define the average of the surviving coefficients by

$$\bar{u}_L = \int dr_{12} 4\pi r_{12}^2 u_{LL0}(r_{12}) g(r_{12})/(2L+1)^{\frac{1}{2}}; \tag{53}$$

to simplify the notation we have introduced the factor of $(2L+1)^{-\frac{1}{2}}$ from eq. (51) into this definition. Collecting these three averages together gives the orientational pseudo-potential as

$$U(\beta) = \sum_L{}' \bar{u}_L \bar{P}_L P_L(\cos\beta), \tag{54}$$

where the prime restricts the summation to even values of L.

The orientational contributions to the thermodynamic functions follow in much the same way as for the Maier-Saupe theory. We begin with the reduced orientational singlet distribution function

$$f(\beta) = \tilde{Z}^{-1} \exp\{-\beta \sum_L{}' \bar{u}_L \bar{P}_L P_L(\cos\beta)\}, \tag{55}$$

where the orientational partition function is

$$\tilde{Z} = \int d\beta \sin\beta \exp\{-\beta \sum_L{}' \bar{u}_L \bar{P}_L P_L(\cos\beta)\}. \tag{56}$$

The orientational internal energy is obtained from eqs. (55) and (16) as

$$U = (N/2) \sum_L{}' \bar{u}_L \bar{P}_L^2 \tag{57}$$

and eq. (19) gives

$$S = (N/T) \sum_L{}' \bar{u}_L \bar{P}_L^2 + Nk\ln\tilde{Z}, \tag{58}$$

for the orientational entropy. These combine to give the orientational contribution to the Helmholtz free energy as

$$A = (N/2) \sum_L{}' \bar{u}_L \bar{P}_L - NkT\ln\tilde{Z}. \tag{59}$$

The order parameters needed to evaluate A are obtained from the consistency conditions

$$\bar{P}_L = \tilde{Z}^{-1} \int d\beta \sin\beta P_L(\cos\beta) f(\beta). \tag{60}$$

Finally we note that this form of the free energy is a minimum with respect to variations in the order parameters $\bar{P}_2$ and $\bar{P}_4$, as required by eq. (20).

If the expansion of the pseudo-potential is truncated after the term with L = 2 then the equations which we have just derived reduce to those of the Maier-Saupe theory. We see, therefore, that it is not necessary to assume that dispersion forces make the sole contribution to the intermolecular potential [6]. Indeed all anisotropic forces contribute to the coefficient ε occurring in the Maier-Saupe theory. Unfortunately this effectively prevents any attempt at predicting the transition temperature, for although there is considerable uncertainty in the molecular polarizability the expansion coefficients $u_{LL0}(r_{12})$ are virtually unknown for nematogenic molecules.

To use the mean field results obtained from the general form of the anisotropic pair potential we must decide how many terms to retain in the pseudo-potential. The need to minimise the number of adjustable parameters in the theory suggests that the series would be truncated after the term with L = 4. Calculations for simple diatomics [40] certainly show that the expansion (48) of the pair potential does converge rapidly for large intermolecular separations but not for small r_{12}. However, even though the pair potential may not converge rapidly for all separations the pseudo-potential may be convergent because, for nematics, the order parameters $\bar{P}_L$ which are part of the coefficients in $U(\beta)$ do decrease rapidly with increasing L. For these reasons therefore we approximate the general pseudo-potential by

$$U(\beta) = \bar{u}_2\{\bar{P}_2 P_2(\cos\beta)+\lambda\bar{P}_4 P_4(\cos\beta)\}, \tag{61}$$

where λ is $\bar{u}_4/\bar{u}_2$ and is a measure of the relative importance of the fourth rank terms.

To calculate the properties resulting from this truncated pseudo-potential we first fix λ and determine the order parameters $\bar{P}_2$ and $\bar{P}_4$, from the two coupled consistency conditions in eq. (60), as a function of the reduced variable $kT/\bar{u}_2$. These are then used to determine the Helmholtz free energy and hence the nematic-isotropic transition, at constant volume. Finally the entropy of transition may be obtained from the truncated expression for the internal energy in eq. (57) as

$$\Delta S^{(NI)}/Nk = (N/2)(\bar{u}_2/kT_{NI})\{\bar{P}_2^{(NI)^2}+\lambda\bar{P}_4^{(NI)^2}\}. \tag{62}$$

The results of such calculations are listed in table 2 and these show the effects of changing λ on the behaviour of the nematic phase at the transition. The addition of the P_4 term clearly enhances the anisotropy of the pseudo-potential while its subtraction must reduce this anisotropy. We expect therefore that the orientational order at the transition will increase with increasing λ and consequently the entropy of transition as well as the transition temperature will also become larger. These expectations are borne out by the results in table 2. However it is interesting to note that the enhancement of the properties by a positive P_4 term in the pseudo-potential is greater than the reduction caused by a comparable negative P_4 term. Finally we show in figure 5 the dependence of the second rank order parameter on the reduced temperature. The

Table 2. *The dependence of the nematic transitional properties on λ.*

λ	$-kT_{NI}/\bar{u}_2$	$\bar{P}_2^{(NI)}$	$\bar{P}_4^{(NI)}$	$\Delta S^{(NI)}/Nk$
-0.3	0.2164	0.353	0.070	0.282
-0.2	0.2175	0.372	0.081	0.314
-0.1	0.2186	0.398	0.098	0.358
0	0.2203	0.429	0.120	0.417
0.1	0.2222	0.467	0.154	0.496
0.2	0.2250	0.516	0.199	0.606
0.3	0.2285	0.572	0.252	0.755

effect of changing λ is simply to shift the curve either up or down from the Maier-Saupe result (λ = 0) depending on whether λ is positive or negative; the slopes of the curves are however independent of λ.

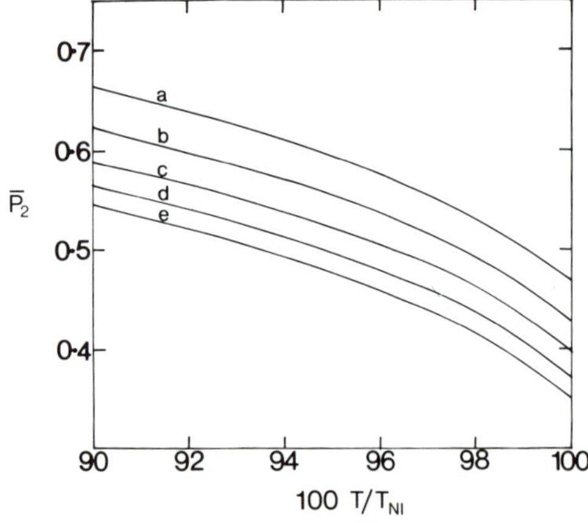

Fig. 5 *The dependence of the order parameter $\bar{P}_2$ on the reduced temperature calculated from the Humphries-James-Luckhurst theory for λ equal to (a) 0.1, (b) 0, (c) -0.1, (d) -0.2 and (e) -0.3.*

Let us consider, briefly, how these predictions of the Humphries-James-Luckhurst theory might usefully be employed. For example, from figure 4 we can see that the order parameter $\bar{P}_2$ for the nematogen 4,4'-diethoxyazoxybenzene (PAP) is significantly greater that that of its lower homologue PAA. In addition the temperature dependence of $\bar{P}_2$ for the two nematics lie on essentially parallel curves. We might argue, therefore, that the pseudo-potential and hence the anisotropic pair potential for PAP is steeper than that for PAA. If this is so then the entropy of transition for PAA should be less than for PAP; this is indeed the case for $\Delta S^{(NI)}/Nk$ is 0.19 for PAA and 0.41 for PAP [41]. Past usages of the theory have attempted to make the analysis quantitative, for example, by fitting theory to the observed temperature dependence of $\bar{P}_2$. Often the agreement between theory and experiment has been quite impressive as we can see from the results in figure 2 for PAA [42]. Despite such excellent agreement it should now be realised that these comparisons may be quite misleading, for the inclusion of the fourth rank term in the pseudopotential compensates not only for the limitations of the pair potential, as intended, but also for the errors introduced by the use of the molecular field approximation.

Non-cylindrically Symmetric Molecules

The variation of the order parameter $\bar{P}_2$ with the structure of the nematogen provides some evidence for the inclusion of fourth rank and possibly higher rank terms in the anisotropic pair potential, but such evidence is indirect. In contrast, there are more compelling reasons to justify the inclusion of terms in the pair potential to allow for deviations from molecular cylindrical symmetry. For example the symmetry of nematogenic molecules is never as high $D_{\infty h}$; indeed the molecules are more lath-like than cylindrical, as we can see from the structures given in Chapter 1. Of course, the lack of molecular symmetry does not preclude the possibility that the pair potential is effectively that for cylindrical particles because the terms which result from deviations from cylindrical symmetry may be small. An indication of the importance of these terms is given by the second rank ordering matrix **S** introduced in Chapter 3, but until recently few experiments have been able to determine the five independent elements of **S**. Indeed most experiments provide a single number whose analysis to yield an order parameter necessitates the assumption of effective cylindrical symmetry. Thus the principal elements of **S** are assumed to be related by

$$S_{xx} = S_{yy} = -S_{zz}/2, \tag{63}$$

so that the experiment yields S_{zz} rather than some unknown combination of S_{zz} and $(S_{xx}-S_{yy})$. However Alben, McColl and Shih [43] have attempted to estimate the difference $(S_{xx}-S_{yy})$ for PAA by combining the results of two experiments and obtained a value of about 0.07. Their analysis is open to question because of their simplistic assumptions concerning the geometry of the phenyl groups in PAA [44]. However the advent of new techniques in NMR spectroscopy has removed the need to make such assumptions and a very recent study of perdeuteriated PAA has shown, quite clearly, that the ordering matrix does deviate from cylindrical symmetry [45]. In addition a proton NMR experiment on partially deuteriated 4-cyano-4'-n-pentylbiphenyl has also revealed that the ordering matrix for the cyano substituted phenyl group is not cylindrical although the deviation is small [46]; this experiment is discussed in Chapter 15.

The evidence now available suggests that the ordering matrix is not cylindrically symmetric and so the pair potential must be modified to allow for deviations from molecular cylindrical symmetry. However, even before these results were known the consequences of molecular symmetry lower than $D_{\infty h}$ were being explored theoretically. For example, Frieser had shown that an ensemble of non-cylindrically symmetric particles should exhibit a transition from a uniaxial to a biaxial nematic phase [47]. This intriguing possibility has since been studied by a variety of theoretical techniques [48-50] but, as yet, the transition has not been discovered for any real nematogen. Presumably the uniaxial nematic phase freezes before the transition to the biaxial mesophase can occur. This should not be a problem in computer simulation experiments and a Monte Carlo calculation for a cubic lattice of biaxial particles does exhibit the predicted transition [51]. None-the-less we shall concentrate here on the effect of molecular biaxiality on the properties of the uniaxial mesophase. The molecular field approximation is used to obtain these properties and so we emphasise again that the predictions cannot be precise but they will be useful in providing a qualitative guide to the consequences of molecular biaxiality. For example, our calculations should enable us to gauge the relative magnitudes of the various terms in the anisotropic pair potential necessary to account for the observed deviation of the ordering matrix from cylindrical symmetry.

We start, as before, with the pair potential and, for molecules of arbitrary shape, this is given by eqs. (40) and (41) of Chapter 2. The potential is

$$U(\mathbf{r}_{12},\Omega_1,\Omega_2) = \sum u^{kk'}_{LL'J}(r_{12})S^{kk'}_{LL'J}(\Omega), \tag{64}$$

where the S-function is defined by

$$S^{kk'}_{LL'J}(\Omega) = (i)^{L-L'-J} \sum_{m,m',M} \begin{pmatrix} L & L' & J \\ m & m' & M \end{pmatrix} D^{L}_{m,k}(\Omega_1)D^{L'}_{m',k'}(\Omega_2)C_{J,M}(\theta,\phi) \tag{65}$$

and $D^{L}_{m,k}$ is an Lth rank Wigner rotation matrix. The various angles employed to define the molecular orientations and that of the intermolecular vector are shown in figure 5 of Chapter 2; we shall take the laboratory z axis in this figure to be parallel to the director. We now obtain the orientational pseudo-potential by evaluating the same three averages as those which we have encountered in the previous sections.

Since we are restricting our attention to the uniaxial mesophase the order parameters occuring in the theory are $\overline{D^L_{0,k}}$, where L is even. The averages are readily evaluated and give the pseudo-potential as [52]

$$U(\beta\gamma) = \sum_{L,k,k'}{}' u_{Lkk'}\overline{D^L_{0,k'}}D^L_{0,k}(\beta\gamma), \tag{66}$$

where the prime restricts the summation to even values of L and $u_{Lkk'}$ is a convenient notation for the average $\bar{u}^{kk'}_{LL}$ defined by

$$\bar{u}^{kk'}_{LL0} = \rho\int dr_{12}4\pi r^2_{12}u^{kk'}_{LL0}(r_{12})g(r_{12})(2L+1)^{-\frac{1}{2}} \tag{67}$$

An analogous expression for the pseudo-potential has been obtained by Straley [50] although he considered only second rank terms.

Thus far our analysis is quite general and we could proceed to determine the orientational order parameters and thermodynamic functions. However eventually we shall need to minimise the number of arbitrary parameters in the theory and so we begin by making certain assumptions, about the molecular symmetry, to simplify the pseudo-potential before calculating any properties of the system. As we saw in Chapter 2 the expansion coefficients $u^{kk'}_{LL'J}(r_{12})$ in the pair potential may be restricted depending on the symmetry of the system and of individual molecules. For example, if the molecules are identical and possess a reflection plane orthogonal to their z axis then (L+k) and (L'+k') must both be even. In the pseudo-potential L is equal L' and even, consequently the summation is further restricted to even values of k and k'. This together with, say, a second reflection plane perpendicular to the y axis means that the averaged coefficients are related by

$$u_{Lkk'} = u_{Lk-k'} = u_{L-kk'} = u_{L-k-k'}. \tag{68}$$

We may now write the pseudo-potential as

$$U(\beta\gamma) = \sum_{\substack{L,|k|\\|k'|}}{}' u_{Lkk'}\overline{d^L_{0,k'}\cos k\gamma}\, d^L_{0,k}(\beta)\cos k\gamma/(1+\delta_{0k})(1+\delta_{0k'}), \tag{69}$$

where we have introduced the reduced Wigner rotation matrices defined in the appendix of Chapter 3. If we set k and k' equal to zero then we obtain the Humphries-James-Luckhurst pseudo-potential given in eq. (54); consequently the terms with k and k' ≠ 0 represent the contributions which stem from deviations from cylindrical symmetry.

There are clearly still too many unknowns in the expansion of the pseudo-potential and so we restrict the summation to second rank terms; this gives

$$U(\beta\gamma) = (u_{200}\overline{d^2_{0,0}}+2u_{220}\overline{d^2_{0,2}\cos 2\gamma})d^2_{0,0}(\beta)$$

$$+ (2u_{220}\overline{d^2_{0,0}}+4u_{222}\overline{d^2_{0,2}\cos 2\gamma})d^2_{0,2}(\beta)\cos 2\gamma. \quad (70)$$

Here $\overline{d^2_{0,0}}$ is the usual order parameter $\bar{P}_2$ which we equate with the z component of **S** in its principal axis system; the other order parameter $\overline{d^2_{0,2}\cos 2\gamma}$ is related to the x and y components by

$$\overline{d^2_{0,2}\cos 2\gamma} = (6)^{-\frac{1}{2}}(S_{xx}-S_{yy}). \quad (71)$$

(n.b. Our assumptions concerning the molecular symmetry result in the molecular frame being the principal axis system for **S**). In order to simplify the notation for the thermodynamic functions it is convenient to denote the coefficients of the angular functions in eq. (70) by

$$a = -\beta(u_{200}\overline{d^2_{0,0}}+2u_{220}\overline{d^2_{0,2}\cos 2\gamma}) \quad (72)$$

and

$$b = -\beta(2u_{220}\overline{d^2_{0,0}}+4u_{222}\overline{d^2_{0,2}\cos 2\gamma}). \quad (73)$$

The orientational internal energy is then

$$U = -(NkT/2)(a\overline{d^2_{0,0}}+b\overline{d^2_{0,2}\cos 2\gamma}) \quad (74)$$

and the entropy is

$$S = -Nk(a\overline{d^2_{0,0}}+b\overline{d^2_{0,2}\cos 2\gamma})+Nk\ln\tilde{Z}. \quad (75)$$

The single particle partition function is readily evaluated as

$$\tilde{Z} = 2\pi\int d\beta\sin\beta I_0\{(bd^2_{0,2}(\beta)\}\exp\{ad^2_{0,0}(\beta)\}, \quad (76)$$

where $I_n(x)$ is an nth order modified Bessel function

$$I_n(x) = \pi^{-1}\int^{\pi} d\phi\cos n\phi\exp(x\cos\phi). \quad (77)$$

The orientational Hemlholtz free energy is then

$$A = (NkT/2)(ad^2_{0,0}+b\overline{d^2_{0,2}\cos 2\gamma})-NkT\ln\tilde{Z} \quad (78)$$

and the order parameters are

$$\overline{d^2_{0,0}} = 2\pi\tilde{Z}^{-1}\int d\beta\sin\beta d^2_{0,0}(\beta)I_0\{bd^2_{0,2}(\beta)\}\exp\{ad^2_{0,0}(\beta)\} \quad (79)$$

and

$$\overline{d^2_{0,2}\cos 2\gamma} = 2\pi\tilde{Z}^{-1}\int d\beta\sin\beta d^2_{0,2}(\beta)I_1\{bd^2_{0,2}(\beta)\}\exp\{ad^2_{0,0}(\beta)\}. \quad (80)$$

Finally we note that this expression for the Helmholtz free energy is a minimum with respect to both order parameters, as demanded by eq. (20).

These equations still contain three unknowns although as before one of them (u_{200}) will be combined with the temperature to produce a reduced temperature scale. This leaves two and it is desirable to find a relationship between these; we therefore adopt the combining rule

$$u_{220} = (u_{200}u_{222})^{\frac{1}{2}}. \tag{81}$$

Such a rule is expected when the molecules interact via dispersion forces [52] and although other combinations are possible [50] eq. (81) is sufficient for our semi-quantitative study. The equations for a and b must now reduce to

$$a = -(\beta u_{200})(\overline{d^2_{0,0}}+2\delta\overline{d^2_{0,2}\cos 2\gamma}) \tag{82}$$

and

$$b = 2\delta a, \tag{83}$$

where δ is the ratio u_{220}/u_{200} which is related to the deviation from cylindrical symmetry. The numerical calculations proceed as before; δ is fixed at a particular value and the order parameters are calculated from eqs. (79) and (80) as a function of the reduced variable kT/u_{200}. This is, in fact, assigned negative values to ensure that only solutions for which the molecular z axis tends to align parallel to the director are obtained. The Helmholtz free energy is then evaluated as a function of kT/u_{200} in order to locate the isotropic-nematic transition; as before the volume change at the transition is taken to be zero. Since the Helmholtz free energy at the transition is zero eqs. (75), (78) and (83) may be combined to give the entropy of transition as

$$\Delta S^{(NI)} = -Nka^{(NI)}(\overline{d^2_{0,0}}^{(NI)}+2\delta\overline{d^2_{0,2}\cos 2\gamma}^{(NI)}). \tag{84}$$

The results for the properties of the nematic phase at T_{NI} are listed in table 3 for increasing deviations from molecular cylindrical symmetry; here we have reverted to our original notation for the primary order parameters $\overline{d^L_{0,0}}$, namely $\bar{P}_L$. Only positive values of δ have been used because reversing the sign of δ is equivalent to an interchange of the x and y axes; consequently the only parameter to be affected is $\overline{d^2_{0,2}\cos 2\gamma}$ and this simply changes sign. We begin with the variation in the transition temperature with increasing δ; for constant u_{200} we see that T_{NI} increases with the molecular biaxiality. This result is, at first sight, somewhat surprising but is readily understood by considering eq. (50) for the pseudo-potential. Since u_{200} is fixed, increasing δ has the effect of increasing u_{220} and u_{222}; the orientational energy of a molecule in the mean field therefore becomes more negative and this

Table 3. *The order parameters and entropy change at the nematic-isotropic transition*

δ	$-kT_{NI}/u_{200}$	$\bar{P}_2^{(NI)}$	$\overline{d^2_{0,2}\cos 2\gamma}^{(NI)}$	$\bar{P}_4^{(NI)}$	$\Delta S^{(NI)}/Nk$
0	0.2203	0.429	0	0.120	0.417
0.1	0.2220	0.408	0.017	0.109	0.384
0.2	0.2280	0.341	0.035	0.075	0.275
0.3	0.2404	0.207	0.041	0.029	0.112

results in the increase in the range of the nematic mesophase. It would be difficult to test this prediction with real nematogens because it is hard to conceive of an alteration in the structure which would keep u_{200} constant and yet change δ.

The value of the second rank order parameter at T_{NI} decreases with increasing biaxility. This trend reflects the fact that for large δ the molecular z axis will tend to be orthogonal to the director as we can see from eq. (70). In this limit the order parameter $\bar{P}_2$ becomes $-1/2$ consequently as δ increases $\bar{P}_2$ must decrease from its value when δ is zero. Similarly the fourth rank order parameter $\bar{P}_4$ must also decrease because in the limit of large δ $\bar{P}_4$ tends to 3/8 which should be compared with the other limit of 1 when the z axis is parallel to the director. The most important order parameter in this theory is $\overline{d^2_{0,2}\cos 2\gamma}$ because this measures the molecular biaxility. As δ increases so $\overline{d^2_{0,2}\cos 2\gamma}$ is expected to increase and this behaviour is shown by the results in table 3. Finally the entropy of transition is seen to decrease quite markedly with increasing biaxility and from eq. (84) we see that this trend may be traced to the reduction in $\bar{P}_2^{(NI)}$.

The temperature dependence of the second and fourth rank order parameters are shown in figure 6. It is interesting and may be of importance to note that increasing δ affects the slope of the $\bar{P}_2$ curve near the phase transition. This behaviour contrasts with that found by including higher rank terms in the pseudo-potential, as shown in figure 5. The influence of δ on $\bar{P}_4$ is less dramatic and simply reduces $\bar{P}_4$ by a constant amount, at least in the reduced temperature range exhibited by real nematics.

These departures from the Maier-Saupe theory produced by allowing for deviations from molecular cylindrical symmetry are reminiscent of those observed for nematics. Indeed one of the objectives of such theoretical developments was to account quantitatively for the major properties

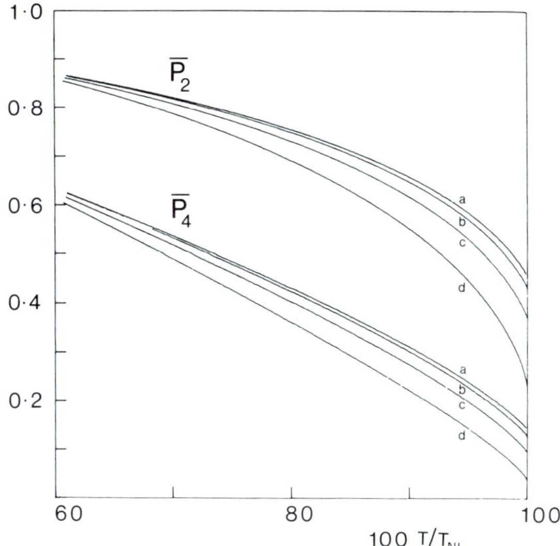

Fig. 6. *The variation of the order parameters $\bar{P}_2$ and $\bar{P}_4$ with the reduced temperature T/T_{NI} calculated for δ equal to (a) 0, (b) 0.1, (c) 0.2 and (d) 0.3.*

of the nematic mesophase [52]. We now appreciate that this is an impossible task for any theory based on the molecular field approximation. Indeed most of the discrepancies between experiment and the Maier-Saupe theory can be attributed to the limitations of this approximation. There is however one notable exception, for clearly any theory developed for molecules with $D_{\infty h}$ symmetry can never explain a non-zero value for the order parameter $\overline{d^2_{0,2}\cos 2\gamma}$. To see whether the results for this order parameter obtained for PAA [43, 52] may be reasonably understood by this mean field theory we have plotted $\overline{d^2_{0,2}\cos 2\gamma}$ against $\bar{P}_2$ in figure 7. The predicted dependence is also shown as the curves for different

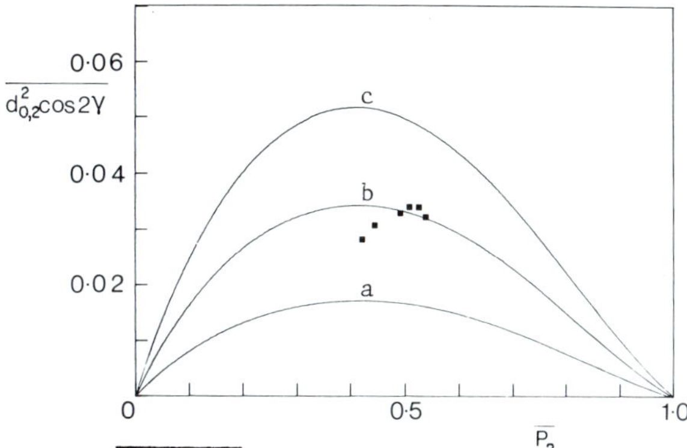

Fig. 7. *The dependence of $\overline{d^2_{0,2}\cos 2\gamma}$ on $\bar{P}_2$ for PAA. The curves are predicted by the mean field theory with δ equal to (a) 0.1, (b) 0.2 and (c) 0.3.*

molecular biaxiality. The uncertainty in the values for $\overline{d^2_{0,2}\cos 2\gamma}$ is high but the points do scatter reasonably well about the curve for $\delta = 0.2$; it is clear that results spanning a wider range of $\bar{P}_2$ are required. However the value for δ is reasonable both with respect to the anisotropy in the molecular plarizability and indeed in the shape [50]. In addition the theory suggests that the molecular biaxility required to explain the magnitude of $d^2_{0,2}\cos 2\gamma$ for PAA will have a pronounced effect on the other properties of the nematic mesophase. As a consequence any theory of nematics based on the assumption of molecular cylindrical symmetry is likely to be quantitatively inadequate.

References

1. M. Born, *Sitz. Phys.-Math.*, **25**, 614 (1916); M. Born and F. Stumpf, *ibid.*, **27**, 1043 (1916).
2. F. London, *Z. Physik*, **63**, 245 (1930); *Z. Physik. Chem. B*, **11**, 221 (1930).
3. W. Maier and A. Saupe, *Z. Naturforsch.*, **13a**, 564 (1958); **14a**, 882 (1959); **15a** 287 (1960).
4. P.G. de Gennes, *The Physics of Liquid Crystals*, Oxford University Press, (1974).
5. H.E. Stanley, *Introduction to Phase Transitions and Critical Phenomena*, Oxford University Press, (1971).
6. R.L. Humphries, P.G. James and G.R. Luckhurst, *J. Chem. Soc. Faraday Trans. II*, **68**, 1031 (1972).
7. T.D. Schultz, *Liquid Crystals 3*, eds. G.H. Brown and M.M. Labes, Gordon and Breach, p. 263 (1972). C. Zannoni, Thesis, University of Southampton, (1975).
8. A. Wulf, *J. Chem. Phys.*, **55**, 4512 (1971).
9. L. Shen, H.K. Sim, Y.M. Shih and C.-W. Woo, *Mol. Cryst. Liq. Cryst.*, **39**, 229 (1977); M.A. Lee and C.-W. Woo, *Phys. Rev. A* **16**, 750 (1977): V.T. Rajan and C.-W. Woo, *ibid.*, **17**, 382 (1978).
10. J.P. Hansen and I.R. McDonald, *Theory of Simple Fluids*, Academic Press, (1976).
11. C.N. Yang, *Phys. Rev. A* **5**, 808, (1952).
12. S. Chandrasekhar, *Liquid Crystals*, Cambridge University Press, (1977).
13. M.E. Rose, *Elementary Theory of Angular Momentum*, Wiley, (1957).
14. T.J. Krieger and H.M. James, *J. Chem. Phys.*, **22**, 796 (1954).
15. P.A. Lebwohl and G. Lasher, *Phys. Rev. A* **6**, 426 (1972).
16. H. Meirovitch, *Chem. Phys.* **21**, 251 (1976).
17. H.J.F. Jansen, G. Vertogen and J.G.J. Ypma, *Mol. Cryst. Liq. Cryst.*, **38**, 87 (1977).
18. C. Zannoni, to be published.
19. G.R. Luckhurst and S. Romano, to be published.
20. J.I. Kaplan and E. Drauglis, *Chem. Phys. Letters*, **9**, 645 (1971).
21. A. Wulf, *J. Chem. Phys.*, **64**, 104 (1976).
22. S. Chandrasekhar and N.V. Madhusudana, *Acta Cryst.*, **27**, 303 (1971).
23. J.R. McColl and C.S. Shih, *Phys. Rev. Letters*, **29**, 85 (1972).
24. I. Penchev and I. Dozov, *Phys. Letters A*, **60**, 34 (1977).
25. G.R. Luckhurst and R.N. Yeates, *J. Chem. Soc. Faraday Trans. II*, **72**, 996 (1976).
26. G.R. Luckhurst and R.N. Yeates, *Mol. Cryst. Liq. Cryst. Letters*, **34**, 57 (1976).
27. M. Kohli, K. Otnes, R. Pynn and T. Riste, *Z. Phys. B*, **24**, 147 (1976).
28. M.A. Cotter, *Mol. Cryst. Liq. Cryst.*, **39**, 173 (1977).
29. B. Widom, *J. Chem. Phys.*, **39**, 2808 (1963).
30. H.C. Anderson, D. Chandler and J.D. Weeks, *Adv. Chem. Phys.*, **34**, 105 (1976).
31. P. Sheng, *J. Chem. Phys.*, **59**, 1942 (1973).
32. G.R. Luckhurst and C. Zannoni, *Nature*, **267**, 412, (1977).
33. A.J. Leadbetter, R.M. Richardson and C.N. Colling, *J. de Phys.*, **36**, C1 - 37 (1975).
34. N. Nimura, *Mol. Cryst. Liq. Cryst.*, **31**, 123 (1975).
35. M. Evans, M. Davies and I. Larkin, *J. Chem. Soc. Faraday Trans II* **69**, 1011 (1973).
36. D.H. Chen, P.G. James and G.R. Luckhurst, *Mol. Cryst. Liq. Cryst.*, **8**, 71 (1969).
37. W.A. Steele, *J. Chem. Phys.*, **39**, 3197 (1963).
38. L. Blum and A.J. Torruella, *J. Chem. Phys.*, **56**, 303 (1972).
39. S. Chandrasekhar, D. Krishnamurti and N.V. Madhusudana, *Mol. Cryst. Liq. Cryst.*, **8**, 45 (1969).

40. J.R. Sweet and W.A. Steele, *J. Chem. Phys.,* **47**, 3022 (1967).
41. H. Arnold, *Z. Phys. Chem.,* **226**, 146 (1964).
42. R.L. Humphries and G.R. Luckhurst, *Chem. Phys. Letters,* **17**, 514 (1972).
43. R. Alben, J.R. McColl and C.S. Shih, *Solid State Commun.,* **11**, 1081 (1972).
44. G.R. Luckhurst, *Mol. Cryst. Liq. Cryst.,* **21**, 125 (1973).
45. J.W. Emsley, S.K. Khoo and G.R. Luckhurst, *Mol. Phys.,* **37**, 959 (1979).
46. J.W. Emsley, G.R. Luckhurst, G.W. Gray and A. Mosley, *Mol. Phys.,* **35**, 1499 (1978).
47. M.J. Freiser, *Phys. Rev. Letters,* **24**, 1041 (1970); *Liquid Crystals 3,* eds. G.H. Brown and M.M. Labes, Gordon and Breach, p.281 (1972).
48. C.S. Shih and R. Alben, *J. Chem. Phys.,* **57**, 307 (1972).
49. R. Alben, *Phys. Rev. Letters,* **30**, 778 (1973).
50. J.P. Straley, *Phys. Rev. A* **10**, 1881 (1974).
51. G.R. Luckhurst and S. Romano, to be published.
52. G.R. Luckhurst, C. Zannoni, P.L. Nordio and U. Segre, *Mol. Phys.,* **30**, 1345 (1975).

Chapter 5

A MOLECULAR FIELD THEORY OF THE CHOLESTERIC LIQUID CRYSTAL STATE

H. SCHRÖDER

Fachbereich Physik, Universität Konstanz,
7750 Konstanz, West Germany

Introduction

In this Chapter, we outline a molecular statistical theory for cholesteric liquid crystals. The cholesteric phase can be considered as an inhomogeneous nematic phase, where, on average, the long axes of the molecules rotate in space in such a way that, in a local frame, nematic order is essentially preserved. Depending on the particular interaction between chiral molecules, a rotation with respect to each of the principal axes on the average cannot be excluded. The possibility of such highly non-uniform phases has been pointed out by Brazovskii and Dmitriev [1] within the framework of de Gennes' theory of cholesterics [2,4]. The subject of this Chapter is the one-dimensional cholesteric phase with the long molecular axes rotating on the average in a plane, as one proceeds perpendicular to the plane. As long as there is no external field, the amplitudes of the order parameters remain constant, whereas the phase is subject to a spatial variation. The temperature dependent behaviour of the cholesteric pitch is quite complex. Near the cholesteric-smectic transition, the pitch shows a drastic increase in the immediate neighbourhood of T_{SCh} [5]. For higher temperatures the increase of the pitch is less dramatic. In both regions, the slope of the inverse pitch is virtually a constant [6,7]. In a single component system the pitch tends, in general, to increase with decreasing temperature. The only exception is cholesteryl 2-(2-ethoxyethoxy)ethyl carbonate [8], where the pitch increases with increasing temperature. Binary mixtures, for example, cholesteryl nonanoate/cholesteryl chloride [6], may exhibit a decrease in pitch with decreasing temperature near the cholesteric-isotropic phase transition. On cooling, the tendency is eventually reversed accompanied by a transition to the smectic phase. Deviations from this scheme are observed in compensated mixtures, where the inverse pitch may pass through zero and change sign as a function of the temperature, as well as of the concentration of the components [6,9].

A helicoidal phase may also be obtained from mixtures of achiral with chiral molecules. An example has been reported by Hanson *et al.* [7] with mixtures of cholesteric cholesteryl chloride and nematic 4-ethoxybenzylidene -4'-n-butylaniline and 4-[(4'-ethoxybenzylidene)amino] benzonitrile. This induced phase chirality can probably be well explained with a statistical generalization of the Samulskis' theory [10].

In the development of theories explaining the cholesteric phase, the main emphasis has naturally been placed on predictions of the pitch. Starting with Keating's [11] and Goossens' [12] work serious effort has been devoted to explaining the spontaneous twist. Since then, several attempts have been made to develop molecular statistical models for binary mixtures [13,14], as well as for single component systems [14-19]. However, a satisfactory analysis of the temperature dependence of the cholesteric pitch is still not available. A critical examination shows that the more or less phenomenological molecular statistical models [13,19] deal only with one of two possible descriptions of molecular chirality which allows only a certain structure for the pair potential. A group theoretical analysis employed in this Chapter shows that both descriptions are formally allowed, but seem to lead to opposite and contradictory physical interpretations of the cholesteric phase. Therefore a decision on physical grounds is possible and necessary. The crucial differences have been noticed only recently and will be discussed here.

In the second section, the pair potential in the form of an invariant expansion for arbitrarily shaped molecules is introduced and its symmetry properties are discussed. In the third section, the mean field approximation is employed to define order parameters and the pseudopotential. The free energy, the equilibrium conditions for the order parameters and the pitch, as well as the stability conditions are discussed in the fourth section. In the final section, the molecular symmetry properties and their influence on the properties of the phase are studied. For this purpose a Landau expansion of the cholesteric free energy is developed.

The Intermolecular Potential

The existence of a liquid crystal mesophase is closely related to the shape of the constituent molecules. That is, the anisotropy of the molecules has to be sufficiently strong to prevent the system from freezing before an orientational order is stabilized. Therefore, the strength of the anisotropic part of the interaction has to be comparable with the condensation energy of the liquid system, which is related to the isotropic part of the interaction. We assume this to be the case when considering the orientational potential alone. We begin therefore with a discussion of the intermolecular potential and pay particular attention to those terms responsible for the helical structure of the cholesteric phase.

Although the interaction of quite complex molecules is not known, it is formally possible, as we saw in Chapter 2, to obtain the isotropic and anisotropic parts of the pair potential for arbitrarily shaped molecules by separation of the angular and distance dependent contributions. This is achieved with an invariant expansion of the potential which has been taken from the excellent work of Blum and Torruella [20] as well as that by Steele [21]. The interaction for elongated molecules is, in general, anisotropic with respect to both the orientation Ω_i of the molecules and the intermolecular vector $\mathbf{r}_{ij}$.

The position of a molecule is denoted by $\mathbf{r}_i = (x, y, z)_i$ and its orientation, in the laboratory frame, is described by the Euler angles $\Omega_i = (\alpha, \beta, \gamma)_i$; both the variables are represented by $\mathbf{X}_i=(\mathbf{r}_i,\Omega_i)$. The pair potential can be written as

$$U(\mathbf{X}_1,\mathbf{X}_2) = \sum_{LL'J} \sum_{q,q'} u^{qq'}_{LL'J} S^{qq'}_{LL'J}(\Omega_1,\Omega_2,\Omega_{12}) \tag{1}$$

with the rotational invariant

$$S^{qq'}_{LL'J}(\Omega_1,\Omega_2,\Omega_{12}) = \sum_{p,p'M} \begin{pmatrix} L & J & L' \\ p & M & p' \end{pmatrix} D^{L}_{p,q}(\Omega_1) D^{L'}_{p',q'}(\Omega_2) D^{J}_{M,0}(\Omega_{12}), \tag{2}$$

where

$$\Omega_{12} = (\mathbf{r}_1-\mathbf{r}_2)/|\mathbf{r}_1-\mathbf{r}_2|,$$

describes the orientation of the intermolecular vector. The summation over J in eq. (1) is restricted according to the triangular condition $|L-L'| \leqslant J \leqslant (L+L')$. Translational invariance has already been established with the use of $\mathbf{r}_{12} = \mathbf{r}_1-\mathbf{r}_2$. The second basic property of a pair potential for identical particles is invariance under permutation of the molecules,

$$U(\mathbf{X}_1,\mathbf{X}_2) = U(\mathbf{X}_2,\mathbf{X}_1). \tag{3}$$

For the D^J matrix, we have

$$D^{J}_{M,0}(\Omega_{12}) = (-)^J D^{J}_{M,0}(\Omega_{21}). \tag{4}$$

Recalling the behaviour of Wigner 3-j symbols under an odd permutation

$$\begin{pmatrix} L & J & L' \\ p & M & p' \end{pmatrix} = (-)^{L+J+L'} \begin{pmatrix} L' & J & L \\ p' & M & p \end{pmatrix} \tag{5}$$

we find, with the aid of eq. (4), that

$$u_{LL'J}^{qq'} = (-)^{L'+L} u_{L'LJ}^{q'q} , \tag{6}$$

as a consequence of eq. (3). Similarly, we derive from the reality condition

$$U(\mathbf{X}_1,\mathbf{X}_2) = \{U(\mathbf{X}_1,\mathbf{X}_2)\}^* \tag{7}$$

that

$$(u_{LL'J}^{qq'})^* = (-)^{L+J+L'+q+q'} u_{LL'J}^{-q-q'} . \tag{8}$$

The orientational energy of two arbitrarily shaped, rigid molecules is given according to three degrees of freedom each of the molecules possesses. The parametrization is, in principle, arbitrary and for convenience sets of Euler angles have been chosen. Each set of Euler angles expresses the rotation of the three axes of the molecular frame with respect to the three axes of the laboratory frame. The choice of the molecular axes is a matter of definition. This means that, if the pair potential is known for a given set of independent coefficients $u_{LL'J}^{qq'}$ which may be defined with respect to the molecular principal axes, then the pair potential is known for any other set of independent coefficients $u_{LL'J}^{qq'}$, which is obtained through a unitary transformation of the individual molecular coordinates. This is guaranteed by the completeness of the set of basis functions $D(\alpha,\beta,\gamma)$. It is clear that in general the different terms of the expansion will depend on the choice of the molecular frame, though the transformation will yield an equivalent representation of the pair potential. A rotation in the molecular frame is performed by Wigner rotation matrices, which define rotations of irreducible tensors $T^{(L,p)}$

$$T^{(L,p)}(\Omega) = \sum_q D_{p,q}^{(L)}(\Omega) T^{(L,q)}(0) ; \tag{9}$$

this employs Edmond's convention [22] and so differs from that introduced in Chapter 3. The effect of a rotation R in the molecular frame is therefore

$$\sum_q D_{p,q}^{(L)}(\Omega)[R T^{(L,q)}(0)] = \sum_{q,n} D_{p,n}^{(L)}(\Omega) D_{n,q}^{(L)}(\Omega_R) T^{(L,q)}(0) .$$

Hence

$$\sum D_{p,q}^{(L)}(\Omega)[R u_{LL'J}^{qq'}] = \sum_{q,n} D_{p,n}^{(L)}(\Omega) D_{n,q}^{(L)}(\Omega_R) u_{LL'J}^{qq'} . \tag{10}$$

Similar results hold for the second molecule, so that the use of either $u_{LL'J}^{qq'}$ or

$$\hat{u}_{LL'J}^{nn'} = \sum_{n,n'} D_{n,q}^{L}(\Omega_R) D_{n',q'}^{L'}(\Omega_{R'}) u_{LL'J}^{qq'} , \tag{11}$$

in the expansion (1) yields equivalent expressions for the pair potential. Generally the choice of the molecular frame will be dictated by the geometry of the molecule. The requirement

$$\hat{u}^{nn'}_{LL'J} = u^{nn'}_{LL'J} , \tag{12}$$

defines a symmetry operation with respect to L or L', or both, which restricts the number of independent coefficients $u^{qq'}_{LL'J}$. Considering molecules with a twofold rotational symmetry about the molecular z axis, for example, eliminates coefficients with q and q' odd,

$$\left.\begin{aligned} u^{nq'}_{LL'J} &= \sum_{q} \delta_{n,q} \, e^{i\pi q} u^{qq'}_{LL'J} \\ u^{qn'}_{LL'J} &= \sum_{q'} \delta_{n,q'} e^{i\pi q'} u^{qq'}_{LL'J} \end{aligned}\right\} \quad . \tag{13}$$

For linear molecules we find $q = q' = 0$.

Since the symmetry of the mesophase is reflected in the molecular symmetry, a given orientational order in a many body system requires the presence of certain coefficients $u^{qq'}_{LL'J}$ in the interaction, supporting the ordering effect, or the absence of others which may suppress the order. In the present case it is our aim to identify those terms in the pair potential which stabilise the cholesteric phase. This can be achieved in quite a general manner. The only restriction is that the constituent molecules of a cholesteric phase obviously need to be chiral. A chiral molecule has the property that a reflection at an arbitrary plane yields its mirror image, which is not identical with the original molecule and cannot be obtained by a rotation. This eliminates any reflection as a symmetry operation. Therefore the pair potential cannot be invariant under inversion of any one of the molecular axes, if it is to describe the interaction between chiral molecules. Under this condition we call the interaction chiral. The next step is, therefore, to establish a test for molecular chirality, including a classification of tensors under inversion. There are two kinds of objects, those transforming under inversion I according to

$$I \, T^{(L,q)} = (-)^{L} T^{(L,q)} \tag{14}$$

and those transforming according to

$$I \, T^{(L,q)} = (-)^{L+1} T^{(L,q)} . \tag{15}$$

Quantities transforming according to eq. (15) are called pseudotensors. The rotationally invariant expansion of the pair potential consequently allows for two types of coefficients, which cannot be reduced to each other. Both types of coefficients are members of an irreducible representation of the roto-reflection group, but contain different physical information and lead to different physical statements, as we shall show.

In order to elaborate the crucial differences, a separate discussion of the possible cases is opportune. Since the irreducible tensors $T^{(L,q)}$ are isomorphic with products of spinor components [22], it is easy to derive the behaviour of $T^{(L,q)}$ under transformations S which are combinations of a rotation R and inversion I

$$S = RI = IR.$$

Consider a reflection in the molecular frame with respect to a plane perpendicular to the unit vector $\mathbf{n}$. Then

$$D^{(\frac{1}{2})}_{p,q}(\pm\boldsymbol{\sigma}\mathbf{n}) = \pm(\boldsymbol{\sigma}\mathbf{n})_{pq}, \tag{16}$$

where

$$\boldsymbol{\sigma}\mathbf{n} = \begin{pmatrix} -n_3 & n_1 - in_2 \\ n_1 + in_2 & n_3 \end{pmatrix},$$

can be chosen as the representation of the reflection operator S for $L = \frac{1}{2}$ [23]. Note that

$$\left.\begin{array}{l} \det(\pm\boldsymbol{\sigma}\mathbf{n}) = -1 \\ (\pm\boldsymbol{\sigma}\mathbf{n})(\pm\boldsymbol{\sigma}\mathbf{n}) = 1 \end{array}\right\}. \tag{17}$$

and

Under a reflection S tensors transform according to

$$S\ T^{(L,p)} = \sum_q D^L_{p,q}(\boldsymbol{\sigma}\mathbf{n})T^{(L,q)} \tag{18}$$

and pseudotensors according to

$$S\ T^{(L,p)} = -\sum_q D^L_{p,q}(\boldsymbol{\sigma}\mathbf{n})T^{(L,q)}. \tag{19}$$

Note that in both cases the invariance of $T^{(L,q)}$ to S^2 is guaranteed by eqs. (17). Molecular chirality requires

$$S\ T^{(L,q)} \neq T^{(L,q)} \tag{20}$$

for any S. As independent operations we only need to consider reflections of the three conditions of type (20). Proceeding by analogy to eq. (10), we obtain:

reflection of the molecular x axis:

$$\boldsymbol{\sigma}\mathbf{n} = \sigma_x, \tag{21a}$$

$$\sum_q D^L_{p,q}(\Omega)\{S_x u^{qq'}_{LL'J}\} = \pm\sum_{q,n} D^L_{p,n}(\Omega) D^L_{n,q}(\sigma_x) u^{qq'}_{LL'J},$$

$$= \pm\sum_{q,n} D^L_{p,n}(\Omega)\delta_{n,-q} u^{qq'}_{LL'J};$$

reflection of the molecular y axis:

$$\sigma n = \sigma_y, \tag{21b}$$

$$\sum_q D^L_{p,q}(\Omega)\{S_y u^{qq'}_{LL'J}\} = \pm\sum_{q,n} D^L_{p,n}(\Omega) D^L_{n,q}(\sigma_y) u^{qq'}_{LL'J},$$

$$= \pm\sum_{q,n} D^L_{p,n}(\Omega)(-1)^q \delta_{n,-q} u^{qq'}_{LL'J};$$

reflection of the molecular z axis:

$$\sigma n = \sigma_z, \tag{21c}$$

$$\sum_q D^L_{p,q}(\Omega)\{S_z u^{qq'}_{LL'J}\} = \pm\sum_{q,n} D^L_{p,n}(\Omega) D^L_{n,q}(\sigma_z) u^{qq'}_{LL'J},$$

$$= \pm\sum_{q,n} D^L_{p,n}(\Omega)(-1)^{L+q}\delta_{n,q} u^{qq'}_{LL'}.$$

The minus sign in front of the right hand side of eqs. (21) has to be chosen if $u^{qq'}_{LL'J}$ is a pseudo-quantity. For the case of *true tensors* we find immediately that S_x and S_y represent symmetry operations if q = 0. In addition, S_z is a symmetry operation if L alone is even. In order to guarantee molecular chirality at least one term with *q ≠ 0, L odd* and *L + q odd* must be present in the pair potential. In lowest order this is fulfilled for L = 3 and q = ± 2, or L' = 3 and q' = ± 2, respectively. In this case any reflection operation yields the *optical isomer* by reversing the handedness of the molecule.

For the case of *pseudotensors* L + q must be *even,* but q may be *zero* so as to describe molecular chirality. In the lowest order approximation this is fulfilled for L = 2 or L' = 2, respectively. The fact that q = 0 is not ruled out, is equivalent to the statement that linear molecules could exhibit optical activity, which is in striking contrast to the observation that optical activity is associated with an asymmetric carbon atom.

Another important feature of the pair potential is given by its behaviour under inversion of all coordinates. The energy of a physical system is independent of the choice of the coordinate frame. This is reflected in the law of conservation of parity and guaranteed by the vanishing commutator of the hamiltonian *H* and inversion operator *J*, acting on all coordinates:

$$[H,J] = 0. \tag{22}$$

The physical system may consist of just a single molecule or a system of interacting molecules; the energy must be a scalar. Eq. (22) is trivially fulfilled for the electrostatic interaction which is the origin of molecular interactions and therefore represents the only case of interest (cf. Chapter 2). The inversion of a physical system is the geometrical operation of reflecting each point of its configuration (mass and charge distribution) through the origin of the laboratory frame. If the constituent molecules are chiral, their handedness will be reversed automatically by this process. The inversion of a two particle system consists of an inversion of the molecules in their individual frame and in the inversion of the intermolecular vector. Eq. (22) requires:

$$J\ U(\mathbf{X}_1,X_2) = U(\mathbf{X}_1,X_2). \tag{23}$$

Note that the inversion operation with respect to a molecule can be written as:

$$I = S_x S_y S_z,$$

$$I\ T^{(L,q)} = \pm \sum_{m,nq} D^L_{p,m}(\sigma_z) D^L_{m,n}(\sigma_y) D^L_{n,q}(\sigma_z) T^{(L,q)}. \tag{24}$$

Since

$$D^J_{M,0}(\Omega_{12}) = (-)^J D^J_{M,0}(-\Omega_{12}),$$

we find the following conditions for the expansion coefficients from eq. (23)

$$u^{qq'}_{LL'J} = \eta^{(\pm)} u^{qq'}_{LL'J}\ . \tag{25}$$

For tensorial quantities we have

$$\eta^{(+)} = (-)^{L+L'+J}, \tag{26}$$

so that L + L' + J must be even. If $u^{qq'}_{LJL'}$ is a pseudo-quantity, we find

$$\eta^{(-)} = (-)^{L+J+L'+1}, \tag{27}$$

so that L + L' + J must be odd.

The different behaviour under parity shows that the two competing representations in question, not only lead to incompatible descriptions of molecular chirality but, as well, to completely different structures of the interaction itself. If only tensors appear in the pair potential, L + L' + J odd is completely ruled out by conservation of parity. Also, if pseudotensors are involved, then L + J + L' odd is allowed. This means that the expansion of a pair potential, which does not conserve parity, is formally indistinguishable from the corresponding expression where the terms with L + L' + J odd are pseudo-quantities.

With the aid of results from eq. (21) and eq. (25) we can easily specify the lowest order contribution of the pair potential to the chiral interaction U_c. Omitting pseudotensors, we find:

$$U_c = - \sum_{p,M,p'} \sum_{q,q'} \sum_{J} u^{qq'}_{23J} \begin{pmatrix} 2 & J & 3 \\ p & M & p' \end{pmatrix} D^2_{p,q}(\Omega_1) D^3_{p',q'}(\Omega_2) D^J_{M,o}(\Omega_{12}), \quad (28)$$

where $q = 0, \pm 2$, $q' = \pm 2$ and $J = 1,3,5$. (One can always find a transformation of the type of eq. (11) so that terms with $q,q' = \pm 1$ vanish [24]). Allowing for pseudotensors, we have

$$U_c = - \sum_{p,M,p'} \sum_{J} u^{00}_{22J} \begin{pmatrix} 2 & J & 2 \\ p & M & p' \end{pmatrix} D^2_{p,0}(\Omega_1) D^2_{p',0}(\Omega_2) D^J_{M,o}(\Omega_{12}), \quad (29)$$

where $J = 1,3$ and U_c vanishes for $p = M = p' = 0$.

The expansion of the pair potential is physically meaningless as long as it is impossible to identify these coefficients with those of a known interaction or to derive them from first principles. This holds for either representation. Since we deal with molecular interactions, we have to consider electrostatic and induced interactions (cf. Chapter 2). For electrostatic interactions the situation is clear, since pseudotensors cannot appear. Dispersion forces lead to the identification of the expansion coefficients with electronic transition matrix elements, corresponding to induced multipole moments. The usual procedure is to treat the electrostatic multipole expression as a perturbation operator and to calculate the dispersion energy in second order. The coefficients $u^{qq'}_{LL'J}$ are then products of electronic multipole matrix elements.

This is the foundation of Goossens' theory [12], which shows that the lowest order contribution to dispersion forces for a chiral interaction originates in the induced dipole-quadrupole interaction. Since molecules are essentially assemblies of charges, their symmetry is determined by the symmetry group of the respective ionic skeletons. The transition matrix elements depend, in general, on all nuclear positions. Goossens has demonstrated that the interaction of optically active molecules, which do not have any plane of symmetry, requires a non-vanishing electrostatic dipole-quadrupole transition matrix element, which is a cartesian third rank tensor. This is the lowest order contribution to molecular chirality. Depending on the symmetry group of the ionic skeleton, arbitrarily high order contributions from true tensors may be obtained. If a chiral molecule possesses an axis of n-fold rotational symmetry, the lowest order tensor element accounting for chirality is of the type $T^{(n+1,\pm n)}$ if n is even and $T^{(2n+1,\pm n)}$ if n is odd.

Other molecular theories for the cholesteric phase use a pair potential in which the chiral part is an odd-parity term [1-4, 13-17]. However, nothing is said about the possible microscopic origin of those terms, neither is it explained whether the term with L + L' + J odd (L = L' = 2, J = 1) is meant to be parity violating or to be a member of the pseudo-representation (except for [2]). This is certainly a serious defect, because inspection does not tell us the difference between a coefficient $u^{qq'}_{LL'J}$ which behaves under inversion as a tensor, and one that behaves as a pseudotensor. A representative example for such a pair potential is:

$$U_{12} = -J_{12}(\mathbf{a}_1.\mathbf{a}_2)^2 - K_{12}(\mathbf{a}_1.\mathbf{a}_2)\{(\mathbf{a}_1\wedge\mathbf{a}_2).\mathbf{a}_{12}\}, \tag{30a}$$

where $\mathbf{a}_1$ and $\mathbf{a}_2$ are the orientations of linear molecules and $\mathbf{a}_{12}$ is the intermolecular vector. The first term on the right hand side is the usual nematic ordering term, the second is identical with that in eq.(29). Beside the fact that this potential implies that linear molecules may be optically active, it has a very disturbing property. Expression (30a) may either describe the interaction between two chiral molecules or, under violation of parity, the interaction between two molecules with $D_{\infty h}$ symmetry. However, since optical activity is the property of a single molecule and non-conservation of parity a property of space, there is no equivalence and eq. (30a) can only describe either phenomenon. This problem can be resolved with the aid of Goossens' expression for the chiral interaction (eq. (6) in [12]).

Making simple assumptions concerning the transition matrix elements as suggested by van der Meer and Vertogen, we obtain from eq. (5) and (6) in [12]:

$$U_{12} = - J_{12}(\mathbf{a}_1.\mathbf{a}_2)^2 - K_{12}(\mathbf{a}_1.\mathbf{a}_2)\{(\mathbf{a}_1.\mathbf{c}_2)(\mathbf{b}_2.\mathbf{a}_{12}) - (\mathbf{a}_1.\mathbf{b}_2)(\mathbf{c}_2.\mathbf{a}_{12})\}$$

which correctly reads:

$$U_{12} = -J_{12}(\mathbf{a}_1.\mathbf{a}_2) - K_{12}(\mathbf{a}_1.\mathbf{a}_2)(\mathbf{a}_1\wedge\mathbf{a}_{12})(\mathbf{c}_2\wedge\mathbf{b}_2), \tag{30b}$$

where $\mathbf{a}_2$, $\mathbf{b}_2$ and $\mathbf{c}_2$ are the directions of the axes of the frame attached to molecule 2. K_{12} is a scalar

$$K_{12} = 3 \sum_{\nu,\nu'} (\xi^1_\nu \; \chi^1_{\nu'}) \;/(E_{\nu\nu,oo}),$$

with $\xi_\nu^1 = (p_z^1|p_z^1)_\nu \qquad \chi_\nu^2 = (p_x^2|q_{yz}^2) + c.c. = -(p_y^2|q_{xz}^2) - c.c.$.

U_{12}, in the form of eq. (30b), is indeed invariant under inversion of all coordinates and therefore parity conserving. A reflection of any of the molecular axes of molecule 2 yields the optical isomer with a change in sign in front of K_{12}. Replacing $\mathbf{c}_2 \wedge \mathbf{b}_2$, which is an axial vector, by $\mathbf{a}_2$, which is a polar vector with orientation and magnitude of $\mathbf{c}_2 \wedge \mathbf{b}_2$, indeed leads to parity violation in eq. (30a) whereas $\mathbf{c}_2 \wedge \mathbf{b}_2$ expresses a single particle property. The potential in the form of eq. (30b) can be constructed in the following way and has been used in this manner by van der Meer and Vertogen.

The philosophy is to create a chiral molecule by superposition of anisotropically polarizable particles. The mutual orientational interaction of such particles is the induced dipole-dipole interaction [25],

$$U_{ij} \sim P_2(\chi_{ij}) r_{ij}^{-6}, \tag{31}$$

which alone would tend to align the molecules parallel. However, the development of a nematic phase is suppressed by preventing the particles from taking their parallel alignment along a cylindrical surface. For this purpose an effective pair potential, U_{12}, is defined, simply by restricting the particles along the surface of each of the two cylinders to non-equilibrium orientations followed by summation over all positions:

$$U_{12} = \sum_{i,j} U_{ij} \sim \sum_{i,j} P_2(\chi_{ij}) / |\mathbf{r}_i - \mathbf{r}_j + \mathbf{a}_{12}|^6. \tag{32}$$

The summation is evaluated by taking a double Taylor expansion of r_{ij}^{-6} in powers of the cylinders' radii over their separation $\mathbf{a}_{12}$. The relative orientation between the anisotropically polarizable particles on the cylinder and the cylindrical axis is constant and defines a screw sense. In lowest order the potential (31) is reproduced and is identical with the first term of eq. (30). The next order yields the second term of eq. (30), which is identical with eq. (29). Apart from the definition of the interaction, eq. (32), it is clear that it has nothing in common with Goossens' expression. Indeed the cholesteric phase obtained from eq. (32) is rather imposed by boundary conditions, which keep the particles on the cylinders in non-equilibrium orientations relative to each other, than derived from a microscopic interaction. Although this model seems to be plausible for chiral molecules, the procedure is incorrect, simply because van der Waals forces do not generally superpose. In addition an identification of the expansion coefficients requires a comparison with a known interaction. Electrodynamic forces are non-additive, because multiple scattering processes are responsible for this interaction. Therefore the interaction energy of two molecules, as used by van der Meer and Vertogen, cannot be obtained by a summation over all polarizabilities in each of the molecules. (For a detailed treatment of retarded and non-retarded van der Waals forces see McLachlan [26,27]). In addition the existence of molecules behaving according to eq. 30 should be shown.

The Mean Field Approximation

In the mean field approximation, the variable $D^L_{p,q}(\Omega_i)$ is replaced by its thermal average $\overline{D^L_{p,q}}$ which means that the quantity $\{D^L_{p,q}(\Omega_i)-\overline{D^L_{p,q}}\}$ is neglected in comparison with the order parameter. We obtain the effective potential acting on the reference molecule j as

$$U(\mathbf{X}_j) = \sum_{L,L',J} \sum_{q,q'} u^{qq'}_{LL'J} \sum_{p,p'M} \begin{pmatrix} L & J & L' \\ p & M & p' \end{pmatrix} \sum_{<i>} \overline{D^L_{p,q}} D^L_{p',q'}(\Omega_j) D^J_{M,o}(\Omega_{ij}). \quad (33)$$

The next step is to perform the summation over all neighbours contributing to the mean field potential at molecule j, i.e. $\sum_{<i>} \overline{D^L_{p,q}} D^L_{M,o}(\Omega_{ij})$. In the cholesteric phase the order parameters $\overline{D^L_{p,q}}$, expressed in a laboratory frame, are expected to vary in space because of the helical arrangement of the director. For convenience, we choose the z direction of the laboratory frame as the *helical axis*, so that the projection of the *director* rotates in the xy plane. If the relative *twist angle* between the *director* at two points $\mathbf{r}_i$ and $\mathbf{r}_j$ is ϕ_{ij}, we can write

$$\begin{aligned} \overline{D^L_{p,q}(\Omega_i)} &= \overline{D^L_{p,q}}(\alpha_i\beta_i\gamma_i), \\ &= \overline{D^L_{p,q}}(\alpha_j+\phi_{ij},\beta_j\gamma_j), \\ &= e^{ip\phi_{ij}}\, \overline{D^L_{p,q}}(\alpha_j\beta_j\gamma_j), \\ &= e^{ip\phi_{ij}}\, \overline{D^L_{p,q}}(\Omega_j). \end{aligned} \quad (34)$$

It should be noted that, at this stage, the orientation of the director is still arbitrary and not restricted to the xy plane.

In the absence of any inhomogeneities, the twist angle is a linear function of the distance

$$\phi_{ij} = \kappa(z_i - z_j), \quad (35)$$

where κ is the inverse pitch. Eq. (35) states that the spontaneous twist is translationally invariant. This is an assumption and need not necessarily be fulfilled. For example, the effect of a symmetry breaking magnetic or electric field unwinding the helical structure is a torque which depends on the orientation of the individual molecule in the laboratory frame. Only counting nearest neighbour contributions to the molecular field, the summation $\sum_{<i>}$ is to be taken over the solid angle Ω_{12} of the intermolecular vector $\mathbf{r}_{ij} = (a, \psi_{ij}, \gamma_{ij})$ with $\cos\psi_{ij} = (z_i - z_j)/a$, where a is the average distance between two neighbours. This summation implies a homogeneous distribution of the molecular centres of mass throughout the liquid. Owing to the spatial variation of order parameters in the cholesteric phase, the averaging process includes both the intermolecular vector and the order parameters. Replacing the sum by an integral, and using the following identity for spherical Bessel functions $j_L(x)$,

$$j_L(x) = (1/2)(-i)^L \int d\cos\psi\ e^{ix\cos\psi} D^L_{0,0}(0,\psi,0),$$

we find

$$\sum_{<i>} \overline{D^L_{p,q}}(\Omega_i) D^J_{p,q}(\Omega_{ij}) = i^J \delta_{M,o}\ j_J(\kappa ap) \overline{D^L_{p,q}}(\Omega_j). \tag{36}$$

Inserting this result into eq. (33) yields the effective potential

$$U(\Omega) = -\sum_{L,L',J} \sum_{q,q'} u^{qq'}_{LL'J} i^J \sum_p j_J(\kappa ap) \begin{pmatrix} L & J & L' \\ p & 0 & -p \end{pmatrix} \overline{D^L_{p,q}} D^{L'}_{-p,q'}(\Omega). \tag{37}$$

Self-consistency is established by

$$\overline{D^L_{p,q}} = \int d\Omega\ D^L_{p,q}(\Omega)\exp\{-\beta U(\Omega)\}/\int d\Omega\ \exp\{-\beta U(\Omega)\}. \tag{38}$$

This is the usual definition of self-consistency as employed in previous calculations for orientational order parameters. Restrictions on the number of independent order parameters are obtained from phase and particle symmetries as discussed in the fifth section and in Chapter 3. According to eq. (38), the order parameters are defined in the local frame so that they are independent of a constant phase shift

$$\phi_i \to \phi_i + \phi\ ;\ \phi_j \to \phi_j + \phi$$

which leaves the potential, eq. (37), invariant.

Discussion of the Free Energy

For the sake of completeness, we add field terms $H^L_{p,q}$ to the interaction, eq. (37), which are conjugate variables with respect to the order parameters $\overline{D^L_{p,q}}$. The field components $H^L_{p,q}$ play the role of external fields as given by magnetic or electrical fields, or of boundary conditions expressed by fields (cf. [28]). The generalized expressions for the order parameters are

$$\overline{D^L_{p,q}} = \int d\Omega\, D^L_{p,q}(\Omega)\exp\{\sum_{\lambda',\nu,\nu'} \mathcal{L}^{\lambda'}_{\nu,\nu'} D^{\lambda'}_{\nu,\nu'}(\Omega)\}$$

$$\times[\int d\Omega\, \exp\{\sum_{\lambda',\nu,\nu'} \mathcal{L}^{\lambda'}_{\nu,\nu'} D^{\lambda'}_{\nu,\nu'}(\Omega)\}]^{-1}, \tag{39}$$

with

$$\mathcal{L}^{\lambda'}_{\nu,\nu'} = \beta\{\sum_{L,J} \sum_{q} u^{q\nu'}_{L\lambda' J} i^J j_J(\kappa a\nu) \begin{pmatrix} L & J & \lambda' \\ \nu & 0 & -\nu \end{pmatrix} \overline{D^L_{\nu,q}}(\Omega) + H^{\lambda'}_{\nu,\nu'}\}. \tag{40}$$

Any self-consistency relation, eq. (39), is an equation of the form

$$\overline{D^L_{p,q}} = L^L_{p,q}(\ldots,\mathcal{L}^{\lambda'}_{\nu,\nu'}, \ldots), \tag{41}$$

where the $\mathcal{L}^{L'}_{\nu,\nu'}$ are the inverse functions of the $L^L_{p,q}$,

$$\mathcal{L}^L_{p,q} = (L^L_{p,q})^{-1}(\ldots, D^{\lambda}_{\nu,\nu'}, \ldots), \tag{42}$$

so that the equation of state is obtained by formal inversion of eq. (39),

$$H^{\lambda'}_{\nu,\nu'} = -\sum_{L,J} \sum_{q} u^{q\nu'}_{L\lambda' J} i^J j_J(\kappa a\nu) \begin{pmatrix} L & J & \lambda' \\ \nu & 0 & -\nu \end{pmatrix} D^L_{\nu,q} + \beta^{-1}\mathcal{L}^{\lambda'}_{\nu,\nu'}. \tag{43}$$

The conjugate field is just the derivative of the Gibbs free energy, eq. (46), with respect to the corresponding order parameter,

$$\partial A/\partial \overline{D^{\lambda'}_{\nu,\nu'}} = H^{\lambda'}_{\nu,\nu'}. \tag{44}$$

Consequently, the free energy is an integral of the system of equations given by (43). The relation

$$\partial H^{\lambda'}_{\nu\nu'}/\partial \overline{D^L_{p,q}} = \partial H^L_{p,q}/\partial \overline{D^{\lambda'}_{\nu,\nu'}}, \tag{45}$$

guarantees integrability. Formally, we obtain

$$A = U - TS \tag{46}$$

with

$$U - -\tfrac{1}{2} \sum_{L,J,L'} \sum_{q,q'} u^{qq'}_{LJL'} i^J \sum_{p} j_J(\kappa ap) \begin{pmatrix} L & J & L' \\ p & 0 & -p \end{pmatrix} \overline{D^L_{p,q}}\, \overline{D^{L'}_{-p,q'}} \tag{47}$$

and

$$\partial S/\partial \overline{D^L_{p,q}} = -k\mathcal{L}^L_{p,q}. \tag{48}$$

Eq. (48) states that the entropy S is a function of the order parameters only. The dependence on the inverse pitch, κ, appears explicitly only in the internal energy U.

In the absence of any external field, the equilibrium values for the order parameters which minimize the free energy are determined by

$$H^{L}_{p,q} = 0 \tag{49}$$

and the condition that

$$\det|\partial^2 A/\partial\overline{D^{L}_{p,q}}\,\partial\overline{D^{\lambda}_{\nu,\nu'}}| \tag{50}$$

and its principal minors be positive, where

$$\partial^2 A/\partial\overline{D^{L}_{p,q}}\,\partial\overline{D^{\lambda'}_{\nu,\nu'}} = \sum_{J=|L-\lambda'|}^{L+\lambda'} u^{q\nu'}_{L\lambda' J}\, i^{J} j_{J}(\kappa a\nu)\begin{pmatrix} L & J & \lambda' \\ p & 0 & \nu \end{pmatrix}\delta_{p,-\nu} + \beta^{-1}\partial L^{\lambda'}_{\nu,\nu'}/\partial\overline{D^{L}_{p,q}}. \tag{51}$$

Vanishing fields $H^{L}_{p,q}$ mean that the free energy is minimized by order parameter values satisfying eq. (38). Simultaneously, the free energy becomes minimized with respect to κ. For simplicity, we write

$$A(\kappa,T) = -(1/2)\sum_{\mu,\nu} A_{\mu\nu}(\kappa)\sigma_{\mu}\sigma_{\nu} - TS, \tag{52}$$

where $\sigma(\mu = 1, \ldots, n)$ is a set of order parameters depending on κ and T. According to eq. (49), we have at equilibrium

$$0 = -\sum_{\nu} A_{\mu\nu}(\kappa)\sigma_{\nu} + \beta^{-1}L_{\mu}. \tag{53}$$

Differentiation with respect to κ yields

$$0 = \sum_{\nu}\{\chi^{-1}_{\mu\nu}\,\partial\sigma_{\nu}/\partial\kappa - (\partial A_{\mu\nu}(\kappa)/\partial\kappa)\sigma_{\nu}\} \tag{54}$$

with the inverse susceptibility

$$\chi^{-1}_{\mu\nu} = -A_{\mu\nu}(\kappa) + \beta^{-1}\partial L_{\nu}/\partial\sigma_{\mu}, \tag{55}$$

so that

$$\partial\sigma_{\mu}/\partial\kappa = -\sum_{\nu}\chi_{\mu\nu}\partial^2 U/\partial\kappa\partial\sigma_{\nu}. \tag{56}$$

With the aid of eq. (56), we now show that the equilibrium value of the inverse pitch is only determined by the explicit κ dependence of the internal energy. The first derivative with respect to κ is

$$dA/d\kappa = \sum_{\mu}(\partial A/\partial\sigma_{\mu})(\partial\sigma_{\mu}/\partial\kappa) + \partial A/\partial\kappa. \tag{57}$$

The second derivative is

$$d^2A/d\kappa^2 = \sum_{\mu,\nu}(\partial^2 A/\partial\sigma_{\mu}\partial\sigma_{\nu})(\partial\sigma_{\mu}/\partial\kappa)(\partial\sigma_{\nu}/\partial\kappa) + \sum_{\mu}(\partial A/\partial\sigma_{\mu})(\partial^2\sigma_{\mu}/\partial\kappa^2)$$
$$+ \sum_{\mu}(\partial^2 A/\partial\sigma_{\mu}\partial\kappa)(\partial\sigma_{\mu}/\partial\kappa) + \partial^2 A/\partial\kappa^2. \tag{58}$$

Inserting eq. (56) into eq. (58), using $H_\mu = 0$ and $\partial^2 A/\partial\sigma_\mu \partial\sigma_\nu = \chi^{-1}_{\mu\nu}$, we find

$$dA/d\kappa = \partial U/\partial\kappa \; ; \; d^2A/d\kappa^2 = \partial^2 U/\partial\kappa^2 . \tag{59}$$

Since the pitch is usually of the order of several hundred molecular distances, where we consider the average distance as introduced previously, an expansion in the neighbourhood of $\kappa = 0$ is justified,

$$A(\kappa,T) \approx A(\kappa{=}0,T) + \kappa(\partial U/\partial\kappa)_{\kappa=0} + (1/2)\kappa^2(\partial^2 U/\partial\kappa^2)_{\kappa=0} . \tag{60}$$

The equilibrium value of κ is then

$$\kappa_o = -\sum_{\mu,\nu}(\partial A_{\mu\nu}/\partial\kappa)\sigma_\mu\sigma_\nu / \sum_{\mu,\nu}(\partial^2 A_{\mu\nu}/\partial\kappa^2)\sigma_\mu\sigma_\nu . \tag{61}$$

Note that the entropy is not involved in eq. (61). $\partial^2 U/\partial\kappa^2 > 0$ guarantees that $A(\kappa_o,T)$ is a minimum, whereas κ_o is, in general, a function of temperature. In this approximation, the variation of the order parameters due to κ being small is

$$\delta\sigma_\mu = \kappa \sum_{\nu,\lambda} \chi_{\mu\nu}(\partial A_{\lambda\nu}/\partial\kappa)\sigma_\lambda . \tag{62}$$

This is in accordance with the view of a cholesteric system as a quasi-nematic system with a small twist distortion.

Discussion of Phase Properties

Cholesteric systems consist of either right or left handed chiral molecules so that the pitch is either right or left handed. A mixture of equal parts of right and left handed and otherwise like molecules (e.g. racemic mixture) exhibits a purely nematic phase. Consequently, the change of nothing but the handedness of chiral molecules in a cholesteric system of identical particles can only result in a reversed pitch. Therefore, a simultaneous change of the handedness of the molecules and the sign of the pitch must leave the free energy invariant. This yields conditions for the coefficients $u^{qq'}_{LL'J}$, which are automatically satisfied if the criteria of molecular chirality, according to eqs. (21), are fulfilled and parity is observed. We assume that this is the case and find for the inverse pitch in lowest order ($J = 0, 1$) from eq. (61):

$$\kappa_o a = 2i \sum_{L,L'} \sum_{q,q'} u^{qq}_{LL'1} \sum_p \begin{pmatrix} L & 1 & L' \\ p & 0 & -p \end{pmatrix} \overline{D^L_{p,q}}\; \overline{D^{L'}_{-p,q'}}$$

$$\times\Big\{ \sum_{L,L'} \sum_{q,q'} u^{qq'}_{LL'0} \sum_p p^2 \begin{pmatrix} L & 0 & L' \\ p & 0 & p \end{pmatrix} \overline{D^L_{p,q}}\; \overline{D^{L'}_{-p,q}} \Big\}^{-1} . \tag{63}$$

Substitution of the coefficients $u^{qq'}_{LL'J}$ by the respective coefficients for the molecular mirror images yields the mirror image of the phase

$$\text{a)}\quad u^{qq'}_{LL'J} \rightarrow (-)^{L+L'} u^{qq'}_{LL'J} \;.$$

In the denominator L + L' is always even; hence $\kappa_o \rightarrow -\kappa_o$. Therefore a chiral term in the internal energy, eq. (47), consists of a product of an even rank tensor order parameter with an odd rank tensor order parameter

$$\overline{D^{L}_{p,q}}\;\overline{D^{L'}_{-p,q}}\;,\quad L + L' \text{ odd.}$$

Any finite order parameter $\overline{D^{L}_{p,q}}$ with $q \neq 0$ is the mean field for a hindered rotation of the molecules with respect to their longitudinal axes, where $q \neq 0$ states that the molecules cannot undergo uncorrelated rotations. If we consider pseudo-quantities as being responsible for chirality, q and q' may be zero. We can write for the inverse pitch in this case:

$$\kappa_o a = -2 \sum_{L,L'} u^{00}_{LL'0} \sum_{p} \begin{pmatrix} L & 1 & L' \\ p & 0 & -p \end{pmatrix} \mathrm{Im}\{\overline{D^{L}_{p,0}}\;\overline{D^{L'}_{-p,0}}\}$$

$$\times \left[\sum_{L,L'} u^{00}_{LL'0} \sum_{p} p^2 \begin{pmatrix} L & 0 & L' \\ p & 0 & p \end{pmatrix} \mathrm{Re}\{\overline{D^{L}_{p,0}}\;\overline{D^{L'}_{-p,0}}\}\right]^{-1} . \qquad (65)$$

Here the molecular rotation is uncorrelated with respect to the longitudinal axes. According to the mirror operation

$$\text{b)}\quad u^{qq'}_{LL'J} \rightarrow (-)^{L+L'+1} u^{qq'}_{LL'J}$$

L + L' must be even. Therefore a chiral term in the pseudo-representation consists of a product of two even or two odd rank tensor order parameters:

$$\overline{D^{L}_{p,0}}\;\overline{D^{L'}_{-p,0}}\;,\quad L + L' \text{ even.}$$

After the discussion of several properties of the pair potentials we shall conclude this Chapter with a demonstration of phase properties as they are predicted by the two different conceptions of chirality. The theoretical background has been prepared in the previous sections, so that we can present the essential results without going into detail. A relatively simple solution is obtained for the pair potential in the form of eq. (30). In terms of tensor notation the internal energy, according to eq. (47), is

$$U = -(1/2) \sum_{J=0}^{1} u_{22J}^{00} \; i^J \sum_{p=-2}^{2} j_J(\kappa ap)\begin{pmatrix} 2 & J & 2 \\ p & 0 & -p \end{pmatrix}\bar{D}_{p,0}^{2}\bar{D}_{-p,0}^{2}. \quad (66)$$

The expansion for small κa yields the following contributions:

$$U/U_o = -(1/2)\bar{S}_2^2 + (1/2)(\kappa a)^2(\bar{S}_{2,1}^2 + \bar{S}_{2,2}^2) - \kappa aQ(\bar{S}_{2,1}^2 + \bar{S}_{2,2}^2), \quad (67)$$

where we have defined

$$U_o = u_{220}^{00}/\sqrt{5}, \; Q = i\, u_{221}^{00}/\sqrt{6}\, u_{220}^{00}.$$

The variables are

$$S_{2,0} = P_2(\cos\beta),$$

$$S_{2,1} = \sin\beta\cos\beta\cos\alpha$$

and

$$S_{2,2} = \sin^2\beta\cos 2\alpha.$$

From the transformation properties of spherical tensors it follows that

$$\bar{S}_2^2 = \bar{S}_{2,0}^2 + 3\,\bar{S}_{2,1}^2 + (3/4)\bar{S}_{2,2}^2$$

is an invariant, where $\bar{S}_2$ is the nematic order parameter in the local frame. For $\kappa = 0$ the free energy depends only on $\bar{S}_2$ and T, that means it is degenerate with respect to the orientations of the director. This degeneracy is removed in the cholesteric phase. From $\partial U/\partial\kappa = 0$ we find:

$$\kappa a = Q; \quad (68)$$

hence

$$U/U_o = -(1/2)\bar{S}_2^2 - (1/2)Q^2(\bar{S}_{2,1}^2 + \bar{S}_{2,2}^2). \quad (69)$$

Since $\bar{S}_{2,1}^2 + \bar{S}_{2,2}^2$ is not an invariant, the free energy depends on the average orientation $\omega = \cos\theta$ of the director relative to the helical axis. In addition the director is rotating uniformly about the helical axis creating the cholesteric structure. The average orientation ω_o minimises the free energy. With

$$\bar{S}_{2,1} = \sin\theta\cos\theta\bar{S}_2$$

$$\bar{S}_{2,2} = \sin^2\theta\bar{S}_2,$$

we find

$$U/U_o = -(1/2)\bar{S}_2^2 - (1/2)Q^2\bar{S}_2^2 \;(1-\omega^2), \quad (70)$$

$$\omega_o = 0.$$

Finally

$$U/U_o = -(1/2)\bar{S}_2^2 \; (1 + Q^2). \tag{71}$$

Eq. (68) is the lowest order approximation for the pitch of the type of eq. (65). For the use of linear molecules in the pseudo-representation a temperature independent pitch is typical, as shown in [15] and [18]. Since the condensation energy is lowered by the constant value Q^2, the question arises as to how Q is to be removed in the completely unwound smectic phase, if such a model is used. It should also be noted that the angle between the director and the helical axis is typically 90°.

The situation is much more involved if pseudotensors are ignored and the expansion coefficients are identified with true irreducible tensor elements. Here we need molecules with a deviation from cylindrical symmetry and third rank tensors to account for molecular chirality. The general self-consistent potential from eq. (37) is, in this case,

$$\begin{aligned} U(\Omega) = &- \sum_{q,q'} u^{qq'}_{220} \sum_p j_o(\kappa ap)\begin{pmatrix} 2 & 0 & 2 \\ p & 0 & -p \end{pmatrix}\bar{D}^2_{p,q} D^2_{-p,q'}(\Omega) \\ &- i \sum_{q,q'} u^{qq'}_{231} \sum_p j_1(\kappa ap)\begin{pmatrix} 2 & 1 & 3 \\ p & 0 & -p \end{pmatrix}\bar{D}^2_{p,q} D^3_{-p,q'}(\Omega) \\ &- i \sum_{q,q'} u^{qq'}_{321} \sum_p j_1(\kappa ap)\begin{pmatrix} 3 & 1 & 2 \\ p & 0 & -p \end{pmatrix}\bar{D}^3_{p,q} D^2_{-p,q'}(\Omega). \end{aligned} \tag{72}$$

The first term is purely nematic and the dominating one. On the energy scale the difference between the nematic and the cholesteric state is extremely small. Apart from the helical structure the basic properties of the cholesteric phase are therefore governed by the nematic potential. For the present purpose it is sufficient therefore to study the phase behaviour for $\kappa = 0$, and to obtain chiral effects from a linear response theory. For $\kappa = 0$ the first part of eq. (72) is identical with the mean field potential for biaxial molecules [29] and the assumption of dispersion forces allows us to find relations which reduce the number of independent coefficients. Here we have

$$\lambda^2 = u^{22}_{220}/u^{00}_{220} = (u^{20}_{220}/u^{00}_{220})^2, \tag{73}$$

where λ is a parameter describing the deviation from molecular cylindrical symmetry. We may write

$$u^{qq'}_{LL'J} = A^J_{LL'} T^{(L,q)}(0) T^{(L',q')}(0), \tag{74}$$

where the $T^{(L,q)}(0)$ are the molecular tensor components from eq. (9)

with

$$T^{(2,\pm2)}(0) = \lambda,\ T^{(2,0)}(0) = 1,\ T^{(2,\pm1)}(0) = 0.$$

The internal energy reduces to

$$U/U_o = -(1/2)\{\overline{S}^2_{2,0} + 3\ \overline{S}^2_{2,1} + (3/4)\overline{S}^2_{2,2}\}, \tag{75}$$

with $U_o = A^0_{22}/\sqrt{5}$. The variables are

$$S_{2,0} = P_2(\chi) + (3/2)^{\frac{1}{2}}\lambda(1-\chi^2)\cos2\gamma$$

$$S_{2,1} = \chi(1-\chi^2)^{\frac{1}{2}}\cos\alpha-(2/3)^{\frac{1}{2}}\lambda\{\chi(1-\chi^2)^{\frac{1}{2}}\cos\alpha\cos2\gamma-(1-\chi^2)^{\frac{1}{2}}\sin\alpha\sin2\gamma\},$$

$$S_{2,2} = (1-\chi^2)\cos2\alpha+(2/3)^{\frac{1}{2}}\lambda\{(1+\chi^2)\cos2\gamma\cos2\alpha-2\chi\sin2\alpha\sin2\gamma\}, \tag{76}$$

with $\chi = \cos\beta$. To investigate the ordered phase we use a Landau expansion of the free energy eq. (46) in powers of the order parameter components. The evaluation of the free energy requires a knowledge of the equation of state, eq. (43), which we obtain from an expansion of the orientational partition function Z,

$$Z = \int d\Omega\ \exp\{L_{1,0}S_{2,0}+L_{2,1}S_{2,1}+L_{2,2}S_{2,2}\}. \tag{77}$$

Here we have defined

$$L_{2,0} = \beta(\bar{S}_{2,0}+H_{2,0}),$$

$$L_{2,1} = 3\beta(\bar{S}_{2,1}+H_{2,1}),$$

$$L_{2,2} = (3/4)\beta(\bar{S}_{2,2}+H_{2,2}) \tag{78}$$

and

$$A^0_{22}\ \sqrt{5}\ k = 1.$$

The order parameters are the logarithmic derivatives of the partition function

$$\bar{S}_{2,p} = \ln Z/\partial L^{(2)}_p, \tag{79}$$

$$\bar{S}_{2,0} = L^{(2)}_0(L^{(2)}_0, L^{(2)}_1, L^{(2)}_2),$$

$$\bar{S}_{2,1} = L^{(2)}_1(L^{(2)}_0, L^{(2)}_1, L^{(2)}_2,$$

$$\bar{S}_{2,2} = L_{2,2}(L^{(2)}_0, L^{(2)}_1, L^{(2)}_2). \tag{80}$$

This system of non-linear equations can be inverted and solved for the $L^{(2)}_p$ and $H^{(2)}_p$, respectively.

We find for lnZ, including fifth order contributions

$$
\begin{aligned}
\ln Z = {} & (1/10)(1+2\lambda^2)\{L_0^{(2)2}+(1/3)L_1^{(2)2}+(4/3)L_2^{(2)2}\} \\
& +(1/105)(1-6\lambda^2)[L_1^{(2)2}\{(1/2)L_0^{(2)}+L_2^{(2)}\}+L_0^{(2)}(L_0^{(2)2}-4L_2^{(2)2})] \\
& -(1/700)(1+2\lambda^2)^2\{L_0^{(2)2}+(1/3)L_1^{(2)2}+(4/3)L_2^{(2)2}\}^2 \\
& -(1/1925)(1+2\lambda^2)(1-6\lambda^2)\{L_0^{(2)2}+(1/3)L_1^{(2)2}+(4/3)L_2^{(2)2}\} \\
& \times[L_2^{(2)2}\{(1/2)L_0^{(2)}+L_2^{(2)}\}+L_0^{(2)}(L_0^{(2)2}-4L_2^{(2)2})]. \qquad (81)
\end{aligned}
$$

The inversion of eqs. (80) is straightforward, but tedious. Integration of the $H_p^{(2)}$ finally yields the Landau free energy for the nematic phase:

$$
\begin{aligned}
F_N = {} & (1/2)\{(5T-1-2\lambda^2)/(1+2\lambda^2)\}(\bar{S}_{2,0}^2+3\,\bar{S}_{2,1}^2+(3/4)\bar{S}_{2,2}^2) \\
& -TB(\lambda)\{(9/2)\bar{S}_{2,1}(\bar{S}_{2,0}+(3/2)S_{2,2})+\bar{S}_{2,0}(\bar{S}_{2,0}^2-(9/4)\bar{S}_{2,2}^2)\} \\
& +TC(\lambda)(\bar{S}_{2,0}^2+3\,\bar{S}_{2,1}^2+(3/4)\bar{S}_{2,2}^2)^2 \\
& -TD(\lambda)(\bar{S}_{2,0}^2+3\,\bar{S}_{2,1}^2+(3/4)\bar{S}_{2,2}^2) \\
& \times[(9/2)\bar{S}_{2,1}^2(\bar{S}_{2,0}+(3/2)\bar{S}_{2,2})+\bar{S}_{2,0}\{\bar{S}_{2,0}^2-(9/4)\bar{S}_{2,2}^2\}]. \qquad (82)
\end{aligned}
$$

The coefficients are

$$
\begin{aligned}
B(\lambda) &= 25(1-6\lambda^2)/21(1+2\lambda^2), \\
C(\lambda) &= 25\{7(1+2\lambda^2)^3+10(1-6\lambda^2)^2\}/196(1+2\lambda^2)^5, \qquad (83) \\
D(\lambda) &= 125(1-6\lambda^2)\{119(1+2\lambda^2)^3+165(1-6\lambda^2)^2\}/44 \times 196(1+2\lambda^2)^7.
\end{aligned}
$$

The combinations of the order parameter components appearing in eq. (82) are independent of the biaxiality parameter, λ. For $\lambda = 0$ the free energy reduces to that of Maier and Saupe for axially symmetric molecules, in which case it is usually calculated for only one of the order parameter components, namely $\bar{S}_{2,0}$. However, the choice of the director with respect to a laboratory frame is arbitrary. From the structure of the free energy F_N, it is self-evident that this holds for biaxial molecules as well as for uniaxial molecules in the Maier and Saupe model. It seems therefore to be superfluous to carry all three components. Nevertheless, it is necessary to use a rotational invariant form of the free energy, because imposing the orientation of the director with respect to the laboratory frame on the system does not allow us to study the symmetry breaking effect of the pitch in a possible cholesteric phase, if we add the chiral contribution to eq. (82). We can use again the invariant $\overline{S_2}$, and $\omega = \cos\theta$ as the orientation between the molecular and the laboratory z axis. Inserting

$$\bar{S}_{2,0} = \bar{S}_2\ P_2(\omega),$$
$$\bar{S}_{2,1} = \bar{S}_2\omega(1-\omega^2)\ , \tag{84}$$
$$\bar{S}_{2,2} = \bar{S}_2\ (1-\omega^2),$$

into eq. (82), we find that the free energy

$$F_N = (1/2)\{(5T-1-2\lambda^2)/(1+2\lambda^2)\}\bar{S}_2^2 \quad - TB(\lambda)\bar{S}_2^3$$
$$+ TC(\lambda)\bar{S}_2^4 - TD(\lambda)\bar{S}_2^5\ , \tag{85}$$

in all orders, depends only on the invariant $\bar{S}_2$ and T as thermodynamical variables. The only possible transition is that from the isotropic to the uniaxial nematic state described by the single order parameter

$$\bar{S}_{2,0} = \bar{P}_2(\chi) + (3/2)^{\frac{1}{2}}\ \lambda\overline{(1-\chi^2)\cos 2\gamma},$$

which is just the invariant $\bar{S}_2$ taken in the local frame. F_N, in the form of eq. (85), is stable against a transition to a biaxial nematic state, whereas the free energy, eq. (82), allows for such a transition. As discussed by Freiser [24] in a similar fashion, the biaxial transition would appear at low temperatures for small λ, and in the neighbourhood of the isotropic phase for $\lambda^2 \approx 1/6$. Exactly at $\lambda^2 = 1/6$ the phase transition from the isotropic into the biaxial nematic state is of second order. Obviously λ is small enough for nematogenic molecules, so that biaxiality within the temperature range of the nematic has not been observed, yet. The lack of phase biaxiality in the nematic state suggests that the biaxiality of the cholesteric phase is induced and not the cause of the helical twist.

As the stability of a spontaneous cholesteric pitch requires molecular chirality with $u_{231}^{qq'} \neq 0$ we must include therefore the second and third term of eq. (72) in the energy balance. From eq. (74) we have

$$u_{231}^{qq'} = A_{23}^1\ T^{(2,q)}(0)T^{(3,q')}(0); \tag{86}$$

molecular chirality is expressed by a non-vanishing imaginary part of $T^{(3,q')}(0)$. However, a complex $T^{(3,q')}(0)$ allows for only one twofold rotation axis. It is reasonable to assume that this is an axis perpendicular to the long molecular axis. For convenience we choose the easy representation for which the components of $T^{(3,q')}(0)$ with q' odd vanish. The twofold axis of rotation is therefore parallel to the z axis of the molecular frame, whereas the long molecular axis is assumed to be parallel to the x axis of the molecular frame. Then the molecular tensor components are

$$T^{(3,\pm3)}(0) = T^{(3,\pm1)}(0) = 0,$$

$$T^{(3,0)}(0) = 1,\ T^{(3\pm2)}(0) = \varepsilon'\pm i\varepsilon''.$$

This particular choice of the molecular frame does not affect the coordinate free definition of the invariant $\bar{S}_2$.

Assuming small κ, we find for the internal energy, including the cholesteric contribution,

$$U = -(1/2)\{\bar{S}_{2,0}^2+3\ \bar{S}_{2,1}^2+(3/4)\bar{S}_{2,2}^2\}+(\kappa a/2)^2(\bar{S}_{2,1}^2+\bar{S}_{2,2}^2)$$
$$+\ (\kappa a/2)\Delta\{(2/5)\overline{S_{2,1}}\,\overline{S_{3,1}}+\overline{S_{2,2}}\,\overline{S_{3,2}}\}, \tag{87}$$

with

$$\Delta = (25/21)^{\frac{1}{2}}(A_{23}^1/A_{22}^0).$$

The third rank tensor variables are

$$S_{3,1} = (5\chi^2-1)(1-\chi^2)^{\frac{1}{2}}\sin\alpha \tag{88}$$
$$-(40/3)^{\frac{1}{2}}\varepsilon'\{P_2(\chi)(1-\chi^2)^{\frac{1}{2}}\sin\alpha\cos2\gamma+\chi(1-\chi^2)^{\frac{1}{2}}\cos\alpha\sin2\gamma\}$$
$$+(40/3)^{\frac{1}{2}}\varepsilon''\{P_2(\chi)(1-\chi^2)^{\frac{1}{2}}\sin\alpha\sin2\gamma-\chi(1-\chi^2)^{\frac{1}{2}}\cos\alpha\cos2\gamma\},$$

$$S_{3,2} = (1-\chi^2)\chi\ \sin2\alpha$$
$$+\ (8/15)^{\frac{1}{2}}\varepsilon''\{(2\chi^2-1)\cos2\alpha\sin2\gamma+P_2(\chi)\chi\ \sin2\alpha\cos2\gamma\}$$
$$+\ (8/15)^{\frac{1}{2}}\varepsilon''\{(2\chi^2-1)\cos2\alpha\cos2\gamma-P_2(\chi)\chi\ \sin2\alpha\sin2\gamma\}.$$

The equilibrium value of the inverse pitch {eq. (61)} is obtained as:

$$\kappa_0 a = -\Delta\{(2/5)\bar{S}_{2,1}\bar{S}_{3,1}+\bar{S}_{2,2}\bar{S}_{3,2}\}/(\bar{S}_{2,1}+\bar{S}_{2,2}). \tag{89}$$

The problem is obviously solved, when we know $\bar{S}_{3,1}$ and $\bar{S}_{3,2}$. In principle, this involves the solution of the following five non-linear equations:

$$\bar{S}_{2,p} = L_p^{(2)}(\mathcal{L}_0^{(2)},\mathcal{L}_1^{(2)},\mathcal{L}_2^{(2)},\mathcal{L}_1^{(3)},\mathcal{L}_2^{(3)}),\ p = 0,\ 1,\ 2$$
$$\bar{S}_{3,p} = L_p^{(3)}(\mathcal{L}_0^{(2)},\mathcal{L}_1^{(2)},\mathcal{L}_2^{(2)},\mathcal{L}_1^{(3)},\mathcal{L}_2^{(3)}),\ p = 1,\ 2. \tag{90}$$

We have defined

$$\mathcal{L}_0^{(2)} = \beta(\overline{S_{2,0}}+H_0^{(2)}),$$
$$\mathcal{L}_1^{(2)} = 3\beta[\{1-(1/6)\kappa^2a^2\}\overline{S_{2,1}}+\ (2/15)\kappa a\Delta\ \overline{S_{3,1}}\ +\ H_1^{(2)}], \tag{91}$$
$$\mathcal{L}_2^{(2)} = (3/4)\beta[\{1-(2/3)\kappa^2a^2\}\overline{S_{2,2}}+(4/3)\kappa a\Delta\ \overline{S_{3,2}}+H_2^{(2)}],$$

$$L_1^{(3)} = \beta\{-(2/5)\Delta\kappa a\ \bar{S}_{2,1} + H_1^{(3)}\},$$
$$L_2^{(3)} = \beta(-\Delta\kappa a\ \bar{S}_{2,2} + H_2^{(3)}). \tag{92}$$

The present approximation consists essentially in a linearisation of eqs. (92), which is useful in the high temperature region of the cholesteric phase. We obtain the linearised form of eqs. (92) from an expansion of the partition function including the linear and quadratic contribution of $L_1^{(3)}$ and $L_2^{(3)}$. The logarithmic derivatives yield

$$\overline{S_{3,1}} = (8/21)(1+2\varepsilon^2)^{-1}L_1^{(3)} + \varepsilon''\lambda(1/63\sqrt{5})\{1-(2/3)\lambda^2\}(L_1^{(2)3} - 8L_2^{(2)2}L_2^{(2)2} - 12L_0^{(2)}L_1^{(2)}L_2^{(2)}),$$

$$\overline{S_{3,2}} = (4/105)(1+2\varepsilon^2)^{-1} + \varepsilon''\lambda(4/35\sqrt{5})\{1-(2/3)\lambda^2\}[\{L_0^{(2)2} - (4/9)L_2^{(2)2}\}L_2^{(2)} - (1/12)\{L_0^{(2)} - (2/3)L_2^{(2)}\}L_1^{(2)2}]. \tag{93}$$

Neglecting the term proportional to $\kappa a\Delta$ ($|\kappa a\Delta| \ll 1$) we find for $\overline{S_{3,1}}$ and $\overline{S_{3,2}}$ in the high temperature limit,

$$\overline{S_{3,1}} = \varepsilon''\ (375/7\sqrt{5})\{1-(2/3)\lambda^2\}(1+2\lambda^2)^{-3}\{\bar{S}_{2,1}^2 - \bar{S}_{2,0}\ \bar{S}_{2,2} - (1/2)\bar{S}_{2,2}^2\}\bar{S}_{2,1} \tag{94}$$

and

$$\overline{S_{3,2}} = \varepsilon''\lambda(75/7\sqrt{5})(1-(2/3)\lambda^2(1+2\lambda^2)^{-3}[\{\bar{S}_{2,0}^2 - (1/4)\bar{S}_{2,2}^2\}\bar{S}_{2,2} - \{\bar{S}_{2,0} - (1/2)\bar{S}_{2,2}\}\bar{S}_{2,1}]. \tag{95}$$

The same approximation applied to eqs. (91) simply reproduces the system of eqs. (80) which has been solved already. According to eq. (60) the contribution to the free energy, caused by a small inverse pitch, is obtained by combination of eq. (87) and eq. (89)

$$\delta U_c = -\Delta^2\{(2/5)\bar{S}_{2,1}\bar{S}_{3,1} + \bar{S}_{2,2}\bar{S}_{3,2}\}^2/4(\bar{S}_{2,1}^2 + \bar{S}_{2,2}^2). \tag{96}$$

This term removes the degeneracy of the free energy with respect to the director. Since we have chosen the z axis as the direction of the helix axis, the total free energy depends on the relative orientation between the molecular axes and the z axis of the laboratory frame, where the average orientation minimizes δU_c. In contrast to the result of eq. (70) here the average orientation between the long molecular axis and the laboratory z axis is not necessarily 90°, but a function of temperature, as is the azimuthal orientation. For the general solution of this problem higher order contributions of $\bar{S}_{3,1}$ and $\bar{S}_{2,2}$ must be considered, because the isotropic-cholesteric transitions so far observed are first order transitions and successive phases with different molecular orientations relative to the pitch axis cannot be excluded. For example, a solution yielding an optically isotropic modification of the cholesteric phase seems possible, where the average birefringence,

$$\Delta\bar{n} = \bar{n}_{||} - \bar{n}_{\perp} ,$$

vanishes, because the average angle between the long molecular axis and the pitch axis is 54.74^{o}, i.e. the magic angle.

The present approximation is sufficient to give a special equilibrium solution, which holds in the neighbourhood of $\lambda^2 = 1/6$; this solution is obtained for $\bar{S}_{2,1} = 0$. From eqs. (94) and (95) we find

$$\bar{S}_{3,1} = 0,$$

$$\bar{S}_{3,2} = \varepsilon''(225/56\sqrt{30})\{\bar{S}^2_{2,0} - (1/4)\bar{S}^2_{2,2}\}\bar{S}_{2,2}. \tag{97}$$

The inverse pitch is

$$\kappa_o a = -(75/56)(5/14)^{\frac{1}{2}}\mathrm{Im}u^{22}_{231}(\bar{S}^2_{2,0}-(1/4)\bar{S}^2_{2,2})/u^{00}_{220} \tag{98}$$

and the cholesteric contribution to the free energy

$$\delta U_c = -5\mathrm{x}75^2\mathrm{x}56^{-3}(\mathrm{Im}u^{22}_{231}/u^{00}_{220})\{\overline{S^2_{2,0}}-(1/4)\overline{S^2_{2,2}}\}^2\overline{S^2_{2,0}}; \tag{99}$$

δU_c is minimized for

$$\bar{S}_{2,0} = \pm(3/2)\bar{S}_{2,2}.$$

This is exactly the solution which would be obtained from F_N, eq. (82), for $\lambda^2 = 1/6$, if the sixth order term is included. Introducing

$$\bar{S}^2 = \bar{S}^2_{2,0} + (3/4)\bar{S}^2_{2,2}$$

we obtain for the inverse pitch,

$$\kappa_o a = -(25/28)(5/14)^{\frac{1}{2}}(\mathrm{Im}u^{22}_{231}/u^{00}_{220})\overline{S^2} , \tag{100}$$

which is proportional to the square of the local nematic order parameter $\bar{S}$. This result shows some resemblance to the behaviour of the elastic constants, which are also proportional to the square of the order parameter. The order parameter itself is readily obtained from eq. (85). With $\lambda^2 = 1/6$ we find

$$F_N = (1/2)\{(15/4)T-1\}\overline{S}^2 + (225/448)T\ \overline{S}^4, \tag{101}$$

where we have for $T \approx T_c = 4/15$,

$$\overline{S}^2 = (28/15)(1-T/T_c). \tag{102}$$

Hence the inverse pitch is proportional to the temperature

$$\kappa_o a = -(5/3)(5/14)^{\frac{1}{2}}(\mathrm{Im}u_{231}^{22}/u_{220}^{00})(1-T/T_c). \tag{103}$$

This result for strongly biaxial molecules certainly does not allow us to draw special conclusions for other cases, but we can close with some general statements.

Non-vanishing order parameters $\overline{S}_{3,1}$ and $\overline{S}_{3,2}$ induce phase biaxiality (δU_c is finite). A coupling between $\overline{S}_{3,1}$ and $\overline{S}_{3,2}$ on the one hand and the second rank tensor order parameters on the other requires $\mathrm{Im}u_{22}^{231} \neq 0$ (see eqs. (94) and (95)), which again is only possible for chiral molecules. Since a change in sign of $\mathrm{Im}u_{22}^{231}$ yields the optical isomer, a racemic mixture cannot be biaxial as long as λ^2 is not in the neighbourhood of 1/6 or the temperature is sufficiently low. It is an interesting question whether the two corresponding regions in a cholesteric are separated by a phase transition. We hope that this matter as well as the possibility of cholesteric phases with different molecular orientations relative to the pitch axis will be the subject of future communications.

Acknowledgements

Parts of this work have been accomplished at the IBM Zürich Research Laboratory, Rüschlikon. I acknowledge the technical support and discussions I had with Dr. E. Courtens. Furthermore, I am grateful to Dr. K. Froböse, Universität Konstanz, for his stimulating arguments. Finally I want to thank Dr. W.J.A. Goossens for his friendly assistance and a critical reading of the manuscript.

References

1. S.A. Brazovskii and S.G. Dmitriev, *Zh. Eksp. Teor. Fiz.* **69**, 979 (1975).
2. P.G. de Gennes, *Mol. Cryst. Liq. Cryst.* **7**, 325 (1969).
3. P.G. de Gennes, *Mol. Cryst. Liq. Cryst.* **12**, 193 (1971).
4. P.G. de Gennes, *The Physics of Liquid Crystals* (Clarendon, Oxford, 1974).
5. R.S. Pindak, C.C. Huang and J.T. Ho, *Solid State Commun.* **14**, 821 (1974).
6. W.U. Müller, Dissertation, Technical University of Berlin, unpublished, (1974).
7. H. Hanson, A.J. Dekker and F. van der Woude, *Mol. Cryst. Liq. Cryst.* **42**, 16 (1977)
8. T. Harada and P. Crooker, *Mol. Cryst. Liq. Cryst.* **30**, 79 (1974).
9. H. Hanson, A.J. Dekker and F. van der Woude, *J. Chem. Phys.* **62**, 1941 (1975).
10. T.V. Samulski and E.T. Samulski, *J. Chem. Phys.* **67**, 824 (1977).
11. P.N. Keating, *Mol. Cryst. Liq. Cryst.* **8**, 315 (1969).
12. W.J.A. Goossens, *Mol. Cryst. Liq. Cryst.* **12**, 237 (1971).
13. A. Wulf, *J. Chem. Phys.* **60**, 3994 (1974).

14. Y.R. Lin-Liu, Yu Ming Shih and Chia-Wei Woo, *Phys. Rev. A* **15**, 2550 (1977).
15. Y.R. Lin-Liu, Yu Ming Shih, Chia-Wei Woo and H.T. Tan, *Phys. Rev. A* **14**, 445 (1976).
16. R.G. Priest and T.C. Lubensky, *Phys. Rev. A* **9**, 893 (1974).
17. A.G. Khachaturyan, *J. Phys. Chem. Solids* **36**, 1055 (1975).
18. B.W. van der Meer, G. Vertogen, A.J. Dekker and J.G.J. Ypma, *J. Chem. Phys.* **65**, 3935 (1976).
19. B.W. van der Meer and G. Vertogen, *Phys. Letters* **59A**, 4 (1976).
20. L. Blum and A.J. Torruella, *J. Chem. Phys.* **56**, 303 (1972).
21. W.A. Steele, *J. Chem. Phys.* **39**, 3197 (1963).
22. A.R. Edmonds, *Angular Momentum in Quantum Mechanics* (Princeton U.P., Princeton, N.J. 1957).
23. S.S. Schweber *An Introduction to Relativistic Quantum Field Theory* (Harper & Row, New York, 1961).
24. M.J. Freiser, *Phys. Rev. Letters,* **24**, 1041 (1970).
25. W. Maier and A. Saupe, *Z. Naturforschung* 13a, 564 (1958), **14a**, 882 (1958).
26. H. Schröder, *J. Chem. Phys.* **67**, 16 (1977).
27. G.R. Luckhurst, C. Zannoni, P.L. Nordio and V. Segre, *Mol. Phys.* **30**, 1345 (1975).
28. W.J.A. Goossens, paper presented at the 7th International Liquid Crystal Conference, Bordeaux (1978).

Chapter 6

A MOLECULAR MODEL FOR THE CHOLESTERIC MESOPHASE

B.W. VAN DER MEER

Solid State Physics Laboratory and Institute for Theoretical Physics, Melkweg 1, University of Groningen, Groningen, The Netherlands.

and

G. VERTOGEN

Institute for Theoretical Physics, Toernooiveld, Catholic University of Nijmegen, Nijmegen, The Netherlands.

Introduction

The cholesteric mesophase is simply a helically distorted nematic mesophase. Consequently an understanding of the nematic phase is a prerequisite to understand the cholesteric phase. We begin, therefore, by dealing briefly with the essential characteristics of the theory of nematics which is discussed in greater detail in Chapter 4.

The difference between the nematic mesophase and the isotropic fluid phase lies in the appearance of long-range orientational order of the long molecular axes. In the nematic phase these axes tend to align along a common direction, called the director **n**. An explanation for this tendency to align parallel is provided by the Maier-Saupe theory [1]. The anisotropic interaction between two molecules is described, in this theory, essentially by the simple interaction

$$U_{ij} = -JP_2(\mathbf{a}_i \cdot \mathbf{a}_j), \tag{1}$$

where J is a coupling constant depending on the intermolecular separation r_{ij}, $\mathbf{a}_i$ is a unit vector along the long axis of molecule i and P_2 is the second Legendre polynomial. In the original presentation of Maier and Saupe it was assumed that J is due entirely to van der Waals forces. Consequently J, which is proportional to the nematic-isotropic transition temperature T_{NI}, is found to be proportional to the product of the anisotropy of the polarizabilities of both interacting molecules. This interpretation of J is not consistent with all the existing experimental data on T_{NI}; e.g. it disagrees with those for the cyclohexylcyclohexanes [2]. Contributions from other forces, particularly from repulsive forces, cannot be neglected. According to Gelbart and Gelbart the strength of J is due mainly to the coupling between isotropic attractions and anisotropic hard core repulsions [3]. Suggestions concerning the eventual unraveling of the contributions of the various forces, can be found in references [3] and [4]. Treatment of the pair potential in eq. (1) via the mean field approximation gives rise to a free energy, which contains only one order parameter $\bar{P}_2$, the thermal average of $P_2(\mathbf{a}\cdot\mathbf{n})$ [1]. We will refer to this mean field analysis as the Maier-Saupe theory which has many satisfactory features. For instance, the predicted temperature dependence of the order parameter $\bar{P}_2$ and its value at the nematic-isotropic transition are found to be in fairly good agreement with experiment [5]. The success of the Maier-Saupe theory suggests that the following two statements, which enter as essential ingredients of the theory, may be made:

(1) the deviation from cylindrical symmetry of the molecules is only a secondary effect; and

(2) to a first approximation only one order parameter, $\bar{P}_2$, is needed to describe the nematic state.

In the cholesteric mesophase the spatial long-range order is also absent and the molecular orientations prefer to align along the director. However, **n** is not constant in space. Instead it varies as a function of the position, **r**, in the same way as the director of a nematic that has been twisted uniformly about an axis perpendicular to **n**. In cholesterics the director **n** describes a helix, i.e. the director rotates around an axis. This rotation can be represented by an axial vector (pseudovector) along the helix axis. In order to clarify this behaviour we imagine a very small observer proceeding along the helix axis. He would see a

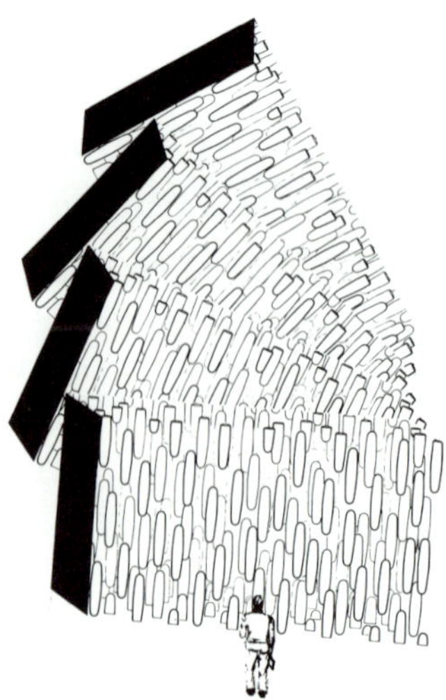

Fig. 1. *A small observer is proceeding along the helix axis of a cholesteric liquid crystal. The cholesteric in the figure is right-handed. While going ahead, he observes that the preferred direction rotates clockwise. (The slices are drawn for convenience; in reality the director rotates continuously).*

director rotating around his path. According to common convention the rotation vector is parallel (antiparallel) to his movement, if he observes a right-handed (left-handed) rotation. The magnitude of the rotation vector is the angle of rotation per unit length. The inner product of the rotation vector with the unit vector along the helix axis in the direction of the observer's moment equals $2\pi/p$, where $|p|$ denotes the pitch of the helix, i.e. the length of the vector describing the translation of the observer along the helix axis over a complete turn of the director. It is clear that the pitch, being inversely proportional to an inner product of an axial vector and a polar vector, is a pseudoscalar; i.e. the pitch changes sign under an inversion operation I, that reverses the sign of all cartesian coordinates. According to common convention the pitch of a right-handed (left-handed) helix is termed positive (negative).

The presence of chiral molecules is a prerequisite for the appearance of the cholesteric mesophase. In other words there must be molecules present which differ from their mirror image. In fact, this requirement forms the basis of the molecular theory developed by Goossens [6] and the phenomenological theory provided by de Gennes [5(chapters 3 and 6)]; the latter utilizes the concept of long-range distortions induced by chiral molecules in a nematic matrix. De Gennes also worked out a very powerful continuum theory based on the work of Frank, who derived a general expression for the free energy density of a cholesteric using only symmetry arguments [5(chapters 3 and 6), 7]. Goossens was the first to put forward a molecular statistical theory of the cholesteric mesophase. Extending the idea of Maier and Saupe he calculated the dispersion energy between two molecules taking into account, not only the dipole-dipole interaction, but also the dipole-quadrupole interaction. The statistical part of his theory was based firmly on the belief, held at that time, that the molecules in cholesteric liquid crystals are arranged in layers [8]. However, such an arrangement turned out to be incorrect; a cholesteric is locally uniaxial and is just a spontaneously twisted nematic. Consequently Goossens' treatment is partly incorrect; notably his claim that cylindrically symmetric molecules do not give rise to a cholesteric state, cannot be justified. Nevertheless the main point of his paper, namely the presentation of a molecular model for the cholesteric mesophase, is of interest. As for the model, we note that it is probably too limited, because it does not take into account the effect of repulsive interactions on the pitch.

It is clear that one has to resort to molecular models in order to gain insight into the behaviour of cholesterics, e.g. the physical origin of the helix. Obviously it can also be claimed that the resulting molecular statistical theory should be in agreement with the phenomenological theory of de Gennes and that it should also be an extension of the Maier-Saupe theory in the sense that, if the chirality of the molecules is removed, the theory reduces to that of Maier and Saupe. However, it shoul be noted that the introduction of a planar structure for cholesterics pro hibits a reduction to the Maier-Saupe theory.

The main purpose of this Chapter is to elucidate the appearance of the cholesteric mesophase in terms of a very simple model for a chiral molecule. After deriving the interaction between chiral molecules we discuss a theory of the cholesteric state exhibiting the features previously mentioned. Given the fact that a cholesteric is simply a spontaneously twisted nematic, it is quite natural to accept statements (1) and (2) which we have formulated not only for nematics, but for cholesterics as well. Thus, in a first approximation, we shall be content with a one order parameter theory of the cholesteric state; our starting point is an interaction potential that ignores the deviation from molecular cylindrical symmetry. The interaction model, which was introduced in references [9] and [10] is

$$U_{ij} = -J\, P_2(\mathbf{a}_i.\mathbf{a}_j) - K(\mathbf{a}_i.\mathbf{a}_j)(\mathbf{a}_i \wedge \mathbf{a}_j.\mathbf{u}_{ij}), \tag{2}$$

where $\mathbf{u}_{ij}(\equiv \mathbf{r}_{ij}/r_{ij})$ is a unit vector pointing from the centre of mass of molecule i towards that of j and K is a coupling constant that vanishes if both molecules are achiral, i.e. identical with their mirror images. This pair potential is, therefore, a direct extension of that employed in the Maier-Saupe theory. The constant K, like J, depends on the intermolecular separation. The potential can easily be treated within the mean field approximation; as we shall show the resulting Helmholt: free energy and distortion free energy have the required properties.

The potential in eq. (2) is not generally accepted as correct for the interaction between two chiral molecules. For example Schröder, in a paper presented at the 7th International Liquid Crystal Conference (Bordeaux 1978), concludes from symmetry arguments, in contrast to Frank, that potentials for uniaxial molecules are inadequate to describe the cholesteric state. It seems relevant to refute Schröder's arguments before proceeding with our own analysis. Schröder's theory is based upon the groundless application of the following law describing the transformation of an irreducible tensor **T** with spherical components $T^{(L,m)}$ under the action of the inversion operator I

$$I T^{(L,m)} = (-1)^L T^{(L,m)}. \tag{3}$$

This transformation law is not correct and a counter example can be easily constructed. Consider an angular momentum vector **L** with spherical components

$$L^{(1,0)} = L_z$$

and

$$L^{(1\pm 1)} = \pm (1/\sqrt{2})(L_x \mp i\, L_y),$$

where L_x, L_y and L_z are cartesian components. This vector is an axial vector (pseudovector) and is *invariant* under inversion

$$I L^{(1,m)} = L^{(1,m)}. \qquad (4)$$

It might be that Schröder was motivated by an incorrect statement in the literature that the twist term in the free energy of the cholesteric state should be a pseudoscalar [11]. He studied the behaviour of the system under the parity operator in order to formulate a theory of the cholesteric state. His analysis is based on an expansion of the potential in the basis functions of the rotation group, although the basis functions of the full rotation-inversion group are the natural candidates. We will try to proceed in the spirit of Feyman's remark, that group theory is suitable to deal with symmetry "but usually you can figure out what you want with common sense" [12]. Application of the inversion operator I to the terms of expression (2), which are both present in the general expansion used by Schröder, yields:

(1) $IJ\, P_2(\mathbf{a}_i.\mathbf{a}_j) = J\, P_2(\mathbf{a}_i.\mathbf{a}_j)$ if J is a scalar; consequently $J\, P_2(\mathbf{a}_i.\mathbf{a}_j)$ is a scalar;

(2) $IJ\, P_2(\mathbf{a}_i.\mathbf{a}_j) = -J\, P_2(\mathbf{a}_i.\mathbf{a}_j)$ if J is a pseudoscalar; consequently $J\, P_2(\mathbf{a}_i.\mathbf{a}_j)$ is a pseudoscalar in this case;

(3) $IK(\mathbf{a}_i.\mathbf{a}_j)(\mathbf{a}_i \wedge \mathbf{a}_j.\mathbf{u}_{ij}) = -K(\mathbf{a}_i.\mathbf{a}_j)(\mathbf{a}_i \wedge \mathbf{a}_j.\mathbf{u}_{ij})$ if K is scalar; consequently $K(\mathbf{a}_i.\mathbf{a}_j)(\mathbf{a}_i \wedge \mathbf{a}_j.\mathbf{u}_{ij})$ is pseudoscalar;

(4) $IK(\mathbf{a}_i.\mathbf{a}_j)(\mathbf{a}_i \wedge \mathbf{a}_j.\mathbf{u}_{ij}) = K(\mathbf{a}_i.\mathbf{a}_j)(\mathbf{a}_i \wedge \mathbf{a}_j.\mathbf{u}_{ij})$ if K is a pseudoscalar; consequently $K(\mathbf{a}_i.\mathbf{a}_j)(\mathbf{a}_i \wedge \mathbf{a}_j.\mathbf{u}_{ij})$ is a scalar in this case.

Parity conservation requires that terms referring to an energy have to be scalars; i.e. J is a scalar and K is a pseudoscalar. Schröder excluded beforehand the possibility that some of his coefficients $u^{qq'}_{LL'J}$ could be pseudoscalars, a mathematically and physically untenable procedure. Consequently he arrives at the wrong conclusion that parity conservation requires that $K(\sim u^{00}_{212})$ should be zero and that the present model is forbidden.

Why is K a pseudoscalar? This question is the subject of the next section, where we show that chirality of the molecules imperatively results in K being a pseudoscalar. This is done on the basis of a symmetry argument very similar to that given by Lin-Liu *et al.* [10]. In the third section we introduce simple models for chiral molecules and present a derivation of the interaction involved between these molecules. The interpretation of the coupling constants in terms of molecular properties is discussed. In the fourth section the relation between the interaction model, the Helmholtz free energy and the distortion free energy of the cholesteric state is explained. Extensions of the simple theory of cholesterics are dealt with briefly in section five; they are quite analogous to extensions of the Maier-Saupe theory described in Chapter 4. Finally, in the last section, the results are briefly discussed and compared with experiment.

The Symmetry of the Chiral Interaction

Consider a pair of molecules i and j subject to the anisotropic interaction

$$U_{ij} = -J\, P_2(\mathbf{a}_i.\mathbf{a}_j) - K(\mathbf{a}_i.\mathbf{a}_j)(\mathbf{a}_i \wedge \mathbf{a}_j.\mathbf{u}_{ij}). \tag{5}$$

Next apply an inversion operation, that is reflection in a point, taken as the origin. This is an active transformation, where each point of the molecules is inverted, such that the coordinate system is unchanged; the transformation is depicted in figure 2. The molecules i and j

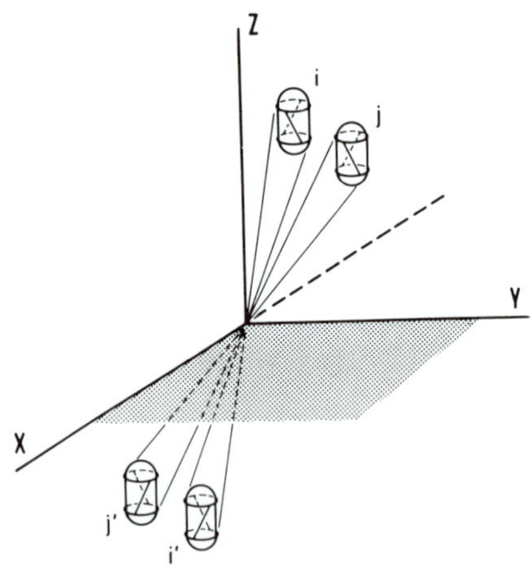

Fig. 2. *The inversion transformation. The molecules i and j transform into i' and j' respectively. The handedness of i and j is opposite to that of i' and j'.*

transform into their mirror images i' and j', and the coordinates of i' and j' are the inverse of i and j, e.g.

$$\mathbf{a}_i \rightarrow \mathbf{a}_{i'} = -\mathbf{a}_i,$$
$$\mathbf{a}_j \rightarrow \mathbf{a}_{j'} = -\mathbf{a}_j,$$
$$\mathbf{u}_{ij} \rightarrow \mathbf{u}_{i'j'} = -\mathbf{u}_{ij}. \tag{6}$$

The interaction $U'_{i'j'}$ between i' and j' is in general different from the original interaction U_{ij}; in fact

$$U'_{i'j'} = -J'P_2(\mathbf{a}_{i'}.\mathbf{a}_{j'})-K'(\mathbf{a}_{i'}.\mathbf{a}_{j'})(\mathbf{a}_{i'} \wedge \mathbf{a}_{j'}.\mathbf{u}_{i'j'}) + \text{other terms.} \quad (7)$$

Parity conservation, however, requires that the interaction energy, being a scalar, must satisfy the relation

$$U_{ij} = U'_{i'j'}. \quad (8)$$

Using eqs. (5), (6) and (7) it follows immediately that J' = J and K' = -K, i.e.

$$U'_{i'j'} = -J\ P_2(\mathbf{a}_{i'}.\mathbf{a}_{j'}) + K(\mathbf{a}_{i'}.\mathbf{a}_{j'})(\mathbf{a}_{i'} \wedge \mathbf{a}_{j'}.\mathbf{u}_{i'j'}) \quad (9)$$

and all remaining terms are zero. If the molecules are achiral, the molecules are identical with their mirror immage. In that case the molecules themselves are not affected by the inversion operation; only their coordinates are inverted. Consequently the interaction is form-invariant under the transformation. This means U' = U, or K = 0 for achiral molecules. Concluding we can state:

(1) K = 0 for achiral molecules;

(2) K changes sign, i.e. K is a pseudoscalar, if the handedness of the molecules is changed.

The Interaction Model

In the preceding sections we discussed symmetry properties of the interaction. Here we present a simple molecular model that produces such an interaction. In this way the coupling constants J and K can be interpreted in terms of molecular properties such as the polarizability, shape and chirality. First we start, for the sake of clarity, in the spirit of the original approach of Maier and Saupe and consider the molecules as anisotropic polarizabilities practically neglecting their shape. We shall indicate the influence of the hard core repulsive forces on the coupling constants later.

The first version of our model consists of N identical linear polarizabilities on a twisted cylindrical surface, as shown in figure 3a, and we consider the limit as $N \to \infty$, i.e. we obtain a cylindrically symmetric molecule. The lines represent harmonic oscillators (polarizabilities) whose equilibrium positions lie on a circle with radius s. In order to describe the position and orientation of a given polarizability m situated on molecule i, we attach a cartesian coordinate system to the molecule with its origin in the centre of mass, as in figure 4. The coordinate system is given by the unit vectors $\mathbf{a}_i$, $\mathbf{b}_i$ and $\mathbf{c}_i$, such that $\mathbf{a}_i$ points along the cylindrical axis, $\mathbf{a}_i.\mathbf{b}_i = \mathbf{b}_i.\mathbf{c}_i = \mathbf{a}_i.\mathbf{c}_i = 0$ and $\mathbf{b}_i \wedge \mathbf{c}_i = \mathbf{a}_i$. The direction of the oscillator is given by $\mathbf{e}_{im}$ and its equilibrium position by $s_i\mathbf{d}_{im}$. The unit vectors $\mathbf{d}_{im}$ and $\mathbf{e}_{im}$ are

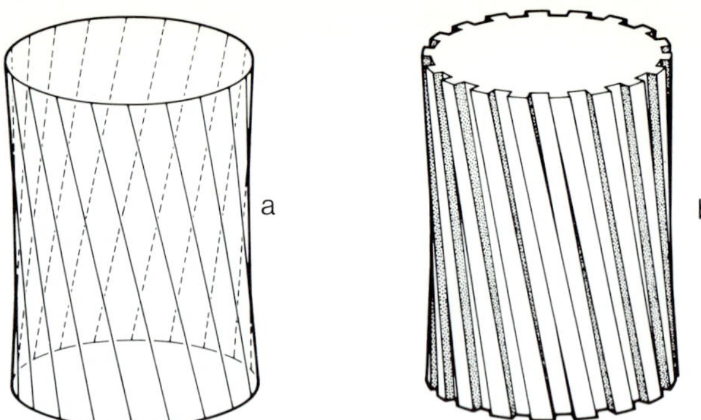

Fig. 3. *(a) A model for a chiral molecule. The lines denote linear polarizabilities which may be thought to represent interatomic bonds. (b) A chiral cog-wheel as a model for an enantiomorphic molecule.*

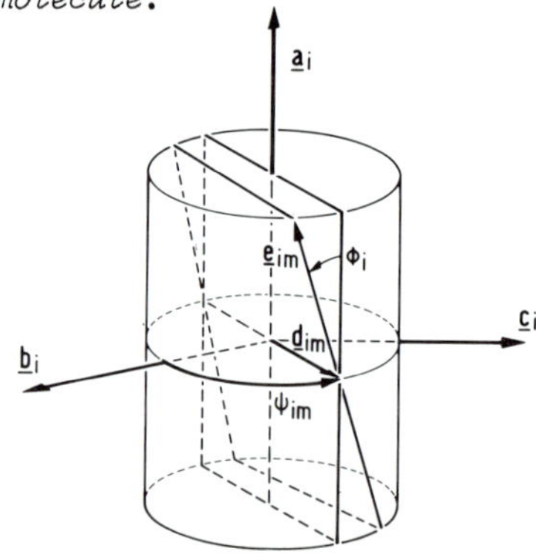

Fig. 4. *The geometry of the chiral molecule i. Only the oscillator m is shown here.*

$$\mathbf{d}_{im} = \mathbf{b}_i \cos\psi_{im} + \mathbf{c}_i \sin\psi_{im}, \tag{10}$$

$$\mathbf{e}_{im} = \mathbf{a}_i \cos\phi_i + \mathbf{b}_i \sin\phi_i \sin\psi_{im} - \mathbf{c}_i \sin\phi_i \cos\psi_{im}. \tag{11}$$

The magnitude of the angle ϕ_i between $\mathbf{e}_{im}$ and $\mathbf{a}_i$ is a measure of the deviation from inversion symmetry; the sign of ϕ_i gives the handedness of the molecule. We call the molecule right-handed ($\phi_i > 0$) if by going from the cylinder axis to the polarizability along their connection line, we have to rotate the axis clockwise in order to be parallel with the polarizability. The polar angle ψ_{im} defines the equilibrium position of the oscillator. It holds that

$$A_i = P_2(\mathbf{a}_i.\mathbf{e}_{im}) = P_2(\cos\phi_i) \tag{12}$$

and

$$B_i = (\mathbf{a}_i.\mathbf{e}_{im})(\mathbf{a}_i \wedge \mathbf{e}_{im}.\mathbf{d}_{im}) = (1/2)\sin\phi_i \tag{13}$$

are molecular properties independent of m. The pseudoscalar B_i describes the chirality of molecule i.

To calculate the orientational interaction energy between two molecules i and j with polarizabilities α_i and α_j we have to use the dipole-dipole dispersion energy. It is quite instructive, however, to assume a Maier-Saupe type of interaction between their constituent linear polarizabilities, namely,

$$-(E\alpha_i\alpha_j/N^2)(\mathbf{r}_{ij}-s_i\mathbf{d}_{im}+s_j\mathbf{d}_{jn})^{-6}P_2(\mathbf{e}_{im}.\mathbf{e}_{jn}), \tag{14}$$

where m and n are oscillators situated at i and j respectively, $\mathbf{r}_{ij}$ is the vector between the centres of mass of the molecules pointing from i to j and E is a coupling constant. It should be remarked here that the replacement of the dipole-dipole dispersion interaction by the Maier-Saupe interaction can be made without loss of the essential features. The total orientational interaction energy U_{ij} between molecules i and j is obtained by summing over n and m. Expanding in powers of $s_i/|\mathbf{r}_{ij}|$ and $s_j/|\mathbf{r}_{ij}|$ and retaining only the first two terms we obtain in the limit of a continuous distribution of oscillators

$$U_{ij} = -E\alpha_i\alpha_j(2\pi)^{-2}r_{ij}^{-6}\int_0^{2\pi} d\psi_{im}\int_0^{2\pi} d\psi_{jn}\, XP_2(Y), \tag{15}$$

with

$$X = 1 + 6s_i r_{ij}^{-1}\mathbf{b}_i.\mathbf{u}_{ij}\cos\psi_{im} + 6s_i r_{ij}^{-1}\mathbf{c}_i.\mathbf{u}_{ij}\sin\psi_{im} - 6s_j r_{ij}^{-1}\mathbf{b}_j.\mathbf{u}_{ij}\cos\psi_{jn} - 6s_j r_{ij}^{-1}\mathbf{c}_j.\mathbf{u}_{ij}\sin\psi_{jn}, \tag{16}$$

$$\begin{aligned} Y = {} & \mathbf{a}_i.\mathbf{a}_j\cos\phi_i\cos\phi_j + \mathbf{a}_i.\mathbf{b}_j\cos\phi_i\sin\phi_j\sin\psi_{jn} - \mathbf{a}_i.\mathbf{c}_j\cos\phi_i\sin\phi_j\cos\psi_{jn} \\ & + \mathbf{b}_i.\mathbf{a}_j\sin\phi_i\sin\psi_{im}\cos\phi_j + \mathbf{b}_i.\mathbf{b}_j\sin\phi_i\sin\psi_{im}\sin\phi_j\sin\psi_{jn} \\ & - \mathbf{b}_i.\mathbf{c}_j\sin\phi_i\sin\psi_{im}\sin\phi_j\cos\psi_{jn} - \mathbf{c}_i.\mathbf{a}_j\sin\phi_i\cos\psi_{im}\cos\phi_j \\ & - \mathbf{c}_i.\mathbf{b}_j\sin\phi_i\cos\psi_{im}\sin\phi_j\sin\psi_{jn} + \mathbf{c}_i.\mathbf{c}_j\sin\phi_i\cos\psi_{im}\sin\phi_j\cos\psi_{jn}. \end{aligned} \tag{17}$$

Carrying out the integrations over ψ_{im} and ψ_{jn} we obtain

$$U_{ij} = -J\,P_2(\mathbf{a}_i.\mathbf{a}_j) - K(\mathbf{a}_i.\mathbf{a}_j)(\mathbf{a}_i\wedge\mathbf{a}_j.\mathbf{u}_{ij}), \tag{18}$$

where

$$J = E\,r_{ij}^{-6}(\alpha_i A_i)(\alpha_j A_j) \tag{19}$$

and the pseudoscalar K

$$K = 9E\,r_{ij}^{-7}\{(s_i\alpha_i B_i)(\alpha_j A_j) + (s_j\alpha_j B_j)(\alpha_i A_i)\}. \tag{20}$$

Essentially the same results can be obtained by considering the dipole-dipole and dipole-quadrupole dispersion energy between freely rotating molecules [9].

Up to now we have neglected the effect of shape completely. The hard core repulsive forces act in two different ways. First they give rise to packing effects. These will be discussed briefly in the fifth section. Secondly they couple to attractive forces in a way explained by Gelbart and Gelbart [3]. The hard core repulsive forces determine the minimal nearest neighbour distance or shortest distance of approach ξ_{ij} of two elongated molecules i and j. The distance ξ_{ij} is a complicated function of the molecular dimensions and the orientational coordinates. A simplified description in the case of achiral molecules is

$$\xi_{ij} = B_M + (1/3)(L_M - B_M)\{P_2(\mathbf{a}_i.\mathbf{u}_{ij}) + P_2(\mathbf{a}_j.\mathbf{u}_{ij}) + 1\}, \tag{21}$$

where B_M and L_M are the molecular breadth and length respectively. In the case of chiral molecules ξ_{ij} is more complicated, because in general the chirality will show up in the molecular shape as well. To illustrate this we have drawn a model of a chiral molecule in figure 3b. It is the same as that in figure 3a except that the linear polarizabilities are replaced by cogs; a related model is Straley's threaded rod [13]. In order to model the shortest distance of approach of two such chiral cogwheels, terms like $(\mathbf{a}_i.\mathbf{a}_j)(\mathbf{a}_i \wedge \mathbf{a}_j.\mathbf{u}_{ij})$ and $(\mathbf{a}_i.\mathbf{u}_{ij})(\mathbf{a}_j.\mathbf{u}_{ij})(\mathbf{a}_i \wedge \mathbf{a}_j.\mathbf{u}_{ij})$ with coefficients proportional to the thickness of the cogs have to be added to eq. (21). The Gelbart-Gelbart coupling between repulsive and attractive forces can be formulated as follows. The major part of the attractive interaction is isotropic, $U_{att} \approx U_0(r_{ij})$. The hard core repulsive interactions prevent the molecules from overlapping. This means that the effective interaction is approximately equal to $U_{ij} = U_0(r_{ij})\theta(r_{ij}-\xi_{ij})$, where θ denotes the step function. Expanding U_{ij} in spherical harmonics and retaining only the second rank terms, we obtain again essentially the interaction (2). In this case the coupling constants J and K are determined by the isotropic interactions and the molecular shape.

Above we have discussed two contributions to the interactions between elongated chiral molecules, the interaction between anisotropic polarizabilities (the polarizability contribution) and the combination of hard core repulsive forces and attractive forces (the shape contribution). It seems reasonable to assume, that in the general case both contributions are present so that the coupling constants J and K can be split in two parts $J = J_s + J_p$ and $K = K_s + K_p$, where J_s and K_s are determined by the molecular shape and J_p and K_p denote the contributions due to the dispersion forces between anisotropic polarizabilities. A characteristic feature of the polarizability contribution to the twist interaction of chiral molecules is that the principal axes of the polarizability tensor vary over the molecule in some chiral way. This molecular property is also the basic ingredient of the molecular theory of Goossens [6]

As to de Gennes' model of a chiral object in a nematic matrix it is easy to see that the chiral molecules presented here induce the same twist distortions.

The Free Energy of the Cholesteric State

The thermodynamic properties of our model can be easily calculated in the mean field approximation. In a state with complete orientational order the long axis of a molecule i, $\mathbf{a}_i$, would align everywhere to the local director $\mathbf{n}_i \equiv \mathbf{n}(\mathbf{r}_i)$. In practice, however, the orientational order is never ideal and the orientation $\mathbf{a}_i$ fluctuates about $\mathbf{n}_i$. The mean field approximation assumes that all the necessary information about this disorder is contained in the single orientational distribution function f and that this function exhibits cylindrical symmetry with respect to the director, i.e. $f = f(\mathbf{a}_i.\mathbf{n}_i)$. In general the director is a function of position. The distortions of the director field are very small and we can, therefore, neglect the dependence of the distribution function on the gradients of the director. The orientational free energy in the mean field approximation can be written as

$$A = U - TS_r + A_d, \tag{22}$$

where U is the internal energy in the absence of distortions,

$$U = (N\rho/2) \int d\mathbf{r}_{ij} \int d\mathbf{a}_i \int d\mathbf{a}_j U(i,j) f(\mathbf{a}_i.\mathbf{n}_i) f(\mathbf{a}_j.\mathbf{n}_i) = -(N\rho/2) J_0 \bar{P}_2^2 \tag{23}$$

and

$$J_0 = \int d\mathbf{r}_{ij} J(r_{ij}).$$

The orientational entropy S_r is given by

$$S_r = -Nk \int d\mathbf{a}_i f(\mathbf{a}_i.\mathbf{n}_i) \ln[4\pi f(\mathbf{a}_i.\mathbf{n}_i)], \tag{24}$$

while the distribution function is

$$f(\mathbf{a}_i.\mathbf{n}_i) = \exp\{\beta\rho J_0 \bar{P}_2 P_2(\mathbf{a}_i.\mathbf{n}_i)\} / \int d\mathbf{a}_i \exp\{\beta\rho J_0 \bar{P}_2 P_2(\mathbf{a}_i.\mathbf{n}_i)\}; \tag{25}$$

here ρ denotes the density, and $\beta = 1/kT$. The orientational order parameter $\bar{P}_2$ is the stable solution of

$$\bar{P}_2 = \int d\mathbf{a}_i P_2(\mathbf{a}_i.\mathbf{n}_i) f(\mathbf{a}_i.\mathbf{n}_i). \tag{26}$$

The distortion free energy A_d is given by

$$A_d = (\rho^2/2) \int d\mathbf{r}_i \int d\mathbf{r}_{ij} \int d\mathbf{a}_i \int d\mathbf{a}_j U_{ij} f(\mathbf{a}_i.\mathbf{n}_i)\{f(\mathbf{a}_j.\mathbf{n}_j) - f(\mathbf{a}_j.\mathbf{n}_i)\}$$

$$= (1/2) \int d\mathbf{r}_i \{K_{11}(\nabla.\mathbf{n})^2 + K_{22}(\mathbf{n}.\nabla\wedge\mathbf{n})^2 + K_{33}(\mathbf{n}\wedge\nabla\wedge\mathbf{n})^2 + Q(\mathbf{n}.\nabla\wedge\mathbf{n})\}, \tag{27}$$

with

$$K_{11} = K_{22} = K_{33} = \rho^2\bar{P}_2^2 J_2/2; \quad J_2 = \int d\mathbf{r}_{ij} r_{ij}^2 J(r_{ij}); \tag{28}$$

$$Q = \rho^2\bar{P}_2^2 K_1/3; \quad K_1 = \int d\mathbf{r}_{ij} r_{ij} K(r_{ij}). \tag{29}$$

The expression (27) is obtained by expanding the components of $\mathbf{n}_j = \mathbf{n}_i(\mathbf{r}_i+\mathbf{r}_{ij})$ in a Taylor series with respect to $\mathbf{r}_{ij}$ (see Appendix). We have shown here that the simple interaction potential (2) is equivalent to the one elastic constant approximation of the cholesteric state. In the mean field approximation the elastic constants of this model are proportional to the square of the product of the density and the order parameter. Note that, if the gradients of the director are zero, the free energy reduces to the Maier-Saupe free energy. This is not, however, the state of lowest free energy. In the special case of uniform deformation, where the director depends only on the cartesian coordinate z, the state with the lowest energy is a state of pure twist. With reference to a right-handed coordinate system this state can be described by $n_x = \cos qz$, $n_y = \sin qz$, $n_z = 0$ and

$$q = 2\pi/\text{pitch} = Q/2K_{22} = K_1/3J_2.$$

For the sake of completeness we note again that the term $Q(\mathbf{n}.\nabla\wedge\mathbf{n})$ is a scalar. It is a product of two pseudoscalars; Q, which describes the chirality of the molecules and $\mathbf{n}.\nabla\wedge\mathbf{n}$, which describes the helicity of the phase.

Extensions of the Theory

In this section we indicate briefly extensions of the simple theory.

(1) In order to obtain a temperature dependent pitch the interaction potential (2) has to be extended by allowing for couplings between $\mathbf{a}_i$ and $\mathbf{a}_j$ of higher rank; this is discussed in references [9,10]. The temperature dependence of the pitch is determined by $(\bar{P}_4/\bar{P}_2)^2$ in this theory where $\bar{P}_4$ is the thermal average of the fourth Legendre polynomial $P_4(\mathbf{a}.\mathbf{n})$.

(2) The molecules forming liquid crystals are lath-like rather than rod-like. This deviation from cylindrical symmetry also gives rise to a temperature dependent pitch. Two order parameters appear, the Maier-Saupe order parameter $\bar{P}_2$ and $\bar{D}$, the thermal average of $\{P_2(\mathbf{b}.\mathbf{n}) - P_2(\mathbf{c}.\mathbf{n})\}$ where **b** and **c** denote the short molecular axes. The temperature dependence of the pitch is determined by $\bar{D}/\bar{P}_2$ [13].

(3) Experiments of McColl and Shih may be taken to demonstrate that steric effects are important in nematics [15]. These packing effects will also influence the behaviour of cholesterics.

A phenomenological method to include packing in a way consistent with a more fundamental approach [16] (cf. Chapter 8) is to renormalize the shape contribution according to

$$J_s \rightarrow J_s(1 + T/T_p) \tag{30}$$

and

$$K_s \rightarrow K_s(1 + T/T_p), \tag{31}$$

where T_p may be called the packing temperature. Note that the inclusion of packing leads to a temperature dependent pitch if $J_p/J_s \neq K_p/K_s$.

(4) A very important extension concerns the behaviour of the elastic constants. Experimentally the elastic constants are different, e.g. $K_{11}:K_{22}:K_{33}$ = 1:0.63:1.35 for 4-methoxybenzylidene-4'-butylaniline at T/T_{NI} = 0.96 [17], while the simple potential results in identical elastic constants. In order to improve this situation we have to extend the potential as done by Priest [18]. We indicate here the way to apply his treatment to cholesterics. Expanding the interaction potential in spherical harmonics we obtain

$$U_{ij} = - \Sigma\, J^{L\,J\,L'}\, T^{L\,J\,L'}(\mathbf{a}_i, \mathbf{u}_{ij}, \mathbf{a}_j), \tag{32}$$

where $J^{L\,J\,L'}$ are coupling constants depending on the intermolecular separation and $T^{L\,J\,L'}(a_i, u_{ij}, a_j)$ are rotational invariants. The function $T^{L\,J\,L'}(\mathbf{a}_i, \mathbf{u}_{ij}, \mathbf{a}_j)$ containing $\mathbf{a}_i$ to the Lth power, $\mathbf{a}_j$ to the L'th power and $\mathbf{u}_{ij}$ to the Jth power, is proportional to $S^{00}_{LL'J}(\mathbf{a}_i, \mathbf{a}_j, \mathbf{u}_{ij})$ as being defined by Blum and Torruella as described in Chapter 2. Explicit expressions for the relevant $T^{L\;\;L''}(\mathbf{a}_i, \mathbf{u}_{ij}, \mathbf{a}_j)$ are given in the Appendix. The simplest truncation of the expansion (32) that is consistent with experimental data on the elastic constants is to retain the terms (1) L=L'=2, J =0,1,2,3,4 (2) L=2, L'=4, J =2,3,4,5,6 (3) L=4, L'=2, J =2,3,4,5,6 and to neglect all higher order terms. Note that J^{202} is equal to J and J^{212} is equal to K. Using results derived in the Appendix we arrive at the same expression for the distortion free energy in the mean field approximation as mentioned in expression (27). The elastic constants, however, are different:

$$K_{11} = \bar{K}(1 + \Delta - 3\Delta'\, \bar{P}_4/\bar{P}_2), \tag{33}$$

$$K_{22} = \bar{K}(1 - 2\Delta - \Delta'\bar{P}_4/\bar{P}_2), \tag{34}$$

$$K_{33} = \bar{K}(1 + \Delta + 4\Delta'\bar{P}_4/\bar{P}_2), \tag{35}$$

$$Q = 2\,\bar{K}\,\sigma, \tag{36}$$

where

$$\bar{K} = \frac{1}{2}\,\rho^2 \bar{P}^2\, J_2^{202}, \tag{37}$$

$$\Delta = \frac{1}{5}\, J_2^{222}/J_2^{202}, \tag{38}$$

$$\Delta' = \frac{1}{6}\,(J_2^{422}+J_2^{224})/J_2^{202}, \tag{39}$$

$$\sigma = \frac{1}{3}\, J_1^{212}/J_2^{202} \tag{40}$$

and

$$J_N^{LJL'} = \int d\mathbf{r}\, r^N\, J^{LJL'} \tag{41}$$

The helical wave number q is given by

$$Q/2K_{22} = \sigma/(1-2\Delta-\Delta'\bar{P}_4/\bar{P}_2). \tag{42}$$

Expressions (33), (34) and (35) have been obtained before by Priest [18]. In this case the temperature dependence of the pitch is determined by $\bar{P}_4/\bar{P}_2$.

(5) Near the smectic-cholesteric transition one can imagine that fluctuating smectic droplets are present in the cholesteric liquid [19]. Therefore additional elastic energy is required to distort these droplets, i.e.

$$K_{ii} = K_{ii}^{\text{intrinsic}} + K_{ii}^{\text{fluctuating}}.$$

The contribution $K_{ii}^{\text{fluctuating}}$ diverges at a temperature T^*_{SCh} where T^*_{SCh} is equal to or below the transition temperature. Consequently the pitch is composed of two contributions,

$$p = p^{\text{int}} + p^{\text{fluct}};$$

the intrinsic pitch and elastic constants were discussed earlier. In order to describe the divergence of the pitch near the smectic-cholesteric transition, that is $p^{\text{fluct}} \sim K_{22}^{\text{fluct}}$, the interaction model has to be extended to account for the smectic state. In this way the influence of the pre-smectic fluctuations on the temperature dependent behaviour of the pitch can be calculated in terms of the random phase approximation [20].

(6) To calculate pretransitional phenomena near the cholesteric-isotropic transition short-range orientational order has to be taken into account. This can be done in terms of a two-site cluster approximation [21] and the random phase approximation [22], as for nematics.

(7) To calculate the concentration dependence of the pitch in binary cholesteric mixtures the model has to be extended in the same way as

has been done by Humphries *et al.* [23] for nematics.

Conclusions

In this Chapter we have presented and discussed a simple interaction potential constructed in order to facilitate the understanding of the cholesteric phase. The assumption that the molecules behave as if they were cylindrically symmetric, has been shown to be consistent with relevant properties of the cholesteric mesophase. This result is of special significance, because other authors have arrived, erroneously, at the opposite conclusion.

The relative importance of the various contributions to the intermolecular interaction, that determines the stability of the liquid crystalline phases, are not yet fully understood. As for cholesterics the underlying model might be a valuable tool to analyze experimental results and thus to gain insight into the contribution of shape and polarizability effects. As an example we consider the cholesterogen represented in figure 5. First we remark that a molecule can be left-handed

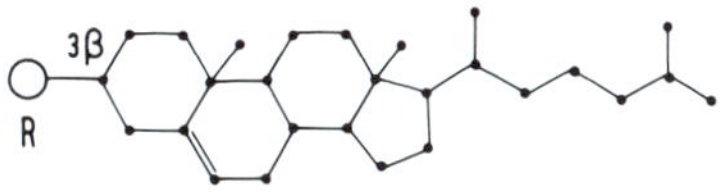

Fig. 5. *The chemical structure of the cholesteryl compounds. The black dots represent carbon atoms and the lines denote bonds; the hydrogen atoms and the carbon-hydrogen bonds are not shown in the figure. The circle R is a group attached to the 3β bond of the cholesteryl skeleton; e.g. R is Cl or O_2CCH_3. A three dimensional model of this structure resembles a twisted rod.*

with respect to its shape but right-handed with respect to its polarizability or *vice versa*. There is no reason why a molecule should possess the same handedness with respect to both contributions. If the length of the 3β chain is increased, the shape chirality will change but the polarizability chirality will remain unaltered. Cholesteryl compounds with small 3β chain length can, therefore, be right-handed and the others left-handed, in agreement with the results of Leder [24]. A similar argument was also presented by Kimura [27]. Assuming that the polarizability of the cholesteryl compounds is left-handed and that the cogs of the nematic and the cholesteryl compounds do not match with each other, it follows that the shape contribution to K is absent in the interaction between a nematogenic molecule and a molecule of the cholesterogen. Consequently the helix is always left-handed in a nematic-rich mixture, whereas it may be right-handed in a cholesteric-rich mixture. This might explain the helix inversion observed in some binary mixtures of a nematic compound and a cholesteryl compound [27, 28]. Note that, in our analysis, the assumption has been made that shape chirality tends to lose its influence with increasing 3β chain length. This might be understood to originate from an effective increase in the distance between the cogs of neighbouring molecules. Other explanations of the occurrence of helix inversion have also been offered [27, 28].

Careful comparison between experiment and the extensions listed in the previous section might contribute to deeper insights into the origin of the cholesteric helix. It would be interesting, for instance, to study the helix at constant volume analogously to the experiments of McColl and Shih [15], i.e. to study the influence of steric effects on the helix.

Finally it should be remarked that an interesting prediction can be inferred from the extension (4) of the preceding section, namely that the helical wave number of the intrinsic pitch is an increasing function of the temperature {i.e. $(T/q)dq/dT > 0$} if $K_{33} < K_{11}$ and a decreasing function {i.e. $(T/q)dq/dT < 0$} if $K_{33} > K_{11}$}.

Acknowledgements

The authors wish to thank Dr. M. Winnink and Dr. H.J.F. Jansen for helpful discussions on group theory, Dr. H. Hanson and Dr. F. Leenhouts for interesting discussions on elastic constants, and Professor A.J. Dekker for his stimulating interest in our work.

Appendix

The relevant generalizations of $P_2(\mathbf{a}_i.\mathbf{a}_j)$ and $(\mathbf{a}_i.\mathbf{a}_j)(\mathbf{a}_i\wedge\mathbf{a}_j.\mathbf{u}_{ij})$ are the following. Putting $\mathbf{x}=\mathbf{a}_i$, $\mathbf{y}=\mathbf{a}_j$ and $\mathbf{z}=\mathbf{u}_{ij}$, we define

$$T^{000}(\mathbf{x},\mathbf{z},\mathbf{y}) = 1,$$

$$T^{202}(\mathbf{x},\mathbf{z},\mathbf{y}) = P_2(\mathbf{x}.\mathbf{y}),\ T^{220}(\mathbf{x},\mathbf{z},\mathbf{y}) = T^{202}(\mathbf{x},\mathbf{y},\mathbf{z}),\ T^{022}(\mathbf{x},\mathbf{z},\mathbf{y}) = T^{202}(\mathbf{z},\mathbf{x},\mathbf{y})$$

$$T^{212}(\mathbf{x},\mathbf{z},\mathbf{y}) = (\mathbf{x}.\mathbf{y})(\mathbf{x}\wedge\mathbf{y}.\mathbf{z}),$$

$$T^{222}(\mathbf{x},\mathbf{z},\mathbf{y}) = (9/2)(\mathbf{x}.\mathbf{y})(\mathbf{x}.\mathbf{z})(\mathbf{y}.\mathbf{z})-(3/2)(\mathbf{x}.\mathbf{y})^2-(3/2)(\mathbf{x}.\mathbf{z})^2-(3/2)(\mathbf{y}.\mathbf{z})^2+1,$$

$$T^{232}(\mathbf{x},\mathbf{z},\mathbf{y}) = \{5(\mathbf{x}.\mathbf{z})(\mathbf{y}.\mathbf{z}) - \mathbf{x}.\mathbf{y}\}\ (\mathbf{x}\wedge\mathbf{y}.\mathbf{z}),$$

$$T^{242}(\mathbf{x},\mathbf{z},\mathbf{y}) = (35/8)(\mathbf{x}.\mathbf{z})^2(\mathbf{y}.\mathbf{z})^2-(5/2)(\mathbf{x}.\mathbf{y})(\mathbf{y}.\mathbf{z})(\mathbf{z}.\mathbf{x})+(1/4)(\mathbf{x}.\mathbf{y})^2 - (5/8)(\mathbf{x}.\mathbf{z})^2-(5/8)(\mathbf{y}.\mathbf{z})^2+ 1/8,$$

$$T^{422}(\mathbf{x},\mathbf{z},\mathbf{y}) = T^{242}(\mathbf{z},\mathbf{x},\mathbf{y})\ ,\ T^{224}(\mathbf{x},\mathbf{z},\mathbf{y}) = T^{242}(\mathbf{x},\mathbf{y},\mathbf{z}),$$

$$T^{432}(\mathbf{x},\mathbf{z},\mathbf{y}) = \{7(\mathbf{x}.\mathbf{z})^2(\mathbf{x}.\mathbf{y}) - 2(\mathbf{x}.\mathbf{z})(\mathbf{z}.\mathbf{y}) - \mathbf{x}.\mathbf{y}\}(\mathbf{x}\wedge\mathbf{y}.\mathbf{z})\ ,$$

$$T^{234}(\mathbf{x},\mathbf{z},\mathbf{y}) = T^{432}(\mathbf{y},\mathbf{z},\mathbf{x}),$$

$$T^{442}(\mathbf{x},\mathbf{z},\mathbf{y}) = (147/8)(\mathbf{x}.\mathbf{y})(\mathbf{x}.\mathbf{z})^3(\mathbf{y}.\mathbf{z}) - (63/8)(\mathbf{x}.\mathbf{z})^2(\mathbf{y}.\mathbf{z})^2- (49/8)(\mathbf{x}.\mathbf{z})^4 - (63/8)(\mathbf{x}.\mathbf{y})^2(\mathbf{x}.\mathbf{z})^2- (27/8)(\mathbf{x}.\mathbf{y})(\mathbf{y}.\mathbf{z})(\mathbf{z}.\mathbf{x}) + (51/8)(\mathbf{x}.\mathbf{z})^2 + (9/8)(\mathbf{y}.\mathbf{z})^2 + (9/8)(\mathbf{y}.\mathbf{x})^2- 6/8,$$

$$T^{244}(\mathbf{x},z,\mathbf{y}) = T^{442}(\mathbf{y},z,\mathbf{x}),$$

$$T^{452}(\mathbf{x},z,\mathbf{y}) = \{(21/2)(\mathbf{x}.z)^3(\mathbf{y}.z)-(7/2)(\mathbf{x}.z)^2(\mathbf{x}.\mathbf{y})-(7/2)(\mathbf{x}.z)(\mathbf{y}.z) + (1/2)(\mathbf{x}.\mathbf{y})\}(\mathbf{x}\wedge\mathbf{y}.z),$$

$$T^{254}(\mathbf{x},z,\mathbf{y}) = T^{452}(\mathbf{y},z,\mathbf{x}),$$

$$T^{462}(\mathbf{x},z,\mathbf{y}) = (231/16)(\mathbf{x}.z)^4(\mathbf{y}.z)^2-(168/16)(\mathbf{x}.\mathbf{y})(\mathbf{x}.z)^3(\mathbf{y}.z) - (126/16)(\mathbf{x}.z)^2(\mathbf{y}.z)^2-(21/16)(\mathbf{x}.z)^4+ (28/16)(\mathbf{x}.\mathbf{y})^2(\mathbf{x}.z)^2 + (56/16)(\mathbf{x}.\mathbf{y})(\mathbf{x}.z)(\mathbf{y}.z)+(14/16)(\mathbf{x}.z)^2+(7/16)(\mathbf{y}.z)^2 - (4/16)(\mathbf{x}.\mathbf{y})^2 - 1,$$

$$T^{264}(\mathbf{x},z,\mathbf{y}) = T^{462}(\mathbf{y},z,\mathbf{x}).$$

The functions $T^{LJL'}(\mathbf{x},z,\mathbf{y})$ have the following property:

$$\int d\mathbf{x} \int dy\, T^{LJL'}(x,z,y)f(x.n_i)f(y.n_j) = \int dx \int dy\, T^{LJL'}(\mathbf{n}_i,z,\mathbf{n}_j)P_L(\mathbf{n}_i.\mathbf{x}),$$

$$P_{L'}(\mathbf{n}_j.\mathbf{y})f(\mathbf{x}.\mathbf{n}_i)f(\mathbf{y}.\mathbf{n}_j) = \bar{P}_L\bar{P}_{L'}T^{LJL'}(\mathbf{n}_i,z,\mathbf{n}_j).$$

The expression $\int d\mathbf{r}_{ij}J^{LJL'}(r_{ij})\,T^{LJL'}(\mathbf{n}_i,z,\mathbf{n}_j)$ can be evaluated using a Taylor series expansion of $\mathbf{n}_j=\mathbf{n}(\mathbf{r}_i+\mathbf{r}_{ij})$ with respect to $\mathbf{r}_{ij}$. Discarding terms higher than quadratic in the gradients of $\mathbf{n}_i$ and ignoring surface terms, we obtain

$$\int d\mathbf{r}_{ij}J^{202}(r_{ij})T^{202}(\mathbf{n}_i,z,\mathbf{n}_j)=J_0^{202}-(1/2)J_2^{202}\{(\nabla.\mathbf{n}_i)^2+(\mathbf{n}_i.\nabla\wedge\mathbf{n}_i)^2+(\mathbf{n}_i\wedge\nabla\wedge\mathbf{n}_i)^2\},$$

$$\int d\mathbf{r}_{ij}J^{212}(r_{ij})T^{212}(\mathbf{n}_i,z,\mathbf{n}_j) = -(1/3)J_1^{212}\mathbf{n}_i.\nabla\wedge\mathbf{n}_i,$$

$$\int d\mathbf{r}_{ij}J^{222}(r_{ij})T^{222}(\mathbf{n}_i,z,\mathbf{n}_j) = -(1/10)J_2^{222}\{(\nabla.\mathbf{n}_i)^2-2(\mathbf{n}_i.\nabla\wedge\mathbf{n}_i)^2+(\mathbf{n}_i\wedge\nabla\wedge\mathbf{n}_i)^2\},$$

$$\int d\mathbf{r}_{ij}J^{232}(r_{ij})T^{232}(\mathbf{n}_i,z,\mathbf{n}_j) = \int d\mathbf{r}_{ij}J^{242}(r_{ij})T^{242}(\mathbf{n}_i,z,\mathbf{n}_j) = 0,$$

$$\int d\mathbf{r}_{ij}J^{422}(r_{ij})T^{422}(\mathbf{n}_i,z,\mathbf{n}_j) = (1/12)J_2^{422}\{3(\nabla.\mathbf{n}_i)^2+(\mathbf{n}_i.\nabla\wedge\mathbf{n}_i)^2-4(\mathbf{n}_i\wedge\nabla\wedge\mathbf{n}_i)^2\},$$

$$\int d\mathbf{r}_{ij}J^{224}(r_{ij})T^{224}(\mathbf{n}_i,z,\mathbf{n}_j = (1/12)J_2^{224}\{3(\nabla.\mathbf{n}_i)^2+(\mathbf{n}_i.\nabla\wedge\mathbf{n}_i)^2-4(\mathbf{n}_i\wedge\nabla\wedge\mathbf{n}_i)^2\},$$

$$\int d\mathbf{r}_{ij}J^{4J2}(r_{ij})T^{4J2}(\mathbf{n}_i,z,\mathbf{n}_j) = \int d\mathbf{r}_{ij}J^{2J4}(r_{ij})T^{2J4}(\mathbf{n}_i,z,\mathbf{n}_j) = 0 \text{ if } J \neq 2.$$

References

1. W. Maier and A. Saupe, *Z. Naturforsch.* **14a** 882 (1959); **15a** 287 (1960).
2. L. Pohl, R. Eidenschink, J. Krause and G. Weber, *Phys. Lett.* A **65** 169 (1978).
3. W.M. Gelbart and A. Gelbart, *Mol. Phys.* **33** 1387 (1977).
4. G.R. Luckhurst and C. Zannoni, *Nature* **267** 412 (1977).
5. See, for example, P.G. de Gennes, *The Physics of Liquid Crystals*, Clarendon, Oxford Ch.2. (1974).
6. W.J.A. Goossens, *Mol. Cryst. Liq. Cryst.* **12** 237 (1971).
7. F.C. Frank, *Disc. Faraday Soc.* **25** 19 (1958).
8. See, for example, J.L. Fergason, *Scientific American* **211** 76 (1964).
9. B.W. van der Meer, G. Vertogen, A.J. Dekker and J.G.J. Ypma, *J. Chem. Phys.* **65** 3935 (1976).

10. Y.R. Lin-Liu, Yu Ming Shih, Chia-Wei Woo and H.T. Tan, *Phys. Rev.* A **14** 445 (1976); *Phys. Rev.* A **15** 2550 (1977).
11. P.G. de Gennes, *Mol. Cryst. Liq. Cryst.* **12** 210 (1971); R. Albèn, *ibid* **20** 233 (1973); M.J. Stephan and J.P. Straley, *Rev. Mod. Phys.* **46** 668 (1974).
12. R.P. Feynman, R.B. Leighton and M. Sands, *The Feynman Lectures on Physics II* Addison-Wesley, Reading, p.30-7 (1969).
13. J.P. Straley, *Phys. Rev.* A **14** 1835 (1976).
14. B.W. van der Meer and G. Vertogen, *Phys. Lett. A* **59** 279 (1976).
15. J.R. McColl and J.S. Shih, *Phys. Rev. Lett.* **29** 85 (1972).
16. H. Kimura, *J. Phys. Soc. Japan.* **36** 1280 (1974); M.A. Cotter, *J. Chem. Phys.* **66** 1098, (1977); J.G.J. Ypma and G. Vertogen, *Phys. Lett.* A **61** 45 (1977), *Phys. Rev.* A **17** 149 (1978); J.G.J. Ypma, Thesis, Groningen (1977).
17. F. Leenhouts, H.J. Roebers, A.J. Dekker and J.J. Jonker, *J. Phys.* (Paris), C3-291 (1979).
18. R.G. Priest, *Phys. Rev.* A **7**, 720 (1973).
19. W.L. McMillan, *J. Phys. (Paris)* C-1 105 (1975); P.G. de Gennes, *Mol. Cryst. Liq. Cryst.* **21** 49 (1973); R.S. Pindak, C.C. Huang and J.T. Ho, *Solid State Comm.* **14** 821 (1974).
20. B.W. van der Meer and G. Vertogen, *paper presented at the 7th International Liquid Crystal Conference*, Bordeaux (1979).
21. J.G.J. Ypma, G. Vertogen and H.T. Coster, *Mol. Cryst. Liq. Cryst.* **37** 57 (1976); P. Sheng and P.J. Wojtowicz, *Phys. Rev.* A **14** 1883 (1976).
22. G. Vertogen and B.W. van der Meer, *paper presented at the 7th International Liquid Crystal Conference*, Bordeaux (1978); *Phys. Rev. A.*
23. R.L. Humphries, P.G. James and G.R. Luckhurst, *Symp. Faraday Soc.* **5** 107 (1971); *J. Chem. Faraday Trans. 11*, **68** 1931 (1971); R.L. Humphries and G.R. Luckhurst, *Proc. R. Soc. Lond.* A **352** 41 (1976).
24. L.B. Leder, *J. Chem. Phys.* **55** 2649 (1971).
25. H. Kimura, *paper presented at the 7th International Liquid Crystal Conference*, Bordeaux (1978).
26. H. Stegemeyer and H. Finkelmann, *Chem. Phys. Lett.* **23** 227 (1973).
27. H. Hanson, A.J. Dekker and F. van der Woude, *Mol. Cryst. Liq. Cryst.* **42** 15 (1977).
28. T.V. Samulski and E.T. Samulski, *J. Chem. Phys.* **67** 824 (1977).

Postscript

After completion of this Chapter we received the final draft of Schröder's Chapter. We saw that he now understood that the twist term $(\mathbf{a}_i.\mathbf{a}_j)(\mathbf{a}_i\wedge\mathbf{a}_j.\mathbf{u}_{ij})$ cannot be excluded on formal grounds. His rejection of this type of twist interaction, i.e. his rejection of the continuum theory of cholesterics, is now based upon, what he calls, physical grounds. An important argument to sustain his point of view is that the induced dipole-quadrupole energy is identical with the simplest version of his pair potential. This conjecture is incorrect as can be easily shown in the following way.

The dipole-quadrupole dispersion interaction between molecules i and j reads, in the notation of Goossens [6],

$$U_{ij}^{pq} = \sum_{\nu,\nu'} (p_\alpha^i|p_{\beta'}^i)_\nu (p_\beta^j|q_{\alpha'\gamma'}^j)_{\nu'} (E_{\nu\nu',oo})^{-1} C_{\alpha\beta}^{ij} D_{\alpha'\beta'\gamma'}^{ij} (r_{ij})^{-7}$$

$$+ \sum_{\nu,\nu'} (p_\alpha^i|p_{\alpha'}^i)_\nu (p_\beta^j|q_{\gamma\gamma}^j)_{\nu'} (E_{\nu\nu',oo})^{-1} C_{\alpha\beta}^{ij} u_{ij\alpha} (r_{ij})^{-7} + \text{c.c.}, \quad (P1)$$

where summation over repeated Greek indices is implied and

$$(p^i_\alpha|p^i_\beta)_\nu = <0|p^i_\alpha|\nu><\nu|p^i_\beta|0>,$$

$$(p^j_\beta|q^j_{\alpha\gamma})_\nu = <0|p^j_\beta|\nu><\nu|q^j_{\alpha\gamma}|0>.$$

Here $|0>$ denotes the groundstate and $|\nu>$ an excited eigenstate; p^i_α denotes the α-component of the electric dipole moment of molecule i and $q^j_{\alpha\beta}$ the $\alpha\beta$-component of the electric quadrupole moment of molecule j; $E_{\nu\nu'00}$ is the difference in energy between the groundstate and an excited state of the system. Further we have

$$C^{ij}_{\alpha\beta} = \delta_{\alpha\beta} - 3u_{ij\alpha}u_{ij\beta},$$

$$D^{ij}_{\alpha\beta\gamma} = (3/2)u_{ij\alpha}(5u_{ij\beta}u_{ij\gamma} - 2\delta_{\beta\gamma}),$$

where $\delta_{\alpha\beta}$ is a Kronecker delta, r_{ij} is the distance between the centres of mass of the molecules i and j and $u_{ij\alpha}$ denotes the α-component of the unit vector $\mathbf{u}_{ij}$ pointing from the centre of mass of molecule i towards that of molecule j.

Next we demonstrate the untenability of Schröder's argument by selecting a particularly simple representation of the induced dipole-quadrupole model. We choose a cartesian coordinate system, defined by the unit vectors $\mathbf{a}_i$, $\mathbf{b}_i$ and $\mathbf{c}_i$, along the principal inertial axes of molecule i. The vector $\mathbf{a}_i$ (z axis) is along the long molecular axis and $\mathbf{b}_i$ (x axis) and $\mathbf{c}_i$ (y axis) are along the remaining principal axes, where $\mathbf{b}_i \wedge \mathbf{c}_i = \mathbf{a}_i$. A simple version of the dipole-quadrupole model is obtained by putting

$$(p^i_z|p^i_z)_\nu = \xi^i_\nu, \quad (p^j_x|q^j_{yz})_{\nu'}+c.c = -(p^j_y|q^j_{xz})_{\nu'}-c.c = \chi^j_{\nu'},$$

and taking all remaining matrix elements to be zero. Substitution yields

$$U^{pq}_{ij} = -K_{ij}\{(\mathbf{a}_i.\mathbf{a}_j)-2(\mathbf{a}_i.\mathbf{u}_{ij})(\mathbf{a}_j.\mathbf{u}_{ij})\}\{(\mathbf{b}_j.\mathbf{u}_{ij})(\mathbf{c}_j.\mathbf{a}_i)-(\mathbf{c}_j.\mathbf{u}_{ij})(\mathbf{b}_j.\mathbf{a}_i)\},$$

$$= -K_{ij}\{(\mathbf{a}_i.\mathbf{a}_j)-2(\mathbf{a}_i.\mathbf{u}_{ij})(\mathbf{a}_j.\mathbf{u}_{ij})\}(\mathbf{a}_i\wedge\mathbf{a}_j.\mathbf{u}_{ij}), \quad \text{(P2)}$$

with

$$K_{ij} = 3\sum_{\nu\nu'} \xi^i_\nu\chi^j_{\nu'}(E_{\nu\nu'oo})^{-1} \quad \text{(P3)}$$

and where we used the relation

$$\mathbf{v}\wedge\mathbf{w}.\mathbf{a}_j = (\mathbf{v}\wedge\mathbf{w}).(\mathbf{b}_j\wedge\mathbf{c}_j) = (\mathbf{v}.\mathbf{b}_j)(\mathbf{w}.\mathbf{c}_j)-(\mathbf{v}.\mathbf{c}_j)(\mathbf{w}.\mathbf{b}_j).$$

The general derivation of a term like (P2) for freely rotating chiral molecules has been given in an earlier paper [9]. Schröder's interpretation of a private communication and reference [9] is a mystery to us.

In terms of Schröder's notation (P2) can be written as

$$U_{ij}^{pq} = \sum_{pMp'} \sum_{J=1,3} u_{2J2}^{00} \begin{pmatrix} 2 & J & 2 \\ p & M & p' \end{pmatrix} D_{p,0}^{2}(\Omega_i) D_{p',0}^{2}(\Omega_j) D_{M,0}^{J}(\Omega_{ij}), \tag{P4}$$

where the coefficients u_{2J2}^{00} J = 1,3) are proportional to K_{ij}. The expression (P4), however, does not belong to Schröder's potential. Consequently the simplest version of his theory is not identical with the induced dipole-quadrupole model as claimed. This strongly suggests that Schröder's model is physically meaningless according to his own criterion.

Chapter 7

HARD PARTICLE THEORIES OF NEMATICS

MARTHA A. COTTER

Department of Chemistry, Rutgers, The State University, New Brunswick, New Jersey 08903.

Introduction

Hard particle theories of nematics are based on a model system of rodlike molecules which are infinitely impenetrable but otherwise do not interact. This hard rod fluid, as it is commonly called, is clearly not a realistic model for a nematic liquid crystal, where there are intermolecular attractions as well as repulsions, and the latter are at least somewhat weaker and longer ranged than hard particle repulsions. Why, then, consider the hard rod fluid? One justification sometimes given is that it is a simple, well-defined, *relatively* tractable system which still mimics certain features of real nematogens. In my opinion, a better answer lies in the theory of ordinary isotropic liquids, where it has been shown that the structure of a simple liquid far from its critical point is largely determined by very short ranged intermolecular repulsions, which can be approximated satisfactorily by hard sphere repulsions [1]. This is the basis of (1) the rather successful van der Waals theory of simple liquids [2], in which the model system consists of hard spheres in a uniform, negative mean field, and (2) the still more successful thermodynamic perturbation theories [3], in which the attractive part of the intermolecular pair potential is treated as a perturbation on the repulsive part, and the properties of the unperturbed or reference system are approximated by those of hard spheres with an effective diameter, $d(\rho)$ or $d(\rho,T)$. If one assumes that very short ranged intermolecular repulsions are of comparable importance in nematics, where they must be highly anisotropic, and that a van der Waals or perturbation theory might therefore prove useful, then the hard rod fluid would seem to be the reference system with which to begin.

The Hard Rod Fluid

Let us consider a fluid of N rodlike molecules in a volume V, with intermolecular pair potential

$$U_{ij} = U^*(\mathbf{r}_{ij},\Omega_i,\Omega_j) = \begin{cases}\infty, & \text{if i and j would overlap} \\ 0, & \text{otherwise}\end{cases}$$

where $\mathbf{r}_{ij} = \mathbf{r}_j - \mathbf{r}_i$ is the vector connecting the centres of molecules i and j; Ω_i and Ω_j represent the sets of Euler angles ($\alpha\beta\gamma$) describing the orientations of the respective molecules relative to a space-fixed coordinate system. The configurational partition function Q_N for this system is

$$Q_N = \{1/N!(8\pi^2)^N\}\int\{dX^N\}\exp(-\beta U_N),$$

$$= \{1/N!(8\pi^2)^N\}\int\{d\mathbf{r}^N\}\int\{d\Omega^N\}\exp(-\beta U_N), \tag{1}$$

where

$$U_N = \sum_{1<i<j<N} U^*(\mathbf{r}_{ij},\Omega_i,\Omega_j).$$

This differs from the partition function introduced in Chapter 3 by the normalisation factor $(8\pi^2)^{-N}$. The angular integrations in eq. (1) can be approximated to arbitrary accuracy by dividing the single particle orientational phase space into n arbitrarily small sections of volume $\Delta\Omega = 8\pi^2/n$ and summing over all possible orientational distributions $(N_1,N_2,\ldots,N_n)$, where N_σ is the number of molecules with the σth orientation (i.e. with Ω in the σth section centred about Ω_σ) and

$$\sum_{\sigma=1}^{n} N_\sigma = N.$$

This yields

$$Q_N = \{1/N!(8\pi^2)^N\}\sum_{N_1}\cdots\sum_{N_n}\{N!(\Delta\Omega)^N/(N_1!N_2!\ldots N_n!)\}\int d\mathbf{r}^N\exp\{-\beta U_N(\mathbf{r}^N,N_1,N_2\ldots N_n!)\}, \tag{2}$$

$$\approx (\Delta\Omega/8\pi^2)^N(\prod_{\sigma=1}^{n}\tilde{N}_\sigma!)^{-1}\int d\mathbf{r}^N\exp\{-\beta U_N(r^N,\tilde{N}_1,\ldots,\tilde{N}_n)\}, \tag{3}$$

where the tildes denote the set $(N_1,\ldots,N_n)$ corresponding to the maximum term in eq. (2). Taking the natural logarithm of both sides of eq. (3),

$$\ln Q_N = N\ln(\Delta\Omega/8\pi^2)-\sum_{\sigma=1}^{n} N_\sigma\ln\tilde{N}_\sigma+\tilde{N}+\ln[\int d\mathbf{r}^N\exp\{-\beta U_N(\mathbf{r}^N,\tilde{N}_1,\ldots,\tilde{N}_n)\}] \tag{4}$$

or

$$\beta A_c/N = -N^{-1}\ln Q_N,$$

$$= \sum_{\sigma=1} s_\sigma\ln(ns_\sigma) + \ln\rho + \beta\phi_N(\tilde{N}_1,\tilde{N}_2,\ldots\tilde{N}_n)/N - 1, \tag{5}$$

where $s_\sigma = \tilde{N}_\sigma/N$, $\rho = N/V$, A_c is the configurational Helmholtz free energy, and

$$\phi_N(\tilde{N}_1,\tilde{N}_2,\ldots,\tilde{N}_n) = -kT \ln[V^{-N}\int d\mathbf{r}^N \exp\{-\beta U_N(\mathbf{r}^N,\tilde{N}_1,\ldots,\tilde{N}_n)\}].$$

As was first noted by Zwanzig [4], $\phi_N(\tilde{N}_1,\tilde{N}_2,\ldots,\tilde{N}_n)$ is the excess Helmholtz free energy of a fluid of $\tilde{N}_1$ hard rods with fixed orientation Ω_1, $\tilde{N}_2$ hard rods with fixed orientation $\Omega_2\ldots$, and $\tilde{N}_n$ hard rods with fixed orientation Ω_n, relative to an ideal n-component mixture. Writing $s_\sigma = f(\Omega_\sigma)\Delta\Omega$ and converting from sums back to integrals yields, finally

$$\begin{aligned}\beta A_c/N &= -N^{-1}\ln Q_N,\\ &= \ln\rho + \int d\Omega f(\Omega)\ln[8\pi^2 f(\Omega)] + \beta\phi_N\{f(\Omega)\}/N - 1,\end{aligned} \tag{6}$$

where $f(\Omega)$ is the singlet orientational distribution function [5].

Clearly, the major task confronting us is to derive the excess free energy function $\phi_N(N_1,N_2,\ldots,N_n)$. $A_c(N,V,T)$ can then be obtained by (i) substituting ϕ_N in eq. (5), (ii) converting from sums to integrals, and (iii) determining $f(\Omega)$ by minimizing the right-hand side of eq. (6). The two most common approaches used to obtain ϕ_N are the cluster expansion technique of Onsager [6] and scaled particle theory. These will be discussed in turn.

The Cluster Expansion Approach

As we saw, the fluid with excess Helmholtz free energy $\phi_N(\tilde{N}_1,\tilde{N}_2,\ldots,\tilde{N}_n)$ is an n-component thermodynamic system, each component consisting of molecules with a particular orientation. As is the case for any n-component system, the excess free energy can be expanded in a cluster or virial expansion in the total density, ρ, when this is sufficiently small; that is

$$\begin{aligned}\beta\phi_N/N &= \beta A^{\text{excess}}/N,\\ &= \sum_{m\geq 2}(m-1)^{-1}B_m(s_1,\ldots,s_n)\rho^{(m-1)},\end{aligned} \tag{7}$$

where B_m is called the mth virial coefficient. Since the total configurational potential energy U_N is pair-wise additive, each B_m can be written as a sum of integrals over products of Mayer functions $f_{ij}^{\sigma\sigma'}$, where

$$f_{ij}^{\sigma\sigma'} = \exp(-\beta U_{ij}^{\sigma\sigma'})-1$$

and $U_{ij}^{\sigma\sigma'}$ is the pair potential between molecules i and j of types σ and σ', respectively (i.e. with fixed orientations Ω_σ and $\Omega_{\sigma'}$, respectively). Expressions for B_2 and B_3 are

$$B_2 = -(1/2)\sum_{\sigma'=1}^{n}\sum_{\sigma=1}^{n} s_\sigma s_{\sigma'}\int d\mathbf{r}_{ij}\, f_{ij}^{\sigma\sigma'}(\mathbf{r}_{ij}) \tag{8}$$

and

$$B_3 = -(1/6)\sum_{\sigma''=1}^{n}\sum_{\sigma'=1}^{n}\sum_{\sigma=1}^{n} s_\sigma s_{\sigma'} s_{\sigma'}\int d\mathbf{r}_{ij} d\mathbf{r}_{ik} f_{ij}^{\sigma\sigma'} f_{ik}^{\sigma\sigma''} f_{jk}^{\sigma'\sigma''} . \tag{9}$$

The expressions for B_4, B_5,...contain increasingly lengthy sums of integrals.

Truncating eq. (7) after the first term, $B_2\rho$, and substituting in eq. (5), gives

$$\beta A_c/N = \sum_{\sigma=1}^{\tilde{n}} s_\sigma \ln(ns_\sigma)+\ln\rho-1-(\rho/2)\sum_{\sigma'=1}^{\tilde{n}}\sum_{\sigma=1}^{\tilde{n}} s_\sigma s_{\sigma'}\int d\mathbf{r}_{ij} f_{ij}^{\sigma\sigma'}(\mathbf{r}_{ij}). \tag{10}$$

For hard rods, $f_{ij}^{\sigma\sigma'}$ can have only two values, -1 when $U_{ij}^{\sigma\sigma'} = \infty$ and 0 (when $U_{ij}^{\sigma\sigma'} = 0$). Eq. (10) therefore reduces to

$$\beta A_c/N = \sum_{\sigma=1}^{\tilde{n}} s_\sigma \ln(ns_\sigma)+\ln\rho-1+(\rho/2)\sum_{\sigma'}\sum_{\sigma} s_\sigma s_{\sigma'} V(\Omega_\sigma,\Omega_{\sigma'}), \tag{11}$$

where $V(\Omega_\sigma\ \Omega_{\sigma'})$ is the mutual exclusion volume (or covolume) of two rods with orientations Ω_σ and $\Omega_{\sigma'}$, respectively. More precisely, $V(\Omega_\sigma,\Omega_{\sigma'})$ is the volume excluded to the centre of a rod with orientation $\Omega_{\sigma'}$ by the presence of a fixed rod with orientation Ω_σ and *vice versa*. Converting from sums to integrals then gives

$$\beta A_c/N = \int d\Omega f(\Omega)\ln[8\pi^2 f(\Omega)]+ \ln\rho-1+(\rho/2)\int d\Omega d\Omega' f(\Omega)f(\Omega')V(\Omega,\Omega'). \tag{12}$$

To obtain the singlet orientational distribution function, we must minimize the functional $A_c\{f(\Omega)\}\beta/N$ subject to the constraint

$$\int d\Omega f(\Omega) = 1. \tag{13}$$

Thus, $f(\Omega)$ is a solution to the variational equation

$$0 = \delta\{\beta A_c/N - \nu\int d\Omega f(\Omega)\}/\delta f,$$

$$= \ln\{8\pi^2 f(\Omega)\}+1-\nu+\rho\int d\Omega' f(\Omega')V(\Omega,\Omega'), \tag{14}$$

where ν is a Lagrange multiplier, to be evaluated from the normalization condition eq. (13).

To proceed further, we must specify the shape of the rod-like molecules. The most convenient such shape, used by Onsager and most subsequent workers, is a spherocylinder; that is a right circular cylinder capped on each end by a hemisphere of the same radius. In this case [6],

$$V(\Omega,\Omega') = 32\pi a^2/3+8\pi a\ell^2+4a\ell^2|\sin\gamma(\Omega,\Omega')|,$$

$$= 8v_o+ 4a\ell^2|\sin\gamma(\Omega,\Omega')|, \tag{15}$$

where a, ℓ, and v_o are the radius, cylindrical length, and volume, respectively, of a spherocylinder and $\gamma(\Omega,\Omega')$ is the angle between the cylinder axes of two molecules with orientations Ω and Ω'. Eq. (14) then becomes

$$\ln\{8\pi^2 f(\Omega)\} = \nu-1-8v_o\rho-4a\ell^2\rho\int d\Omega' f(\Omega')|\sin\gamma(\Omega,\Omega')|, \quad (16)$$

a nonlinear integral equation which must be solved numerically for $f(\Omega)$. All the equilibrium thermodynamic properties of the system can then be determined from eq. (12) and its derivatives.

An approximate variational solution to eq.(16) was given by Onsager [6]; more accurate numerical solutions were later obtained by Lasher [7]. According to the latter, when the quantity $\lambda^*(=4a\ell^2\rho)$ is less than 8.88, there exists only the isotropic solution $f(\Omega) = (1/8\pi^2)$ (using our definition of $f(\Omega)$ [5]). When $\lambda^*>8.88$, however, there are two anisotropic solutions as well, and the free energy is minimized by the $f(\Omega)$ corresponding to nematic ordering of the rods. The isotropic to nematic phase transition is first order with $\lambda^*_{nem} = 10.6$, $\lambda^*_{iso} = 8.4$, $\Delta\rho/\rho_{nem} = 0.21$, and $\bar{P}_2 = 0.784$ in the large (ℓ/a) limit.

The truncation of the cluster expansion eq. (7) after the linear term $B_2\rho$ is valid only when ρ is sufficiently small. Unfortunately, the isotropic to nematic phase transition occurs at such low densities only when the hard rod length-to-breadth ratio x ($x = 1+\ell/2a$ for spherocylinders) is quite large. Based on estimates of the ratio $\rho_c B_3/B_2$ (ρ_c is the smallest density for which anisotropic solutions to eq. (16) exist), Straley [8] argued that the truncation can be justified quantitatively only for very long rods ($x > 100$), although it should produce qualitatively reliable results for considerably shorter particles ($x \approx 40$, for example). Whatever the accuracy of these numerical estimates, it is clear that at molecular length-to-breadth ratios relevant to thermotropic nematic mesophases ($x \approx 3$ to 10 at most) this truncation is totally invalid [8-10]. It must be emphasized, therefore, that the use of the Onsager approach (i.e. eqs. (12) and (16)) to study the isotropic-nematic phase transition in systems of relatively short hard rods ($x<10$) is completely unjustified and any conclusions based on such a calculation must be discounted. On the other hand, this approach can be applied, quite properly, to models of the lyotropic nematic or cholesteric mesophases which occur in dilute solutions of long rigid polymer molecules, (e.g. poly-γ-benzyl-L-glutamate in dioxane). In this case, $v_o\rho$ is the volume fraction of polymer, $U^{\sigma\sigma'}_{ij}$ is a two-body potential of mean force, and B_2 is the osmotic second virial coefficient, as was clearly stated by Onsager [6].

The cluster expansion method could be extended to shorter hard rods if the higher order virial coefficients B_3, B_4, etc. could be evaluated. Due to the mathematical and computational difficulties it would entail, this is extremely unlikely. An alternative, more tractable approach to the statistical mechanics of short hard rods is therefore needed. Scaled particle theory, which we shall now discuss, provides such an alternative.

The Scaled Particle Approach

Scaled particle theory provides a means of evaluating the equilibrium thermodynamic properties of a *hard particle* fluid without first determining the pair distribution function. More specifically, it permits the derivation of (approximate) expressions for the pressure and the chemical potential by evaluating (approximately) the reversible work required to build a molecule-sized cavity in the fluid. It was originally developed by Reiss, Frisch, and Lebowitz [11] to treat the hard sphere fluid and has since been applied to a number of other systems. It is exact in one dimension and gives equations of state for hard discs and hard spheres which are in quite satisfactory agreement with computer simulations.

The scaled particle approach has been applied to hard rod fluids by Cotter and Martire [12], Lasher [7], Timling [13] and Cotter [14,15]. Although the different treatments have produced somewhat different free energy expressions, equations of state, etc., their numerical predictions differ very little. (In each instance, the hard rods were taken to be spherocylinders). The version of the theory to be presented here is that given in ref. [15]. It is the simplest formulation which yields self-consistent thermodynamics for the hard rod fluid with fixed orientations [16] (i.e. the multi-component system whose excess Helmholtz free energy is $\phi_N(N_1,\ldots,N_n)$) *and* expressions for μ_c, P, A_c, etc. which reduce to previously reported scaled particle expressions for hard spheres, isotropic hard spherocylinders, and hard spherocylinders with perfect orientational order in the appropriate limits.

We want to derive an expression for the excess free energy function $\phi_N(\tilde{N}_1,\tilde{N}_2,\ldots,\tilde{N}_n)$, which can then be substituted in eq. (5), and $\beta A_c\{f(\Omega)\}/N$ and $f(\Omega)$ determined as before. For this purpose, let us consider a fluid of N hard spherocylindrical molecules with radius a, cylindrical length ℓ, and fixed orientations in a volume V, where $\tilde{N}_1,\tilde{N}_2$, ... and $\tilde{N}_n$ are the numbers of spherocylinders with orientations Ω_1,Ω_2 .... and Ω_n, respectively. The principal quantity in the scaled particle theory of this system is the work function $W_\sigma(\alpha,\lambda,\rho)$, defined as the reversible work of adding a scaled spherocylinder of radius αa, cylindrical length $\lambda\ell$, and fixed orientation Ω_σ to the fluid at some arbitrary fixed point. W_σ is related to the thermodynamic properties of the fluid through the exact relation

$$\beta G_c/N = \sum_{\sigma=1} s_\sigma\{\ln(s_\sigma\rho)+\beta W_\sigma(1,1,\rho)\}, \tag{17}$$

where G_c is the configurational Gibbs free energy and $s_\sigma = \tilde{N}_\sigma/N$ is the mole fraction of molecules of type σ.

When the scaling parameters α and λ are both sufficiently small so that three ordinary or unscaled spherocylinders cannot simultaneously be in contact with the scaled spherocylinder, it can be shown [12] that

$$\exp(-\beta W_\sigma)=1-\rho \sum_{\sigma'=1} s_{\sigma'}V_{\sigma\sigma'}(\alpha,\lambda)+(\rho^2/2)\sum_{\sigma'\sigma''}\sum s_{\sigma'}s_{\sigma''}\int_{V\sigma\sigma'}d\mathbf{r}_i\int_{V\sigma\sigma''}d\mathbf{r}_j g_{\sigma'\sigma''}(\mathbf{r}_{ij}),$$

$$\equiv 1-\rho\sum_{\sigma'} s_{\sigma'}V_{\sigma\sigma'}(\alpha,\lambda) + A_2^{(\sigma)}(\alpha,\lambda,\rho), \tag{19}$$

where $V_{\sigma\sigma'}(\alpha,\lambda)$ is the volume excluded to the centre of an unscaled molecule with orientation $\Omega_{\sigma'}$ by the presence of the scaled particle (with orientation Ω_σ) at some fixed point; $g^{(2)}_{\sigma'\sigma''}(\mathbf{r}_{ij})$ is the pair correlation function for two unscaled molecules i and j with fixed orientations $\Omega_{\sigma'}$ and $\Omega_{\sigma''}$, respectively; the integrations over $\mathbf{r}_i$ and $\mathbf{r}_j$ are over all positions excluded to the centre of the particular molecule by the presence of the scaled particle. As was shown by Onsager [6],

$$V_{\sigma\sigma'}(\alpha,\gamma) = (4/3)\pi a^3(1+\alpha)^3+\pi a^2\ell(1+\alpha)^2(1+\lambda)+2a\ell^2(1+\alpha)\lambda|\sin\gamma(\Omega_\sigma,\Omega_{\sigma'})|, \tag{20}$$

where $\gamma(\Omega_\sigma,\Omega_{\sigma'})$ is again the angle between the long axes of two molecules with orientations Ω_σ and $\Omega_{\sigma'}$ respectively. When $\alpha = 1$ and $\lambda = 1$, eq. (20) reduces, of course, to eq. (15). Moreover, it can be shown [17] that

$$A_2^{(\sigma)}=\partial A_2^{(\sigma)}/\partial\alpha=\partial A_2^{(\sigma)}/\partial\lambda=\partial^2 A_2^{(\sigma)}/\partial\alpha\partial\lambda=\partial^2 A_2^{(\sigma)}/\partial\alpha^2=\partial^3 A_2^{(\sigma)}/\partial\alpha^2\partial\lambda= 0 \tag{21}$$

when $\alpha = 0$ and $\lambda = 0$, that is when the scaled particle shrinks to a point. At the opposite extreme, when α and λ are both very large, $W_\sigma(\alpha,\lambda)$ must approach the reversible PV work required to create a macroscopic spherocylindrical cavity in the fluid. Hence,

$$\lim_{\substack{\alpha\to\infty\\ \lambda\to\infty}} W_\sigma = [\pi(\alpha a)^2\lambda\ell + \pi(4/3)(\alpha a)^3]P.$$

These considerations suggest that

$$W_\sigma(\alpha\ \lambda) = C_{00}^{(\sigma)}+C_{10}^{(\sigma)}\alpha+C_{01}^{(\sigma)}\lambda+C_{11}^{(\sigma)}\alpha\lambda+C_{20}^{(\sigma)}\alpha^2+(\pi a^2\ell P)\alpha^2\lambda+(4\pi a^3P/3)\alpha^2 \tag{22}$$

might be a good approximation for all $\alpha > 0$, $\lambda > 0$, where the coefficients $C_{ij}^{(\sigma)}$ are to be obtained from eqs. (19-21), and

$$C_{ij}^{(\sigma)} = (i!j!)^{-1}\partial^{(i+j)}W_\sigma/\partial\alpha^i\partial\lambda^j ; \tag{23}$$

eq. (22) is the central assumption of the theory. It involves, in essence, an interpolation between very large and very small values of α and λ.

After evaluating the coefficients, $C_{ij}^{(\sigma)}(\rho,s_1,\ldots,s_n)$, $W_\sigma(1,1,\rho)$ can be obtained from eq. (22) and then substituted in eq. (17), to give

$$\beta G_c/N = \sum_{\sigma=1}^{n} s_\sigma \ln s_\sigma + \ln\rho\{\rho/(1-v_o\rho)\} + 6v_o\rho(1+2/3r\Gamma)/(1-v_o\rho)$$
$$+ 2v_o^2\rho^2(2+q)(1-q/4+r\Gamma)^2/(1-v_o\rho)^2 + 2\beta Pv_o, \tag{24}$$

where

$$v_o = \pi a^2\ell + 4\pi a^3/3,$$
$$q = 4\pi a^3/3v_o = 2/(3x-1),$$
$$r = a\ell^2/v_o = 6(x-1)^2/\pi(3x-1),$$
$$x = 1 + \ell/2a$$

and

$$\Gamma = \sum_{\sigma=1}^{n}\sum_{\sigma'=1}^{n} s_\sigma s_{\sigma'}|\sin\gamma(\Omega_\sigma,\Omega_{\sigma'})|.$$

The pressure, P, can then be determined from the Gibbs-Duhem equation in the form

$$\beta(\partial P/\partial\rho)_{s_1,\ldots,s_n} = 1+\rho\beta\sum_{\sigma=1}^{n} s_\sigma(\partial W_\sigma(1,1)/\partial\rho)_{s_1,\ldots s_n}. \tag{25}$$

Inserting the expression for $\partial W_\sigma(1,1)/\partial\rho$ obtained from eq. (22) and re-arranging gives

$$(1-v\ \rho)(\partial(\beta P)/\partial\rho)_{s_1,\ldots s_n} = 1+v_o\rho/(1-v_o\rho)$$
$$+ 6v_o\rho(1+2/2r\Gamma)/(1-v_o\rho)^2$$
$$+ 4v_o^2\rho^2(2+q)(1-q/4+r\Gamma)/(1-v_o\rho)^3. \tag{26}$$

Dividing both sides of eq. (26) by $(1-v_o\rho)$ and integrating then gives

$$\beta Pv_o = v_o\rho[1+v_o\rho(1+2r\Gamma)+2v_o^2\rho^2\{1+q-q^2/2+(1+2q)r\Gamma\}/3]. \tag{27}$$

We can now obtain $\phi_N(\tilde{N}_1,\tilde{N}_2,\ldots,\tilde{N}_n)$; i.e.

$$\beta\phi_N/N = \beta A_c/N - (\beta A_c/N)_{ideal},$$
$$= \beta A_c/N - \sum_\sigma s_\sigma \ln s_\sigma - 1,$$
$$= \beta G_c/N - \beta P/\rho - \sum_\sigma s_\sigma \ln s_\sigma - 1,$$
$$= -\ln(1-v_o\rho) + 3v_o\rho/(1-v_o\rho) + [(4+q-q^2/2)v_o^2\rho^2$$
$$+ 2rv_o\rho\{3-(1-q)v_o\rho\}\Gamma]/3(1-v_o\rho)^2. \tag{28}$$

Substituting this expression in eq. (5) and converting from sums to integrals then yields our final free energy expression:

$$\beta A_c/N = \underline{(-1/N)\ln Q_N},$$

$$= \overline{\ln\{8\pi^2 f(\Omega)\}} + \ln\{\rho/(1-v_o\rho)\} + 3v_o\rho/(1-v_o\rho)$$

$$+ [(4+q-q^2/2)v_o^2\rho^2 + 2rv_o\rho\{3-(1-q)v_o\rho\}\overline{|\sin\gamma|}]/3(1-v_o\rho)^2, \quad (29)$$

where

$$\overline{\ln 8\pi^2 f(\Omega)\}} = \int d\Omega f(\Omega)\ln\{8\pi^2 f(\Omega)\}$$

and

$$|\overline{\sin\gamma}| = \int d\Omega d\Omega' f(\Omega) f(\Omega')\,|\sin\gamma(\Omega,\Omega')|.$$

All that remains is to minimize the functional $\beta A_c\{f(\Omega)\}/N$ subject to the normalization condition eq. (13). This leads to the integral equation

$$\nu - 1 = \ln\{8\pi^2 f(\Omega)\} + [4rv_o\rho\{3-(1-q)v_o\rho\}]\{3(1-v_o\rho)^2\}^{-1}\int d\Omega'\,|\sin\gamma(\Omega,\Omega')|\,f(\Omega') \quad (30)$$

which must be solved for $f(\Omega)$; (ν is again a Lagrange multiplier). Expressions for other thermodynamic functions can be obtained from eq. (29) and evaluated with the aid of eq. (30). For example,

$$S_c/Nk = \beta(U-A_c)/N = -\beta A_c/N, \quad (31)$$

$$\beta P v_o = -\beta v_o(\partial A_c/\partial V)_{N,T} = v_o\rho^2\partial(\beta A_c/N)/\partial\rho)_{N,T},$$

$$= v_o\rho^2(\partial(\beta A_c/N)/\partial\rho)_{N,T,f(\Omega)} + \int d\Omega\{\delta(\beta A_c/N)/\delta f\}(df/d\rho)$$

$$= v_o\rho\{1 + v_o\rho + (2/3)(1+q-q^2/2)v_o^2\rho^2 + 2rv_o\rho[1+(1+2q)v_o\rho/3]\overline{|\sin\gamma|}\}/(1-v_o\rho)^3 \quad (32)$$

since $\delta(\beta A_c/N)/\delta f = 0$ and

$$\beta\mu_c = \beta G_c/N(\partial(\beta A_c)/\partial N)_{V,T} = (\partial(\beta A_c)/\partial N)_{V,T,f(\Omega)},$$

$$= \overline{\ln\{8\pi^2 f(\Omega)\}} + \ln\{\rho/(1-v_o\rho)\} + 6v_o\rho/(1-v_o\rho)$$

$$+ 4v_o\rho\{(1+q/4-q^2/8)v_o\rho + r(1+qv_o\rho/2)\overline{|\sin\gamma|}\}/(1-v_o\rho)^2 + \beta P v_o, \quad (33)$$

where S_c and μ_c are the configurational entropy and chemical potential, respectively.

Eq. (30) has the same mathematical form as the Onsager integral equation (16); namely,

$$\ln\{8\pi^2 f(\Omega)\} + \lambda^*(\rho,a,\ell)\int d\Omega'\,|\sin\gamma(\Omega,\Omega')|\,f(\Omega') = C, \quad (34)$$

where C does not depend on Ω. (λ^* is, of course, a different function of ρ, a, and ℓ in the two cases). From Lasher's results, therefore, we know that the Helmholtz free energy is minimized by the isotropic solution $f(\Omega) = 1/8\pi^2$ when the quantity $\lambda^* = 4rv_o\rho\{3-(1-q)v_o\rho\}/3(1-v_o\rho)^2$ is less than 8.88, but by the nematic solution when $\lambda^* > 8.88$. In order to locate the first-order transition between the isotropic (iso) and nematic (nem) phases for rods of length-to-breadth ratio $x = 1+\ell/2a$, it is necessary to solve the simultaneous equations

$$\beta Pv\ (x,v_o\rho_{nem},f_{nem}(\Omega)) = \beta Pv_o(x,v_o\rho_{iso},1/8\pi^2) \qquad (35)$$

and

$$\beta\mu_c(x,v_o\rho_{nem},f_{nem}(\Omega)) = \beta\mu_c(x,v_o\rho_{iso},1/8\pi^2), \qquad (36)$$

where ρ_{iso}, ρ_{nem} and $f_{nem}(\Omega)$ are the density of the isotropic phase, the density of the nematic phase, and the orientational distribution function in the nematic phase at the transition, and the functions βPv_o and $\beta\mu_c$ are defined by eqs. (32) and (33), respectively. Instead of fixing x and solving eqs. (30), (35) and (36) for $f_{nem}(\Omega)$, $v_o\rho_{nem}$ and $v_o\rho_{iso}$, it is more convenient to fix λ^* (and therefore f_{nem}) and to solve eqs. (35), (36) and

$$\lambda^* = 4rv_o\rho_{nem}\{3-(1-q)v_o\rho_{nem}\}/3(1-v_o\rho_{nem})^2, \qquad (37)$$

numerically for x, $v_o\rho_{nem}$ and $v_o\rho_{iso}$. The results of these computations are summarized in the figure, where the densities of the coexisting phases $v_o\rho_{nem}$ and $v_o\rho_{iso}$, and the nematic order parameter, $\bar{P}_2$, [18] at the transition are plotted versus x, the molecular length-to-breadth ratio.

The behaviour shown in the figure is typical of that predicted by the hard rod calculations to date; namely increases in x are accompanied by substantial decreases in the transition densities, substantial increases in the relative density discontinuity, $\Delta\rho/\rho_{nem}$, and smaller increases in the order parameter at the transition, $\bar{P}_2^{(NI)}$ Only for very short spherocylinders ($x < 2$) is $\Delta\rho/\rho_{nem}$ less than 1%, as appears to be the case for real nematogens. The predicted values of $\bar{P}_2^{(NI)}$ are considerably larger than experimental values, which are typically in the range 0.3 - 0.5. Moreover, the hard rod order parameter is independent of T at constant ρ, in disagreement with experiment [19].

Clearly, then, the hard rod fluid is not a very satisfactory model for nematogenic systems. This is not surprising, given such idealizations as (i) the absence of intermolecular attractions, (ii) the approximation of intermolecular repulsions by hard rod exclusions, (iii) the neglect of molecular flexibility, and (iv) the assumption of cylindrically symmetric molecules in the hard rod system. It is also probable that some of the

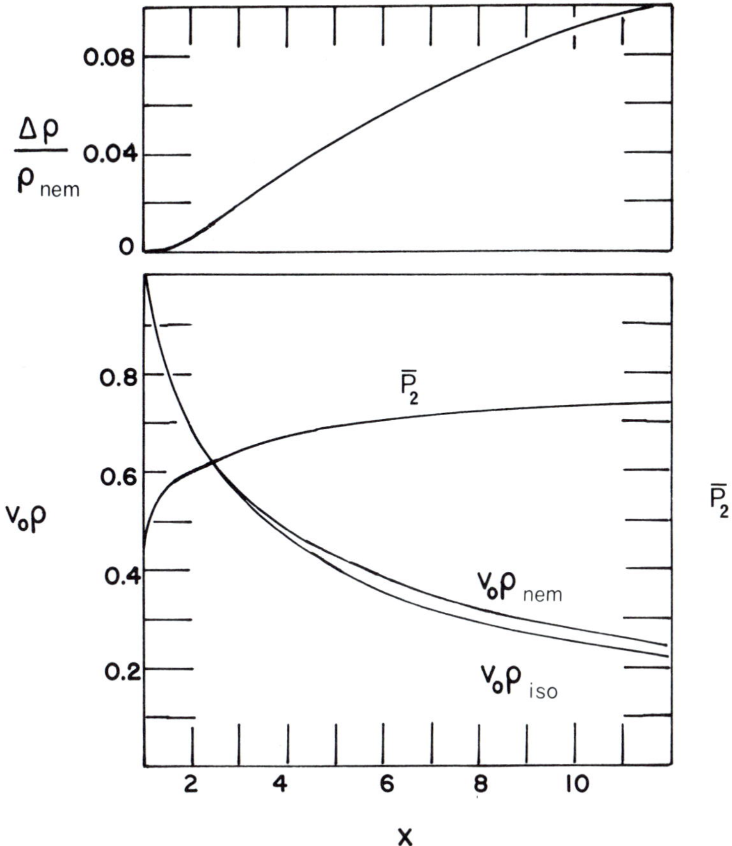

Fig. 1 *Reduced densities and order parameters,* $\bar{P}_2$*, at the nematic to isotropic phase transition as a function of the molecular length-to-breadth ratio x.*

discrepancy between theoretical predictions and experimental results is due to deficiencies in the scaled particle treatment, particularly the apparent neglect of short-range orientational correlations. Unfortunately, the only available check on the scaled particle method is Vieillard-Baron's [10] Monte Carlo calculation of the equation of state of an isotropic fluid of hard spherocylinders with x = 3. For this system, the agreement between the scaled particle and Monte Carlo predictions is quite good at low densities, not so good at relatively high densities. (At the highest value of $v_o\rho$ used in the Monte Carlo calculation, 0.54, scaled particle theory gives $(\beta P v_o)$ 24% higher than that obtained in the simulation). Due to difficulties in obtaining convergence, the Monte Carlo simulation could not be extended to higher densities; thus no nematic phase was observed.

Other Approaches

The hard rod fluid has also been considered by Alben [20], using an ingenious, although admittedly crude, free volume argument to obtain ϕ_N.

The results of this approach are qualitatively very similar to those of scaled particle theory.

Finally, systems of hard rods have also been studied using the Flory-DiMarzio lattice model [21], in which a rod-like molecule of length-to-breadth ratio x occupies x adjacent sites on a regular three-dimensional lattice. This approach is discussed in Chapter 11.

References

1. For a recent discussion of the roles of repulsive and attractive forces in liquids, see H.C. Andersen, D. Chandler, and J.D. Weeks, *Adv. Chem. Phys.* **34**, 105 (1976).
2. See, for example, H.C. Longuet-Higgins and B. Widom, *Mol. Phys.* **8**, 549 (1964).
3. For a recent review of thermodynamic perturbation theories (among other things) see J.A. Barker and D. Henderson, *Rev. Mod. Phys.* 48, 587 (1976).
4. R. Zwanzig, *J. Chem. Phys.* **39**, 1714 (1963).
5. More precisely, $f(\Omega)d\Omega$ is the fraction of molecules with orientations between Ω and $\Omega + d\Omega$ where $d\Omega = \sin\beta d\beta d\alpha d\gamma$. According to this definition, $f(\Omega) = 1/8\pi^2$ in an isotropic system since $\int d\Omega = \int_0^{2\pi} d\gamma \int_0^{2\pi} d\alpha \int_0^{\pi} \sin\beta d\beta$. This differs from the convention used in most discussions of the hard rod fluid, where it is assumed that the particles are cylindrically symmetric and only two Euler angles α and β (frequently called ϕ and θ, respectively) are used to specify molecular orientations. In this case, $\int d\Omega = \int_0^{2\pi} d\alpha \int_0^{\pi} \sin\beta d\beta = 4\pi$ and $f(\Omega) = 1/4\pi$ in an isotropic phase.
6. L. Onsager, *Ann. N.Y. Aca. Sci.* **51**, 627 (1949).
7. G. Lasher, *J. Chem. Phys.* **53**, 4141 (1970).
8. J.P. Straley, *Mol. Cryst. Liq. Cryst.* **24**, 7 (1973).
9. J.P. Straley, *Mol. Cryst. Liq. Cryst.* 22, 333 (1973).
10. J. Vieillard-Baron, *Mol. Phys.* **28**, 809 (1974).
11. H. Reiss, H.L. Frisch, and J.L. Lebowitz, *J. Chem. Phys.* 31, 369 (1959).
12. M.A. Cotter and D.E. Martire, *J. Chem. Phys.* 52, 1902, 1909 (1970); 53, 4500 (1970).
13. K.M. Timling, *J. Chem. Phys.* **61**, 465 (1974).
14. M.A. Cotter, *Phys. Rev.* A **10**, 625 (1974).
15. M.A. Cotter, *J. Chem. Phys.* **66**, 1098 (1977), Appendix.
16. See references 7, 14 and 15 for discussions of the difficulties in obtaining consistency.
17. M.A. Cotter, Doctoral Dissertation (Georgetown University, 1969), Appendix III; M.A. Cotter, unpublished results.
18. As is customary, θ is the angle between the long axis of a molecule and the nematic director. Assuming that the Z axes of the space-fixed and molecule-fixed coordinate systems coincide with the director and the molecular long axes, respectively, θ is just the Euler angle β. To be quite proper, therefore, $\bar{P}_2 = \int d\Omega P_2(\cos\beta) f(\Omega)$.
19. J.R. McColl and C.S. Shih, *Phys. Rev. Lett.* 29, 85 (1972).
20. R. Alben, *Mol. Cryst. Liq. Cryst.* 13, 193 (1971).
21. For a discussion of this model with references to previous work see M.A. Cotter, *Mol. Cryst. Liq. Cryst.* **35**, 33 (1976).

Chapter 8

THE VAN DER WAALS APPROACH TO NEMATIC LIQUIDS

MARTHA A. COTTER

Department of Chemistry, Rutgers, The State University, New Brunswick, New Jersey 08903

Introduction

In the van der Waals theory of liquids [1], it is assumed that the detailed structure of a liquid far from its critical point is determined largely by very short-ranged intermolecular repulsions (which may be approximated by hard particle repulsions), while the role of the longer-ranged intermolecular attractions is, to a first approximation, merely to provide a negative, spatially uniform mean field in which the molecules move. This approach has been applied to fluids of rod-like molecules by Alben [2], who used an orientation-independent mean field potential, and by Cotter [3] and Gelbart and Baron [4], who used orientation-dependent mean fields. Cotter considered a model system of hard spherocylinders moving in a spatially uniform mean field potential

$$\bar{\psi}(\Omega,\rho) = -\varepsilon_o\rho - \varepsilon_2\rho\bar{P}_2P_2(\cos\theta),$$

where ε_o and ε_2 are positive constants. Gelbart and Baron, in their generalized van der Waals (GVDW) theory of nematogens, considered a model system of rod-like molecules interacting via the pair potential

$$U(\mathbf{r}_{ij},\Omega_i,\Omega_j) = \underset{\text{hard rod}}{U(\mathbf{r}_{ij},\Omega_i,\Omega_j)} + \underset{\text{attractive}}{U(\mathbf{r}_{ij},\Omega_i,\Omega_j)},$$

$$\equiv U^*(\mathbf{r}_{ij},\Omega_i,\Omega_j) + U^{(a)}(\mathbf{r}_{ij},\Omega_i,\Omega_j) \qquad (1)$$

and derived a self-consistent expression for the mean field potential or pseudo-potential $\bar{\psi}(\Omega,\rho)$ in terms of $U^{(a)}$, the attractive part of the pair potential. The GVDW theory was subsequently rederived by Cotter [5] in a more straightforward manner, which is closely related to the usual derivation [6] of the van der Waals equation of state. The more compact formulation will be presented here.

Statistical Mechanics

Our model consists of N rod-like molecules in a volume V with the intermolecular pair potential (1). (The shape of the molecular hard cores will be specified later). The configurational partition function Q_N of this system is given by eq. (1) of Chapter 7, hereafter referred to as (7,1), except that

$$U_N = \sum_{i<j} U^*(r_{ij},\Omega_i,\Omega_j) + \sum_{i<j} U^{(a)}(r_{ij},\Omega_i,\Omega_j),$$
$$= U_N^*(\mathbf{r}^N,\Omega^N) + U_N^{(a)}(\mathbf{r}^N,\Omega^N). \quad (2)$$

With this substitution, moreover, eqs. (7,2) and (7,3) are also valid, i.e. in the maximum term approximation

$$Q_N = (\Delta\Omega/8\pi^2)^N\left(\prod_{\sigma=1}^{n} N_\sigma!\right)^{-1} \int\{d\mathbf{r}^N\}\exp\{-\beta U_N(\mathbf{r}^N,\tilde{N}_1,\tilde{N}_2,\ldots,\tilde{N}_n)\}, \quad (3)$$

where N_σ is the number of molecules with fixed orientation $(\Omega_\sigma\pm\Delta\Omega/2)$, and the tildes again denote the maximum term values. The GVDW approach then consists of (i) approximating $U_N^{(a)}(\mathbf{r}^N,\tilde{N}_1,\ldots,\tilde{N}_n)$ by its average value $\bar{U}_N^{(a)}(\tilde{N}_1,\ldots,\tilde{N}_n)$ in Q_N; i.e., assuming

$$Q_N \approx \Delta\Omega^N\exp(-\beta\bar{U}_N^{(a)})\{(8\pi^2)^N \prod_{\sigma=1}^{n} N_\sigma!\}^{-1}\int\{d\mathbf{r}^N\}\exp\{-\beta U_N^*(\mathbf{r}^N,\tilde{N}_1,\ldots,\tilde{N}_n)\} \quad (4)$$

and (ii) approximating $\bar{U}_N^{(a)}$ by

$$\bar{U}_N^{(a)} = \tfrac{1}{2}V^{-1}\sum_{\sigma\sigma'}\sum \tilde{N}_\sigma\tilde{N}_{\sigma'}\int d\mathbf{r}g_{\sigma\sigma'}^{(2)}(\mathbf{r})u^{(a)}(\mathbf{r},\Omega_\sigma,\Omega_{\sigma'}),$$

$$\approx \tfrac{1}{2}V^{-1}\sum_{\sigma\sigma}\sum \tilde{N}_\sigma\tilde{N}_\sigma \int d\mathbf{r}\ \exp\{-\beta u^*(\mathbf{r},\Omega_\sigma,\Omega_{\sigma'})\}u^{(a)}(\mathbf{r},\Omega_\sigma,\Omega_{\sigma'}), \quad (5)$$

where $\mathbf{r}$ is the separation vector and $g_{\sigma\sigma'}^{(2)}(\mathbf{r})$ the pair correlation function between two molecules with fixed orientations Ω_σ and $\Omega_{\sigma'}$. Substituting $\bar{U}_N^{(a)}$ from (5) into (4) and taking natural logarithms, we obtain

$$-N^{-1}\ \ln Q_N = \ln(8\pi^2\rho/\Delta\Omega) + \sum_\sigma s_\sigma \ln s_\sigma$$
$$- N^{-1}\ln[V^{-N}\int\{d\mathbf{r}^N\}\exp\{-\beta U_N^*\ (\mathbf{r}^N,\tilde{N}_1,\ldots,\tilde{N}_n)\}]$$
$$+ \tfrac{1}{2}\beta\rho\sum_{\sigma\sigma'}\sum\ s_\sigma s_{\sigma'}\int d\mathbf{r}\ \exp\{-\beta U^*(\mathbf{r},\Omega_\sigma,\Omega_{\sigma'})\}U^{(a)}(\mathbf{r},\Omega_\sigma,\Omega_{\sigma'}), \quad (6)$$

where $s_\sigma = \tilde{N}_\sigma/N$. Finally, writing $s_\sigma = f(\Omega_\sigma)\Delta\Omega$ and converting from sums back to integrals yields the GVDW Helmholtz free energy equation, first derived by Gelbart and Baron [4]:

$$\beta A_c/N = -N^{-1}\ln Q_N = \ln\rho + \int d\Omega f(\Omega)\ln\{8\pi^2 f(\Omega)\} - N^{-1}\ln[V^{-N}\int d\mathbf{r}^N\exp\{-\beta U_N^*(\mathbf{r}^N;f(\Omega))\}]$$

$$+ \tfrac{1}{2}\beta\rho\int d\Omega f(\Omega)\int d\Omega' f(\Omega')\int d\mathbf{r}\, \exp\{-\beta U^*(\mathbf{r},\Omega,\Omega')\}U^{(a)}(\mathbf{r},\Omega,\Omega'),$$

$$= \beta A_c^* \{f(\Omega)\}/N + \tfrac{1}{2}\beta\rho\int d\Omega f(\Omega)\int d\Omega' f(\Omega')\int d\mathbf{r}\exp\{-\beta U^*(\mathbf{r},\Omega,\Omega')\}U^{(a)}(\mathbf{r},\Omega,\Omega'), \quad (8)$$

where $A_c^*\{f(\Omega)\}$ is the free energy functional for a fluid of hard rods at temperature T and density ρ, and the orientational distribution function $f(\Omega)$ is to be determined by minimizing the right-hand side of eq. (7) subject to the normalization constraint (7,13).

As is clear from references [4] and [5], the approach just outlined is equivalent to considering a model system of molecules with hard-rod repulsions, each moving in a spatially uniform potential $\bar{\psi}(\Omega,\rho)$, where

$$\bar{\psi}(\Omega,\rho) = \rho\int d\Omega' f(\Omega')\int d\mathbf{r}\, \exp\{-\beta U^*(\mathbf{r},\Omega,\Omega')\}U^{(a)}(\mathbf{r},\Omega,\Omega'). \quad (9)$$

Eq. (8) can, therefore, also be written as

$$\beta A_c/N = \beta A^*_c\{f(\Omega)\}/N + (\beta/2)\int d\Omega f(\Omega)\bar{\psi}(\rho,\Omega). \quad (10)$$

The first derivatives of the free energy can also be expressed quite conveniently in terms of the pseudo-potential $\bar{\psi}$ and the properties of the hard rod fluid (with the orientational distribution function $f(\Omega)$); i.e.,

$$S_c/Nk = -\{\partial(A_c/Nk)/\partial T\}_{V,N,f(\Omega)} = -\{\partial(A_c^*/Nk)/\partial T\}_{V,N,f(\Omega)} = S_c^*/Nk, \quad (11)$$

$$\beta P = \rho^2\{\partial(\beta A_c/N)/\partial\rho\}_{N,T,f(\Omega)} = \beta P^* + (\beta\rho/2)\int d\Omega f(\Omega)\bar{\psi}(\Omega,\rho) \quad (12)$$

and

$$\beta\mu_c = \{\partial(\beta A_c)/\partial N\}_{T,V,f(\Omega)} = \beta\mu_c^* + \beta\int d\Omega f(\Omega)\bar{\psi}(\Omega,\rho). \quad (13)$$

The Pseudo-potential $\bar{\psi}(\Omega,\rho)$

Since $\exp(-\beta U^*)$ can have only two values, 0 (when $U^* = \infty$) and 1 (when $U^* = 0$), eq. (9) can be rewritten as

$$\bar{\psi}(\Omega,\rho) = \rho\int d\Omega' f(\Omega')\int_{\zeta(\Omega,\Omega')} d\mathbf{r}U^{(a)}(\mathbf{r},\Omega,\Omega'), \quad (14)$$

where $\zeta(\Omega,\Omega')$ is the set of all separation vectors $\mathbf{r}$ for which the hard cores of two molecules with orientations Ω and Ω' do not interpenetrate. If we assume, moreover, that

$$U^{(a)}(\mathbf{r}_{ij},\Omega_i,\Omega_j) = U^{(a)}_{iso}(\mathbf{r}_{ij}) + U^{(a)}_{aniso}(\mathbf{r}_{ij},\Omega_i,\Omega_j), \quad (15)$$

then

$$\bar{\psi}(\Omega,\rho) = \rho\int d\Omega' f(\Omega')\int_{\zeta(\Omega,\Omega')} d\mathbf{r}\{U^{(a)}_{iso}(\mathbf{r}) + U^{(a)}_{aniso}(\mathbf{r},\Omega,\Omega')\},$$

$$\equiv \rho\int d\Omega' f(\Omega')\{I_{iso}(\Omega,\Omega') + I_{aniso}(\Omega,\Omega')\}. \quad (16)$$

From eq. (16), it is clear that the isotropic or orientation-independent part of the attractive pair potential $U_{iso}^{(a)}$, as well as the orientation-dependent or anisotropic part $U_{aniso}^{(a)}$, will contribute to the orientation dependence of $\bar{\psi}$. Thus, even if the intermolecular attractions were completely isotropic, the effective potential felt by a single molecule would be orientation-dependent, because the hard rod repulsions build in short-range correlations between **r** and the angle $\gamma(\Omega,\Omega')$ between molecular long axes. This would no longer be true, however, if the rod-like hard cores were replaced by hard spheres, in which case the domain of integration ζ is independent of Ω and Ω' and the orientation dependence of $\bar{\psi}$ comes only from orientation-dependent terms in $U^{(a)}$.

Gelbart and Gelbart [7] recently used eq. (16) to characterize $\bar{\psi}(\Omega)$ for several systems with spherocylindrical hard cores and attractive pair potentials of the form

$$U^{(a)}(\mathbf{r}_{ij},\Omega_i,\Omega_j) = -\{C_{iso} + C_{aniso}\cos^2\gamma(\Omega_i,\Omega_j)\}/r_{ij}^6, \qquad (17)$$

where C_{iso} and C_{aniso} are positive constants and $\gamma(\Omega_i,\Omega_j)$ is the angle between the cylindrical axes of molecules i and j. For this choice of $U^{(a)}$, $f(\Omega)$ is cylindrically symmetric about the nematic director and

$$I_{iso} + I_{aniso} = \{1 + (C_{aniso}/C_{iso})\cos^2\gamma(\Omega,\Omega')\}I_{iso}(\gamma,a,\ell),$$

$$= \{1 + (C_{aniso}/C_{iso})\cos^2\gamma\}\int_{\zeta(\gamma,a,\ell)} d\mathbf{r}(-C_{iso}/r^6), \qquad (18)$$

where a and ℓ are the radius and cylindrical length, respectively, of the spherocylindrical hard cores. $\bar{\psi}(\Omega,\rho)$ can, therefore, be expanded in the familiar manner:

$$\bar{\psi}(\Omega,\rho) = A_o\rho + A_2\rho\bar{P}_2P_2(\cos\theta) + A_4\rho\bar{P}_4P_4(\cos\theta) + \ldots, \qquad (19)$$

where θ is the angle between the long axis of a molecule and the director, $\bar{P}_L = \int d\Omega f(\Omega)P_L(\cos\theta)$, and the coefficients A_L depend only on C_{iso}, C_{aniso}, a, and ℓ. Gelbart and Gelbart characterized $\zeta(\gamma,a,\ell)$ analytically, then evaluated the integral I_{iso} numerically and obtained the A_L's in terms of C_{iso}, for hard cores of length-to-breadth ratios 1 (i.e. hard spheres), 2, 3, and 4.2 and, in each case, $(C_{aniso}/C_{iso}) = 8^{-1}, 50^{-1}$ and 250^{-1}. They also estimated the relative magnitudes of successive terms in eq. (19), using Onsanger's [8] one-parameter representation of $f(\Omega)$; namely,

$$f(\Omega) = \xi\cosh(\xi\cos\theta)/8\pi^2\sinh\xi, \qquad (20)$$

to calculate the order parameters $\bar{P}_L$ (ξ was chosen to give $\bar{P}_2 = 0.52$). From these calculations, they found that (1) $\bar{\psi}$ is rather well approximated by the first two terms in eq. (19); higher order terms ($L \geq 4$) contribute less than 2% over the whole range $\cos\theta = -1$ to 1; (2) the coefficient A_2 is determined largely by the isotropic terms $-C_{iso}/r_{ij}^6$ and (3) A_2/A_o is of order 20% for length-to-breadth ratios in the range 2 to 4, even when $U^{(a)}$ is completely isotropic.

These results can be used (after the fact) to justify the pseudo-potential used in reference [3].

Results

As noted in the Introduction, Cotter [3] applied the van der Waals approach to a fluid of hard spherocylinders (once again with radius a and cylindrical length ℓ) in an anisotropic mean field with pseudo-potential

$$\psi(\Omega,\rho) = -\varepsilon_o\rho-\varepsilon_2\rho\bar{P}_2P_2(\cos\theta), \tag{21}$$

where ε_o and ε_2 are positive energy parameters and $\bar{P}_2 = \int f(\Omega)P_2(\cos\theta)d\Omega$. A_c, S_c, P, and μ_c were obtained from eqs. (10) - (13), using the scaled particle expressions (7,29), (7,31), (7,32), and (7,33) for A_c^*, S_c^*, P^*, and μ_c^* respectively, with one additional approximation, namely,

$$\overline{|\sin\gamma(\Omega,\Omega)|} \approx \pi/4-(5\pi/32)\bar{P}_2^2.$$

This is equivalent to expanding $|\sin\gamma(\Omega,\Omega')|$ in Legendre polynomials $P_L(\cos\gamma)$, averaging term by term, and then truncating the resulting series after its second term. It enormously simplifies subsequent numerical computations {by allowing the derivation of a closed-form expression for $f(\Omega)$} without qualitatively affecting the predictions of the model. (Quantitatively, it lowers the predicted values of the order parameter $\bar{P}_2$ for *hard* spherocylinders by roughly 0.10 to 0.15 while changing the densities of the coexisting phases at the transition only very slightly.) Substituting eq. (22) in eq. (10) and solving the now familiar variational equation

$$\delta\{\beta A_c/N-\nu\int f(\Omega)d\Omega\}/\delta f = 0 \tag{23}$$

yields

$$f(\Omega) = (4\pi^2Z)^{-1}\exp\{\Lambda\bar{P}_2P_2(\cos\theta)\}, \tag{24}$$

where

$$\Lambda(\rho,T) = 5\pi rv_o\rho\{1-(1-q)v_o\rho/3\}\{8(1-v_o\rho)^2\}^{-1} + \varepsilon_2\rho\beta,$$

(cf. (7,24) for the definitions of r, q, and v_o),

$$Z = \int_o^\pi d\theta\sin\theta\exp\{\Lambda\bar{P}_2P_2(\cos\theta)\}$$

and the order parameter $\bar{P}_2$ is calculated (iteratively) from the consistency condition

$$\bar{P}_2 = \int d\Omega f(\Omega) P_2(\cos\theta),$$

$$= Z^{-1} \int_0^{\pi} d\theta \sin\theta \exp\{\Lambda \bar{P}_2 P_2(\cos\theta)\} P_2(\cos\theta). \quad (25)$$

Extensive numerical computations were carried out for a system with molecular length-to-breadth ratio x = 3, hard core volume $v_o = 230$ Å^3, and energy parameters $(v_o k)^{-1} \varepsilon_o = 25000$ K, $(v_o k)^{-1} \varepsilon_2 = 2000$ K. The results were compared with available experimental data for 4,4'-dimethoxy-azoxybenzene (PAA). (x = 3 and $v_o = 230$ Å^3 are estimates for the length-to-breadth ratio and molecular volume of PAA made by Vieillard-Baron [9], using tabulated van der Waals radii and bond lengths). Quite satisfactory qualitative agreement between theory and experiment was obtained; i.e. the model system exhibits order parameter versus temperature curves (at constant P or constant ρ) of roughly the correct shape, nearly linear plots of lnT versus lnρ at constant $\bar{P}_2$ with slopes close to 4 (in agreement with the data of McColl and Shih [10]), a first-order nematic-isotropic phase transition, increases in T_{NI} with increasing pressure of the correct order of magnitude, and large pretransitional increases in the compressibility, expansivity, and specific heat as T_{NI} is approached from below. The only significant effects not predicted are the premonitory phenomena observed experimentally in the isotropic phase above T_{NI}. On the other hand, satisfactory quantitative agreement with experiment was certainly not achieved, particularly with respect to the properties of the NI transition, as can be seen from the table. With the particular choice of ε_o and ε_2 [11], quantitatively correct results were obtained for T_{NI} and for the ratio

$$\Gamma = -(\rho/T)\{(\partial \bar{P}_2/\partial \rho)_T / (\partial \bar{P}_2/\partial T)_\rho\},$$

which measures the relative sensitivity of the order parameter to changes in density versus changes in temperature. However, the predicted values of the relative density discontinuity, entropy of transition, and slope of the P-T coexistence curve are too large by factors of roughly 10, 5, and 3½, respectively. Furthermore, the mean reduced density at the transition $v_o\bar{\rho}$ is too small and $\bar{P}_2^{NI}$ is too large. (It would be larger still were it not for the approximation (22)).

These quantitative deficiencies probably result primarily from two factors:

(1) the use of a rigid spherocylinder for the molecular shape. This ignores the flexibility and asymmetry of real nematogens. It can be compensated for to some extent by decreasing the hard core length-to-breadth ratio x. (In fact, much better, though still not

quantitative, agreement with the PAA data can be achieved using $x = 2$ to 2.5.), and

(2) the inadequate treatment of short-range order. Obviously, short-range orientational and translational correlations resulting from intermolecular attractions are completely neglected. Furthermore, as was noted previously, the effects of short-range *orientational* correlations at least are probably seriously underestimated in the scaled particle treatment of the molecular hard cores. The resulting deficiencies in A_c^*, P^*, etc. will clearly be magnified by applying the mean field potential.

Table. *Comparison of the nematic-isotropic (NI) phase transitions in PAA and in the model system.* $\bar{P}_2^{NI}$ *and* Γ_{NI} *are the values of* $\bar{P}_2$ *and* Γ *in the nematic phase at* T_{NI} *and* $\bar{\rho} = (\rho_{nem} + \rho_{iso})/2$.

Quantity	Predicted Value	Observed[a] Value
T_{NI}	410.4 K	409 K
$\bar{P}_2^{NI}$	0.5424	∿ 0.40
Γ_{NI}	3.90	∿ 4.0
$(dT_{NI}/dP)_{P=1\ bar}$	175 K/kbar	48 K/kbar
$v_o\bar{\rho}$	0.445	0.62
$\Delta\rho_{NI}/\rho_{nem}$	0.040	0.0035
$\Delta S_{NI}/Nk$	0.887	0.17

[a]For the references from which these values were taken, see reference [3]

In addition to this van der Waals calculation, Baron and Gelbart [13] have recently applied their generalized van der Waals theory to systems with spherocylindrical hard cores and attractive forces described by eq. (17) using (7,29) for A_c^* and approximating $f(\Omega)$ by the Onsager representation (20). Qualitatively, they observed some interesting trends in T_{NI}, ΔS_{NI}, the transition densities, etc. as the hard core volume v_o, the length-to-breadth ratio x, and the van der Waals coefficient C_{iso} (cf. 17) were varied. Quantitatively, however, their use of eq. (20) introduces substantial errors, as can be seen by comparing their values of $\bar{P}_2^{NI}$, $v_o\rho_{nem}$, and $v_o\rho_{iso}$ for hard spherocylinders at x = 3 with figure 1 in Chapter 7. In my opinion further calculations using a more accurate representation of f(Ω) are needed. It must be emphasized, however, that the GVDW approach in which $\bar{\psi}$ is determined from the model pair potential, is clearly superior to the use of an essentially phenomenological pseudo-potential such as (21).

Discussion

Despite its quantitative inadequacies, the van der Waals approach does, in my opinion, tell us something about nematic liquid crystals. First and foremost, it indicates that the anisotropy of the short-range intermolecular repulsions plays a major role in determining nematic order and stability and cannot be neglected, even to a first approximation. For model systems with hard cores plus attractions, anisotropic hard cores are clearly necessary in order to explain the behaviour of nematogens in even a qualitatively satisfactory manner. For example, if the spherocylindrical hard cores in references [3] and [13] are replaced by hard spheres, the model cannot explain the following:

(1) the order parameter data of McColl and Shih [10]. A system of hard spheres with a mean field potential

$$\bar{\psi}(\Omega,\rho) = A_o\rho + A_2\rho\bar{P}_2P_2(\cos\theta) + A_4\rho\bar{P}_4P_4(\cos\theta) + \ldots, \quad (26)$$

will predict that the parameter Γ is equal to one, no matter what the values of the A_L's since the coefficient of $\bar{P}_2P_2(\cos\theta)$ is directly proportional to ρ. Moreover, one can show that eq. (26) is the only pseudo-potential of the form

$$\bar{\psi}(\Omega,\rho) = \bar{\psi}_o(\rho) + \bar{\psi}_2(\rho)\bar{P}_2P_2(\cos\theta) + \bar{\psi}_4(\rho)\bar{P}_4P_4(\cos\theta)+\ldots \quad (27)$$

which maintains statistical mechanical self-consistency [14]. Even if one chooses to ignore this consistency requirement, it is necessary to make the physically unreasonable assumption $\bar{\psi}_2(\rho) \propto \rho^4$ in order to obtain Γ = 4. With spherocylindrical hard cores, on the other hand, one can easily obtain Γ = 4 without producing inconsistency or introducing any unphysical assumptions;

(2) the observed values of T_{NI}. As was mentioned before, when the molecular hard cores are spherical, the orientation-dependent part of $\bar{\psi}$, $\bar{\psi}^{(a)}$, arises solely from the orientation dependence of the intermolecular attractive forces, presumably dispersion forces.

From rough estimates of $U^{(a)}_{aniso}/U^{(a)}_{iso}$ for typical nematogens [15,16], however, it appears that the contribution to $\bar{\psi}^{(a)}$ from the anisotropy of the dispersion forces is at least an order of magnitude too small to explain experimental values of T_{NI}. No such problem arises with spherocylindrical hard cores, since Gelbart and Gelbart [7] have shown that the main contribution to $\bar{\psi}^{(a)}$ comes from the isotropic attractions, $U^{(a)}_{iso}$, coupled to the hard rod exclusions, and there is no difficulty in obtaining large enough values of the coefficient A_2 in eq. (26).

Secondly, references [3], [4], [7] and [13], taken together, strongly suggest that the dominant factors governing nematic order and stability are the anisotropy of the short-range intermolecular repulsions and the overall strength of the intermolecular attractions, while the anisotropy of the attractive forces plays a somewhat secondary role.

References

1. See, for example, M. Kac, G.E. Uhlenbeck, and P.C. Hemmer, *J. Math. Phys.* 4, 216 (1963); J.L. Lebowitz, G. Stell, and S. Baer, *J. Math. Phys.* 6, 1282 (1965); and H.C. Longuet-Higgins and B. Widom, *Mol. Phys.* **8**, 549 (1964). In the last cited article, the theory was used very successfully to predict the properties of argon at its triple point.
2. R. Alben, *Mol. Cryst. Liq. Cryst.* 13, 193 (1971) (section on the "Steric Model", pp. 214-217).
3. M.A. Cotter, *J. Chem. Phys.* 66, 1098 (1977).
4. W.M. Gelbart and B.A. Baron, *J. Chem. Phys.* 66, 207 (1977).
5. M.A. Cotter, *J. Chem. Phys.* 66, 4710 (1977).
6. M. Kac, G.E. Uhlenbeck, and P.C. Hemmer, *J. Math. Phys.* 4, 216 (1963).
7. W.M. Gelbart and A. Gelbart, *Mol. Phys.* 33, 1387 (1977).
8. L. Onsager, *Ann. N.Y. Acad. Sci.* 51, 627 (1949).
9. J. Vieillard-Baron, *Mol. Phys.* 28, 809 (1974).
10. J.R. McColl and C.S. Shih, *Phys. Rev. Lett.* 29, 85 (1972).
11. ε_0 and ε_2 were chosen primarily to reproduce the observed T_{NI} and Γ_{trans}. However, $(v_0k)^{-1}\,\varepsilon_0$ = 25000 K is essentially what one would guess *a priori* from the estimated heat of vaporization of PAA (76 kJ mol^{-1}[12]).
12. J.F. Solsky and E. Grushka, *J. Phys. Chem.* **78**, 275 (1974).
13. B.A. Baron and W.M. Gelbart, *J. Chem. Phys.* 67, 5795 (1977).
14. M.A. Cotter, *Mol .Cryst. Liq.Cryst.* 39, 173 (1977); *J.Chem.Phys.* **67**, 4268 (1977). In the latter, it is shown that $\bar{\psi}(\Omega,\rho)$ must equal $\rho(\partial\bar{\psi}/\partial\rho)_{T,f(\Omega)}$ in order to achieve statistical mechanical consistency, whatever the shape of the hard cores or the assumed orientation dependence of $\bar{\psi}$.
15. J.I. Kaplan and E. Drauglis, *Chem. Phys. Lett.* 9, 645 (1971).
16. A. Wulf, *J. Chem. Phys.* 64, 104 (1976).

Chapter 9

COMPUTER SIMULATIONS

CLAUDIO ZANNONI

Department of Chemistry, The University,
Risorgimento 4, 40136 Bologna, Italy

Introduction

The problem of calculating the thermodynamic observables of a fluid from a given intermolecular potential is so complex that, apart from a few exceptional cases, there are only two possibilities open: one is to use approximate theories, the other to resort to computer simulations [1-6]. These latter techniques, which are the subject of this Chapter, consist of numerical solutions to the problem of many interacting particles. Let us start by discussing some of the advantages and limitations of this numerical approach before becoming involved in the details of the two techniques (Monte Carlo [1] and molecular dynamics [2]) used almost universally in computer simulations.

In many respects simulations represent a theoretician's dream of what an experiment should be. A potential is fed into the computation and comparison can then be made between the essentially exact results produced by the simulation with those produced by theory. One can use convenient potentials such as those with hard cores with their discontinuities, which are obviously absent in real systems. In addition, with computer simulations we have the possibility of calculating quantities not readily accessible or perhaps not available at all from experiment. For instance we can calculate not only thermodynamic observables such as energy, specific heat *etc.* but also distribution and correlation functions. In some cases we can also evaluate separately the contributions provided by different terms in the intermolecular potential. We can also control certain experimental conditions such as temperature and volume with great accuracy. We might also extend these experimental conditions to extreme values difficult to attain in the laboratory.

However, nothing is ever perfect, and there are some problems associated with simulation techniques. Perhaps the major one is that only a limited number of particles, typically from a few hundred to a few thousand in the most favourable cases, can be studied. Now several hundred particles does seem a small number indeed when compared to the number, of the order of 10^{23}, present in a macroscopic system. A legitimate question is then if we can compare at all the results obtained via computer simulations with those of real experiments. This problem has, of course, received considerable attention. In particular, simple fluids and spin systems have been studied in detail. It turns out that in most practical cases bulk properties can be obtained from samples composed of a few tens to a few hundred molecules [4] if, as we shall see later, free surfaces are avoided by using periodic boundary conditions.

The situation is slightly more complicated if we want to study the properties of a system near a phase transition [5]. One of the problems is that for a finite number of particles there can be no true phase transition (cf. Chapter 3). In addition, since near a critical point the range of correlations diverges we can expect fluctuations of correlation length comparable to the dimensions of the sample itself. Thus in the proximity of a transition, finite size effects might be particularly important. Apart from this, simulations near a critical point are in many respects similar to real experiments, where a long time is needed to reach equilibrium. Hysteresis phenomena are also difficult to avoid, the system being susceptible to small amounts of impurities, *etc*. Fortunately, the existence of exactly soluble models in one and two dimensions affords a valuable way of checking the reliability of computer simulations even in these circumstances. Extensive studies of the Ising model with a varying number of particles have been performed [5,6]. If, for example, the change of phase is monitored by the temperature dependence of the specific heat, the simulations show a peak which, as the number of particles, N, increases, tries to mimic the divergence predicted by Onsager [7]. In contrast the simulation of a one dimensional system of rotors, where it is known [8] that no transition exists gives results for the specific heat that do not show this sharpening of the peak as N increases [9]. By the strength of these and other experiences [5] we feel justified in saying that, taking the due precautions, computer simulations can be a useful aid in studying properties not only far from a phase transition but also in its neighbourhood.

To turn now to the specific problems of computer simulations in liquid-crystalline systems it must be said that these are still very much in their infancy and the number of papers published to date [10-15] on the subject does not indicate its importance. We think that this is due in part to the unfamiliarity of many people involved in liquid crystal research with computer simulation techniques and, *vice versa*, because computer simulators are unaware of the important problems in liquid crystal physics. Here we wish to make an attempt to bridge the gap between the two parties. We shall spend, therefore, some time illustrating the Monte Carlo and molecular dynamics techniques and also discuss the calculation of some of the quantities of interest in liquid crystal theory. Some results will be given for the order parameters and correlations discussed in Chapters 3 and 4.

The Monte Carlo Method

Monte Carlo methods can be defined in general as "that branch of experimental mathematics which is concerned with experiments on random numbers"(Hammersley and Handscomb [16]). In liquid state physics, however, the term is now universally reserved for the technique devised by Metropolis *et al.* [1] to evaluate statistical averages. Consider a system whose potential energy is known, say

$$U_N = (1/2) \sum_{i \neq j} U(X_i, X_j) \; ; \; i,j = 1,\ldots,N. \tag{1}$$

In the canonical ensemble (constant N, V, T) any time independent configurational property of interest can be written as the average

$$\langle A \rangle = \int \{d\mathbf{X}^N\} A \exp(-\beta U_N)/Z_N, \tag{2}$$

where we use the notation established in Chapter 3. Since the intermolecular potential is assumed to be known, the problem of calculating $\langle A \rangle$ is reduced to performing a 6N dimensional integration. Now imagine a real system and some way of taking configurational photographs in configurational space. From the j-th configuration, i.e. from the set of positions and orientations specifying the state of the system in this j-th photograph, we could calculate the value of property A; call this $A^{(j)}$. We could repeat the process M times, say. It is clear that the average (2) would now be

$$\langle A \rangle = (1/M) \sum_{j=1}^{M} A^{(j)}, \tag{3}$$

where, of course, M should be large enough to reduce the statistical uncertainty in $\langle A \rangle$ to acceptable values. In this *gedanken* experiment the substitution of integral (2) with the simple sum (3) is, clearly, possible since in nature every configuration appears automatically weighted with a probability given by the Boltzmann factor $\exp(-\beta U_N)$. The Metropolis *et al.* technique consists of introducing a stochastic process in which asymptotically (for indefinitely long chains) each configuration recurs with a frequency proportional to the Boltzmann factor for that state. To see how this can be possible let us now recall briefly some results from the theory of discrete Markov processes [17] that is, stochastic processes with memory of only the preceding step. The probability that a system, whose evolution is Markovian, is in a state k at time step t depends only on the state at time (t-1). Thus a discrete Markov process is completely described in terms of a transition probability matrix whose elements p_{ij} have the following intuitive properties:

$$p_{ij} \geq 0, \tag{4}$$

$$\sum_j p_{ij} = 1. \tag{5}$$

An n-step transition probability is defined by the recurrence relation

$$p_{ij}^{(n)} = \sum_k p_{ik}^{(n-1)} p_{kj}. \tag{6}$$

It is known that if every state can be reached from any other state with a certain sequence of steps (the system is ergodic) then the limit

$$\lim_{n\to\infty} p_{ij}^{(n)} = w_j \tag{7}$$

exists for every pair i,j and is independent of the starting state i. In addition it is found that the asymptotic frequency factors w_j obey the relations

$$w_j > 0, \tag{8}$$

$$\sum_j w_j = 1, \tag{9}$$

and, from eqs. (6) and (7)

$$w_j = \sum_k w_k p_{kj}. \tag{10}$$

In our case we know what the asymptotic frequency factors should be; if every configuration represents a state of a Markov chain, then we require

$$w_j = \exp(-\beta U^{(j)})/Z_N, \tag{11}$$

where $U^{(j)}$ is the energy of the j-th configuration.

Eqs. (8) and (9) are automatically satisfied by this choice of w_j and our problem is to select a transition matrix p_{kj} so as to obey eq. (10). One possible way of achieving this is to impose the condition of microscopic reversibility, that is

$$w_k\, p_{kj} = w_j\, p_{jk}, \tag{12}$$

whatever j and k. In this case eq. (10) follows from

$$\sum_k w_k\, p_{kj} = \sum_k w_j\, p_{jk} = w_j \sum_k p_{jk},$$

$$= w_j,$$

where the last equality is obtained with the aid of eq. (9). From eq. (12) we find now the constraint on the transition probabilities

$$p_{jk}/p_{kj} = \exp\{-\beta(U^{(k)}-U^{(j)})\}. \tag{13}$$

Eq. (13) still does not specify a unique Markov process. The prescription proposed by Metropolis *et al.* [1] is to choose

$$p_{jk} = a_{jk}\,, \quad \text{if } U^{(k)} \leqslant U^{(j)},$$

$$p_{jk} = a_{jk}\, \exp\{-\beta(U^{(k)}-U^{(j)}\}, \text{ if } U^{(k)} > U^{(j)}$$

and

$$p_{jj} = 1 - \sum_{k\neq j} p_{jk}, \tag{14}$$

where a_{jk} are constants and $a_{jk} = a_{kj}$. It is easily verified that this transition probability obeys eqs. (11-13).

In practice the process is realized rather simply by moving one particle at a time in the following way. A starting configuration is chosen, typically with the N particles in a box with periodic boundary conditions which are described later and the energy of this configuration, $U^{(j)}$, is calculated. One particle is then chosen, either sequentially or at random, and a new configuration is generated by giving this particle a random displacement δX. The energy, $U^{(k)}$, of this trial configuration is calculated, and if $(U^{(k)}-U^{(j)}) \leqslant 0$ the new configuration is accepted. If $(U^{(k)}-U^{(j)}) > 0$, the move is accepted with the relative probability $\exp\{-\beta(U^{(k)}-U^{(j)})\}$. In practice a random number uniformly distributed between 0 and 1 is generated, if this is less than $\exp\{-\beta(U^{(k)}-U^{(j)})\}$ the move is accepted, if not the original configuration is restored and counted again. This procedure is then repeated, typically for a few hundred thousand to a few million times to equilibrate the system. When equilibrium has eventually been reached the new configurations generated can be used to calculate averages according to eq. (3).

It is worth stressing, perhaps, that the time involved in proceeding along the Monte Carlo Markov chain has nothing to do with real time. In the Monte Carlo method the true trajectory of the system in phase space is replaced by an artificial Markovian trajectory chosen by us. However, it is sometimes useful to point out the formal analogy of the Monte Carlo method with an (artificial) dynamic method [5]. For example, we shall be able to use the same methods of calculation for equilibrium thermodynamic quantities: order parameters *etc.* both for the Monte Carlo method and the molecular dynamics technique to be described later. The difference with the molecular dynamics method is that there the true dynamics ar followed while in Monte Carlo all we can say is that, by construction, the process will, after a sufficiently large number of steps, lead to equilibrium, in the sense that configurations will occur with a frequency proportional to their Boltzmann factors. In practice it is hardly possible to gauge *a priori* how large this number of steps will be, (see, however, Ref.[5]). We expect the convergence to depend in some way on how efficiently we sample the configurational space and so it is clear that we want to reach some balance between the number of configurations accepted and rejected. This acceptance ratio can in turn be affected by the magnitude of the random displacement, $\delta\mathbf{X}$, given to the particle. Since the maximum jump length $\delta\mathbf{X}$ is not dictated by the method it can be adjusted to speed up convergence. As a common rule of thumb the maximum displacement is chosen [1b] so that approximately half of the configurations are accepted and half rejected.

We have now reached a stage where it would be beneficial to consider some aspects of the calculations in detail. Thus next we shall comment on some general points such as periodic boundary conditions, the choice of the initial configuration and the calculation of thermodynamic observables. We shall then discuss the calculation of order parameters and orientational pair correlation functions.

Boundary Conditions

Even though we are always forced to study a finite and relatively small system of interacting particles, we are not normally interested in the properties of that small sample itself but rather in predicting those of the corresponding macroscopic system. In doing this we are hampered not only by having to deal with the effect of a limited number of particles but also with surface effects. In order to reduce the latter we do not normally use free boundary conditions, which correspond to an isolated system of N particles, but instead employ the common artefact of periodic boundaries where the sample box is surrounded by exact replicas of itself. Every particle at a position (x,y,z) in a box of side length λ will have 26 ghost images at $(x\pm\lambda,y,z)$, $(x,y\pm\lambda,z)$, $(x,y,z\pm\lambda)$, *etc.* These are in turn surrounded by similar images, *ad infinitum*. The space filling system obtained in this way does not have free surfaces at all so we have remedied that part of the problem. Notice, however, that a spurious periodic correlation between particles has been introduced.

An appreciation of this extra correlation factor can help us in judging the effect of various boundary conditions on a phase transition. Consider, for example, [6], a lattice model where the correlation between sites is obviously greatest for periodic boundary conditions and a minimum for a system with free surfaces. We expect the infinite lattice case to lie somewhere in between these two extremes. Consequently periodic and free boundaries should give, respectively, an upper and lower bound for the transition temperature T_K:

$$T_{free} < T_K < T_{periodic}.$$

In the same way we would expect order parameters obtained with periodic boundaries to be greater than those calculated for a system with free surfaces at the same temperature. The predicted trend is actually borne out by Monte Carlo simulations on the two and three dimensional Ising model [5,6]. Increasing the number of particles has the effect of restricting the upper and lower bounds; thus for instance, the peak in the specific heat shifts to higher temperatures for the isolated system and to lower temperatures for periodic boundaries.

When using periodic conditions the distance between two different particles, i and j, is usually taken as the distance between i and the nearest image of j (minimum image convention). Thus every particle is the centre of an identical box and accordingly a given molecule i interacts only with the image of another molecule j which is the nearest. If the range of the molecular interactions is less than $\lambda/2$ this comprises all interactions. If the intermolecular potential is very long range, however, the minimum image convention for calculating the interaction energy is often abandoned [18]. Instead, although we shall not discuss it here, the complete, space filling system of the box and all its images is considered and the Ewald techniques [18], originally developed for lattice summations, employed. Still another method in use for long-range interactions is to truncate the intermolecular potential at a given cut-off distance and to take into account the long-range tail of the interaction using perturbation theory.

Influence of the Initial Configuration

We have shown that, whatever the initial configuration, the Monte Carlo procedure will, after an infinite number of steps, converge to equilibrium. Thus in principle any configuration e.g. a completely ordered, completely disordered or an intermediate case, can be chosen as a starting point. However, since we always perform a finite number of steps it is unwise to overlook completely the choice of starting configuration. For instance, in simulations of simple fluids it is thought [19-21] that one cannot normally start from an isotropic fluid and generate a crystal structure, even if this would be the thermodynamically stable state at the temperature and density chosen. If we want our system to reach a given region of phase space in a reasonable number of steps and hence computer time then the starting point is not immaterial. To visualize this, let us consider the situation [20] sketched in figure 1 and imagine that the system finds itself in region A of the configuration space, while we would like it to reach region B. This is clearly not impossible, since there exists an open path from A to B, but it is easy to convince ourselves that the process may well require a very long time due to the bottleneck shape of the pathway. A non-equilibrium state like A having a relatively long lifetime is sometimes called a metastable state. The problem can be particularly important in simul-

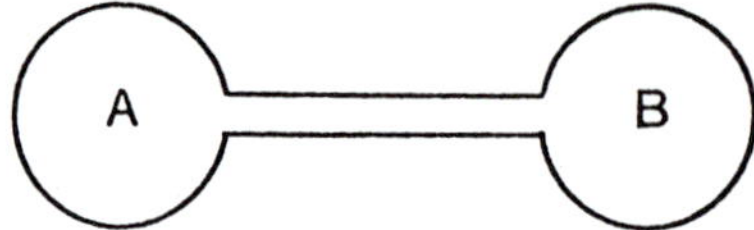

Fig. 1 *Two hypothetical regions of configurational space joined by a bottleneck pathway.*

ations of liquid crystals formed of hard elongated particles such as spherocylinders. In this case configurations like the one shown in figure 2 may prove difficult to unlock.

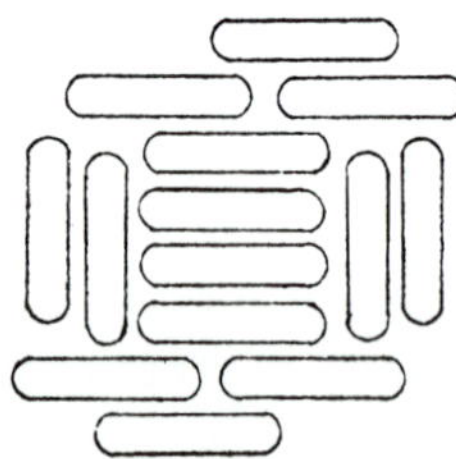

Fig. 2 *A relatively stable configuration for a system of elongated particles.*

Thermodynamic Observables

The thermodynamic properties can be calculated in principle from their expressions as statistical averages introduced in Chapter 3. The energy is the simplest to calculate since its evaluation is already part of the Monte Carlo prescription. The pressure can be obtained from the average virial for differentiable potentials or from the pair distribution at contact point for hard cores. We now wish to comment briefly on some other observables: the specific heat, the free energy and the order parameters.

Specific Heat The constant volume specific heat, C_V, is defined as

$$C_V = (\partial U/\partial T)_V. \tag{15}$$

Thus we can calculate the internal energy for a series of temperatures, interpolate and perform a numerical differentiation of U(T) to find $C_V(T)$. An alternative procedure consists in determining average energy fluctuations. From Chapter 3 we have

$$C_V/k = \beta^2\{<U^2>-<U>^2\},$$

or [6]

$$C_V/k = \beta^2[(1/M)\sum U^{(j)^2}-\{(1/M)\sum U^{(j)}\}^2]. \tag{16}$$

The dimensionless quantity on the right hand side of eq. (16) can be obtained directly from the simulation. However since it is a fluctuation quantity we expect it to be affected by large errors.

Free Energy In the canonical ensemble the relevant free energy is the Helmholtz function

$$A = U - TS.$$

This is a rather difficult quantity to evaluate, because of the presence of the entropy term. The fact is that quantities easily calculated with Monte Carlo methods are average quantities and we do not have such an expression for A. None-the-less we can actually make some progress in this direction. Given a quantity F we can write its average as

$$<F> = (1/Z_N)\int\{dX^N\}F(\{X^N\})\exp[-\beta U(\{X^N\})]. \tag{17}$$

Thus, if we choose $F \equiv \exp[+\beta U(\{X^N\})]$ we have [22]

$$Z_N = (\int\{dX^N\})/<\exp[+\beta U(\{X^N\})]>. \tag{18}$$

The partition function can therefore be calculated, in principle, as an average and the free energy obtained from

$$A = -kT(\ln Z_N-\ln N!). \tag{19}$$

Unfortunately only very poor estimates of Z_N can be obtained by the method when the system is somewhat ordered. In this case, in fact, important contributions to the summation come from configurations of high energy, with a small Boltzmann factor, which are very poorly sampled by the Monte Carlo procedure. This results in very slow convergence except when the system is relatively disordered.

A common method of evaluating the free energy exploits an interpolating relation obtained from the Gibbs-Helmholtz equation

$$U(\beta) = \{\partial(\beta A)/\partial\beta\}_V, \tag{20}$$

which gives on integration

$$\int_{\beta_1}^{\beta_2} d\beta U(\beta) = \beta_2 A(\beta_2) - \beta_1 A(\beta_1), \tag{21}$$

that is

$$A(\beta_2) = (\beta_1/\beta_2)A(\beta_1) + (1/\beta_2)\int_{\beta_1}^{\beta_2} d\beta U(\beta). \tag{22}$$

Thus if the free energy at a temperature $T_1 = (1/k\beta_1)$ is known together with the temperature dependence of the internal energy in the interval $T_1 \leqslant T \leqslant T_2$ the free energy at the new temperature T_2 can be calculated from eq. (22). This method was used by Lebwhol and Lasher in their simulation of the Maier-Saupe lattice model [11]. They started the interpolation on the two sides of the transition by a high and low temperature analytic series expansion, while of course $U(\beta)$ is obtained from the Monte-Carlo calculation. It might be possible in some cases to use the method previously described to start the iteration procedure at high temperature without having to recur to series expansions.

Similarly, one gets for the free energy as a function of density

$$A(\rho_2) = A(\rho_1) + N\int_{\rho_2}^{\rho_1} d\rho \rho^{-2} P(\rho). \tag{23}$$

The free energy is calculated first at a very low density from, say, a virial expansion. Thermodynamic integration of the density dependence of the pressure can then be used to extrapolate the free energy to the high densities of interest.

Other methods have been proposed recently by Valleau *et al.* [18] for calculating free energy differences between the system of interest, characterized by a configurational potential energy U, and a reference system with potential energy U_o; that is

$$\beta(A-A_o) = -\ln\langle\exp\{-\beta(U-U_o)\}\rangle_o,$$

where the angular brackets indicate the canonical ensemble average over the reference system.

Orientational order parameters

The calculation of orientational order parameters is of particular importance in computer simulations of model liquid crystals. Here we consider for simplicity a uniaxial mesophase formed of cylindrically symmetric particles. The second rank order parameter $\overline{P}_2$ is given by

$$\overline{P}_2 = \int_0^{\pi} d\beta \sin\beta f(\beta) P_2(\cos\beta) / \int_0^{\pi} d\beta \sin\beta f(\beta), \tag{24}$$

where β is the angle between the molecular symmetry axis and the director. Because the order parameter is not a scalar and since we do not work in a director frame but in an arbitrary laboratory frame the calculation of $\bar{P}_2$ and of the other order parameters may well not be a trivial matter. It is helpful to distinguish three possible situations:

(a) the director orientation is known. As an example the director could be fixed along the z direction by an external field. This can be applied by adding to the potential energy a term

$$U_{ext} = -\zeta \sum P_2(\cos\beta_i), \tag{25}$$

where the positive coupling parameter ζ measures the strength of interaction with the field and β_i measures the angle between the axis of the i-th molecule and the field. In this rather special case $\bar{P}_2$ can be simply calculated from an average over M equilibrium configurations. We have $\bar{P}_2 = \langle P_2 \rangle$,

$$\langle P_2 \rangle = (1/M) \sum_{j=1}^{M} P_2^{(j)}, \tag{26}$$

where

$$P_2^{(j)} = (1/N) \sum_{i=1}^{N} P_2(\cos\beta_i) \tag{27}$$

is the order parameter computed for the j-th configuration and we use angular brackets to indicate the Monte Carlo estimate of $\bar{P}_2$. In this particular case it may be even simpler to calculate a histogram for the singlet orientational probability $f(\beta)$ and subsequently determine all the desired order parameters $\bar{P}_L$ by integration. However, for a system of a few thousand particles it is likely that the symmetry-breaking field in eq. (25) has to be relatively large to have an appreciable effect on the director. Unfortunately a large field does in turn affect the transition, in the sense of stabilizing the ordered phase. This effect has been observed by Viellard-Baron [12] in his simulation of a two dimensional system of hard ellipses;

(b) the orientation of the director is unknown but director fluctuations are negligible. The order is calculated by maximizing the expression

$$P_2' = (1/N) \sum_{i=1}^{N} P_2(\mathbf{q}_i \cdot \mathbf{n}) , \tag{28}$$

with respect to the unit vector $\mathbf{n}$. The unit vector

$$\mathbf{q}_i \equiv (\sin\beta_i \cos\alpha_i, \sin\beta_i \sin\alpha_i, \cos\beta_i),$$

defines the orientation of the i-th molecule in the laboratory frame. The director $\mathbf{n}$ is the unit vector which renders P_2' a maximum. It is convenient to proceed and determine $\mathbf{n}$ in cartesian coordinates [12]. To do this, we rewrite eq. (28) as

$$P_2' = (3/2)\langle \mathbf{n}.\mathbf{Q}.\mathbf{n}\rangle, \tag{29}$$

where the tensor **Q** is

$$Q = (1/N)\sum_i \mathbf{q}_i\mathbf{q}_i - (1/3)\mathbf{1}, \tag{30}$$

e.g.

$$Q_{xy} = (1/N)\sum_i q_{ix}q_{iy}.$$

Notice that, by definition, **Q** is symmetric and traceless. In the special case that **n** is parallel to the z axis it is immediately seen that

$$\langle Q\rangle = \begin{pmatrix} -(1/3)\langle P_2\rangle+\xi & 0 & 0 \\ 0 & -(1/3)\langle P_2\rangle-\xi & 0 \\ 0 & 0 & (2/3)\langle P_2\rangle \end{pmatrix}, \tag{31}$$

if the nematic is homogeneous. The biaxiality parameter ξ will tend to zero if the mesophase has uniaxial symmetry. It is now obvious that the rotation diagonalizing $\langle Q\rangle$ defines the orientation of the director frame in terms of our laboratory frame. The director itself is defined by the eigenvector corresponding to the largest eigenvalue, λ_{max}, of $\langle Q\rangle$. The second rank order parameter $\langle P_2\rangle$ is obtained from the largest eigenvalue as

$$\langle P_2\rangle = (3/2)\lambda_{max}. \tag{32}$$

Once the rotation Ω connecting the laboratory to the director frame has been found, the order parameters calculated in the old frame can be transformed to the new one. In general we can show that

$$\langle D^L_{m,n}\rangle_{\text{new frame}} = \sum_n D^{L*}_{q,m}(\Omega)\langle D^L_{q,n}\rangle_{\text{old frame}}, \tag{33}$$

if Ω is the rotation carrying the old frame into the new one;

(c) director fluctuations are not negligible. In this case we can still define a **Q** tensor for every configuration, say $Q^{(j)}$ for the j-th one. By diagonalizing $\mathbf{Q}^{(j)}$, we obtain an order parameter $P_2^{(j)}$ and a director $\mathbf{n}^{(j)}$. However, the director can change from one configuration to the next. In practice, we could divide the Monte Carlo chain into subchains so that the director will fluctuate appreciably from one subchain to another but not within a subchain. Since $P_2^{(j)}$ is obtained as an eigenvalue and the eigenvalues of a matrix are rotationally invariant, we then calculate

$$\langle P_2\rangle = (3/2)\sum_{\alpha=1}^{M_\alpha}(\lambda_{max})_\alpha/M_\alpha, \tag{34}$$

where $(\lambda_{max})_\alpha$ is the largest eigenvalue of the average tensor in the α-th subchain $\langle Q\rangle_\alpha$. The calculation of the orientational distribution with respect to the director strictly involves transforming the orientations, after each diagonalization, to the new director frame. An alternative is to transform the order parameters to the director frame after every diagonalization.

Statistical Errors

We now wish to examine the error involved in the calculation of an observable A (e.g. the energy or the order parameter) using the Monte Carlo method. Since we use a finite chain of M steps instead of an infinite one, we expect that our estimate,

$$\langle A\rangle = (1/M)\sum A^{(j)},$$

will deviate from the true expectation value $\langle A\rangle_{true}$. It is known from the theory of Markov chains that this deviation has an asymptotically normal distribution [5]. Now, to estimate the standard deviation in a computer experiment we divide the chain of states into a number of subchains. We then calculate averages of the quantity A in every subchain α, $\langle A\rangle_\alpha$ together with the usual average $\langle A\rangle$ over the complete chain. The statistical error on $\langle A\rangle$ is then estimated as the standard deviation from the average:

$$\sigma(A) \simeq [\{1/M_\alpha(M_\alpha-1)\}\sum_{\alpha=1}^{M_\alpha}(\langle A\rangle_\alpha-\langle A\rangle)^2]^{\frac{1}{2}}, \tag{35}$$

where M_α is the number of sub-averages. For this estimate to be reliable we need the subchains to be long enough so that they are statistically uncorrelated. Compatibility with this restriction obviously requires the number of sub-averages, M_α, to be as large as possible. In general the Monte Carlo average, $\langle A\rangle$, tends to the true value as the number of configurations, M, increases according to [18]

$$\langle A\rangle \simeq \langle A\rangle_{true} + O(M^{-\frac{1}{2}}). \tag{36}$$

It is useful to examine the behaviour of the subchain averages to decide if convergence to equilibrium has been achieved. If this is the case the sub-averages should simply oscillate about the average and not exhibit any systematic drift.

Pair Distribution Function

There are two commonly used procedures for evaluating the reduced pair distribution $G(r_{12},\Omega_1,\Omega_2)$ in a computer simulation experiment.

One consists of producing it as a multidimensional histogram [23]. This is achieved by first dividing the range of separations and orientations into a number of intervals. Then, after a suitable number of Monte Carlo steps the orientations and separations $(\mathbf{r}_{12},\Omega_1,\Omega_2)$ of every pair of molecules are sorted into the appropriate volume elements and the relative counters increased. The histogram of the pair distribution is obtained by repeating the sorting process for a sufficient number of equilibrium configurations and then normalizing. The limitations of this procedure are in the large amount of computer storage required which in turn limits the resolution that can be obtained in the histogram.

An alternative procedure is to evaluate the coefficients for the expansion of $G(r_{12},\Omega_1,\Omega_2)$ in a product basis of Wigner rotation matrices [24] (cf. Chapter 3). For example, consider the case of a reduced pair distribution independent of the orientation of the intermolecular vector, or $G(r_{12},\Omega_1,\Omega_2)$. We have shown in Chapter 3 that the most general, rotationally invariant form of $G(r_{12},\Omega_1,\Omega_2)$ is

$$G(r_{12},\Omega_1,\Omega_2) = G(r_{12},\Omega_{12}).$$

Thus the reduced pair distribution is, in this limit, a function of relative orientations Ω_{12}. For cylindrically symmetric particles we have just $G(r_{12},\cos\beta_{12})$ which can be expanded in Legendre polynomials as

$$G(r_{12},\Omega_{12}) = (1/64\pi^4)\sum(2L+1)G_L^{00}(r_{12})P_L(\cos\beta_{12}). \tag{37}$$

Here

$$G_L^{00}(r_{12}) \equiv G_0^{00}(r_{12})G_L(r_{12}), \tag{38}$$

where

$$G_0^{00}(r_{12}) = \int d\Omega_1 d\Omega_2 G(r_{12},\Omega_{12}) \tag{39}$$

is the centre of mass pair distribution and

$$G_L(r_{12}) = \{1/G_0^{00}(r_{12})\}\int d\Omega_1 d\Omega_2 G(r_{12},\Omega_{12})D_{0,0}^L(\Omega_{12}), \tag{40}$$

$$= \overline{P_L[\cos\beta_{12}(r_{12})]}$$

represents an L-th rank angular correlation as discussed in Chapter 3. The average Legendre polynomials $G_L(r_{12})$ can be calculated during the course of a simulation using the following procedure. First the range of interparticle separations is divided into a number of intervals (buckets) of width Δr each labelled by an integer. Thus, to every separation r_{12} we can assign an integer number labelling one of the buckets. Then, for a given configuration, a particle, i, is chosen as an origin and the quantity of interest, $P_L(\cos\beta_{ij})$, is computed and added for every pair i,j into the bucket corresponding to the separation r_{ij}. The process is then repeated choosing another particle as an origin and so on. Normalization is achieved by dividing the content of every bucket by the number of pairs it holds. The sorting process is then repeated, as usual, for sufficiently many equilibrium configurations to obtain the Monte Carlo estimate of $G_L(r_{12})$.

The usefulness of expansion (37) as a representation of $G(r_{12},\Omega_{12})$ relies on its rate of convergence, which in general can only be verified *a posteriori* from the simulation results. However, if we are only interested in calculating the average of a two-particle quantity $F(r_{12},\Omega_{12})$ whose angular dependence can be written as a terminating expansion in Legendre polynomials

$$F(r_{12},\Omega_{12}) = \sum_{L=0}^{J} F_L(r_{12}) D^L_{0,0}(\Omega_{12}), \tag{41}$$

then only the $G_L(r_{12})$ of rank L up to J are needed. In fact

$$\begin{aligned} \langle F(r_{12},\Omega_{12})\rangle &= (\rho/N)\int d\mathbf{r}_{12} d\Omega_1 d\Omega_2 G(r_{12},\Omega_{12}) F(r_{12},\Omega_{12}), \\ &= (4\pi\rho/N) \sum_{L=0}^{J} \int dr_{12} r_{12}^2 F_L(r_{12}) G_0^{00}(r_{12}) G_L(r_{12}). \end{aligned} \tag{42}$$

Lattice Models

Lattice models play an important role in the theory of phase transitions. Here we shall discuss such models for liquid crystals in general and examine in the next section some results for the Maier-Saupe-Lasher model [10,11].

It may seem paradoxical, at first, to choose a model of liquid crystals where the molecules are constrained on lattice sites, since a characteristic of nematics is that of being ordered fluids in which orientational order coexists with translational freedom. The answer to the paradox is that when choosing a lattice model the aim is not to try and reproduce the properties of a real liquid crystal (4,4'-dimethoxyazoxybenzene say). In any case this would certainly be overambitious if we recall that the most complicated systems usually studied by computer simulations are rare gases or diatomics (water is a notable exception). Quite aside from the predictable difficulties and computing time demanded by such a calculation there would also be the problem of finding a realistic intermolecular potential for such large molecules. On the other hand, even with model systems one could allow translational freedom [12b]. The main reason for studying lattice models lies therefore in their simplicity.

This in turn means that they can be studied in more detail and e.g. a larger number of particles can be used. In this perspective the problem is not if a lattice model is a reasonable model of liquid crystals but rather if it can be satisfactory in simulating its orientational properties, such as ordering and the nematic-isotropic phase transition. Numerical experiments in this and similar three dimensional systems can hopefully play a role somewhat similar to that of exactly soluble models in two dimensions. One can choose to study a simple system e.g. the Maier-Saupe lattice model [10, 11, 15] and, in the absence of an analytic solution, perform computer experiments with which to compare results from approximate theories. Possibly the conclusions reached for the model system will then be applicable even to real systems, particularly near the transition, if the essential physics is contained in the model. There is some hope that this is the case. In particular the behaviour near a phase transition (critical exponents, etc.) appears to be independent of the details, such as the exact nature of intermolecular interactions and lattice structures [25]. For example, an extremely simple representation of the gas-liquid transition is obtained with the lattice gas model. Here each lattice site can be empty or contain a particle and each particle interacts only with its nearest neighbours. The agreement between the lattice gas coexistence curve and experiments for simple gases (Ne, Ar, Kr, Xe, N_2, O_2) [25, 26] is remarkable, bearing in mind the simplicity of the model. It is, of course, necessary that however simple the model, the basic physics is present. For instance, stric lattice models where orientations are quantized seem to present some conceptual shortcomings [27]. In particular changing the continuous rotational symmetry seems a severe alteration of the problem [10], if we are interested in just orientational properties.

Results for the Maier-Saupe-Lasher model

Here we wish to consider in some detail the lattice version of the Maier-Saupe model [28] proposed by Lasher [10,11]. In this model molecules are replaced by centres of interaction located at the sites of a simple cubic lattice and interacting with a pair potential

$$U_{ij} = U(X_i, X_j),$$

$$= -\varepsilon_{ij} P_2(\cos\beta_{ij}), \tag{43}$$

where ε_{ij} is a positive constant, ε, for neighbouring sites i and j and zero otherwise. β_{ij} is the relative orientation of particles i and j.

It is interesting to consider this relatively simple model since it has been investigated by a number of people using a variety of techniques. For example it offers an opportunity of comparison [29] with the mean field theory considered in Chapter 4 as well as the two site cluster theory [30-32].

The model has been studied [11, 15, 29] using the Metropolis *et al.* Monte Carlo technique. Periodic boundary conditions are imposed to minimize surface effects and an initial orientation can be chosen arbitrarily. For the soft potential given by eq. (43) it is possible to reach virtually any ordered or disordered state from an arbitrary starting configuration. In practice it is convenient to store end configurations of the runs already performed for various temperatures and then start a new calculation from an equilibrium configuration at a nearby temperature. This increases the rate of approach to equilibrium, even though the same final state can be reached from any initial configuration.

It has been found by various authors [11, 14, 15] that the model exhibits a first-order orientational phase transition. This has been located by the change in the slope of the free energy-temperature plot [11] and by the peak in the graph of specific heat versus temperature [15]. Samples of various sizes have been studied: Lebwohl and Lasher [11] and Jansen *et al.* [15] have examined lattices from 10 x 10 x 10 to 20 x 20 x 20 with the Monte Carlo technique and Meirovitch, samples as large as 50 x 50 x 50 with a different technique [14]. In the Table we compare their

Table *The reciprocal transition temperature* ε/KT_{NI} *and the jump in the long range order parameter and energy at the phase transition in the Maier–Saupe–Lasher lattice model.*

	MF	S-W	M	L-L	J-Y-V	
ε/kT_{NI}	0.7569	0.8622	0.90±0.003	0.89±0.005	0.894±0.001	0.75
$\Delta\bar{P}_2$	0.429	0.382	0.27±0.04	0.33±0.04	0.333±0.009	0.82
$\Delta U/\varepsilon$	0.55	0.327	0.12±0.04	0.11	0.10±0.02	1.50

Here MF stands for Mean Field and S-W for the Sheng-Wojtowicz constant coupling theory [30]. The column marked M shows results obtained by a recently proposed approximate simulation technique, the Stochastic Models method, by Meirovitch [14]. The other columns show Monte Carlo results by Lebwhol and Lasher [11] (L-L) and Jansen-Ypma-Vertogen [15] (J-Y-V). In the last column Lasher, dodecahedral model (DODEC) results are reproduced for comparison.

results for the thermodynamic properties at the transition. We also report the results from the mean field treatment of the same model and for Lasher's dodecahedral model which has the same interaction potential (43) but where only 6 different orientations are allowed [10]. It is seen that fairly good agreement is obtained between the various computer simulations [11, 14, 15] although the discrete dodecahedral model gives significantly different results.

We have recently studied the Maier-Saupe-Lasher [29] model to obtain in addition to the basic thermodynamic properties such as energy, specific heat etc. the order parameters and pair correlation. Let us examine very briefly some of the results obtained for a 10 x 10 x 10 lattice using the Monte Carlo method.

Angular Correlation We have computed the angular co-relation coefficients $G_L(r_{12}) = \overline{P_L[\cos\beta_{12}(r_{12})]}$. As we have discussed in Chapter 3 the tail of these coefficients $G_L(r_{12})$, for even L and relatively large separations, should give a measure of the long range order. Thus $G_L(r_{12})$ should level off to essentially $\bar{P}_L{}^2$. In figure 3 we show the results for $G_2(r)$ at temperatures kT/ε well above, well below, and near the orientational phase transition. It is apparent that above the transition the correlation decays rapidly to zero, corresponding to the absence of long-range order.

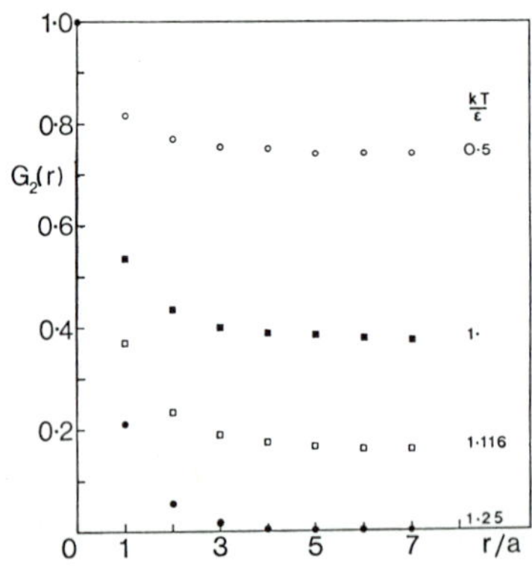

Fig.3 *The angular pair correlation of rank two $G_2(r_{12}) = \overline{P_2\{\cos\beta_{12}(r_{12})\}}$ for the Maier-Saupe-Lasher model as a function of the interparticle separation expressed in units of the lattice constant a. The results, shown with different symbols for the four temperatures kT/ε reported, were obtained [29] from Monte Carlo simulations on a 10 x 10 x 10 lattice with periodic boundary conditions, averaging over at least 2×10^6 configurations. The histograms of $G_2(r)$ were calculated, with a resolution of 0.1a, in the range $0<r/a<(75)^{\frac{1}{2}}$, where $(75)^{\frac{1}{2}}$ is the maximum significant separation (minimum image convention used). Only values corresponding to integer separations r/a are shown here.*

Short-range Order We can define short-range order parameters as

$$\sigma_L \equiv G_L(a), \tag{44}$$

where a is the nearest neighbour distance or lattice constant. In particular σ_2 is the short-range order parameter obtained from two site cluster theories [30-32]. In a mean field treatment there is no distinc tion between short and long-range order. Thus

$$\sigma_2 = \overline{P_2(\cos\beta_{ij})},$$
$$= \overline{P}_2^2, \qquad \text{(mean field).}$$

Therefore the mean field theory predicts $\sigma_2 = 0$ in the isotropic phase. From figure 3 we can see, however, that short range order actually persists above the transition indicating orientational correlation between neighbouring particles. Cluster theories [30] keep approximate account of the existence of such correlations. The predictions of σ_2 from the treatment of Sheng and Wojtowicz [30] are shown in figure 4, as the

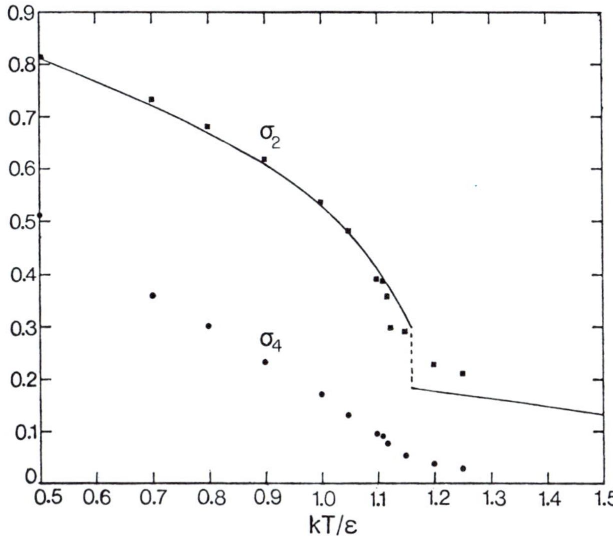

Fig. 4 *The temperature dependence of the short-range order parameters of rank two, σ_2, and four, σ_4, obtained from Monte Carlo simulation [29] of the Maier-Saupe-Lasher model. The continuous line corresponds to the prediction for σ_2 of the constant-coupling theory by Sheng and Wojtowicz [30].*

continuous line, together with our Monte Carlo results for σ_2 and σ_4. It is seen that the two site cluster theory represents a significant improvement over the mean field treatment. However, the short-range correlation in the isotropic phase is still somewhat underestimated. It is worth mentioning that for the Maier-Saupe-Lasher model there exists a simple way of checking the values for the short-range order σ_2. Consider the calculation of the energy for a potential

$$U_{12} = \varepsilon(r_{12})P_2(\cos\beta_{12}).$$

Then

$$\langle U\rangle = (\rho^2/2)\int dX_1 dX_2 G(r_{12},\Omega_{12})U(r_{12},\Omega_{12}),$$

$$= (4\pi N^2/2V)\sum\int dr_{12}r_{12}^2 d\Omega_1 d\Omega_2 \{(2L+1)/64\pi^4\}G_0^{00}(r_{12})G_L(r_{12}) \times D_{0,0}^L(\Omega_{12})\varepsilon(r_{12})D_{0,0}^2(\Omega_{12}), \tag{45}$$

or

$$\langle U\rangle/N = 2\pi\rho\int dr_{12}r_{12}^2 G_0^{00}(r_{12})G_2(r_{12})\varepsilon(r_{12}). \tag{46}$$

Now, for a lattice, the centre of mass pair distribution $G_0^{00}(r)$ or $g(r)$ is just that given in eq. (25) of Chapter 3. Therefore

$$\langle U\rangle = (N/2)\{z_1\varepsilon(r_1)G_2(r_1)+z_2\varepsilon(r_2)G_2(r_2)+..\}. \tag{47}$$

In our case of nearest neighbour interactions, with $r_1 = a$ and $\varepsilon(r_1) = -\varepsilon$, we find

$$-\langle U\rangle = (N/2)\varepsilon z_1\sigma_2. \tag{48}$$

Thus σ_2 can be obtained simply from the energy or, alternatively, in the Maier-Saupe-Lasher model the internal energy is determined completely by the short-range order. The short-range order parameter σ_2 that we obtain from the energy agrees well with the values obtained from $G_2(r)$ at nearest neighbour separation. The maximum differences, when we calculate $G_L(r_{ij})$ from 50 randomly selected particles as origins, are about 2%. Since the energy can be calculated rather accurately ($\sim$ 0.5% except very near the transition) this constitutes a comforting check on the reliability of our $G_L(r)$. Similar internal consistency tests can be obtained from eq. (46) for more complex potentials. In general the test will obviously involve recalculating the energy from the pair correlations instead of the contrary.

Long-range order In figure 5 we show some results for the temperature dependence of the long-range order parameters of rank two and four. We

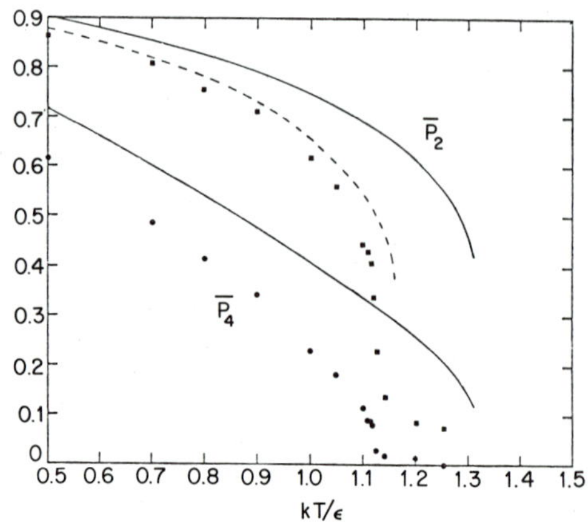

Fig. 5 *Monte Carlo results for the temperature dependence of the long-range order parameters $\bar{P}_2$ (■) and $\bar{P}_4$ (●) in the Maier-Saupe-Lasher model. The continuous lines show the mean field predictions for $\bar{P}_2$ and $\bar{P}_4$. The broken line gives the long-range order parameter $\bar{P}_2$ according to Sheng-Wojtowicz [30].*

have evaluated the long-range order parameter $\overline{P}_2 = <P_2>$ for a range of temperatures kT/ε by averaging the largest eigenvalue of **Q** every 10^3 configurations, as described for the general case where the director is not known and its fluctuations are not negligible. We have also estimated $\overline{P}_2$ from the square root of the limiting average plateau value of $G_2(r)$. The two results agree to less than a few percent. We feel justified then in adopting the same procedure to estimate $\overline{P}_4$ from the tail of $G_4(r)$. The results are shown in figure 5 together with the mean field and Sheng-Wojtowicz predictions [30].

It should be mentioned that averages over about 2×10^6 configurations are sufficient to obtain a good estimate of $\overline{P}_2$, with a standard deviation smaller than 1%, except in a narrow temperature range near the transition. In this critical region very large fluctuations occur and the system spends part of its time in a nematic-like and part in an isotropic-like situation. This is shown for example by the histogram of the frequency of occurrence of the second rank order parameter [11]. The bistability shows, convincingly, the first order character of the transition but complicates the interpretation of the results and contrasting views have appeared in the literature [11, 15]. We shall not discuss these problems further, except to stress once more that if accurate values of critical properties are desired, very large samples should be studied (lattices 20 x 20 x 20 [11, 15] and even 50 x 50 x 50 [14] have been used).

Molecular Dynamics

Molecular dynamics simulations of liquid crystals are, even more than Monte Carlo ones, in an early stage of development. In this section we shall outline the method [2, 33] and develop the equations necessary to treat orientational problems.

Even though the molecular dynamics technique is more complicated to implement than the Monte Carlo procedure, the idea behind it is, in principle, even simpler. It consists of setting up and solving numerically the equations of motion for a system of N molecules contained in a box of a given volume. Since the system is isolated its total energy, linear and angular momentum will be conserved, while of course the potential and kinetic energy, and therefore the temperature fluctuate. Periodic boundary conditions are used, as in Monte Carlo, to minimize surface effects. By solving the set of coupled equations of motion for the N particles the real trajectory of the system in phase space is followed. Thus the great advantage of the molecular dynamics technique is that dynamic as well as equilibrium properties can, in principle, be calculated. We now start by deriving the equation of motion for a rotor.

Equations of Motion

The kinetic energy K of a rotor of mass m and inertia tensor **I** is the sum of the translational and rotational contributions K_t and K_r:

$$K_r = \tfrac{1}{2}m\mathbf{v}.\mathbf{v}, \tag{49}$$

$$K_r = \tfrac{1}{2}\boldsymbol{\omega}.\mathbf{I}.\boldsymbol{\omega}. \tag{50}$$

Here $\mathbf{v} \equiv \dot{\mathbf{r}}$ is the linear velocity of the centre of mass of the rotor. The angular velocity, $\boldsymbol{\omega} = \dot{\boldsymbol{\Omega}}$, is the time derivative of the rotation vector $\boldsymbol{\Omega}$ which has modulus Ω and direction along the rotation axis. Following the convention of Rose [34] for the Euler angles we have

$$\Omega = \alpha\mathbf{Z}+\beta\mathbf{y}'+\gamma\mathbf{z},$$

where $\alpha\mathbf{Z}$ is a positive, right-handed, rotation about the laboratory Z axis, $\beta\mathbf{y}'$ a rotation of β about the new $\mathbf{y}'$ axis and finally $\gamma\mathbf{z}$ a rotation of γ about the body fixed **z**.

The intermolecular potential acting on the rotor is, assuming effective pairwise additivity [2b],

$$U \equiv U_i = \sum U(X_i,X_j),\ j \neq i. \tag{51}$$

The summation is extended in principle to all the rotors except the one considered, even though in practice only those within a given range are taken into account. We do not enter here into the nature of the effective two body potential but we shall assume it is a continuous and differentiable function of positional and orientational coordinates as well as time independent.

The dynamic evolution of the rotor is regulated by Lagrange's equations [35]

$$\partial(\partial L/\partial\dot{q}_\alpha)/\partial t-\partial L/\partial q_\alpha = 0 \quad , \quad \alpha = 1,2,\ldots, n_f, \tag{52}$$

where L = K - U is the lagrangian function for the rotor under consideration, q_α a generalized coordinate component and n_f the number of degrees of freedom. Thus, for translational degrees of freedom, $q_\alpha = r_x,r_y,r_z$, we obtain at once Newton's equation

$$m\dot{\mathbf{v}} = \mathbf{F}, \tag{53}$$

where $F \equiv \mathbf{F}_i \equiv -\partial U_i/\partial r_i$ is the force acting on the particle with centre of mass at $\mathbf{r}_i$. Similarly we can deduce the rotational equation of motion in the laboratory frame

$$d(\mathbf{I}.\boldsymbol{\omega})/dt = \mathbf{N}, \tag{54}$$

where $\mathbf{N} \equiv N_i \equiv -\partial U_i/\partial\Omega$ is the torque produced on the rotor, under consideration, by all the others. It is convenient to rewrite eq. (56) in a frame moving with the molecule by using the vector formula for the time derivative of a vector **A** attached to a frame moving with angular velocity ω_R with respect to a reference frame:

$$d\mathbf{A}_{Ref.}/dt = d\mathbf{A}_{Mov.}/dt + \boldsymbol{\omega}_R \wedge \mathbf{A} \ . \tag{55}$$

Thus we find from eq. (54) the so-called Euler equation describing the reorientation of the particle

$$\mathbf{I}.\dot{\boldsymbol{\omega}}+\boldsymbol{\omega}\wedge(\mathbf{I}.\boldsymbol{\omega}) = \mathbf{N}. \tag{56}$$

If our molecule fixed coordinate system is the principal frame of the inertia tensor we have explicitly

$$I_{xx}\dot{\omega}_x-\omega_y\omega_z(I_{yy}-I_{zz}) = N_x, \tag{57}$$

$$I_{yy}\dot{\omega}_y-\omega_x\omega_x(I_{zz}-I_{xx}) = N_y \tag{58}$$

and

$$I_{zz}\dot{\omega}_z-\omega_y\omega \ (I_{xx}-I_{yy}) = N_z. \tag{59}$$

Here I_{xx}, I_{yy}, I_{zz} are the principal moments of inertia of the particle and ω_x, ω_y, ω_z and N_x, N_y, N_z are respectively the components of the angular velocity and of the torque around the principal axes. As an example, again for the i-th rotor

$$\begin{aligned} N \equiv N_i &= \mathbf{r}_i \wedge \mathbf{\Gamma}_i, \\ &\equiv -\mathbf{r}_i \wedge \nabla U_i. \end{aligned} \tag{60}$$

Since we often express intermolecular potentials in terms of Euler angles it is helpful to give also the expressions for the angular velocities and the torques in terms of $(\alpha\beta\gamma)$ and their derivatives. We find, for the molecule fixed components of the angular velocity,

$$\omega_x = -\sin\beta\cos\gamma\dot{\alpha}+\sin\gamma\dot{\beta}, \tag{61}$$

$$\omega_y = \sin\beta\sin\gamma\dot{\alpha}+\cos\gamma\dot{\beta}, \tag{62}$$

$$\omega_z = \cos\beta\dot{\alpha}+\dot{\gamma} \tag{63}$$

and for those of the torque

$$N_x = \cos\gamma\cot\beta N_\gamma+\sin\gamma N_\beta-(\cos\gamma/\sin\beta)N_\alpha, \tag{64}$$

$$N_y = -\sin\gamma\cot\beta N_\gamma+\cos\gamma N_\beta+(\sin\gamma/\sin\beta)N_\alpha, \tag{65}$$

$$N_z = N_\gamma, \tag{66}$$

where

$$N_\alpha \equiv -\partial U/\partial\alpha, N_\beta \equiv -\partial U/\partial\beta \text{ and } N_\gamma \equiv -\partial U/\partial\gamma \ .$$

We can also derive the equations of motion for the Euler angles. For a symmetric top molecule we obtain, after a fair amount of algebra,

$$\ddot{\alpha} = \{-(A-B)\dot{\alpha}\dot{\beta}\cos\beta\sin\beta+(A+B)\dot{\beta}\dot{\gamma}\sin\beta+N_\alpha-N_\gamma\cos\beta\}A\sin^2\beta, \tag{67}$$

$$\ddot{\beta} = \{-(A+B)\dot{\alpha}\dot{\gamma}\sin\beta-B(\dot{\alpha})^2\sin\beta\cos\beta+N_\beta\}/A \tag{68}$$

$$\ddot{\gamma} = \{-(A+B)^2\dot{\beta}\dot{\gamma}\cos\beta\sin\beta+(A-B\cos^2\beta)(A+B)\dot{\alpha}\dot{\beta}\sin\beta+(A+B\cos^2\beta)N_\gamma - (A+B)\cos\beta N_\alpha\}/A(A+B)\sin^2\beta. \tag{69}$$

Here $A \equiv I_\perp$ and $B \equiv I_{||} - I_\perp$, where $I_{||}$ and $I_\perp$ are the principal moments of inertia for the rotor. For a linear rotor, where $A = I_\perp = -B$, eqs. (67-69) reduce to

$$\ddot{\alpha} = -2\dot{\alpha}\dot{\beta}\cot\beta+N_\alpha/A\sin^2\beta, \tag{70}$$

$$\ddot{\beta} = (\dot{\alpha})^2\sin\beta\cos\beta+N_\beta/A. \tag{71}$$

For a linear molecule the Euler angles β and α become simply the polar angles θ and ϕ of the particle.

Integration of the Equations of Motion

The set of coupled differential equations derived in the preceding section are solved to generate the trajectory of the system in phase space, that is, to compute positional and angular coordinates and velocities as a function of time. The numerical integration can be performed using one of several algorithms available [4, 33]. We wish to consider here, as an example, the integration of the translational equation of motion. Thus, suppose we have positions $\mathbf{r}_i(t)$ and velocities $\mathbf{v}_i(t)$ for every particle at a time t and that we want to find their value at a later time $(t + \Delta t)$. A straight forward Taylor expansion gives a simple finite difference approximation for the derivatives $\dot{\mathbf{v}}$ and $\dot{\mathbf{r}}$ and yields [19]

$$\mathbf{v}_i(t+\Delta t) = \mathbf{v}_i(t)+(\Delta t/m)\mathbf{F}_i(\{X^N(t)\}+O(\Delta t^2) \tag{72}$$

and

$$\mathbf{r}_i(t+\Delta t) = \mathbf{r}_i(t)+\Delta t\mathbf{v}_i(t+\Delta t), \tag{73}$$

where $\mathbf{F}_i(t) \equiv \mathbf{F}_i(\{\mathbf{X}^N(t)\})$ is the force acting on the i-th particle at time t. More sophisticated algorithms can obviously be used. For example Rahman has employed the following predictor-corrector method [19, 33]

$$\mathbf{r}_i'(t+\Delta t) = \mathbf{r}_i(t-\Delta t)+2\Delta t\mathbf{v}_i(t)+O(\Delta t^3), \tag{74}$$

$$\mathbf{v}_i'(t+\Delta t) = \mathbf{v}_i(t-\Delta t)+(\Delta t/m)\mathbf{F}_i(\{X^N(t)\})+O(\Delta t^3), \tag{75}$$

$$\mathbf{r}_i(t+\Delta t) = \mathbf{r}_i(t)+\tfrac{1}{2}\Delta t\{\mathbf{v}_i(t)+\mathbf{v}_i'(t+\Delta t)\}+O(\Delta t^3) \tag{76}$$

and

$$v_i(t+\Delta t) = v_i(t)+(\Delta t/2m)\{F_i(\{\mathbf{X}^N(t)\})F_i(\{X'^N(t+\Delta t)\})\}+O(\Delta t^3). \quad (77)$$

Notice however that the improved accuracy in estimating the new coordinates and velocities is counteracted by the need for calculating and storing additional quantities, such as higher time derivatives of positions. The loss of computer storage can be a disadvantage if very large systems have to be studied. A useful compromise is given by the algorithm proposed by Verlet [33],

$$r_i(t+\Delta t) = -r_i(t-\Delta t)+2r_i(t)+(\Delta t/m)^2F_i(\{X^N(t)\})+O(\Delta t^4) \quad (78)$$

and

$$v_i(t) = \{r_i(t+\Delta t)-r_i(t-\Delta t)\}/2\Delta t+O(\Delta t^2), \quad (79)$$

obtained by combining the Taylor expansion of $r_i(t+\Delta t)$ and $r_i(t-\Delta t)$.

Similar algorithms can be given for the angular coordinates [2b, 23] by taking finite difference approximations of the Euler equations given before. It is worth stressing that, whatever the integration algorithm used, it has to be stable in the sense that the solution obtained should not diverge exponentially with time from the true one [33]. This requirement also determines the magnitude of the increment Δt. Once the new configuration is obtained at time $(t+\Delta t)$ the procedure is repeated. Thus the continuous trajectory of the system in phase space is replaced by a series of points at time intervals Δt.

The process is normally started from a non-equilibrium configuration for the desired temperature and density. The initial positions and orientations are chosen keeping in mind the arguments developed for Monte Carlo calculations. Velocities can be sampled from a suitable Maxwell distribution to accelerate the approach to equilibrium or they can just be set in some rather arbitrary way (for example all with constant magnitude but random directions). For an arbitrary choice there will be, in general, a net linear and angular momentum **P** and **J** respectively. These can be subtracted so that the box does not, so to speak, move around during the computation.

Rotational and translational temperatures can be obtained from the respective kinetic energies

$$K_r = \tfrac{1}{2}\langle\sum_i \boldsymbol{\omega}_i.\mathbf{I}.\boldsymbol{\omega}_i\rangle = NkT_rn_r/2 \quad (80)$$

and

$$K_t = \tfrac{1}{2}\langle\sum_j m_i|v_i|^2\rangle = NkT_tn_t/2, \quad (81)$$

where n_r and n_t are the number of rotational and translational degrees of freedom respectively. Obviously for a fluid at equilibrium, $T_R=T_t=T$. In the first part of the calculation the temperature needs in general to be adjusted. This can be achieved by multiplying the velocities by a scaling factor $h = (T/T')^{\frac{1}{2}}$ where T' is the actual temperature and T the desired one. Normally a series of these adjustments will be necessary since, of course, the temperature is not a constant of the motion. After this preparation stage the system is left to equilibrate for an adequate number of steps (a few thousands, say). As a check on the reliability of the integration the constants of motion such as energy and angular momentum are monitored. They should not drift systematically, but fluctuate about their average value. The temperature should also vary about its time average. The mean square amplitude of the equilibrium temperature fluctuations is determined by the specific heat of the system This has been shown, for spherical particles, by Lebowitz *et al.* [36], who have generalized the usual canonical ensemble relation [37]

$$\langle(\Delta T)^2\rangle/T^2 = k/C_V \tag{82}$$

to the microcanonical conditions characteristic of a molecular dynamic simulation. They find [36] for the translational contribution

$$\langle(\Delta T)^2\rangle/T^2 = 2(1-3Nk/2C_V)/3N. \tag{83}$$

The requirement of constant total energy in the molecular dynamics ensemble implies that the fluctuations in temperature, and thus in kinetic energy, have to be equal to the potential energy fluctuations. For instance, in molecular dynamics simulation of the Maier-Saupe-Lasher model [38], we find that large fluctuations in temperature mean also large fluctuations in the order parameters since these determine the internal energy.

Calculation of static and dynamic properties

The mean value of a quantity A can be calculated as time averages from the configurations obtained after the system has evolved to equilibrium

$$\langle A\rangle = \lim_{\Theta\to\infty}(1/\Theta)\int_0^\Theta dtA(t), \tag{84}$$

$$\simeq (1/M)\sum_{k=1}^{M} A(k\Delta t). \tag{85}$$

Thus in molecular dynamics we do not calculate ensemble averages, as in the Monte Carlo method and we do not even have to assume the system to be ergodic. From the practical point of view, the calculation of static properties (energy, order parameters, pair distribution etc.) is obtained as an average over M configurations. Considerations similar to those in Monte Carlo calculations can be applied to estimate the error involved in substituting the infinite integral (84) with the finite sum (85).

Dynamic properties which describe the response of the system to some weak perturbation can be written [39] in terms of equilibrium time correlation functions, $\langle A(0)A(\tau)^*\rangle$ and these can be evaluated by molecular dynamics:

$$\langle A(0)A(\tau)^*\rangle = \lim_{\Theta\to\infty}(1/\Theta)\int_0^\Theta A(t)A(t+\tau)^* dt, \tag{86}$$

$$\simeq (M-n)^{-1} \sum_{k=1}^{M-n} A(k\Delta t)A(\{k+n\}\Delta t)^*, \quad \tau = n\Delta t. \tag{87}$$

If $\langle A\rangle \neq 0$ it is often useful to redefine A as $(A-\langle A\rangle)$ so that only correlations of the fluctuations of A about the average are treated. It is also convenient to normalize the correlation function as

$$C_A(t) = \langle A(0)A(t)^*\rangle/\langle A(0)A(0)^*\rangle. \tag{88}$$

With these definitions $C_A(t)$ starts at one at time $t = 0$ and decays to $|\langle A\rangle|^2 = 0$ as $t \to \infty$.

Zwanzig and Ailawadi [40] have estimated the error $\sigma[C_A(t)]$ involved in calculating the normalized autocorrelation function $C_A(t)$ with the finite sum (87). They find, assuming a gaussian model for fluctuations,

$$\sigma[C_A(t)] = (2\tau/\theta)^{\frac{1}{2}}\{1-C_A(t)\}, \tag{89}$$

where τ is an effective decay time for C(t) and θ is the time over which the average is performed. In a typical molecular dynamics simulation $\theta/\tau \simeq 20$ and fairly large errors can be expected. If A is a single particle quantity the situation can be improved by calculating the average

$$N^{-1} \sum_{i=1}^{N} \langle A_i(0)A^*(t)\rangle$$

instead of $\langle A_1(0)A_1^*(t)\rangle$. In this case the error (89) is divided by, roughly $\sqrt{N}$ where N is the number of particles considered [40].

A further limitation of the molecular dynamics technique in evaluating dynamic properties is that the longest meaningful time accessible is of the order of λ/C where λ is the box edge and C the velocity of sound in the system [39]. The reason is that this is the time it takes for a perturbation generated at a given position to travel through the box and to be brought back by the periodic boundaries.

From the point of view of application of molecular dynamics to liquid crystal theory [13, 38, 41] correlation functions of particular interest are those describing the rotational motion. These are the correlation functions of the Wigner rotation matrices (cf. Chapter 18).

$$C^L_{m,n}(t) = \langle D^L_{m,n}(0) D^{L^*}_{m',n'}(t) \rangle,$$

and angular velocity correlation functions [42]. For a symmetric top particle it is interesting to calculate the correlation function of $\boldsymbol{\omega}_\perp$, the component of the angular velocity perpendicular to the orientation of the top axis [42]. In a liquid crystal $C_{\omega_\perp}(t)$ should reflect how hindered the reorientation of the long axis is and how this is affected by the onset of orientational ordering. As an illustration we show in figure 6 some preliminary results for $C_{\omega_\perp}(t)$ in a model liquid crystal of the Maier-Saupe-Lasher type. This consists of a system of symmetric

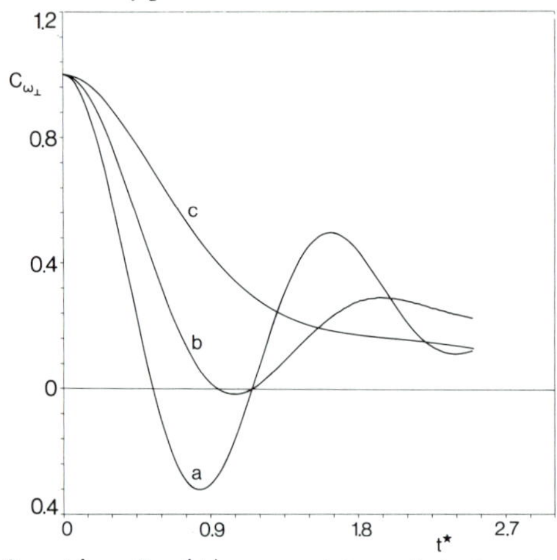

Fig. 6 *The correlation function* $C_{\omega_\perp}(t) = \langle\omega\perp(0).\omega\perp(t)\rangle/\langle\omega\perp(0).\omega\perp(0)\rangle$ *for a 10 x 10 x 10 lattice system of symmetric top rotors, with* $I_{||}/I_\perp = 0.8$ *and interacting with the Maier-Saupe-Lasher potential (43). The correlation function is shown as a function of dimensionless time* $t^* = t(\varepsilon/I_\perp)^{\frac{1}{2}}$ *for three temperatures (a)* $kT/\varepsilon = 0.68$ *(highly ordered nematic), (b)* $kT/\varepsilon = 1.03$ *(nematic but not too far from the transition (c)* $kT/\varepsilon = 1.30$ *(isotropic).*

top rotors, with centre of mass at the sites of a cubic lattice, which interacts through potential (43). We see that in the ordered phase the correlation function for the perpendicular component of the angular velocity exhibits a negative region. This corresponds to a change in the sense of rotation of the particle which in turn indicates some sort of librational motion as the orientational order increases.

Acknowledgements

I wish to thank Dr. M. Guerra, Dr. J. Lewis and especially Professor G.R. Luckhurst for various useful discussions. Dr. M. Guerra has also collaborated in deriving the equations of motion. The simulations reported in this Chapter were run on the CDC 7600 computers of Bologna University (CINECA) and London University (ULCC). Computing time from Laboratorio C.N.R. (Ozzano E.) and the University of Southampton is gratefully acknowledged.

References

1a. N. Metropolis, A.W. Rosenbluth, M.N. Rosenbluth, A.H. Teller and E. Teller, *J. Chem. Phys.* 21, 1087 (1953).
1b. W.W. Wood and F.R. Parker, *J. Chem. Phys.* 27, 720 (1957).
2a. B.J. Alder and T.E. Wainwright, *J. Chem. Phys.* **31**, 459 (1959); **33**, 1439 (1960).
2b. A. Rahman, *Phys. Rev.* **136**, 405 (1964).
3. J.P. Hansen and I.R. McDonald, *Theory of Simple Liquids*, Academic, (1976).
4. J.A. Barker and D. Henderson, *Rev. Mod. Phys.* 48, 587 (1976).
5a. K. Binder, in *Phase Transitions and Critical Phenomena*, Vol. 5B, eds. C. Domb and M.S. Green, Academic, (1976).
5b. K. Binder, *Adv. Phys.*, 23, 917 (1974).
6. C.P. Yang, IBM Research Report No. RC564, (1961) (unpublished).
6a. L.D. Fosdick, in *Methods in Computational Physics*, Vol.1, eds. B. Alder, S. Fernback and M. Rotenberg, Academic, (1963).
6b. A.E. Ferdinand and M.E. Fisher, *Phys. Rev.* 185, 832 (1969).
7. L. Onsager, *Phys. Rev.* 65, 117 (1944).
8. B.C. Freasier and L.K. Runnels, *J. Chem. Phys.* 58, 2963 (1973).
9. J. Denham, R.L. Humphries and G.R. Luckhurst, *Mol.Cryst.Liq.Cryst.Lett*, **41**,67(1977)
10. G. Lasher, *Phys. Rev.* A5, 1350 (1972).
11. P.A. Lebwohl and G. Lasher, *Phys. Rev.* A6, 426 (1972).
12a. J. Viellard-Baron, *J. Chem. Phys.* 56, 4729 (1972).
12b. J. Viellard-Baron, *Mol. Phys.* 28, 809 (1974).
13. J. Kushick and B.J. Berne, *J. Chem. Phys.* 64, 1362 (1976).
14. H. Meirovitch, *Chem. Phys.* 21, 251 (1976).
15. H.J.F. Jansen, G. Vertogen and J.G.J. Ypma, *Mol. Cryst. Liq. Cryst.* 38, 87 (1977).
16. J.M. Hammersley and D.C. Handscomb, *Monte Carlo Methods*, Methuen, (1964).
17. W. Feller, *An Introduction to Probability Theory and its Applications*, Wiley, (1950).
18. J.P. Valleau and S.G. Whittington, in *Modern Theoretical Chemistry*, Vol. 5A, Equilibrium Statistical Mechanics of Fluids, ed. B.J. Berne, Plenum, (1977).
19. F.H. Ree, in *Physical Chemistry, An Advanced Treatise*, Vol.8A, eds. H. Eyring, D. Henderson and W. Jost, Academic, (1971).
20a. L. Verlet in *Statistical Mechanics, Cargese Lectures in Theoretical Physics*, eds. B. Jancovici, Gordon and Breach, (1966).
20b. L. Verlet in *Phase Transitions*, Solvay Institute 14th Chemistry Conference, Wiley, (1971).
21. W.B. Streett, J.H. Raveché and R.D. Mountain, *J. Chem. Phys.* **61**, 1960 (1974).
22. Z.W. Salsburg, J.D. Jacobson, W. Fickett and W.W. Wood, *J. Chem. Phys.* 30, 65 (1959).
23a. B. Quentrec, *Etude de l'Azote en Phase Condensée par la Dynamique Moleculaire Numerique*, These, Orsay, (1974).
23b. J. Barojas, D. Levesque and B. Quentrec, *Phys. Rev. A* 7, 1092 (1973).
24. W. Streett and D. Tildesley, *Proc. Roy. Soc.* 348, 485 (1976).
25. M.E. Fisher, *J. Math. Phys.* 5, 944 (1964).
26. H.E. Stanley, *Introductions to Phase Transitions and Critical Phenomena*, Oxford U.P. (1971).
27. P.G. De Gennes, *The Physics of Liquid Crystals*, Oxford U.P. (1974).
28. M. Maier and A. Saupe, *Z. Naturforsch.* 14A, 882 (1959); **15A**, 287 (1960).
29. C. Zannoni, to be published.
30. P. Sheng and P.J. Wojtowicz, *Phys. Rev.* A14, 1883 (1976).
31. J.G.J. Ypma and G. Vertogen, *Sol. State Comm.* 18, 475 (1976) and *J. Phys.* **(Paris)**, **37**, **557** **(1976)**.
32. N.V. Madhusudana, K.L. Savithramma and S. Chandrasekhar, *Pramana*, 8, 22 (1977).
33a. B.J. Berne and G.D. Harp, *Adv. Chem. Phys.* **22**, 63 (1970).
33b. G.D. Harp and B.J. Berne, *Phys. Rev.* A2, 975 (1970).
34. M.E. Rose, *Elementary Theory of Angular Momentum*, Wiley,(1957).
35. L. Landau and E.M. Lifshitz, *Classical Mechanics*, Pergamon,(1960).
36. J.L. Lebowitz, J.K. Percus and L. Verlet, *Phys. Rev.* 153, 250 (1967).
37. L. Landau and E.M. Lifshitz, *Statistical Physics*, Pergamon, (1969).

38. C. Zannoni and M. Guerra, to be published.
39. B.J. Berne, in *Physical Chemistry, An Advanced Treatise,* Vol. 8B, eds. H. Eyring, D. Henderson and W. Jost, Academic (1971).
40. R. Zwanzig and N. Ailawadi, *Phys. Rev.* **182**,280 (1969).
41. A.L. Tsykalo, A.D. Baymet, *Russian J. Phys. Chem.* **50**, 439 (1976).
42. J. Kushick and B.J. Berne, *J. Chem. Phys.* **59**, 4486 (1973).

Chapter 10

THERMODYNAMICS OF PHASE TRANSITIONS

DANIEL E. MARTIRE

Department of Chemistry, Georgetown University
Washington D.C. 20057

Single Component Systems

Invaluable qualitative and quantitative information is provided about liquid crystals from thermodynamic data at and in the vicinity of their phase transitions. Often the first (and sometimes the best) evidence on molecular structure-phase stability relations comes from detailed thermodynamic studies where the rigid central core and/or the flexible terminal chains are varied systematically. In addition, theories of liquid crystals depend on the availability of reliable thermodynamic data for their definitive testing. Yet, there is a paucity of *complete sets* of such data. For the substances that have been studied extensively, there are sometimes marked discrepancies among the reported values, as we shall see.

This Chapter is not intended to be a comprehensive review. Rather, the primary focus will be on the equilibrium thermodynamics of first-order phase transitions (mainly nematic-isotropic) and their bearing on molecular statistical theories of mesophases. In this first part, single component systems are considered. Experimental quantities of interest are listed and examples of their usage in molecular interpretation and model testing are given. In the second part, the basic thermodynamic relations for phase transitions in dilute non-mesomorphic solute/mesomorphic solvent mixtures are presented. Selected data are then used to illustrate the effects of solute size, shape and flexibility on nematic phase stability.

Order of Phase Transitions

In general, an nth-order phase transition is one in which it is the nth derivative of the chemical potential, μ, with respect to temperature, T, and pressure, P, which first becomes discontinuous at the point of the transition, as we saw in Chapter 3. For an a-b equilibrium phase transition,

$$\mu_a(T,P) = \mu_b(T,P);$$

the zeroth derivative is thus continuous. The $\mu(T,P)$ surfaces of the a and b phases intersect (first order) or make contact (second order) over a range of T and P. For a first-order phase transformation, where $\partial\mu/\partial T$ and $\partial\mu/\partial P$ are discontinuous, changes in enthalpy ΔH (or entropy ΔS) and in volume ΔV are observed. The projection of the line of intersection of the μ surfaces onto the P-T plane provides a consistency relation between dT/dP and $T\Delta V/\Delta H$ through the Clausius-Clapeyron equation. For a second-order transition, both ΔH and ΔV are zero, but discontinuities in $\partial^2\mu/\partial T$, $\partial^2\mu/\partial T\partial P$ and $\partial^2\mu/\partial P^2$ lead to changes in the constant pressure heat capacity, ΔC_p, in the expansion coefficient, $\Delta\alpha$, and in the isothermal compressibility coefficient $\Delta\kappa$, respectively.

With liquid crystals, changes in the order parameters, which were introduced in Chapter 3, must also be considered. At the first-order nematic-isotropic (NI) transition, the orientational order parameter $\bar{P}_2$, which is invariably employed to characterise the orientational order, changes discontinuously. The smectic A phase possesses spatial order and so additional order parameters are required. According to theory [1] both orientational and spatial order parameters may change discontinuously or continuously, depending on whether the smectic A-nematic (S_AN) transition is first or second order, respectively. Most mesophase transitions are termed weakly first-order in nature because ΔH and $\Delta V/V$ are small in comparison with those observed at the melting point and as a consequence large pretransitional effects, reminiscent of second-order behaviour are usually observed in the region of the transition point.

Data Used for Testing Theories

Thermodynamic data most frequently used for testing equilibrium molecular theories of phase transitions are:

1. transition temperatures; these are usually obtained by optical microscopy or by calorimetry;
2. transition enthalpy and entropy changes; these may be obtained by classical adiabatic calorimetry or by dynamic differential scanning calorimetry (DSC) [2];
3. transition densities ρ and fractional density changes $\Delta\rho/\rho$ (or V and $\Delta V/V$); they are traditionally obtained by dilatometry [3];
4. dT/dP which is related to the above is determined by direct measurement of the effect of external pressure on the transition temperature [4];
5. orientational order parameters are of particular importance for it is

long-range orientational order which distinguishes liquid crystals from simple fluids. For the N-I transition, e.g. $\bar{P}_2$ at the normal transition (P = 1 atm) and $(\partial\bar{P}_2/\partial T)_{P=1\ atm}$ in the nematic phase [5] are often measured by nuclear magnetic resonance as we shall see in Chapter 15;

6. ΔC_p, $\Delta\alpha$ and $\Delta\kappa$ and $C_p(T)$, $\alpha(T)$ and $\kappa(T)$ on both sides of the transition; they are of interest in searching for second-order transitions, studying first-order pretransitional effects and determining pseudo-critical exponents [6].

Another useful quantity is

$$\Gamma = -\rho(\partial\bar{P}_2/\partial\rho)_T/T(\partial\bar{P}_2/\partial T)_\rho,$$

$$= (\rho/T)(\partial T/\partial\rho)_{\bar{P}_2};$$

this was initially suggested by Alben [7] as a particularly sensitive probe of the relative importance of attractive and repulsive interactions. It was first determined successfully by McColl and Shih [8]. The plots of $\ln T$ against $\ln\rho$ at constant $\bar{P}_2$ are virtually linear and were found to have a common slope of $\Gamma = 4.0 \pm 0.1$ in the nematic phase of p-azoxyanisole (PAA).

Examination of the recent literature reveals the extent to which theoreticians depend on reliable values for these quantities [1,7,9-12]. Because more or less complete sets of data are available for them, PAA and 4-methoxybenzylidene-4'-butylaniline (MBBA) are the usual materials chosen to investigate the nematic-isotropic transition; presumably these are *typical* nematic liquids. Often, one or more of the data points are used to fit model parameters and the theory is tested by its ability to yield agreement with the remainder of the experimental data. The relative success or failure of such a theory is therefore subject to the accuracy of the data.

Comparison of Data for PAA and MBBA

The table is a compilation of ΔH, $\Delta V/V$ and dT/dP values reported for PAA and MBBA [3,4,8,13-27]. All measurements were apparently done with care; yet, variations of up to 170% (ΔH for MBBA) are evident. Assuming that dT/dP is the most accurate quantity for PAA, the $\Delta V/V$ values support ΔH's in the range 540 to 710 J mol^{-1}. Assuming that $\Delta V/V$ is the most accurate quantity for MBBA, the estimated dT/dP suggests ΔH to be about 290 J mol^{-1}. Parenthetically, we note that a precision of $\pm$ 10% is claimed for the measurement of ΔH by adiabatic calorimetry which gave a value of 623 J mol^{-1} [23]. There are clearly problems with the

Table. *ΔH, ΔV/V and dT/dP for PAA and MBBA at the nematic-isotropic transition.*

PAA(T_{NI} = 409K)

ΔH/J mol^{-1}	Ref	$\Delta V/V \times 10^2$	Ref	(dT/dP)/$Kkbar^{-1}$	Ref
573	[13]	0.35	[17[a]]	48^d	[22]
736	[14]	0.30	[17[b]]	48^d	[4]
628	[15]	0.28	[17[c]]	46^e	[8]
757	[16]	0.36	[18,19]		
607	[3]	0.34	[20]		
		0.35	[21]		
		0.31	[3]		

(a) direct value quoted in ref [21]. (b) extrapolated value quoted in ref [3]
(c) extrapolated value from analysis in ref [4]. (d) up to 1 kbar.
(e) determined indirectly from NMR.

MBBA (T_{NI} = 318K)

ΔH/J mol^{-1}	Ref	$\Delta V/V \times 10^2$	Ref	(dT/dP)/$Kkbar^{-1}$	Ref
623	[23]	0.11	[26]	$\sim 33^a$	[27]
284	[24]	0.11	[3]		
314	[25]				
234	[3]				

(a) up to 2 kbar

PAA and MBBA data. Possible reasons for these discrepancies are: relative sample purity, the minor difficulty of working at relatively high temperatures for PAA, deterioration (through hydrolysis) of MBBA, and the problem of adequately accounting for pretransitional contributions to ΔH [2,4] and ΔV/V [4].

In addition to the requirement of more definitive measurements on PAA and MBBA, what can be suggested at this point? First, although many data exist or are readily obtainable for transition temperatures, enthalpy changes and, to a certain extent, order parameters, there is a need for additional data (e.g., ρ, Δρ/ρ, and dT/dP and Γ) on selected substances and homologous series. The materials chosen should be stable, have relatively low transition temperatures and be pure. Secondly, there should be inter-laboratory comparisons to ensure reproducibility of result

Both classical and scanning calorimetry [2], differential dilatometry [3] and newer density measurement techniques [28,29], should be included. The example of non-electrolytic solution thermodynamicists [29] should be followed in time; *viz*, the eventual use of reference systems for ΔH, ΔV, $\bar{P}_2$ etc., (see Appendix).

Data Usage in Molecular Interpretation and Model Testing

Although the PAA and MBBA results are less than satisfactory, they are good enough to indicate, e.g., that orientationally restricted hard rod models are too high by an order of magnitude in their prediction of $\Delta\rho/\rho$ and that any respectable rigid rod model should yield enthalpy changes of hundreds rather than thousands of J mol^{-1}. Similarly qualitative and even quantitative information has been successfully used and is being used to infer important molecular structure-phase stability relations and to test molecular statistical models. A few prominent examples are described below.

Importance of Molecular Geometry in Nematic Stability Dewar and co-workers [30, 31] undertook a systematic study "to provide further information concerning the relative importance of four features of molecules that have been thought to contribute to mesophase stability, i.e., rigidity, linearity, polarizability and enhancement of polarity by conjugation" [30]. In this classic investigation, also discussed in Chapter 1, the central p-phenylene ring of p-phenylene dianisate was replaced by about twenty other groups, causing variations in these four features. Transition temperatures and ΔH (hence ΔS) were measured and interpreted. It was concluded that the predominant factors in determining mesophase stability were geometric; that is, the essential requirement for nematic phase stability was a "linear rod-shaped molecule of more or less uniform cross-section". These findings support the belief, based on now extensive knowledge of simple fluids [32], that theoretical efforts should be properly directed towards obtaining a good equation of state for hard rods. This school of thought maintains that short-range repulsive interactions determine the structure of liquids, while the longer-range attractive interactions are responsible for the equilibrium density [12].

End-Chain Flexibility in Nematic Liquids Often the first evidence, albeit inferential, of unexpected molecular structural features in mesophases comes from simple thermodynamic measurements. The work of Haller and co-workers [33,34] is a good example of this. In a study of homologous nitrones [33], they observed that the entropies for the mesophase (smectic and nematic) to isotropic transitions, in their systems and others, increased regularly and fairly linearly with the number of chain carbons. For longer chains, the $\Delta S/R$ increments per methylene group ranged from 0.04 to 0.15(NI) and from 0.08 to 0.27(SI), both far less than the increment associated with the melting of all-trans chains which is approximately ln3 or 1.10. This suggested appreciable chain flexibility in both mesophases (slightly more so in nematics). In a subsequent study [34], they compared the molar volume increment per methylene group in nematic stilbene derivatives (16.3 $cm^3\ mol^{-1}$) with the group contributions in normal isotropic liquids (16.4 - 17.1 $cm^3\ mol^{-1}$) and solid state n-alkanes (14.5 $cm^3\ mol^{-1}$). This comparison indicated that the packing of the n-alkyl groups was more liquid-like than solid-like, consistent with their earlier deduction. These findings and the Wulf-de Rocco model [35] stimulated us [36-38] and Marcelja [39] to relax the rigid rod models by allowing for the possibility of end-chain flexibility. Further, this incisive (though qualitative) thermodynamic inference has since been confirmed and placed on a more quantitative level by NMR studies of chain ordering [40-42].

A recent dilatometric investigation [43] may interest those concerned with the physics of smectic phases. For the homologous series studied, methylene volume increments of 14.7 $cm^3\ mol^{-1}$ for the crystal phase (C) and 15.6 - 16.0 $cm^3\ mol^{-1}$ for three smectic phases (designated S_1, S_2 and S_3) were found. This, considered together with aromatic core contributions, led to the suggestion that the aliphatic chains essentially melt in a single step during the $C \rightarrow S_1$ transition, followed by stepwise increases of disorder in the smectic layers from $S_1 \rightarrow S_2$ and $S_2 \rightarrow S_3$.

Second-Order Smectic A-Nematic (S_AN) Transitions McMillan's molecular theory [1,44] predicts that a second-order S_AN transition should occur at the short chain length end of homologous series; more precisely, where $T_{S_AN}/T_{NI} < 0.89$. Recent refinements [45] place this point at $T_{S_AN}/T_{NI} < 0.89$. This interesting prediction has motivated demanding thermodynamic measurements, particularly on 4-cyanobenzylidene-4'-n-octyloxyaniline (CBOOA). In support of a weakly first-order transition, one can cite volumetric [46], calorimetric [47], high pressure [48] and diamagnetic susceptibility [49] evidence. While for a second-order transition, there is Rayleigh light scattering evidence [50] and, more recently, a differential dilatometry study [51] which sets the S_AN volume discontinuity at less than $10^{-5}\ cm^3g^{-1}$!. At present the majority of evidence is in favour of a weakly first-order S_AN transition in CBOOA, although it should be noted that T_{S_AN}/T_{NI} is 0.935 [49].

Better systems have been studied, first by de Jeu [52], using calorimetry and birefringence measurements, Δn, and more recently by the Bordeaux group [49] who also employed calorimetry together with the anisotropy of dimagnetic susceptibility, $\Delta\chi$; these systems are the hexyl to octyl homologues of 4,4'-di-n-alkylazoxybenzene and mixtures thereof. For the hexyl derivative T_{S_AN}/T_{NI}=0.877. Although non-zero transition enthalpies (ΔH is 20 J mol^{-1} for the hexyl) are found for all homologues and their binary mixtures, there appears to be simultaneous convergence towards zero for ΔH, Δn and $\Delta\chi$ at the S_AN transition as T_{S_AN}/T_{NI} approaches about 0.89. Further, the experimental curves [49] are in excellent agreement with the theoretical curves obtained by Lee *et al.* [45]. It would seem, therefore, that the molecular field theory of McMillan, as modified by Lee *et al.* may indeed be correct, at least as to its prediction of the onset of second-order behaviour at $T_{S_A}/T_{NI} \approx 0.89$. Whatever the final outcome, thermodynamic information will continue to be crucial and a dividend of this considerable effort has been increased precision in volumetric and calorimetric measurements.

Binary Mixtures

We shall now consider equilibrium phase transitions in binary non-mesomorphic solute/mesomorphic solvent mixtures. As before, the perspective is a somewhat restricted one, directed here towards providing the thermodynamic relations and experimental results required to examine certain molecular statistical theories of the nematic-isotropic (NI) transition in binary mixtures. The thermodynamics of only dilute solutions, that is, the Henry's law region, are developed. Also, the experimental results presented and discussed appertain to low concentration solute-induced NI transitions, with emphasis on the effects of solute size, shape and flexibility on nematic phase stability.

Thermodynamic Relations

With few exceptions, which we describe, non-mesomorphic solutes depress the normal NI transition temperature (T_{NI}) in liquid crystals [53-61]. Moreover, recent experiments [56, 58-60, 62] have confirmed de Kock's [63] observation that the presence of such impurities leads to the formation of a two-phase region. This is entirely consistent with the laws of thermodynamics and the first-order nature of the nematic-isotropic transition ($\Delta H \neq 0$ [56, 64], $\Delta V \neq 0$ [62] and $\Delta \bar{P}_2 \neq 0$ [54, 60]). The general phase behaviour at low solute mole fraction, x, is illustrated in figure 1, there T^* (=T/T_{NI}) is the reduced temperature. Thus, when a binary nematic system of composition x' is heated, the isotropic phase begins to appear at $T_n{}^*$ and the nematic phase completely disappears at

T_i^*. Alternatively, incremental addition of solute at a fixed temperature $T^{*'}$ to an originally pure nematic material results in initial appearance of the isotropic phase at x_n and final disappearance of the nematic phase at x_i.

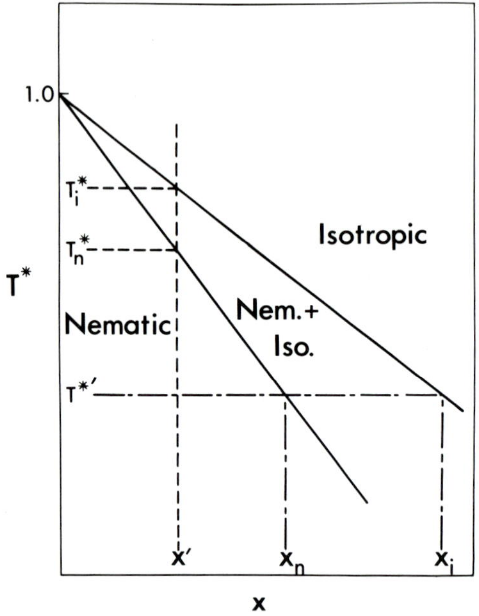

Fig. 1 *Typical reduced temperature (T^*)-solute mole fraction (x) diagram at low solute mole fraction.*

For the two phases in equilibrium at $T^{*'}$, the solute partial pressures must be equal [56, 59]; hence

$$\gamma_n x_n = \gamma_i x_i, \tag{1}$$

where γ_n and γ_i are the solute activity coefficients at x_n and x_i, respectively; they are both based on the Raoult's law convention (i.e., $\gamma \to 1$ as $x \to 1$). For sufficiently dilute solutions ($x \to 0$), the γ's approach limiting or infinite dilution activity coefficients (γ_n^∞ and γ_i^∞), and so

$$\gamma_n^\infty/\gamma_i^\infty = x_i/x_n. \tag{2}$$

The infinite dilution activity coefficient is proportional to the Henry's law constant K. The smaller the value of γ^∞ or K, the more soluble is the solute at fixed partial pressure and temperature; that is, a small γ^∞ implies solute compatibility with the solvent.

Applying the conditions of phase equilibrium for each component and utilizing the Gibbs-Duhem relations for each phase, we obtain, for a first-order transition, the following expressions for the limiting slopes of the lower line or nematic boundary (β_n^∞) and the upper line or isotropic boundary (β_i^∞) in figure 1 [56, 59],

$$\beta_n^\infty = \{(x_n - x_i)/x_n\}(R/\Delta S_{NI}),$$
$$= \{(\gamma_i^\infty - \gamma_n^\infty)/\gamma_i^\infty\}(R/\Delta S_{NI}),$$

$$\beta_i^\infty = \{(x_n - x_i)/x_i\}(R/\Delta S_{NI}),$$
$$= \{(\gamma_i^\infty - \gamma_n^\infty)/\gamma_n^\infty\}(R/\Delta S_{NI}),$$

where ΔS_{NI} is the NI transition entropy for the pure nematic solvent. It follows that

$$\beta_n^\infty/\beta_i^\infty = x_i/x_n,$$
$$= \gamma_n^\infty/\gamma_i^\infty. \quad (5)$$

Thus, at fixed $T^{*\prime}$

$$(x_i - x_n) = (1/\beta_n^\infty - 1/\beta_i^\infty)(1 - T^{*\prime}),$$
$$= (\Delta S_{NI}/R)(1 - T^{*\prime}) \quad (6)$$

and at fixed x'

$$(T_i^* - T_n^*) = (\beta_i^\infty - \beta_n^\infty)x'. \quad (7)$$

Although eqs. (3) - (7) are applied here to nematic-isotropic phase equilibria, they are generally valid for any first-order situation at low solute concentration (e.g. smectic-nematic or smectic-isotropic equilibria).

Eq. (6) clearly shows that, if $\Delta S_{NI} \neq 0$, a two-phase region must exist at fixed $T^{*\prime}$ and that the size of this region depends only on the magnitude of ΔS_{NI}. Eq. (6) is similar in form to the expression for equilibrium between a dilute solution and the pure solid phase of the solvent, i.e., the freezing-point depression relation:

$$x = (\Delta S_m/R)(1 - T/T_m),$$
$$= (\Delta S_m/R)(1 - T_m^*), \quad (8)$$

where x is the solute mole fraction in the solution phase and m denotes melting. Comparison of eqs. (6) and (8) reveals why early proposals [63,65] to use liquid crystal solvents for solute molecular weight determination by transition temperature depression were impractical. Eq. (8) involves a readily determinable absolute solute concentration, while eq. (6) involves the far less accurate determination of a small difference in solute concentrations.

Eq. (7) gives the extent of the two-phase region in terms of temperature at fixed x'. If the solute is reasonably similar to the nematic solvent (i.e. if it is a latent nematic such as a non-mesomorphic Schiff's base [53, 65, 66], the β^{∞} values would be expected to be uniformly small in magnitude; hence, $(T_i^* - T_n^*)$ would also be quite small. It is not surprising that most workers [53, 65, 66] have failed to observe a two-phase region with such mixtures.

We see from eqs. (3) and (4) that β^{∞} is equal to the product of a pure solvent property $(\Delta S_{NI}/R)$ and a solution property determined by the ratio $\gamma_n^{\infty}/\gamma_i^{\infty}$. Both properties influence the magnitude of β^{∞} while the latter property alone determines the sign of β^{∞}. If $\gamma_n^{\infty}/\gamma_i^{\infty} < 1$, the solute elevates the nematic-isotropic transition ($\beta^{\infty} > 0$ and the lines in figure 1 are rotated counterclockwise by 90^{0}); if $\gamma_n^{\infty}/\gamma_i^{\infty} > 1$, the solute depresses the NI transition and $\beta^{\infty} < 0$ as in figure 1.

The smaller the value of $\Delta S_{NI}/R$, the larger is the modulus of β^{∞}. Hence, all else being equal, nematic solvents exhibiting more weakly first-order NI transitions are more easily perturbed by solute impurities. Returning to the solution property, it is important to recognize that the *ratio* $\gamma_n^{\infty}/\gamma_i^{\infty}$ appears in eqs. (3) and (4). This means that β^{∞} depends only on the *relative preference* of the solute for the two phases. If the non-mesomorphic solute is more compatible with the nematic phase, then $\gamma_n^{\infty}/\gamma_i^{\infty} < 1$, $\beta_i^{\infty} > \beta_n^{\infty} > 0$, and the nematic phase is richer in the solute $(x_n > x_i)$. A few cases of this unusual behaviour [61] have been reported in the literature;solutes which form geometrically and energetically favoured charge-transfer complexes with the nematic solvent [57], and certain plate-like solutes [61]; this behaviour is qualitatively consistent with Alben's lattice model predictions [67]. The more usual behaviour is where the solute prefers the isotropic phase. In these cases, the non-mesomorphic solute, present in sufficiently high concentration, disrupts nematic order and induces an NI transition at $T^{*'} < 1$. Accordingly, $\gamma_n^{\infty}/\gamma_i^{\infty} > 1$, $x_i > x_n$ and $\beta_n > \beta_i > 0$, where β will now denote the *modulus* of the slope of β^{∞}. Clearly the larger the value of β, the more disruptive is the solute.

Experimental Results

There is a considerable body of data on nematic-isotropic transition temperatures of binary non-mesomorphic solute/mesomorphic solvent mixtures [53, 55-61, 66]. However, most of these measurements are on latent mesomorphic solutes (rod-like molecules), for which the geometric variations are generally too minor to be of interest here. Some of the remaining studies are suspect, since no two-phase regions are reported, and many are too qualitative, lacking in detail or not sufficiently systematic for our present purposes. Fortunately, a set of systematic measurements recently completed by Oweimreen [68] provides the information required to examine pronounced solute geometric effects and test certain molecular statistical theories.

Oweimreen utilized two different approaches to obtain β_n and β_i. One involved the combination of gas-liquid chromatography (GLC) [69] to determine $\gamma_n^\infty/\gamma_i^\infty$ and DSC to measure $\Delta S_{NI}/R$. Typical GLC results [68] are shown in the $\ln\gamma^\infty$ against T^{-1} plots of figure 2; the dashed vertical line is T_{NI}^{-1} for the pure nematic solvent. The discontinuity in $\ln\gamma^\infty$ at T_{NI} yields $\ln(\gamma_n^\infty/\gamma_i^\infty)$. The problem with this approach, however, is that the

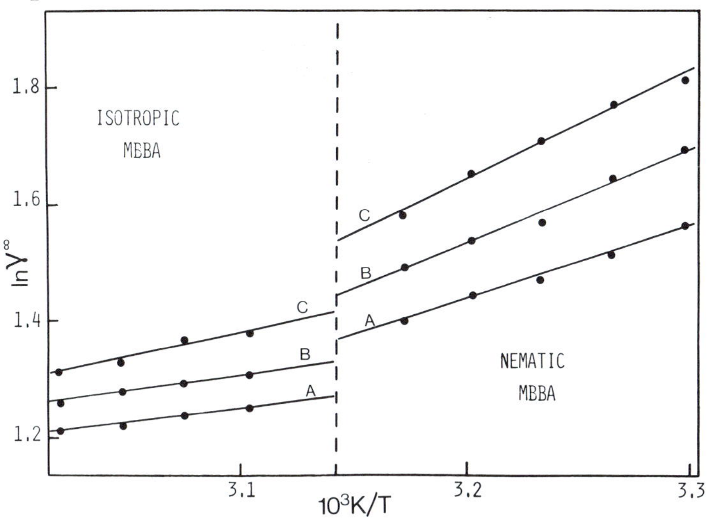

Fig. 2 *Natural logarithm of solute activity coefficient ($\ln\gamma^\infty$) versus reciprocal temperature $10^3K/T$ from GLC measurements* [68]. *(A) n-hexane; (B) 3-ethylpentane; (C) 3,3-diethylpentane.*

combined experimental error in $\gamma_n^\infty/\gamma_i^\infty$ and $\Delta S_{NI}/R$ can lead to uncertainties as high as 20% in β. Thus, although this technique is strictly a semiquantitative one (but an advantageous one for rapid screening of a variety of solutes and nematic solvents), some unequivocally clear trends found *via* this approach [68] will be quoted later.

The other approach is more tedious but far more precise, the typical standard deviation in β being $\sim \pm 0.01$. It is a visual experiment [59], where the appearance and disappearance of the two-phase region are recorded during controlled heating and cooling at fixed composition, that is, moving vertically in figure 1. Normally, solute mole fractions in the range $0.01 < x' < 0.05$ are used. A typical phase diagram so obtained is shown in figure 3, for the system tetra-n-propyl tin (Pr_4Sn) in MBBA for which $\beta_n \approx 1.2$ and $\beta_i \approx 1.0$. For this system, the two-phase region extends over a temperature range of $\Delta T \approx 3^o$ at $x' \approx 0.05$, and over a mole fraction range of $\Delta x \approx 0.008$ at $T^{*'} \approx 0.95$. The overall low-

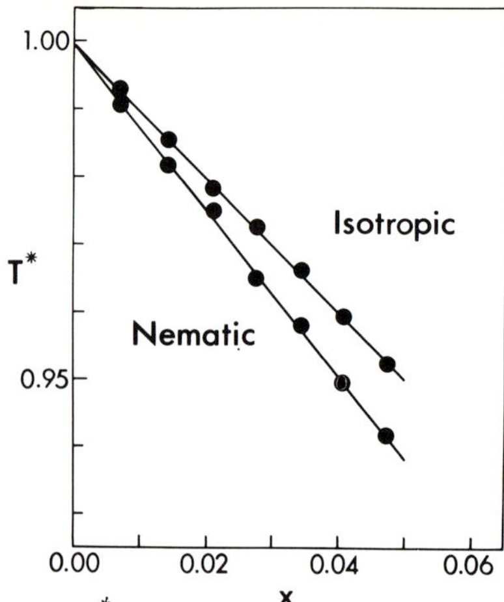

Fig. 3 *Reduced temperature (T^*) - solute mole fraction (x) phase diagram for Pr_4Sn/MBBA.*

ering of the transition temperature at x' ≈ 0.05 is about 19° at the nematic phase boundary line.

Using the visual method, Oweimreen [68] studied a series of five quasi-spherical solutes and a series of seven chain-like solutes in both MBBA and 4'-n-pentyl-4-cyanobiphenyl (5CB). These two nematic solvents have roughly the same length-to-breadth ratio, but different $\Delta S_{NI}/R$ values (0.16 for MBBA and 0.25 for 5CB, determined by DSC [68]). The spherical solutes studied were 3,3-diethylpentane (Et_4C) and several tetra-n-alkyl tins (R_4Sn; R = Me, Et, Pr, Bu); the chain-like solutes were n-alkanes ranging from n-C_8H_{18} to n-$C_{14}H_{30}$. Thus, the surfaces of all solute molecules studied were formed of methyl or methylene groups, leaving only solute size, shape and flexibility as variables.

The experiments, then, were of the solute probe variety, with the attendant advantage of being able to introduce and study the effect of major steric perturbations at low perturbant concentrations. The spherical solute series spans a 2.5-fold range in size and the chain solute series an almost twofold range.

The results for the quasi-spherical solutes in MBBA are summarized in figure 4, where D is the estimated ratio of the hard-core diameter of the solute to that of the solvent (assumed to be a spherocylinder)[59]; similar results were found for 5CB. Also, from GLC and from analysis of the visual β_n/β_i *via* eq. (5), it is found that $\gamma_n^\infty/\gamma_i^\infty$ is larger for solutes in 5CB and increases with increasing D^3 in both solvents [68]. As expected {see eqs. (3) and (4)}, the latter trend parallels the observed increase in β with increasing D^3. That is, as the size of the solute increases, its incompatibility with the nematic phase increases; thus,

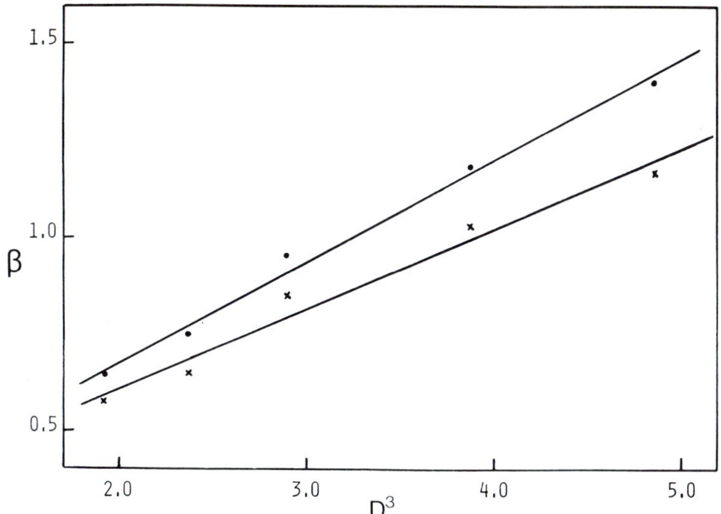

Fig. 4 β *against* D^3 *plot for quasi-spherical solutes in MBBA, where D is a relative diameter defined in the text;* β_n *is denoted by* • *and* β_i *by* x.

the bigger sphere (Bu_4Sn) is roughly twice as disruptive as the smallest sphere (Me_4Sn). Comparing the two solvents, the β values are greater for MBBA, because of its smaller ΔS_{NI} in comparison with 5CB. The size of the two-phase region $(x_i - x_n)$ at fixed $T^{*'}$ is larger for solutes in 5CB which is also consistent with eq. (6).

Figure 5 gives rather interesting results for the n-alkane solutes in MBBA; analogous trends were observed for 5CB. First, we note that, for the same reasons described for the spherical solutes, the β's are greater with MBBA and the difference $(x_i - x_n)$ with 5CB. The next obvious feature is the curious odd-even behaviour, with the even-numbered solutes being

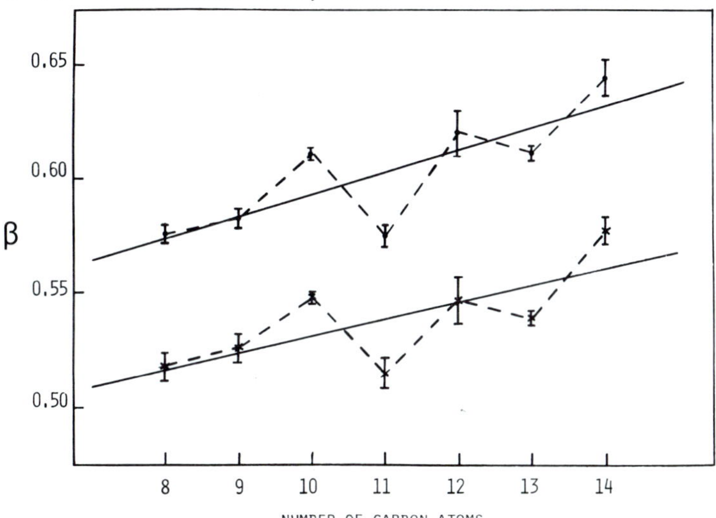

Fig. 5 β *against solute carbon number (n) plot for n-alkanes in MBBA;* β_n *is denoted by* • *and* β_i by x.

more disruptive in MBBA; in contrast the odd alkanes are more disruptive in 5CB. As with most homologous series of pure liquid crystals, the odd carbon members are more stable (higher T_{NI}) in both the MBBA [56] and 5CB [70] series. Given the possibility of segmental correlation between the alkane solute and the liquid crystal end-chain, the reader may speculate about this curious odd-even behaviour.

Neglecting these odd-even effects, we see from the solid lines in figure 5 that almost doubling the size of the alkane solute leads to relatively small changes in the β's. In fact, the β values *increase* by about 10% over the entire range; that is, the longer chains are only slightly more disruptive. (Note that in pure low-temperature liquid crystals, nematic stability usually increases with increasing length of the end-chain [56]). This near constancy of β is confirmed by GLC measurements [68] which indicate that, unlike the spherical solutes, $\gamma_n^\infty/\gamma_i^\infty$ is virtually independent of the alkane solute chain length (n-C_5H_{12} to n-$C_{10}H_{22}$). Further supportive evidence is found in an NMR study by Kronberg *et al.* [60]. They observed that 5 mole percent solutions of n-C_6H_{14} and n-$C_{16}H_{34}$, each dissolved in nematic MBBA, gave nearly superimposable order parameter vs. temperature curves up to the NI transition, although, the n-hexane solutions appear to be slightly more ordered.

Now, if the n-alkanes were completely rigid molecules, we would expect the β values to *decrease* markedly with increasing solute length [71]. That only a slight increase is observed indicates that the dissolved alkane solutes have some degree of flexibility and, perhaps the effective length-to breadth ratio is essentially constant throughout the solute series. A statistical thermodynamic analysis [36] of $\gamma_n^\infty/\gamma_i^\infty$ for such n-alkane solutes indeed suggests that the aligned environment somewhat restricts the conformational freedom of the solute chains, but the chains are more liquid-like than solid-like, and that the contribution to $\gamma_n^\infty/\gamma_i^\infty$ associated with rotation of the solute molecule as a whole is apparently independent of solute chain length.

Figure 6 contains a summary of all the visual data for MBBA; very similar results are observed for 5CB. The dashed and solid lines represent linear least-squares fits of the experimental β results shown in figures 4 and 5; the ratio of solute-to-solvent hard-core volumes, V_2^*/V_1^* is estimated using Bondi's tabulation [59, 68, 72] of group contributions. Figure 6 shows that not only are the β values larger for spheres than for n-alkanes of comparable size, but also that the spherical solutes display a much greater rate of increase with increasing V_2^*. Comparison of $\gamma_n^\infty/\gamma_i^\infty$ for Et_4C and n-C_9H_{20} further confirms that, relative to the isotropic phase, the spherical solute is less compatible with the ordered

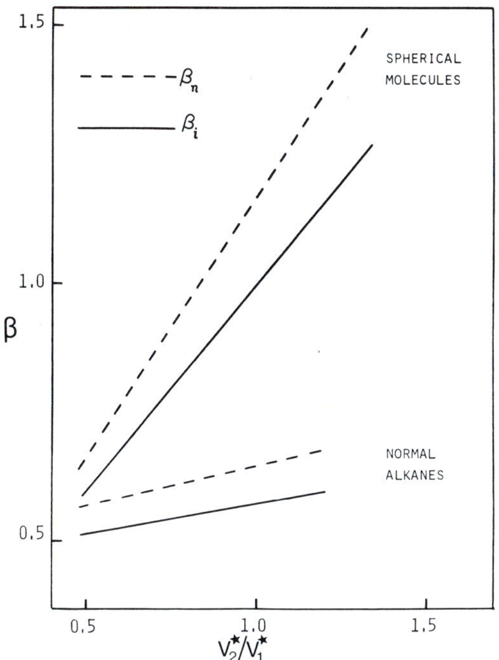

Fig. 6 β *against* V_2^*/V_1^* *plots for quasi-spherical and n-alkane solutes in MBBA;* V_2^* *and* V_1^* *are respectively, solute and solvent hard core volumes.*

nematic phase [68]. Interestingly, the lines for the spherical and chain-like solutes intersect at a V_2^*/V_1^* corresponding to C_5H_{12}. This suggests that n-pentane and neopentane should be equally disruptive. In their NMR work, Kronberg *et al.* [60] found that equivalent amounts of branched and normal hexane produce almost equally ordered nematic solutions, while a highly branched $C_{16}H_{34}$ was clearly more disruptive, with lower solution order parameters, that n-$C_{16}H_{34}$.

Finally, it should be mentioned that the combined GLC-DSC approach [68] and other studies [61] show that, in general, a given solute disrupts shorter rod-like solvents to a greater extent than longer ones. This is intuitively acceptable and also consistent with the observation that ΔS_{NI} generally increases with increasing solvent length-to-breadth ratio.

We shall return to these data, in the following Chapter, when considering the predictions of theories of binary non-mesomorphic solute/mesomorphic solvent mixtures.

Acknowledgement

Research support from the National Science Foundation is gratefully acknowledged. My collaborator in this work, Dr. Ghassan A. Oweimreen, is thanked for several helpful discussions.

Appendix

Possible homologous series for more extensive thermodynamic study of the NI transition are

(1) 4,4'-di-n-alkylazobenzenes [a],
(2) 4,4'di-n-alkoxyazoxybenzes [a],
(3) 4-n-alkyl-4'-cyanobiphenyls [b],
(4) 4-n-alkoxy-4'-cyanobiphenyls [b],
(5) 2-(p-n-alkoxybenzylidene)-aminofluorenones [b].

Potential reference systems could be chosen from the above series (e.g. the pentyl members of series 1 and 3).

Series 1, 3 and 4 have relatively low transition temperatures, thus T_{NI} is between $\sim 15^{o}$ and $\sim 90^{o}C$ and all exhibit the general trend of increasing T_{NI} with increasing chain length (n). Series 2 and 5, however, have higher T_{NI} values in the general range of 120° to $170^{o}C$, but have trends of T_{NI} decreasing or remaining fairly constant, respectively, with increasing n. These series should be of particular interest since there is some controversy over the significance of these trends [b].

It should be noted that Marcelja [39] relied heavily on series 1 and 2 in testing his theory of chain ordering. Also, he, Cotter [c] and other theoreticians have required estimates of attractive energy parameters involving both central core and chain interaction. There would appear to be only one source of such information for nematic liquids: ΔH of vaporization for PAA, determined by Solsky and Grushka [d] from the measurement of vapour pressure as a function of temperature. It would be useful to have such data for other members of series 2.

[a] J. van der Veen, W.H. de Jeu, M.W.M. Wanninkhof and C.A.M. Tienhoven, *J. Phys. Chem.*, **77**, 2153 (1973).

[b] G.W. Gray and A. Mosley, *J. Chem. Soc. Perkin II*, 97 (1976).

[c] M.A. Cotter, *Mol. Cryst. Liq. Cryst.*, **35**, 33 (1976).

[d] J.F. Solsky and E. Grushka, *J. Phys. Chem.*, **78**, 275 (1974).

References

1. W.L. McMillan, *Phys. Rev. A*, 4, 1238 (1971).
2. E.M. Barrall, II and J.F. Johnson in *Plastic Crystals and Liquid Crystals* Vol. 2, ed. by G.W. Gray and P.A. Winsor, Ellis Horwood Ltd, Chichester, 1974, p. 254.
3. see e.g. D. Armitage and F.P. Price, *Phys. Rev. A*, 15, 2496 (1977).
4. see e.g. W. Klement, Jr. and L.H. Cohen, *Mol. Cryst. Liq. Cryst.*, 27, 359 (1974).
5. see e.g. R. Chang, *Mol. Cryst. Liq. Cryst.*, 36, 155 (1976).
6. see e.g. J. Bendler, *Mol. Cryst. Liq. Cryst.*, 38, 19 (1977).
7. R. Alben, *Mol. Cryst. Liq. Cryst.*, 13, 193 (1971).
8. J.R. McColl and C.S. Shih, *Phys. Rev. Lett.*, **29**, 85 (1972); J.R. McColl, *Phys. Lett.*, 38A, 55 (1972).
9. S. Chandrasekhar and N.V, Madhusudana, *Mol. Cryst. Liq. Cryst.*, **24**, 179 (1973).
10. R. Pynn, *J. Chem. Phys.*, 60, 4579 (1974).

11. Y.R. Shih, Y.R. Lin-Liu and C.W. Woo, *Phys. Rev. A*, **14**, 1895 (1976).
12. M.A. Cotter, *J. Chem. Phys.*, **66**, 1098 (1977).
13. H. Arnold, *Z. Phys. Chem.* (Leipzig), **226**, 146 (1964).
14. E.M. Barrall, II, R.S. Porter and J.F. Johnson, *J. Phys. Chem.*, **68**, 2801 (1964).
15. M. Leclercq, J. Billard and J.C. Jacques, *C.r. hebd. Séanc. Acad. Sci.*, Paris, **264**, 1789 (1967).
16. L.C. Chow and D.E. Martire, *J. Phys. Chem.*, 73, 1127 (1969).
17. W. Maier and A. Saupe, *Z. Naturforsch.*, **A15**, 287 (1960).
18. R.S. Porter and J.F. Johnson, *J. Appl. Phys.*, **34**, 51 (1963).
19. E. McLaughlin, M.A. Shakespeare and A.R. Ubbelohde, *Trans. Faraday Soc.*, **60**, 25 (1964).
20. F.P. Price and J.H. Wendorff, *J. Phys. Chem.*, 76, 2605 (1972).
21. B. Bahadur, *Mol. Cryst. Liq. Cryst.*, **35**, 83 (1976).
22. J. Robberecht, *Bull. Soc. Chim. Belges*, 47, 597 (1938).
23. J. Mayer, T. Waluga and J.A. Janik, *Phys. Lett.*, **41A**, 102 (1972).
24. T. Shinoda, Y. Maeda and H. Enokido, *J. Chem. Therm.*, 6, 921 (1974).
25. G.W. Smith and Z.G. Gardlund, *J. Chem. Phys.*, **59**, 3214 (1973).
26. M.J. Press and A.S. Arrott, *Phys. Rev. A*, 8, 1459 (1973).
27. P.H. Keyes, H.T. Weston, W.J. Lin and W.B. Daniels, *J. Chem. Phys.*, **63**, 5006 (1975).
28. E. Gulari and B. Chu, *J. Chem. Phys.*, **62**, 795 (1975).
29. J.R. Goates, J.B. Ott and J.F. Moellmer, *J. Chem. Therm.*, 9, 249 (1977).
30. M.J.S. Dewar and R.M. Riddle, *J. Amer. Chem. Soc.*, **97**, 6658 (1975).
31. M.J.S. Dewar and A.C. Griffin, *J. Amer. Chem. Soc.*, **97**. 6662 (1975).
32. H.C. Anderson, D. Chandler and J.D. Weeks, *Adv. Chem. Phys.*, **34**, 105 (1976).
33. W.R. Young, I. Haller and A. Aviram, *Mol. Cryst. Liq. Cryst.*, 13, 357 (1971).
34. I. Haller, H.A. Huggins, H.R. Lilienthal and T.R. McGuire, *J. Phys. Chem.*, **77**, 950 (1973).
35. A. Wulf and A.G. deRocco, *J. Chem. Phys.*, **55**, 12 (1971).
36. D.E. Martire, *Mol. Cryst. Liq. Cryst.*, 28, 63 (1974).
37. G.I. Ågren and D.E. Martire, *J. Chem. Phys.*, **61**, 3959 (1974).
38. a. F. Dowell, Doctoral Dissertation, Georgetown University, 1977.
    b. F. Dowell and D.E. Martire, *J. Chem. Phys.*, **68**, 1088, 1094 (1978).
39. S. Marcelja, *J. Chem. Phys.*, **60**, 3599 (1974).
40. J.W. Emsley, J.C. Lindon and G.R. Luckhurst, *Mol. Phys.*, **30**, 1913 (1975).
41. B. Deloche, J.Charvolin, L. Liebert and L. Strzelecki, *J. Phys.* (Paris), 36 C1-121 (1975).
42. B. Deloche and J. Charvolin, *J. Phys.* (Paris), **37**, 1497 (1976).
43. D. Guillon and A. Skoulios, *J. Phys.* (Paris), **37**, 797 (1976).
44. W.L. McMillan, *Phys. Rev. A*, 6, 936 (1972).
45. F.T. Lee, H.T. Yan, Y.M. Shih and C.W. Woo, *Phys. Rev. Lett.*, **31**, 1117 (1973); *Phys. Lett.*, **A48**, 68 (1974).
46. S. Torza and P.E. Cladis, *Phys. Rev. Lett.*, **32**, 1046 (1974).
47. D. Djurek, J. Baturic-Rubcic and K. Franulovic, *Phys. Rev. Lett.*, 33, 1126 (1974).
48. W.J. Lin, P.H. Keyes and W.B. Daniels, *Phys. Lett.*, A49, 453 (1974).
49. M.F. Achard, F. Hardouin, G. Sigaud and H. Gasparous, *J. Chem. Phys.*, **65**, 1387 (1976).
50. K.C. Chu and W.L. McMillan, *Phys. Rev. A*, **11**, 1059 (1975).
51. D. Armitage and F.P. Price, *Mol. Cryst. Liq. Cryst.*, **38**, 229 (1977).
52. W.H. de Jeu, *Solid State Commun.*, **13**, 1521 (1973).
53. J.S. Dave and R.A. Vora, in *Plastic Crystals and Liquid Crystals*, Vol.1, ed. by G.W. Gray and P.A. Winsor, John Wiley and Sons, New York, 1974, p153 (and references therein).
54. D.H. Chen and G.R. Luckhurst, *Trans. Faraday Soc.*, **65**, 656 (1969).
55. P.E. Cladis, J. Rault and J.P. Burger, *Mol. Cryst. Liq. Cryst.*, **13**, 1 (1971).
56. H.T. Peterson and D.E. Martire, *Mol. Cryst. Liq. Cryst.*, **25**, 89 (1974).
57. J.W. Park, C.S. Bak and M.M. Labes, *J. Amer. Chem. Soc.*, **97**, 4398 (1975).
58. S.A. Shaya and H. Yu, *J. Phys.* (Paris), **36**, C1-59 (1975).
59. D.E. Martire, G.A. Oweimreen, G.I. Ågren, S.G. Ryan and H.T. Peterson, *J. Chem. Phys.*, **64**, 1456 (1976).
60. B. Kronberg, D.F.R. Gilson and D. Patterson, *J. Chem. Soc., Faraday II*, **72**, 1673 (1976).

61. G. Sigaud, M.F. Achard, F. Hardouin and H. Gasparoux, *Chem. Phys. Lett.*, **48**, 122 (1977).
62. R.A. Orwoll, R.H. Rhyne. Jr., S.D. Christensen and S.N. Young, *J. Phys. Chem.*, **81**, 181 (1977).
63. A.C. de Kock, Z. *Phys. Chem.*, **48**, 129 (1904)
64. E.C.-H. Hsu and J.F. Johnson, *Mol. Cryst. Liq. Cryst.*, 20, 177 (1973).
65. J.S. Dave and M.J.S. Dewar, *J. Chem. Soc.*, 4616 (1954); 4305 (1955).
66. J.M. Lohar and D.S. Shah, *Mol. Cryst. Liq. Cryst.*, 28, 293 (1974).
67. R. Alben, *J. Chem. Phys.*, **59**, 4299 (1973).
68. G.A. Oweimreen, Doctoral Dissertation, Georgetown University, 1977.
69. L.C. Chow and D.E. Martire, *J. Phys. Chem.*, **75**, 2005 (1971).
70. G.W. Gray and A. Mosley, *J. Chem. Soc. Perkin II*, 97 (1976).
71. H.T. Peterson, D.E. Martire and M.A. Cotter, *J. Chem. Phys.*, **61**, 3547 (1974).
72. A. Bondi, *J. Phys. Chem.*, **68**, 441 (1964).

Chapter 11

STATISTICAL MECHANICS OF BINARY MIXTURES

DANIEL E. MARTIRE

Department of Chemistry, Georgetown University
Washington D.C. 20057

Lattice Models

The relevant thermodynamic background and pertinent experimental results having been covered in Chapter 10; we now consider molecular statistical theories for binary non-mesomorphic solute/mesomorphic solvent mixtures. There are two types of models that have proved generally useful for investigating nematic-isotropic transitions in the dilute solution region of such mixtures: lattice models and molecular field theories of the Maier-Saupe variety. The former emphasize the role of anisotropic intermolecular repulsion, that is packing or steric effects, while the latter, in its original form, emphasizes anisotropic intermolecular attraction, that is anisotropy of dispersion forces.

In this first part, a lattice model for a binary mixture of hard rigid-rod solutes (non-mesomorphic) and hard rigid-rod solvents (mesomorphic) is described in some detail, so as to provide an example of the mechanics involved in lattice treatments. We shall discuss briefly lattice models where the solute is a cube, a flexible chain or a semi-flexible chain. Some general results are given for all of these solute systems.

In the second part, an extension of molecular field theory to binary mixtures of both spherical and non-spherical non-mesomorphic solutes in rod-like solvents is presented. Finally, the predictions of both types of models, together with a virial expansion treatment of spheres and spherocylinders, are compared with the experimental results.

Background

Lattice models have been used extensively to investigate the nematic-isotropic (NI) transition in single component systems. Following the pioneering work of Flory [1] and DiMarzio [2], these models have been applied to NI transitions in systems of completely rigid rods [3-5] and systems where the molecules have some degree of flexibility [6-9]; even the possibility of biaxial nematics has been investigated [10]. The reader is referred to Chapter 7 and elsewhere [5, 11] for a general review of various hard-rod models. More recently, lattice models have been developed for mixtures of rod-like solvent molecules and solute molecules which are rigid rods [5, 12], cubes [13], chains [8,14] or plates [15]. With the exception of one study [3], all of these models have utilized the simple cubic lattice which is considered here. Although the advantages, deficiencies, successes and failures of such lattice models are well documented [5], some salient points are reviewed below.

Inherent in simple cubic lattice models is the necessity to restrict the molecular segments to a discrete set of lattice sites and the long axes of the molecular cores to three discrete and mutally orthogonal directions; neither of these assumptions is realistic for liquid crystals. The latter restriction is particularly severe [16], since it does not permit the small angle deviations from the director, which exist in real nematics. As a result of this "either-or" situation, such lattice models lead to larger transition order parameters ($\bar{P}_2^{(K)}$) and fractional density changes ($\Delta\rho/\rho_{nem}$) compared with continuum models [11] where all molecular positions and orientations are allowed. This may be seen by comparing the results given in table 1 with those obtained via scaled

Table 1. *Results for the nematic-isotropic transition in the simple cubic lattice model of hard, rigid rods* [8, 12]

m^a	$\bar{P}_2^{(NI)}$	f_N^b	f_I^b	$\Delta\rho/\rho_{nem}$	$\Delta S/R$
4	0.761	0.902	0.879	0.025	0.156
5	0.799	0.711	0.654	0.080	0.301
6	0.822	0.591	0.520	0.120	0.371
7	0.838	0.507	0.431	0.150	0.415
8	0.849	0.445	0.368	0.173	0.446
9	0.857	0.397	0.321	0.191	0.469
10	0.864	0.359	0.285	0.206	0.487

a. molecular length-to-breadth ratio

b. dimensionless density or fraction of space occupied by the rods. ($f = Nm/L$; see text).

particle theory shown in figure 1 of Chapter 7. It should be noted, though, that the two sets of results are not significantly different for $m \leqslant 5$, a range that contains the vast majority of *truly rigid* molecules.

Another problem, which is not unique to lattice models, is the complete neglect of short-range order. The usual lattice statistics [2] in fact, involve a sort of mean field approximation in the counting procedure, as we shall see. However, as Alben [4] has pointed out, the probability of success in placing a molecule on the lattice is more properly an averaged function of the local order rather than a function of the average order. Furthermore, unlike continuum systems where the expected divergence occurs, the equation of state and free energy remain finite at the close-packed density in lattice systems of hard, rigid rods. This leads to an artefact, namely, no anisotropic phase exists at any density for a simple cubic lattice system where the length-to-breadth ratio (m) of the rigid rod is less than approximately 3.65 [4]. Fortunately, the lack of an ordered phase for small m is an advantage for our present purposes, since we are "modelling" solutions with non-mesomorphic solutes.

In contrast to these drawbacks (notably that of restricted orientation) are the several distinct advantages of lattice models. They are more tractable and, therefore, more widely applicable than more physically appealing approaches such as scaled particle theory [11], which is described in Chapter 7. They can be applied readily to a variety of systems differing in molecular size, shape and flexibility; accordingly, the results can be related directly to the effects of molecular geometry and packing. Also, in the hard particle systems to be considered here, the purely repulsive intermolecular potential is well defined and no adjustable parameters have to be invoked. Moreover, there is no question as to the statistical-thermodynamic consistency of lattice models [17].

A familiarity with lattice and cell theories of polymer and non-electrolytic solutions [18] leads us to expect that the present lattice models might be appreciably more successful for mixtures than for single component systems. The results are seen to fulfil this expectation.

Lattice Model for Mixtures of Hard, Rigid Rods

The model system consists of N_1 hard, rigid solvent rods of length-to-breadth ratio m_1 and N_2 hard, rigid solute rods of length-to-breadth ratio m_2 (with $m_1 > m_2$), placed on a simple cubic lattice of L sites or cells of volume v_o. The rods are hard in that two rod segments cannot occupy the same lattice site; they are rigid in that the m rod segments are co-linear. Also, there are no attractive interactions present. Each rod of type ν occupies m_ν adjacent lattice sites and can point in only one of three mutually orthogonal directions: a, b and c. For such a system, the configurational partition function Q_N is simply equal to the number of distinguishable ways of placing the rods on the lattice, which can be shown to be independent of the order in which they are added [2].

DiMarzio [2] developed the statistical procedure to handle such a counting problem. The basic quantity involved is the conditional probability, $P_r^{(n)}$, that an occupied site is adjoined in direction r by an unoccupied one when n total molecules have already been placed;

$$P_r^{(n)} = (L-mn)(L-mn+B_{-r}^{(n)}), \tag{1}$$

where $B_{-r}^{(n)}$ is the number of ways in which a site can be occupied, given that its neighbour in direction -r is occupied. This counting procedure neglects all local deviations from the mean density and orientational distribution of the rods already placed.

We now wish to obtain an expression for the number of distinct ways,

$$\prod_{\nu=1}^{2} \prod_{r-a}^{c} g_{\nu r}, \tag{2}$$

of placing the N_1 solvent rods and N_2 solute rods on the lattice such that there are N_{1r} and N_{2r} rods pointing in direction r. Consider placing the $(n_{1a}+1)$th rod on the lattice when n_{1a} rods have already been placed in direction a. The number of sites available to the first segment of the $(n_{1a}+1)$th rod is clearly $(L-m_1n_{1a})$. Subsequent adjacent sites in direction a must either be empty or contain the end segment of one of the n_{1a} previously placed rods. According to eq. (1) then, the probability that the next $(m_1 -1)$ co-linear sites, required to constitute the entire rod, are unoccupied is

$$\{P_a^{(n_{1a})}\}^{m_1-1} = [(L-m_1n_{1a})/\{L-(m_1-1)n_{1a}\}]^{m_1-1}. \tag{3}$$

The number of ways of placing the $(n_{1a}+1)$th rod is then the product of $(L-m_1n_{1a})$ and the right hand side of eq. (3):

$$\sigma_{n_{1a+1}} = (L-m_1n_{1a})^{m_1}/\{L-(m_1-1)n_{1a}\}^{m_1-1}. \tag{4}$$

Taking account of the indistinguishability of the rods, the number of distinct ways of placing the N_1 rods such that there are N_{1a} pointing in direction a is

$$g = (1/N_{1a}!) \prod_{n_{1a}=0}^{N_{1a}-1} \sigma_{n_{1a+1}}. \tag{5}$$

Using the relation

$$\prod_{n=0}^{N-1} (F-fn)^f \approx F!/(F-fN)!,$$

we obtain

$$g_{1a} = \{L-(m_1-1)N_{1a}\}!/N_{1a}!(L-m_1N_{1a})!. \tag{6}$$

Let us now consider placement of the $(n_{1b}+1)$th rod in direction b when N_{1a} and n_{1b} rods are already present. The number of ways of placing the first segment is clearly $(L-m_1N_{1a}-m_1n_{1b})$. Subsequent adjacent sites in direction b must either be empty or contain one of the m_1 segments of an N_{1a} rod or the end segment of one of the n_{1b} rods. Hence

$$\{P_b^{(n_{1b})}\}^{m_1-1} = (L-m_1N_{1a}-m_1n_{1b})^{m_1-1}/\{(L-m_1N_{1a}-m_1n_{1b})+(m_1N_{1a}+n_{1b})\}^{m_1-1}, \tag{7}$$

$$g_{1b} = (1/N_{1b}!)\prod_{n_{1b}=0}^{N_{1b}-1}(L-m_1N_{1a}-m_1n_{1b})^{m_1}/\{L-(m_1-1)n_{1b}\}^{m_1-1}, \tag{8}$$

$$= (L-m_1N_{1a})!\{L-(m_1-1)N_{1b}\}!/N_{1b}!L!(L-m_1N_{1a}-m_1N_{1b})!, \tag{9}$$

where we have proceeded as in the formulation of g_{1a}. In a similar manner, we can obtain g_{1c}, g_{2a}, g_{2b} and g_{2c}, and, hence, the required product of eq. (2).

In the limit $N \to \infty$ $(N = N_1 + N_2)$, $L \to \infty$, while N/L remains finite, the partition function Q_N can be replaced by its maximum term Q_N^* (corresponding to the most probable orientational distribution) and eq. (2) yields

$$Q_N^* = \prod_{\nu=1}^{2}\prod_{r=a}^{c} g_{\nu r}^*,$$

$$= \prod_{r=a}^{c}\{L-(m_1-1)\bar{N}_{1r}-(m_2-1)\bar{N}_{2r}\}!/(L-m_1N_1-m_2N_2)!(L!)^2\prod_{r=a}^{c}(\bar{N}_{1r}!\bar{N}_{2r}!), \tag{10}$$

where the $\bar{N}_{\nu r}$'s are chosen to maximize Q_N. Finally, the configurational Helmholtz free energy, A, and configurational entropy, S, are given by

$$-\beta A/N = S/Nk = (1/N)\ln Q_N^*,$$

$$= 1/d[\sum_{r=a}^{c}\{1-(m_1-1)s_rx_1d-(m_2-1)t_rx_2d\}\ln\{1-(m_1-1)s_rx_1d-(m_2-1)t_rx_2d\}$$

$$-(1-m_1x_1d-m_2x_2d)\ln(1-m_1x_1d-m_2x_2d)-d\ln d-d\sum_{r=a}^{c}\{s_rx_1\ln(s_rx_1)+t_rx_2\ln(t_rx_2)\}], \tag{11}$$

where $N = N_1 + N_2$, $x_1 = N_1/N$, $x_2 = N_2/N$, $s_r = \bar{N}_{1r}/N_1$, $t_r = \bar{N}_{2r}/N_2$ and $v_o\rho = N/L$; i.e., x_1 and x_2 are solvent and solute mole fractions, respectively, s_r and t_r are the fractions of solvent and solute rods, respectively, pointing in direction r (in the most probable orientational distribution), and d is the dimensionless density.

The number of dependent variables appearing in eq. (11) can be reduced by arbitrarily selecting direction c as the director and assuming that the orientational distribution is axially symmetric. Hence,

$$s_a = s_b = s; \quad s_c = 1 - 2s,$$

$$\bar{P}_2^{(1)} = 1 - 3s, \tag{12}$$

$$t_a = t_b = t; \quad t_c = 1 - 2s,$$

$$\bar{P}_2^{(2)} = 1 - 3t, \tag{13}$$

where $\bar{P}_2^{(1)}$ and $\bar{P}_2^{(2)}$ are the solvent and solute order parameters, respectively. Also, since pressure P and temperature T do not appear as independent variables in hard, rigid rod lattice models, we define a dimensionless pressure-to-temperature ratio Φ:

$$\Phi = \beta P v_o = -(\partial \beta A/\partial L)_{T,N_1,N_2}. \tag{14}$$

The configurational chemical potentials μ_i are given by

$$\beta\mu_i = (\partial \beta A/\partial N_i)_{L,T,N_j} \tag{15}$$

and the configurational Gibbs free energy by

$$G = x_1\mu_1 + x_2\mu_2. \tag{16}$$

To determine the equilibrium state of the system at fixed temperature, pressure and composition (the usual experimental variables), we must minimize G with respect to s and t at constant Φ and x_2. Since the configurational Gibbs free energy cannot be expressed in closed form as a function of Φ, this minimization cannot be done directly. However, it can be shown [4] that the required minimization can be performed indirectly through A:

$$(\partial G/\partial s)_{N,\Phi,t,x_2} = (\partial A/\partial s)_{d,t,x_2} = 0, \tag{17}$$

$$(\partial G/\partial t)_{N,\Phi,s,x_2} = (\partial A/\partial t)_{d,s,x_2} = 0, \tag{18}$$

provided that the value of d corresponds to that at the chosen Φ. The set of simultaneous non-linear equations, eqs. (14), (17) and (18), can then be solved numerically [8] to determine the values of s, t and d which give the thermodynamically stable state (i.e. absolute minimum in G) at a given value of Φ. As we have chosen the solute rod to be smaller than the solvent rod at the given Φ values, the *pure* solute rods do not form a stable anisotropic state; that is the pure solutes are non-mesomorphic.

Applying this procedure, it can be shown that the model system is anisotropic $\{0<(s,t)<1/3\}$ when x_2 is small and Φ is sufficiently large, while it is isotropic ($s = t = 1/3$) for larger x_2 and/or smaller values of Φ. Moreover, there exists a range of mole fractions for the system where G is minimized by the presence of two phases of different composition in equilibrium, one anisotropic and the other isotropic. In this range, a change in x_2 at constant Φ for the system as a whole will alter the relative amounts of the two phases present, but not the composition of either phase. To determine the solute mole fractions x_i and x_n (dropping the subscript 2) in the coexisting isotropic and anisotropic (nematic) phases, the following phase equilibrium expressions are required

$$\beta\mu_2(\Phi,x_i) = \beta\mu_2(\Phi,x_n) \tag{19}$$

and

$$\beta\mu_1(\Phi,x_i) = \beta\mu_1(\Phi,x_n). \tag{20}$$

The two phase region at a given Φ is then located by determining s,t,d_n and x_n for the coexisting anisotropic phase and d_i and x_i for the coexisting isotropic phase by simultaneously solving (again, numerically) eqs. (14), (17) and (18) for the anisotropic phase; eq. (14) for the isotropic phase {$s = t = 1/3$ is always a solution to eqs. (17) and (18)} and eqs. (19) and (20) for the mixture. The phase transitions are always found to be first order in that $d_n > d_i$.

In order to compare the results of the lattice calculations with the experimental results, the relationship between the experimental reduced temperature ($T^* = T/T_{NI}$) and the model reduced temperature-to-pressure ratio (Φ^*) must be established. Assuming fixed pressure, we obtain from eq. (14)

$$\Phi^* = \Phi_{NI}/\Phi = T/T_{NI} = T^*, \tag{21}$$

where Φ_{NI} applies to the anisotropic-isotropic transition for *pure* solvent rods and Φ is the chosen value. In all of the simple cubic lattice calculations to be discussed, Φ^*'s in the range $0.94 \leqslant \Phi^* \leqslant 1.00$ were used to mimic the experimental range for T^*.

Hence, if a plot of Φ^* against solute mole fraction, x, is constructed from the model results at low x, a virtually linear phase diagram is obtained similar to the plot shown in figure 1 of Chapter 10. Accordingly, the slopes $d\Phi^*/dx_n$ and $d\Phi^*/dx_i$ may be associated with β_n^∞ and β_i^∞, defined in Chapter 10. Since all of the solutes studied are non-mesomorphic in their pure state, the β^∞'s are all negative. Thus, henceforth we shall refer only to the modulus of β^∞ and denote it simply by β.

Lattice Model Results for Various Solute Types

We now present brief descriptions of the lattice models for binary mixtures of hard, rigid-rod solvents and hard solutes of various types. Although the partition functions differ for different solute components, the same general procedure applies as that presented in the preceding section. We shall first summarize salient results specific to the solute type and then some general results. Additional, and more quantitative comparisons will be made in the concluding sections of this Chapter.

Rigid Rod Solutes [12] Solvent rods with $m_1 = 5$ and $m_1 = 10$ and solute rods of $2 \leqslant m_2 \leqslant (m_1-1)$ were studied. It was found that β decreased with increasing m_2 for both $m_1 = 5$ and 10. Also the solute order parameter $\overline{P_2^{(2)}}$ at the transition was independent of Φ^*, but, as expected, increased with increasing m_2.

Cubic Solutes [13] Cubes are the lattice model analogue of spheres. The solute cubes studied were of side dimension D, where $1.0 \leqslant D \leqslant 2.0$; the solvent rods again had $m_1 = 5$ and 10. The configurational partition function Q_N was obtained by a straightforward extension of the DiMarzio counting procedure [3]. However, the maximum value of D considered was 2.0 to ensure that Q_N was independent of the order of placement of cubes and rods on the lattice. Nevertheless, this maximum D value corresponds to eight segments of the solvent molecule. β was found to increase markedly with increasing D^3 for both $m_1 = 5$ and 10.

Flexible and Semi-flexible Chain Solutes [8,14] The lattice statistics here were potentially more complicated, but were greatly simplified, in the mean field spirit, by considering only an average solute chain with an average fraction of bonds in the different directions. Also, the partition function contained a flexibility statistical factor [6,7] related to the number of different ways of distributing the bonds in a given direction. For the completely flexible chains, there was no energy barrier, E_b, to internal rotation, while for the semi-flexible chains E_b was varied to examine the effects of increasing rigidity.

The combinations of E_b, m_1 and m_2 considered are too numerous to summarize here. We do note, though, that solute chains with $m_2 > m_1$ were included, provided that the pure solute was non-mesomorphic. With the flexible chain solutes ($E_b = 0$), for example, combinations of $m_1 = 5$, $2 \leqslant m_2 \leqslant 7$ and $m_1 = 10$, $2 \leqslant m_2 \leqslant 15$ were treated. For the flexible solutes it was found that β increased with increasing m_2, exhibiting a pattern of behaviour closer to cubical solutes than to rigid-rod solutes with the same number of segments. For the semi-flexible solutes β increased or decreased with increasing m_2, depending on the value of E_b, and approached the rigid-rod limit as $E_b \to \infty$.

General Results

The model calculations for rod-like, cubical and chain-like solutes all yield the expected two-phase region. At constant Φ^*, the extent of this region at very low solute mole fraction should be given by

$$(x_i - x_n) = (\Delta S_{NI}/R)(1-\Phi^*), \qquad (22)$$

as we saw in Chapter 10. The anisotropic-isotropic transition entropy $\Delta S_{NI}/R$ of the pure solvent rods was found to be 0.301 for $m_1 = 5$ and 0.487 for $m_1 = 10$ [8, 12]. The lattice model results are in general accord with eq. (22), thus confirming that the β's are essentially limiting values, that is in the Henry's law regime. Moreover the findings that $(x_n - x_i)$ is independent of solute type for a given m_1 and, with all solutes, is larger for $m_1 = 10$ than for $m_1 = 5$ are consistent with eq. (22).

As observed experimentally, the shorter solvent rods (smaller ΔS_{NI}) are more readily disrupted (i.e. have larger β values) than the longer solvent rods by all solute types. At fixed m_1, we may also compare the β values of solutes with the same number of segments, the general trend being: cubes > flexible chains > semi-flexible chains > rigid rods.

Perhaps the most remarkable result of these calculations is the virtual constancy of the solvent order parameter, $\bar{P}_2^{(1)}$, at the transition. For all of the systems studied, $\bar{P}_2^{(1)}$ at the transition was constant to within ± 2% over the entire Φ^* range, independent of the nature of the solute. This is entirely in agreement with experiment [19, 20]. It further suggests that although an anisotropic solvent at fixed Φ^* may respond differently to different types of solute probes (hence, different β values), its minimum level of orientational order is always the same, regardless of the nature of the solute.

Finally, we note that attractive interactions (isotropic and anisotropic) were incorporated into two lattice model treatments of mixtures [5, 14]. Although they are not required to generate the features we have just described, the presence of attractive interactions led to moderately improved absolute agreement between model results and experiment. However, the extent of anisotropy in such attractive interactions was of little consequence [8, 9, 14].

Molecular Field Theory

Here we present an extension of the Maier-Saupe molecular field theory to binary nematic mixtures of unalignable (spherical) and align-able non-mesomorphic solutes in rod-like solvents. This represents a refinement of our previous work and, more prominently, the treatments of Humphries, James and Luckhurst. The quantitative predictions of the lattice models, the molecular mean field theory and a virial expansion treatment of spheres and spherocylinders are then compared with experimental results.

Background

As is the case with lattice models, the Maier-Saupe molecular field theory [21] described in Chapter 4, has its particular strengths and weaknesses. Among the former are its relative simplicity and tractability, qualities which have opened the way to a variety of successful applications too numerous to cite here [21]. Its questionable assumptions and apparent deficiencies have more recently come to light: (a) its assumption of no coupling between radial and orientational correlations (implicit in the factorization of the pair distribution function); [13, 22]; (b) its emphasis of anisotropic attractive interaction, while neglecting angle-dependent, short-range repulsion [22-24]; (c) its statistical-mechanical inconsistency due to the assumed density dependence of the molecular field pseudo-potential [17]; (d) other inadequacies and inconsistencies resulting from these assumptions [17, 22, 24]; and, as we saw in Chapter 17;(e) significant quantitative discrepancies between certain of its theoretical predictions and experiment [25].

In the context of the generalized van der Waals theory [22, 24, 26] described in Chapter 8 (with consistent density dependence of the pseudo-potential), the Maier-Saupe theory of the nematic-isotropic transition is the special case obtained when the transition is assumed to occur at constant density and when rod-like hard cores are replaced by hard spheres. The latter requires that the angle dependence of the pseudo-potential (and, hence, nematic order) derives solely from *anisotropic* attractive interactions. However, explicit inclusion of orientationally dependent hard-rod repulsions in determining the molecular field averaged attractive pseudo-potential leads to the interesting result that the coefficient of the main angle-dependent term is dominated by *isotropic* attractive interactions.

Despite these shortcomings as well as reservations on this author's part [13], application of the modified Maier-Saupe theory to binary nematic mixtures will nevertheless be considered.

Humphries, James and Luckhurst [27] were the first to extend the Maier-Saupe theory of the nematic mesophase to binary mixtures. For each component they formulated an anisotropic pseudo-potential for a single axially symmetric molecule and obtained the orientational Helmholtz free energy of the mixture (A_n) in a manner consistent with the development for single component nematics. Mixtures of both spheres and rods, and rods and rods were treated and theoretical phase diagrams were obtained. Their analysis of A_n and the derived results were unsettling in two respects because no attempt was made to consider phase separation. Thus, it was specified that, at a fixed solute mole fraction A_n for the total system should vanish at a well-defined reduced temperature, T^*, rather than over a small reduced temperature range ($T_n^* - T_i^*$) during the course of the nematic -isotropic transition. This implied lack of phase separation and suggested second-order behaviour would require that $\beta_n^\infty = \beta_i^\infty$ and $\gamma_n^\infty = \gamma_i^\infty$ (cf. Chapter 10 for definitions). This was thermodynamically inconsistent with, e.g., their derived expression for mixtures and spheres and rods:

$$\ln(\gamma_n^\infty/\gamma_i^\infty) = 0.418, \tag{23}$$

where a discontinuity in the infinite dilution activity coefficient is indicative of a first-order phase transition and, hence, phase separation. Secondly, they obtained the physically unrealistic result that $|\beta^\infty| = 1$, regardless of the size of the solute sphere or solvent rod. This result and the absence of a two-phase region were later found to be in clear disagreement with experiment [28, 29].

Their earlier treatment was made thermodynamically consistent and then improved by utilizing a volume fraction weighting [30], rather than a mole fraction weighting, in formulating the anisotropic pseudo-potentials of the two species [29]. This modified weighting allowed for possible size differences between the solute and solvent molecules and led, in a more direct fashion, to the following limiting (infinite dilution) expressions for binary mixtures of solute spheres and solvent rods [29]

$$\gamma_n^\infty/\gamma_i^\infty = \beta_n^\infty/\beta_i^\infty = \exp(0.418\ v_2/v_1), \tag{24}$$

$$|\beta_n^\infty| = \{\exp(0.418\ v_2/v_1)-1\}/0.418, \tag{25}$$

$$|\beta_i^\infty| = \{1-\exp(-0.418\ v_2/v_1)\}/0.418 \tag{26}$$

and

$$(x_i - x_n) = (\Delta S_{NI}/R)(1-T^{*\prime}) = 0.418\ (1-T^{*\prime}), \tag{27}$$

where v_1 and v_2 are, respectively, the solute and solvent molecular volumes, and the other symbols are as defined in Chapter 10. These results have since been confirmed by Humphries and Luckhurst [31]; the plots in figure 5 of ref. [31] follow eqs. (25) and (26) exactly.

We shall now present a further refinement of this recent work [29, 31] and, in so doing, extend the molecular field approach to nematic mixtures where the solute component may now be aligned. The more general equations to be derived here will be seen to reduce to eqs. (24)-(27) for the special case of unaligned (spherical) solutes.

Molecular Field Theory

General Equations We summarize first, the relevant general equations [21] from Chapter 4. The orientational distribution function of component j, $f_j(\cos\theta)$ is given by

$$f_j(\cos\theta) = Z_j^{-1} \exp\{(-\beta U_j(\cos\theta)\}, \tag{28}$$

where θ is the angle between the molecular symmetry axis and the director, axial symmetry being assumed. $U_j(\cos\theta)$ is the single-particle anisotropic pseudo-potential for component j, that is, the orientational energy of a j-type molecule in the molecular field generated by its neighbours and Z_j is the corresponding single-particle orientational partition function, defined by

$$Z_j = \int_0^1 d(\cos\theta)\exp\{-\beta U_j(\cos\theta)\}. \tag{29}$$

Next we have the self-consistency equation for $\bar{P}_2^{(j)}$, the second rank orientational order parameter of component j,

$$\bar{P}_2^{(j)} = Z_j^{-1}\int_0^1 d(\cos\theta)P_2^{(j)}(\cos\theta)\exp\{-\beta U_j(\cos\theta)\}, \tag{30}$$

which is obtained through the minimization of the free energy, A_n, with respect to $\bar{P}_2^{(j)}$ [21]. The average orientational potential energy of a molecule of component j is then

$$U_j = Z_j^{-1} \int_0^1 d(\cos\theta)U_j(\cos\theta)\exp\{-\beta U_j(\cos\theta)\} \tag{31}$$

and the total orientational energy U_n is

$$U_n = (1/2)\sum_j N_j U_j, \tag{32}$$

where N_j is the total number of j-type molecules. Since the average orientational entropy per j-type molecule is given by

$$s_j = -k \int_0^1 d(\cos\theta)f_j(\cos\theta)\ln f_j(\cos\theta), \tag{33}$$

we find

$$s_j = (U_j/T) + k\ln Z_j. \tag{34}$$

The total orientational entropy of the system is therefore

$$S_n = \sum_j N_j s_j,$$
$$= 2(U_n/T) + k\sum_j N_j \ln Z_j. \qquad (35)$$

Finally, the total orientational Helmholtz free energy A_n is found to be

$$A_n = -U_n - \beta^{-1}\sum_j N_j \ln Z_j. \qquad (36)$$

In statistically consistent Maier-Saupe theory of pure nematic mesophases [17], $U_j(\cos\theta)$ has the form

$$U_j(\cos\theta) = -\rho_j b_{jj} \bar{P}_2^{(j)} P_{2(j)}(\cos\theta), \qquad (37)$$

where ρ_j is the number density (Chapter 4). The interaction parameter b_{jj} is a positive constant (independent of ρ_j and T) and is a measure of the strength of intermolecular attractive interactions. Since $\rho_j = v_j^{-1}$ where v_j is the molecular volume, consistent Maier-Saupe theory for a single component nematic mesophase yields [21]

$$U_j = -v_j^{-1} b_{jj} \bar{P}_2^{(j)^2},$$
$$U_n = -(1/2) N_j v_j^{-1} b_{jj} \bar{P}_2^{(j)^2} \qquad (38)$$

and

$$A_n = +(1/2) N_j v_j^{-1} b_{jj} \bar{P}_2^{(j)^2} - \beta^{-1} N \ln Z, \qquad (39)$$

where, at the nematic-isotropic transition when A_n vanishes:

$$b_{jj}/kT_{NI}v_j = 4.5414 \quad ; \quad \bar{P}_2^{NI} = 0.4290 \qquad (40)$$

$$\ln Z_j^{NI} = \Delta S_j^{NI}/R = \Delta U_j^{NI}/RT_{NI} = (\bar{P}_2^{NI})^2 b_{jj}/2kT_{NI}v_j = 0.4179. \qquad (41)$$

Since the transition is assumed to occur at constant density, consistent Maier-Saupe theory (i.e., with the correct density dependence for the pseudo-potential) yields the same numerical results as the original formulation [21]. In this approach neither the scalar part of the pseudo-potential nor anisotropic repulsive interactions play any role in determining these transition quantities.

Binary mixtures The general equations are now extended and applied to nematic-isotropic transitions in binary mixtures. We shall continue to assume that the contribution of the scalar part of the intermolecular potential is the same in both the nematic and isotropic states [31]. Hence, we shall focus on the orientational Helmholtz function of the nematic mixture, A_n, and derive an expression for the corresponding free energy of mixing ΔA_n.

Following Cotter [17], the anisotropic pseudo-potentials of component 1 (the solvent) and component 2 (the solute) in the nematic mixture are derived to be

$$U_1(\cos\theta) = -\rho(x_1 b_{11}\bar{P}_2^{(1)} + x_2 b_{12}\bar{P}_2^{(2)})P_2^{(1)}(\cos\theta) \tag{42}$$

and

$$U_2(\cos\theta) = -\rho(x_1 b_{12}\bar{P}_2^{(1)} + x_2 b_{22}\bar{P}_2^{(2)})P_2^{(2)}(\cos\theta), \tag{43}$$

where the number density is

$$\rho = (N_1+N_2)/(N_1 v_1 + N_2 v_2),$$

the mole fraction is

$$x_j = N_j/(N_1+N_2),$$

the volume fraction is

$$\phi_j = N_j v_j/(N_1 v_1 + N_2 v_2) = x_j \rho v_j$$

and where $\bar{P}_2^{(j)}$ is the order parameter for component j in the mixture at temperature T. Thus

$$U_1(\cos\theta) = -(\phi_1 b_{11} v_1^{-1}\bar{P}_2^{(1)} + \phi_2 b_{12} v_2^{-1}\bar{P}_2^{(2)})P_2^{(1)}(\cos\theta), \tag{44}$$

$$U_2(\cos\theta) = -(\phi_1 b_{12} v_1^{-1}\bar{P}_2^{(1)} + \phi_2 b_{22} v^{-1}\bar{P}_2^{(2)})P_2^{(2)}(\cos\theta). \tag{45}$$

These equations evolve naturally from consistent Maier-Saupe theory [17] and are, in effect, volume-fraction averaged interactions. This sort of averaging has been applied empirically by Marcelja [32] in his molecular field treatment of end-chain ordering in liquid crystals and is commonly used in statistical thermodynamics of polymer solutions [33].

Inserting eqs. (44) and (45), respectively, into eq. (31) {with Z_j and $\bar{P}_2^{(j)}$ given by eqs. (29) and (30)} and the resulting pseudo-potentials into eq. (32), we obtain

$$U_n = -(\phi_1^2 \bar{P}_2^{(1)^2} b_{11} v_1^{-2} + 2\phi_1\phi_2 \bar{P}_2^{(1)}\bar{P}_2^{(2)} b_{12} v_1^{-1} v_2^{-1} + \phi_2^2 \bar{P}_2^{(2)^2} b_{22} v_2^{-2})/2 \tag{46}$$

for the total orientational energy of the nematic mixture, and from eq. (36)

$$A_n = -U_n - N_1\beta^{-1}\ln Z_1 - N_2\beta^{-1}\ln Z_2. \tag{47}$$

The orientational Helmholtz free energy of the pure (unmixed) components (designated by superscript o), at the same temperature as the mixture, is

$$A_n^o = A_1^o + A_2^o = \sum_{j=1}^{2} N_j a_j^o, \tag{48}$$

where the A_j^o's are given by eq. (39) and are written on the right hand side of eq. (48) in terms of molecular quantities (a_j^o) which are independent of N_j. (Note that if the pure solute is isotropic at temperature T, then A_2^o vanishes). The Helmholtz free energy of mixing, ΔA_n, is then obtained by substracting A_n^o from A_n and adding the combinatorial contribution [31, 33]:

$$\Delta A_n = A_n - \sum_{j=1} N_j a_j^o + \beta^{-1} \sum_{j=1} N_j \ln\phi_j, \tag{49}$$

where we shall assume negligible volume of mixing, so that $v_j \approx v_j^o$.

The orientational Helmholtz free energy of the isotropic mixture A_i is zero. Therefore, it follows that the Helmholtz free energy of mixing ΔA_i is

$$\Delta A_i = - \sum_{j=1} N_j' a_j^o + \beta^{-1} \sum_{j=1} N_j' \ln\phi_j', \tag{50}$$

where the primes are used to distinguish isotropic phase variables from those for the nematic phase.

We can now determine the ratio of solute activity coefficients (γ_2/γ_2') in the two coexisting phases at T, which as we saw in Chapter 10, will enable us to construct phase diagrams at very low solute concentrations. From thermodynamics [33]

$$(\partial\beta\Delta A/\partial N_2)_{T,N_1} = \ln\gamma_2 x_2. \tag{51}$$

Thus, from eq. (49):

$$\ln\gamma_2 x_2 = (\partial\beta A/\partial N_2)_{T,N_1} - \beta a_2^o + \ln\phi_2 + \{1-(v_2/v_1)\}(1-\phi_2) \tag{52}$$

and from eq. (50):

$$\ln\gamma_2' x_2' = -\beta a_2^o + \ln\phi_2' + \{1-(v_2/v_1)\}(1-\phi_2'). \tag{53}$$

At equilibrium, $\gamma_2 x_2 = \gamma_2' x_2'$, and from eqs. (52) and (53):

$$\ln(\gamma_2/\gamma_2') = \{\ln(\gamma_2/\gamma_2')\}_{\text{orient.}} + \ln(\gamma_2/\gamma_2')\}_{\text{comb.}}, \tag{54}$$

where the orientational and combinatorial contributions to $\ln(\gamma_2/\gamma_2')$ are

$$\{\ln(\gamma_2/\gamma_2')\}_{\text{orient.}} = (\partial\beta A_n/\partial N_2)_{T,N_1}, \tag{54a}$$

$$\{\ln(\gamma_2/\gamma_2')\}_{\text{comb.}} = \ln(\phi_2 x_2'/\phi_2' x_2) + \{1-(v_2/v_1)\}(\phi_2'-\phi_2) \tag{54b}$$

and where A_n is given by eq. (47), U_n by eq. (46), Z_j by eq. (29), $\bar{P}_2^{(j)}$ by eq. (30) and the pseudo-potentials by eqs. (44) and (45).

Differentiating βA_n with respect to N_2 (note that Z_1 and Z_2 are also functions of N_2) and collecting terms, we find

$$\ln(\gamma_2/\gamma_2') = (v_2/v_1)(\phi_1^2\beta b_{11}\bar{P}_2^{(1)^2}v_1^{-1}+2\phi_1\phi_2\beta b_{12}\bar{P}_2^{(1)}\bar{P}_2^{(2)}v_1^{-1}+\phi_2^2\beta b_{22}\bar{P}_2^{(2)^2})/2$$

$$-\ln Z_2+\{\ln(\gamma_2/\gamma_2')\}_{comb.} \tag{55}$$

To generate phase diagrams at low solute concentrations, we let $x_2\to 0$ such that $x_2'/x_2\neq 0$), in which case $T\to T_{NI}^{(1)}$, $\bar{P}_2^{(1)}\to\bar{P}_2^{NI}$ and, from eq. (41), $\beta b_{11}\bar{P}_2^{(1)^2}/2v_1\to 0.4179$. Thus from eq. (55), we obtain the following expression for the ratio of the infinite-dilution solute activity coefficients (henceforth denoted by $\gamma_n^\infty/\gamma_i^\infty$) in the vicinity of $T_{NI}^{(1)}$ $(T<T_{NI}^{(1)})$:

$$\ln(\gamma_n^\infty/\gamma_i^\infty) = 0.418\ v_2/v_1-\ln Z_2^\infty\ , \tag{56}$$

where the combinatorial term vanishes as $x_2\to 0$ and

$$Z_2^\infty = \int_0^1 d(\cos\theta)\exp\{a\ P_2^{(2)}(\cos\theta)\}, \tag{57}$$

$$\bar{P}_2^{(2)\infty} = (Z_2^\infty)^{-1}\int_0^1 d(\cos\theta)P_2(\cos\theta)\exp\{aP_2^{(2)}(\cos\theta)\}. \tag{58}$$

Here $a = 1.948\ b_{12}/b_{11}$ and b_{12}/b_{11} is a measure of the relative strength of 1-2 and 1-1 anisotropic interactions. Also, $\bar{P}_2^{(2)\infty}$ is a dilute solution value of the solute order parameter at the nematic-isotropic transition. Note that eqs. (56 - 58) are applicable to mesomorphic and non-mesomorphic solutes.

We show in figure 1 plots of $\ln Z_2^\infty$ and $\bar{P}_2^{(2)\infty}$ against b_{12}/b_{11}, the former being virtually parabolic and the latter virtually linear. In fact, to an excellent degree of approximation, the curves are well represented, in the region $0\leq b_{12}/b_{11}\leq 1$, by

$$\ln Z_2^\infty = 0.418(b_{12}/b_{11})^2\ ,\ (\pm\ 0.001) \tag{59}$$

and

$$\bar{P}_2^{(2)\infty} = 0.429(b_{12}/b_{11})\ ,\ (\pm\ 0.002) \tag{60}$$

where the average difference between the exact and approximate expression is given in parentheses. We note in passing that eq. (60) suggests that measurements of $\bar{P}_2^{(2)\infty}/\bar{P}_2^{NI}$ can be used to estimate b_{12}/b_{11}. From eqs. (56) and (57) then,

$$\ln(\gamma_n^\infty/\gamma_i^\infty) = 0.418\{(v_2/v_1)-(b_{12}/b_{11})^2\}. \tag{61}$$

Making the usual assumption [27] that $b_{12} = (b_{11}b_{22})^{\frac{1}{2}}$ and utilizing eq. (40), eq. (61) can be written in the alternative form

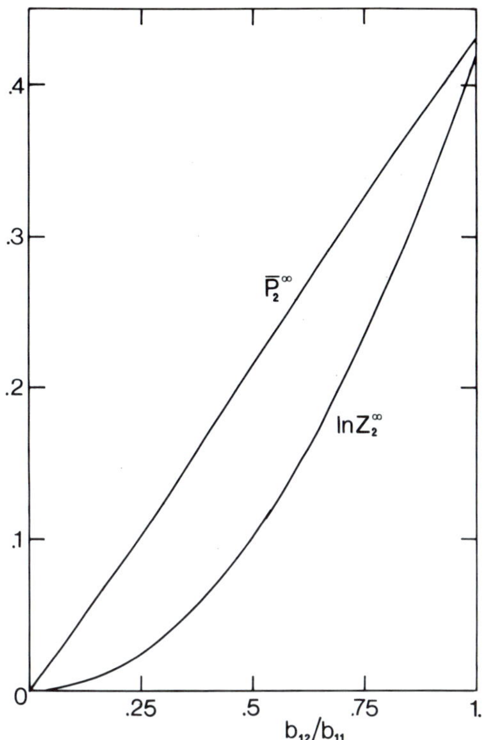

Fig.1 *Dependence of $\ln Z_2^\infty$ and $\bar{P}_2^{(2)\infty}$ on the interaction parameter ratio b_{12}/b_{11}, according to eqs. (57) and (58).*

$$\ln(\gamma_n^\infty/\gamma_i^\infty) = 0.418(v_2/v_1)\{1-T_{NI}^{(2)}/T_{NI}^{(1)}\}, \tag{62}$$

where, depending on whether the pure solute is mesomorphic or non-mesomorphic at the experimental temperature, $T_{NI}^{(2)}$ is an actual or latent nematic-isotropic transition temperature.

With $\gamma_n^\infty/\gamma_i^\infty$ formulated, the limiting slopes β_n^∞ and β_i^∞ can now be determined from

$$\beta_n^\infty = \{1-(\gamma_n^\infty/\gamma_i^\infty)\}/0.418 \tag{63a}$$

and

$$\beta_i^\infty = \{(\gamma_i^\infty/\gamma_n^\infty)-1\}/0.418. \tag{63b}$$

For spherical solutes, $\bar{P}_2^{(2)\infty}$, $\ln Z_2^\infty$, b_{12} and $T_{NI}^{(2)}$ all are zero. Hence from eqs. (56), (61) or (62)

$$\ln(\gamma_n^\infty/\gamma_i^\infty) = 0.418\ v_2/v_1, \tag{64}$$

which, with eq. (63), leads to the results obtained previously [29] for spherical solutes {see eqs. (25-27)}. In a similar manner, eq. (61) or eqs. (56) and (57) (which, for all practical purposes, yield numerically equivalent results) can be used to generate nematic-isotropic phase diagrams of nematic mixtures containing aligned, axially symmetric solutes at low concentrations. We show in figure 2 plots of β_n^∞ and β_i^∞ against

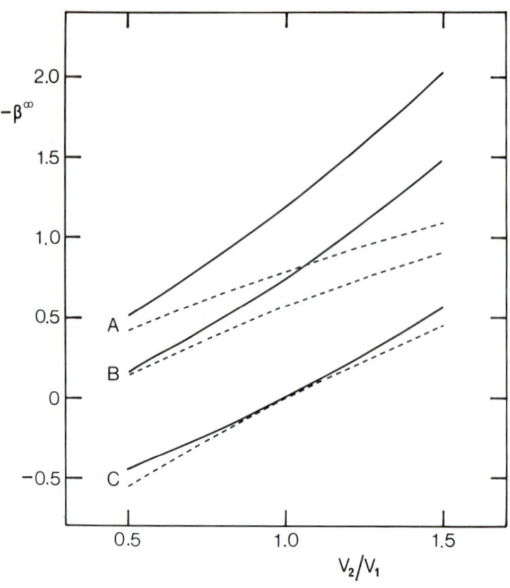

Fig. 2 *Plot of the negative of the limiting slope ($-\beta^\infty$) against the molecular size ratio v_2/v_1 for selected values of b_{12}/b_{11}. Solid curves correspond to nematic phase boundary lines (β_n^∞), dashed curves to isotropic phase boundary lines (β_i^∞). Pairs A, B and C correspond to b_{12}/b_{11} values of 0.2, 0.6 and 1.0 respectively.*

v_2/v_1 for selected values of b_{12}/b_{11} (0.2, 0.6, 1.0). Note the general and physically acceptable trends that smaller and more strongly interacting solutes (i.e. larger b_{12}/b_{11}, hence larger $\bar{P}_2^{(2)\infty}$) are less disruptive and have less negative slopes. In fact, for $b_{12}/b_{11} = 1.0$ and $v_2/v_1 < 1.0$, the solute actually stabilizes the nematic solvent, leading to positive slopes for the limiting nematic and isotropic phase boundary lines. However, in the next section we shall again focus on disruptive solutes ($\gamma_n^\infty/\gamma_i^\infty > 1$, $\beta^\infty < 0$) and shall hereafter refer only to the modulus of β^∞ and denote it by β.

Comparison of all Theories with Experiment

The prediction of the consistent molecular-field model of mixtures developed here may be tested quantitatively only for mixtures containing spherical solutes. However, we note in comparing eq. (64) and eq. (61) or (62) that this model predicts smaller $(\gamma_n^\infty/\gamma_i^\infty)$'s and smaller β's, at comparable v_2/v_1, for solutes which are not spherically symmetric. This is in general agreement with the lattice-model, described earlier in this Chapter, and experimental results. (It would be of interest to carry out quantitative tests of the theory for systems containing aligned non-mesomorphic solutes). On the other hand, the predictions of the lattice treatments of cubical (spherical), chain-like and rod-like solute systems may be readily compared with all of the experimental results given in Chapter 10.

We may begin by considering the size of the two-phase region at constant T*':

$$(x_i - x_n) = (\Delta S_{NI}/R)(1 - T^{*\prime}). \tag{65}$$

Experimentally [34] the coefficient of (1-T*') is 0.18 ± 0.03 for the MBBA systems and 0.26 ± 0.05 for the 5CB systems, both of which are quite close to the DSC $\Delta S_{NI}/R$ values of 0.16 and 0.25, respectively. The lattice models yield coefficients of 0.29 ± 0.02 and 0.48 ± 0.03 for $m_1 = 5$ and $m_1 = 10$, respectively (where m_1 is the length-to-breadth ratio of the solvent), values which are virtually the same as $\Delta S_{NI}/R$ for the pure solvent system, and which are independent of the nature of the solute. Since $m_1 \simeq 3$ for MBBA and 5CB, the agreement with experiment is only reasonable. This agreement may be compared with the molecular field prediction of 0.42, a coefficient fixed by the constant $\Delta S_{NI}/R$ predicted by the Maier-Saupe theory.

Figures 3 and 4 show plots of experimental and theoretical β's against D^3 for spherical solutes. In all but one case, D is the diameter of the

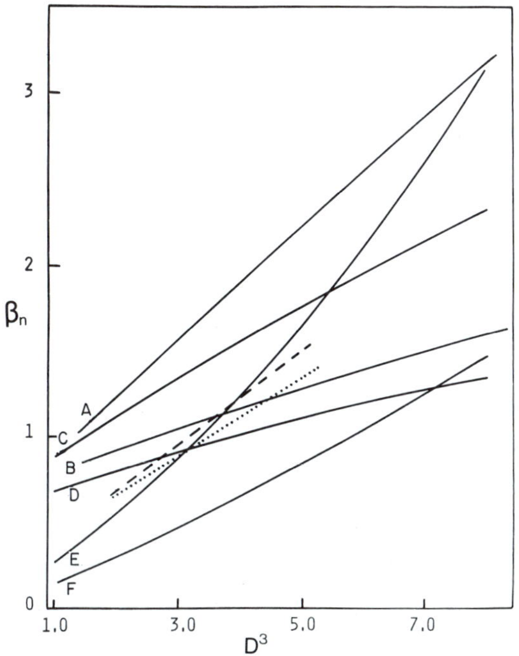

Fig. 3 *Plot of β_n against D^3 for spherical solutes; lattice model: A (m_1=5), B (m_1=10); virial expansion: C (m_1=5), D (m_1=10); molecular field theory: E (m_1=3), F (m_1=5); dashed line: MBBA results; dotted line: 5CB results.*

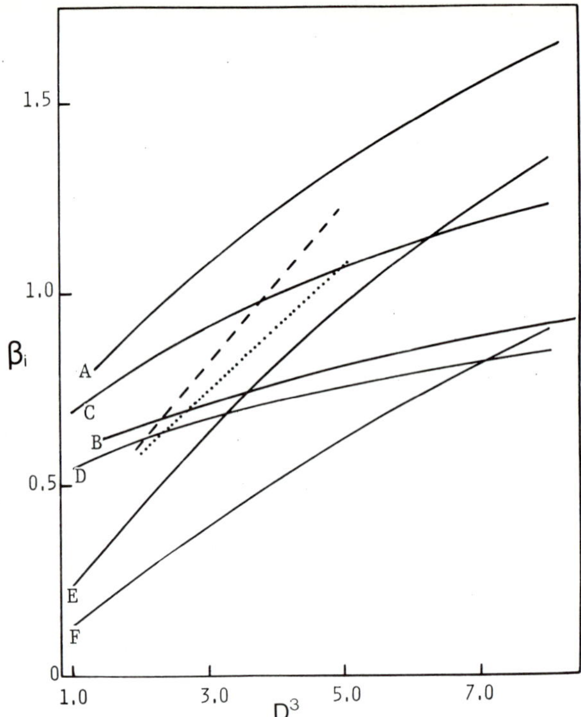

Fig. 4. *Plot of* β_i *against* D^3 *for spherical solutes; lattice model:* $A(m_1=5)$, $B(m_1=10)$; *virial expansion:* $C(m_1=5)$, $D(m_1=10)$; *molecular field theory:* $E(m_1=3)$, $F(m_1=5)$; *dashed line: MBBA results; dotted line: 5CB results.*

spherical solute molecule relative to that of a spherocylindrical solvent molecule. This is the molecular geometry utilized in the virial expansion approach [35] and in the analysis of the experimental results. Accordingly, for the molecular field theory (eqs.(25) and (26)), $v_2/v_1 = 2D^3/(3m_1-1)$. However, for the lattice model plots, D is the side dimension of the cube. (In the lattice model the solvent rod, i.e. a rectangular parallelepiped, has unit thickness). As one can see, the theoretical β values are all within a factor of two of experiment, and the trend of increasing β with increasing D^3 (due to increasing $\gamma_n^\infty/\gamma_i^\infty$ - see the table) is predicted correctly by all three models. Also, all produce the observed trend [34] of decreasing β with increasing m_1. It may also be significant that, to the nearest integer, the virial expansion, lattice model and experimental β_n's correlate best with D^3, while the β_i's correlate best with D^2.

It would appear that, for spherical solutes, the molecular field model gives a marginally better fit to the experimental β's. However, this may be fortuitous. Recall that β_n, for example, is equal to the product of $(\gamma_n^\infty/\gamma_i^\infty-1)$ and $R/\Delta S_{NI}$. As is evident from the table, all models yield rather high $\gamma_n^\infty/\gamma_i^\infty$ values. The lattice model with $m_1 = 5$

Table 2 *Comparison of experimental* $ln(\gamma_n^\infty/\gamma_i^\infty)$ *values for quasispherical solutes with theoretical predictions.*

Solute	Experimental[a]		Theoretical		
	MBBA	5CB	Lattice[b]	Virial[b]	Molecular-Field[c]
Me_4Sn	0.10	0.13	0.30	0.32	0.22
Et_4C	0.12	0.15	0.33	0.35	0.26
Et_4Sn	0.14	0.18	0.37	0.38	0.32
Pr_4Sn	0.16	0.21	0.43	0.44	0.43
Bu_4Sn	0.17	0.23	0.50	0.49	0.53

a. $\ln(\beta_n/\beta_i)$ for visual experiment (cf. Chapter 10).

b. For $m_1 = 5$; from interpolated (β_n/β_i) values

c. From eq. (64)

gives an $R/\Delta S_{NI}$ reasonably close to that of 5CB. Yet, due to the large activity coefficient term, its β_n line is well above that of 5CB (see figure 3). On the other hand, the molecular field model has a relatively small $R/\Delta S_{NI}$, thus partially balancing its large $(\gamma_n^\infty/\gamma_i^\infty)$'s.

Experimentally, the β's for the n-alkane solutes increase by about 10% with roughly a doubling in their chain length. It was argued that this behaviour was not consistent with complete solute rigidity, for which the expected trend would be a marked decrease in β with increasing solute length (cf. Chapter 10). Plotted in figure 5 are lattice model results for β_n as a function of the number of solute segments (m_2) for several values of βE_b, where E_b is the energy barrier to internal rotation. In the light of the experimental results (see figure 6 in Chapter 10), the dramatic difference between curves E and B is self-explanatory and supports the view that the alkane solute chains have some degree of flexibility in the nematic phase. In fact, although the predicted β values would be too high by a factor of two, an $\beta E_b \simeq 1.7$ would produce the same slight increase in β with increasing solute chain length as is found experimentally.

Note also in figure 5 the lattice model behaviour of cubical solutes (line A) relative to chain-like and rod-like solutes. Since the solvent is fixed (a rigid rod with $m_1 = 5$), the absolute β_n values and their progression with increasing m_2 reflect the $\gamma_n^\infty/\gamma_i^\infty$ values and, hence, the mutual compatability of the solute probe and nematic solvent host. As expected the solvent is least tolerant of large globular molecules,

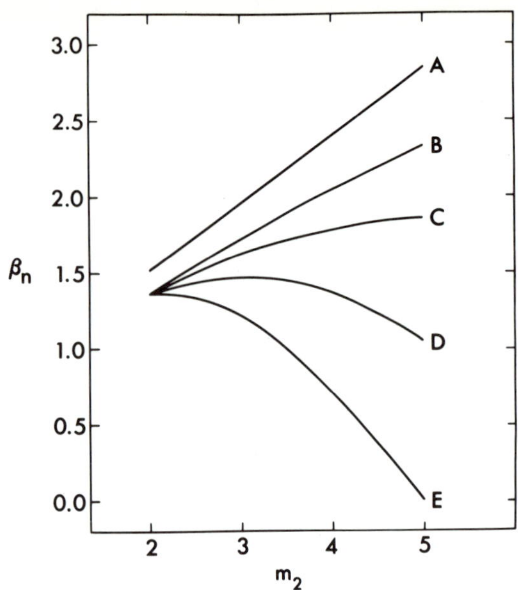

Fig. 5. *Plots of β_n against number of solute segments m_2, from lattice model calculations; m_1=5 for rigid-rod solvent; solute structure: cubes (A), semi-flexible chains (B-D), rigid rods (E); for semi-flexible chains, βE_b: 1.04 (B) 1.38 (C) 2.07 (D).*

most receptive towards long rigid ones, and prefers its chain-like intruders on the inflexible side.

In summary then, considering that we have been dealing with first-generation theoretical models, the overall agreement with experiment is not altogether disappointing. Most of the important trends and some of the quantitative features of the experimental data have been correctly predicted by one or more of the models. The lattice models have proved to be particularly useful in that the effects of solute size, shape and flexibility may be readily studied and then examined in the light of comparable experimental effects.

These lattice models can be improved by allowing for some end-chain flexibility in the solvent molecule. For example, if, instead of having rigid solvent rods with m_1=5, we were to treat mixtures where the solvent rod had four rigid segments and one semi-flexible one, such a minor modification would bring the predicted β values more in line with experiment [8,14]. Also, the molecular field theory may produce better results if deviations from molecular cylindrical symmetry were allowed [36]. Finally, we mention a second-generation model which shows promise, viz, the generalised van der Waals theory (Chapter 8), with anisotropic repulsive interactions and a pseudo-potential dominated by isotropic (rather than anisotropic) attractive interactions. Noting that curves A and F in figures 3 and 4 bracket the experimental lines there is reason to expect that such a balanced model will yield improved agreement. Indeed, preliminary results for such a model mixture of spheres and spherocylinders appear to be in excellent accord with experiment [37]

Acknowledgement

Research support from the National Science Foundation is gratefully acknowledged. Dr. Florrie Dowell is thanked for her valuable collaboration and helpful discussions. Michael W.P. Harbison is acknowledged for computational assistance and useful comments on the molecular field treatment.

References

1. P.J. Flory, *Proc. Roy. Soc.* A234, 73 (1956).
2. E.A. DiMarzio, *J. Chem. Phys.*, 35, 658 (1961).
3. M.A. Cotter and D.E. Martire, *Mol. Cryst. Liq. Cryst.*, 7, 295 (1969).
4. R. Alben, *Mol. Cryst. Liq. Cryst.*, 13, 193 (1971).
5. M.A. Cotter, *Mol. Cryst. Liq. Cryst.*, 35, 33 (1976).
6. A. Wulf and A.G. deRocco, *J. Chem. Phys.*, 55, 12 (1971)
7. G.I. Ågren and D.E. Martire, *J. Chem. Phys.*, 61, 3959 (1974).
8. F. Dowell, Doctoral Dissertation, Georgetown University, 1977.
9. F. Dowell and D.E. Martire, *J. Chem. Phys.* 68, 1088, 1094 (1978).
10. C.S. Shih and R. Alben, *J. Chem. Phys.*, 57, 3055 (1972).
11. M.A. Cotter, *J. Chem. Phys.*, 66, 1098 (1977).
12. H.T. Peterson, D.E. Martire and M.A. Cotter, *J. Chem. Phys.*, 61, 3547 (1974).
13. G.I. Ågren and D.E. Martire, *J. Phys.* (Paris), **36**, 141 (1975).
14. F. Dowell and D.E. Martire, *J. Chem. Phys.*,
15. R. Alben, *J. Chem. Phys.*, 59, 4299 (1973).
16. J.P. Straley, *J. Chem. Phys.*, 57, 3694 (1972).
17. M.A. Cotter, *Mol. Cryst. Liq. Cryst.*, 39, 173 (1977).
18. I. Prigogine (with the collaboration of V. Mathot and A. Bellemans), *The Molecular Theory of Solutions*, North Holland Publ. Co., Amsterdam, 1957.
19. D.H. Chen and G.R. Luckhurst, *Trans. Faraday Soc.*, 65, 656 (1969).
20. B. Kronberg, D.F.R. Gilson and D. Patterson, *J. Chem. Soc., Faraday II*, 72, 1673, (1976).
21. W. Maier and A. Saupe, *Z. Naturforsch.*, 14a, 882 (1959); 15a, 287 (1960); see also, E.B. Priestley, P.J. Wojtowicz and P. Sheng, *Introduction to Liquid Crystals*, Plenum, New York, 1976, Chaps. 3, 4 and 7, and references therein.
22. W.M. Gelbart and B.A. Baron, *J. Chem. Phys.*, 66, 207 (1977).
23. A. Wulf, *J. Chem. Phys.*, 64, 104 (1975).
24. W.M. Gelbart and A. Gelbart, *Mol. Phys.*, **33**, 1387 (1977).
25. S. Jen, N.A. Clark, P.S. Pershan and E.B. Priestly, *J. Chem. Phys.*, **66**, 4635 (1977).
26. M. A. Cotter, *J. Chem. Phys.*, 66, 4710 (1977).
27. R.L. Humphries, P.G. James and G.R. Luckhurst, *Symp. Faraday Soc.*, 5, 107 (1971).
28. H.T. Peterson and D.E. Martire, *Mol. Cryst. Liq. Cryst.*, 25, 89 (1974).
29. D.E. Martire, G.A. Oweimreen, G.I. Ågren, S.G. Ryan and H.T. Peterson, *J. Chem. Phys.*, 64, 1456 (1976).
30. R.L. Humphries and G.R. Luckhurst, *Chem. Phys. Lett.* **23**, 567 (1973).
31. R.L. Humphries and G.R. Luckhurst, *Proc. Roy. Soc.*, A352, 41 (1976).
32. S. Marcelja, *J. Chem. Phys.*, 60, 3599 (1974).
33. T.L. Hill, *An Introduction to Statistical Thermodynamics*, Addison-Wesley Publ. Co. Reading, Mass. (USA), chap. 21, (1960).
34. G. A. Oweimreem, Doctoral Dissertation, Georgetown University, (1977).
35. G. Ågren, *Phys. Rev. A*, 11, 1040 (1975).
36. G.R. Luckhurst, C. Zannoni, P.L. Nordio and U. Segre, *Mol. Phys.* **30**, 1345 (1975).
37. M.A. Cotter, private communication.

Chapter 12

LIQUID CRYSTALS AND MOLECULAR STRUCTURE: SMECTICS

G. W. GRAY

Department of Chemistry, The University, Hull, HU6 7RX, England

Introduction

In considering the effects of change in molecular structure on the smectic properties of mesogens, we have a more difficult subject to tackle, because of smectic polymorphism. There are eight recognised polymorphic smectic modifications, and these are denoted by S_A, S_B, S_C ...S_H. As we will see, each particular smectic polymorphic form can be affected quite differently by a particular alteration in molecular structure. We are therefore concerned with the sensitivity of the thermal stability, or persistence, of each of eight smectic polymorphs to structural change, and also with how structural change affects the incidence of the different polymorphic forms, i.e. their tendency to occur in a given molecular system.

Some of the smectic types are, of course, much less common than others, and as a result, knowledge about the effects of molecular structural change on the different smectic types is very variable.

In a book of this kind, it is not possible to consider in detail how the different smectic types are identified and distinguished from one another. The methods employed involve optical observations using the polarising microscope, thermal analysis, and miscibility studies. A relevant review with photographic illustrations was published recently [1], and we must await with pleasure the publication of a book by Demus dealing far more deeply with this subject. This account is confined, therefore, to a description of the present state of knowledge about the molecular arrangement of the lath-like molecules in each of the smectic types and to illustrating what is known about the way in which molecular structural change affects these molecular arrangements and therefore the phase types themselves.

General Considerations

In relation to the applications of liquid crystals, the very relevant question is frequently asked as to whether a proposed mesogen will exhibit smectic properties (irrespective of their type) and/or nematic/cholesteric properties. Although this basic question can often be answered with some degree of certainty, recent work has shown that quite small changes in molecular structure can have very appreciable effects [2,3,4] on the balance between smectic and nematic/cholesteric tendencies. Greater caution is needed in predicting the type(s) of smectic phase likely to be exhibited; indeed we have a lot to learn about this [5,6,7].

It is however useful to summarise some general points.

(a) If a smectic phase is to be formed, then the lath-like molecules must usually form a layer crystal lattice which will generate the smectic phase on heating. However, even though a suitable molecular arrangement exists in the crystal lattice, if the crystal forces are strong and the melting point of the compound is very high, we may not observe smectic properties if the parallel, stratified arrangement of the molecules is broken down completely (giving an isotropic liquid) or partially, by translation in the direction of the major axes of the molecules (giving a nematic liquid crystal).

(b) Replacement of a terminal ring hydrogen in a mesogen by any substituent which does not destroy the linearity of the molecule or broaden it, enhances T_{NI} values; however, substituents such as -CN, $-NO_2$, CH_3O- which are high in the nematic terminal group efficiency order are low in the smectic order and can suppress [2] smectic properties relative to H. In some instances however, e.g. the long chain 4-n-alkyl- and 4-n-alkoxy-4'-cyanobiphenyls [9,10], the system seems to adapt by forming a S_A phase having an interdigitated bilayer structure [11,12,13], and the smectic properties are then enhanced relative to H. Terminal groups which contribute to the resultant dipole across the long axis, e.g. -CO·O Alkyl, -CH=CH-CO·O Alkyl $-CO\cdot NH_2$ or $-OCF_3$, strongly promote smectic properties, as do ionised functions, e.g. $-CO\cdot O^-M^+$ or $-NH_3^+X^-$. Groups such as -Ph, $-NH\cdot CO\cdot CH_3$, and $-O\cdot CO\cdot CH_3$ strongly promote both smectic and nematic properties, but affect the smectic properties more markedly [14].

(c) Extending the length of a terminal n-alkyl chain increases the smectic tendencies relative to the nematic tendencies of a system [2,8,14]. Eventually a stage is reached when nematic properties are extinguished and the compounds are purely smectic. This behaviour is very general for S_A and S_C phases, and is shown in figure 1 for the 4-(4'-n-alkoxybenzylidene)aminobiphenyls. When S-N transition temperatures rise steeply as in figure 1, no clear alternation of the temperature is distinguishable, but in series where the transition lines are flatter, well

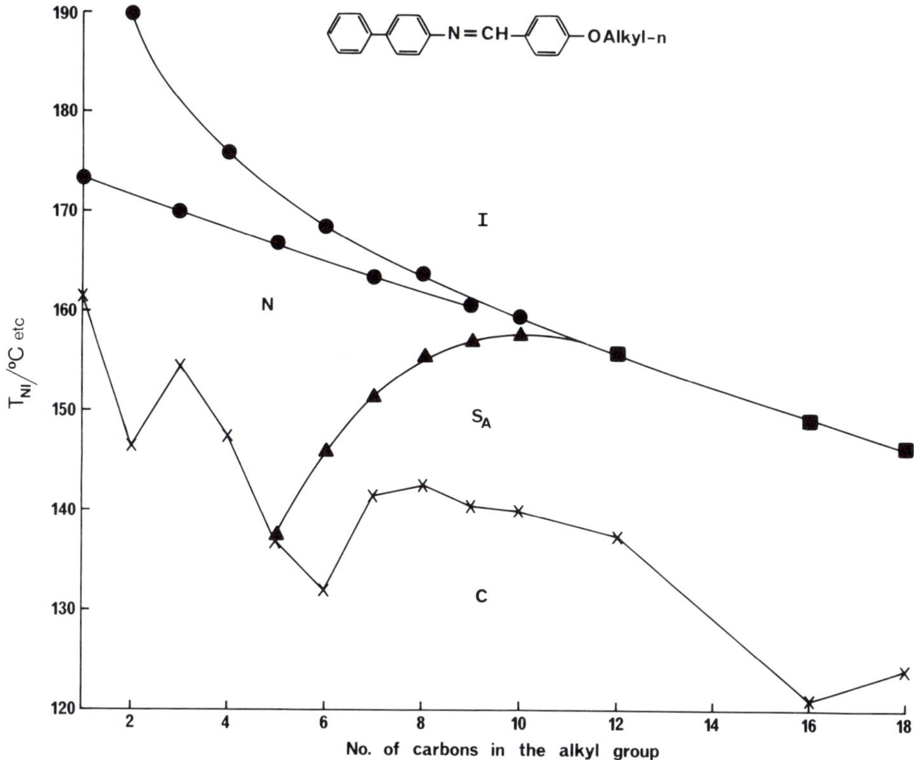

Fig. 1 *Transition temperatures for the 4-(4'-n-alkoxybenzylidene)aminobiphenyls.*

defined alternations of T_{SN} and T_{SI} temperatures occur, similar to those observed for T_{NI}.T_{SAN} and T_{SAI} values alternate in the same sense as T_{NI} values; presumably therefore the same considerations of anistropic molecular polarisability apply to nematic and smectic A phases. However, as we see in figure 2, and shall discuss later, S_B-S_A and S_E-S_B transition temperatures alternate in the opposite sense [14].

Plots of transition temperature against chain length for S_A-N and S_A-I transitions are therefore very similar to those for N-I transitions. Depending on the core structure, the smooth curves can either rise or fall with alternation, as the series is ascended [14], but as we shall see later the types of behaviour vary appreciably with phase type and the molecular system.

(d) Alkyl chain branching on the C_1 of an ester alkyl chain (-CO·O Alkyl) has a much smaller effect on the thermal stability of S_A phases [4,15] than on nematic phases. Moving a branching CH_3 to C_2, C_3 etc. gives progressively smaller decreases. S_B thermal stabilities are not greatly affected by a methyl branch, but S_E thermal stability is enhanced, particularly by a 1-methyl group. The following results illustrate the effects observed.

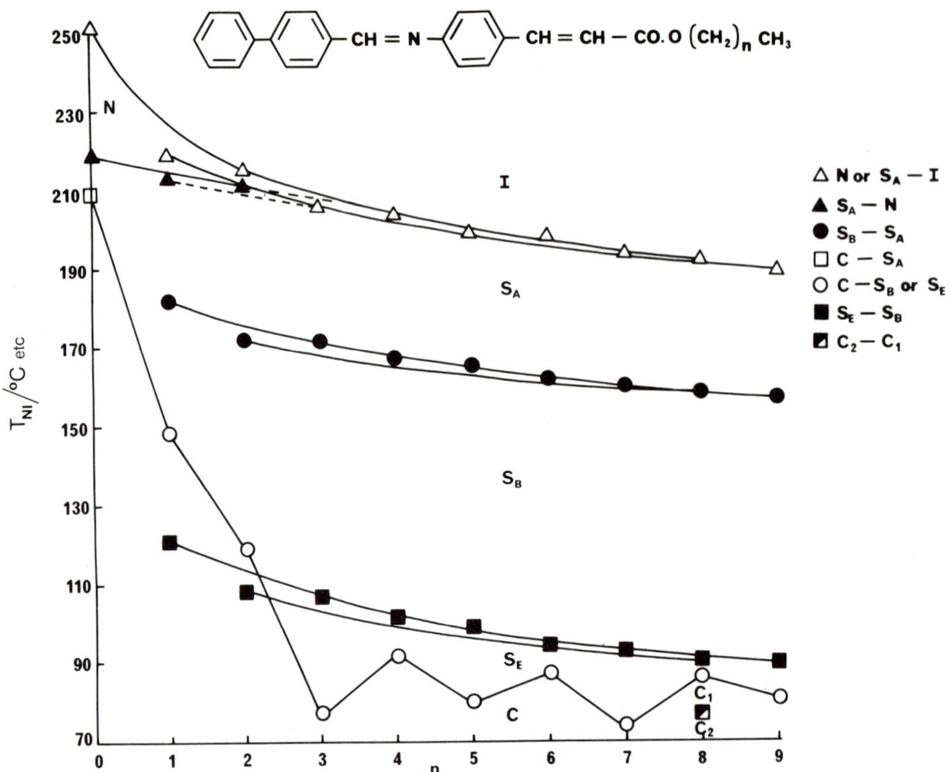

Fig. 2. *Transition temperatures for a series of alkyl esters with the structure indicated.*

C_6H_5–C_6H_4–CH=N–C_6H_4–CH=CH–CO·O–Alkyl

Alkyl	C-S_E	S_E-S_B	S_B-S_A	S_A-I
$CH_2CH_2CH_2CH_2CH_3$	92°	101.5°	168°	204°
$CH(CH_3)CH_2CH_2CH_2CH_3$	70°	128°	168°	180°
$CH_2CH(CH_3)CH_2CH_2CH_3$	68°	113.5°	157°	190.5°
$CH_2CH_2CH(CH_3)CH_2CH_3$	77.5°	109°	167.5°	196°
$CH_2CH_2CH_2CH(CH_3)_2$	93°	108°	168.5°	199°

Slightly different results are obtained for the S_A phase, if the alkyl chain is part of an ether function (-O Alkyl), in that a 1-methyl branch now has a large effect on the S_A-N temperature, but still smaller than that on the N-I temperature [4,14].

(e) The effects of lateral substituents have been studied most fully for system 1 which gives S_C phases when the alkyl chain is sufficiently long [2,8]. Whereas the nematic thermal stabilities decrease in proportion to the size of X, irrespective of its polarity, the S_C-N and S_C-I transition temperatures again decrease, but do so less in relation to the size of X, if X is dipolar. Thus dipole moments do contribute to S_C thermal stabilities e.g. X = CH_3 or Cl has a similar effect in decreasing T_{NI}, but the larger effect on T_{S_CN} is given by X = CH_3. We have then a combined effect of both substituent size and polarity for S_C phases.

Alkyl-O–(C₆H₃X)–(C₆H₄)–CO_2H (1)

As for nematic phases, if the lateral substituent occupies a recess in the structure such that its full broadening effect is not operative, increases in S_C thermal stability can arise [2,14], e.g. in the 5-substituted 6-n-alkoxy-2-naphthoic acids. The increases in T_{S_CN} are larger than those for T_{NI}, showing again that dipole moments play a role in enhancing S_C thermal stability.

The molecular systems in which the steric effects of lateral substituents have been studied [2,14,16] were mainly Schiff's bases of types (2) and (3), which as it happens exhibit S_A and/or N phases. The 2- or 2'-substituents, as we have seen, cause large reductions in T_{NI} because of the additional twisting about the inter-ring bond. For example, a 2-fluoro-substituent in (2) gives $\Delta T_{NI} = 49^o$. Even larger effects on the S_A phases were observed, and $\Delta T_{S_AN} = 71^o$ for a 2-fluoro-substituent in (2). Decreases in N and S_A thermal stability of only 0.5^o and 9°, respectively, have been observed for a fluoro-substituent that exerts no steric effect. Unlike the trends in T_{NI} which followed substituent size, the trends in $T_{S_AN \text{ or } I}$ were irregular, suggesting that dipole moments could again be playing a part.

C₆H₅–C₆H₄–N=CH–C₆H₄–O-Alkyl (2) (positions 2, 2' marked)

Alkyl-O–C₆H₄–CH=N–C₆H₄–C₆H₄–N=CH–C₆H₄–O-Alkyl (3) (position 2 marked)

(f) Finally, in assessing a situation with regard to smectic or nematic tendencies, attention has to be paid to the location in the molecule of particular functions. For example, whereas a *terminal*-CO·O Alkyl group favours smectic properties, the group -CO·O- may be used successfully to link up ring systems and produce strongly nematic materials [17,18].

Thus, benzoate esters such as (4) are purely nematic (C-N, 54°; N-I, 89°), despite the long alkyl chains at each end [17].

$$n\text{-}C_6H_{13}O\text{-}C_6H_4\text{-}C(=O)\text{-}O\text{-}C_6H_4\text{-}OC_8H_{17}\text{-}n \qquad (4)$$

Considerations of Individual Smectic Polymorphic Types

Before considering the individual types of smectic phase, it is useful to look at the transition temperature against alkyl chain length plot, [6] in figure 3, for the homologous series of n-alkyl 4'-n-octyloxybiphenyl-4-carboxylates. The points which emerge stress the extreme

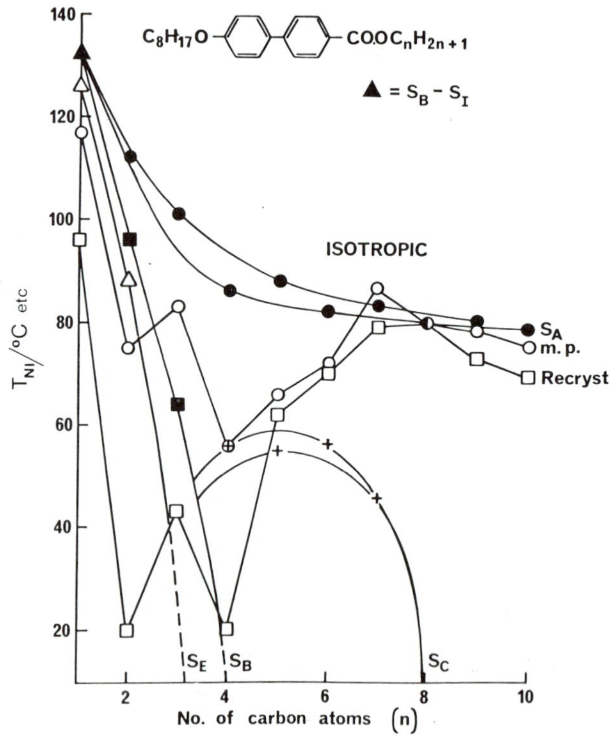

Fig. 3. *A plot of transition temperatures for the n-alkyl 4'-n-octyloxybiphenyl-4-carboxylates.*

sensitivity of both the incidence and thermal persistence of different smectic types to quite minor changes in molecular structure, i.e. the increase in length of a terminal alkyl chain. Moreover, many other examples illustrative of the same effects could be quoted:

(1) the S_A-I transition temperatures fall, steeply at first and then more gradually;

(2) the S_B thermal stability falls away exceedingly steeply, and the last S_B phase occurs at C_3 in contrast to the behaviour [15] shown in figure 2;

(3) the S_E thermal stability falls off even more steeply, and S_E phases are *not* observed after C_2, which contrasts with the behaviour [15] seen in figure 2;

(4) at C_4, a S_C phase is introduced, and virtual S_C-S_A temperatures (obtained by miscibility methods) can be obtained for the C_5, C_6, and C_7 members.

These results make it very clear that *generalisations* relating smectic character to molecular structure should be regarded with caution.

Smectic A Phases

The molecular arrangement in the S_A phase involves a parallel arrangement of the lath-like molecules, with their ends in line, to form layers in which the long axes of the molecules tend to be orthogonal to the layer planes, as we can see in figure 4a. Rotational motion of the molecules is fairly free, but there is no long-range regularity of packing of the centres of gravity of the molecules in the planes of the smectic layers. The layers are therefore liquid-like in nature. The movement of molecules from one layer to another occurs quite freely, and the layers themselves are quite free to slide and move over one another. In the case of unsymmetrical molecules, A-B, there is no evidence for a regular A,A,A and B,B,B ordering of the ends in the layers, and the phases are not ferroelectric.

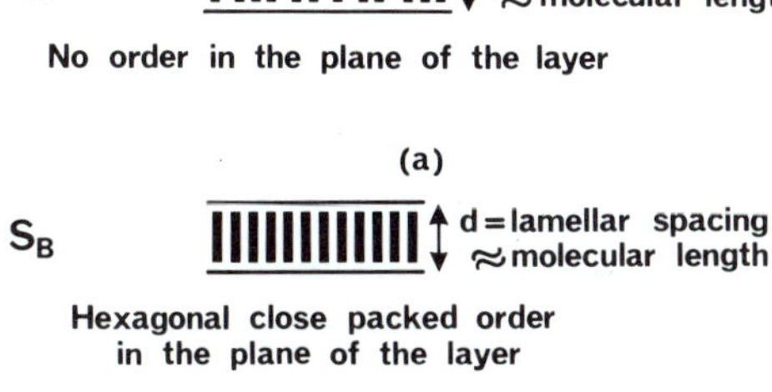

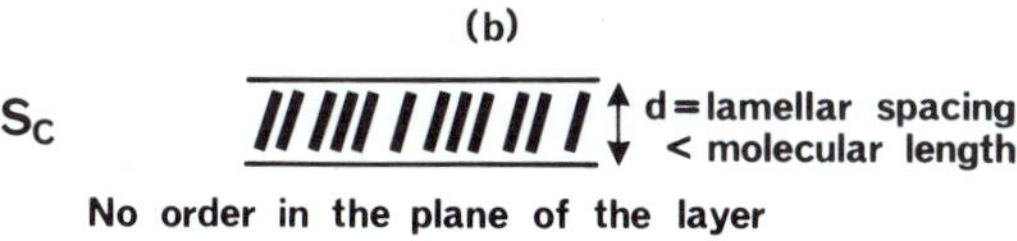

Fig. 4. *Simple representations of the molecular order in the lamallae of S_A, S_B, and S_C phases.*

As we shall see in Chapter 13 the X-ray diffraction pattern of an S_A phase is consistent with this type of arrangement, and has a sharp, inner ring corresponding to the lamellar thickness and a diffuse outer ring reflecting the absence of ordering within the layer planes. The lamellar thickness is often somewhat less than the molecular length. This has variously been attributed to slight interpenetration of the layers which often have liquid-like hydrocarbon chains at the interfaces, or to departures of terminal alkyl chains from their fully extended, *trans*-conformations. An interesting alternative view suggested that the difference in the measured lamellar thickness may arise through a gyroscopic motion of the elongated molecules about some approximately central point [19]. As discussed in subsequent Chapters, this proposal must however be regarded with doubt. A more exceptional form of S_A phase occurs with certain cyano compounds, notably the 4-n-alkyl- and 4-n-alkoxy-4'-cyano biphenyls [11,12]. Here the lamellar spacing for the S_A phase is about 1.5 times the molecular length, and an interdigitated bilayer arrangement must occur (cf. Chapter 13).

We now move on to consider the S_B and S_E phases, after which we can discuss some of the different ways in which the properties of S_A, S_B, and S_E phases are affected by the same changes in molecular architecture.

Smectic B phases

The molecular arrangement for the S_B phase is shown in figure 4(b). The molecules tend to lie orthogonal to the layer planes once more, and the lamellar spacing should therefore approximate to the molecular length. X-ray measurements show that in fact it does. One great difference is that the layers in the S_B phase are highly structured and not unstructured or liquid-like as for S_A phases. The molecular centres lie on a hexagonal net as shown in figure 5(a). Here the molecules are represented as circles to indicate that they are rotating about their long axes. The dimensions of the hexagonal net are however small in relation to the size of the molecules, so that the molecular rotation can hardly be regarded as free, but is probably cooperative. Detailed knowledge

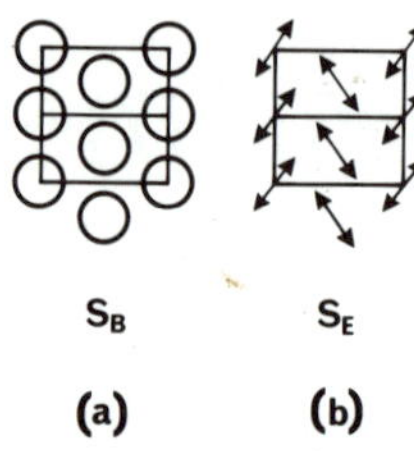

Fig. 5. *Arrangements of the elongated molecules (viewed end on) in the lamellae of (a) the S_B phase (hexagonal) and (b) the S_E phase (orthorhombic).*

of this structural situation for the S_B phase has come from very good X-ray studies [20] on a monodomain sample of the phase formed by heating a single crystal of the solid. The X-ray diffraction pattern consists of a central ring giving the lamellar spacing and a *sharp* outer ring related to the hexagonal order within the lamellae.

The layers of the S_B phase can therefore be regarded as less likely to undergo bending or curving deformations necessitating a splay deformation of the parallel molecular alignment. This has a consequence on the microscopic textures which an S_B phase tends to adopt.

Smectic E Phases

Although the S_E phase may occur on its own, or in conjunction with a higher temperature nematic phase, it is in fact most commonly encountered in the sequence of phases $N \rightarrow S_A \rightarrow S_B \rightarrow S_E$ where the arrows represent decreasing temperature. If the precursor S_B phase is homeotropic (uniaxial), the S_E phase formed is birefringent and therefore biaxial. This distinguishes the S_E phase from both the S_A and S_B phases. The biaxiality of the phase is not however associated with a tilted orientation of the molecules in the layers, for which X-ray studies show that the lamellar spacing is approximately equal to the molecular length. The diffraction pattern shows that the phase is highly ordered, for in addition to the inner ring there are a number of sharp outer rings (cf. Chapter 14). This would be expected of course, as the phase has come from the hexagonally ordered S_B phase by cooling.

As shown by excellent single liquid crystal X-ray studies [21], the position is explained by the chevron-like arrangement of the planes of the molecules viewed end on in the smectic E lamellae, as shown in figure 5b. The system now has an orthorhombic symmetry (positive biaxial) which can be considered as being derived from the hexagonally close packed arrangement of the S_B phase (figure 5a), in which the molecular cross sections are represented as cylinders of rotation. However, recent neutron scattering studies of S_E phases show that the molecules are still rotating quite rapidly [22]. The situation would be explained by jump motions of the molecules about their long axes, such that they are generally restricted to 180° rotations from the positions shown in figure 5b. The cooperative jumps are rapid (10^{9-10} s^{-1}), but the molecular planes can still be regarded as spending a greater proportion of their time in the biaxial chevron arrangement of figure 5b. The molecules are therefore arranged with a two-fold disorder about their long axes as a result of the large amplitude, over-damped librations. Nothing is known about the co-operative nature of the rotations or whether the entire molecules or only the aromatic cores are involved.

It should be mentioned that biaxiality of the S_E phase has not been found for all S_E phases, and that for certain esters of p-terphenyl-4,4"-dicarboxylic acid [25], the S_E phases appear to be uniaxial. Possibly some averaging out of the biaxiality of individual layers to uniaxiality over a number of layers takes place.

Molecular Structural Effects on S_A, S_B, and S_E Phases

It has already been noted that the alternation of S_A-I transition temperatures is opposite to that for S_A-S_B and for S_B-S_E transitions. If dispersion forces, resulting from the anisotropic molecular polarisability are important in determining such transition temperatures, the implication is therefore that interactions between the sides of the molecules in the S_B and S_E phases are more important than in the S_A phase. This would not be inconsistent with the more highly structured situation within the B and E layers.

Other instances in which changes in molecular structure affect the properties of B and E phases differently to A phases have been recorded. The A, B, E sequence of phases occurs quite commonly in systems such as

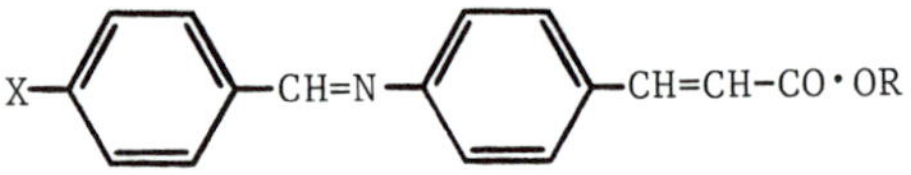

X–⟨C₆H₄⟩–CH=N–⟨C₆H₄⟩–CH=CH–CO·OR

If R is $-(CH_2)_nPh$, as discussed under nematics, then when n changes from even to odd, the Ph group protrudes more beyond the perimeter of the rest of the molecule; the effect is even greater if the terminal Ph also carries a 3- or a 4-substituent [4,15]. These effects give rise to particularly high alternations of T_{NI} values. When the compounds under study exhibit N, S_A, S_B, S_E sequences, e.g. X = Ph, the results for the homologous series can be very interesting, and are exemplified [3] by the cinnamate ester system in figure 6. As n changes from even to odd, there is a very large alternation in the T_{NI} values. However the smectic thermal stabilities are not affected so dramatically, and the result is that we obtain an alternation between materials exhibiting both smectic and nematic and only smectic phases as n increases. Just why the protruding group (which causes a bend in the end of the molecule) affects the smectic properties less is not readily explained [14]. There is, however, an additional effect. Because of the opposite senses of alternation of S_A-I and S_B-S_A transition temperatures, and because in this type of series the alternation is quite pronounced, the two lines meet at n=1 and the S_A properties are extinguished, i.e. we observe a direct S_B-I transition at n = 1. The S_A phase then reappears for n = 2, but for

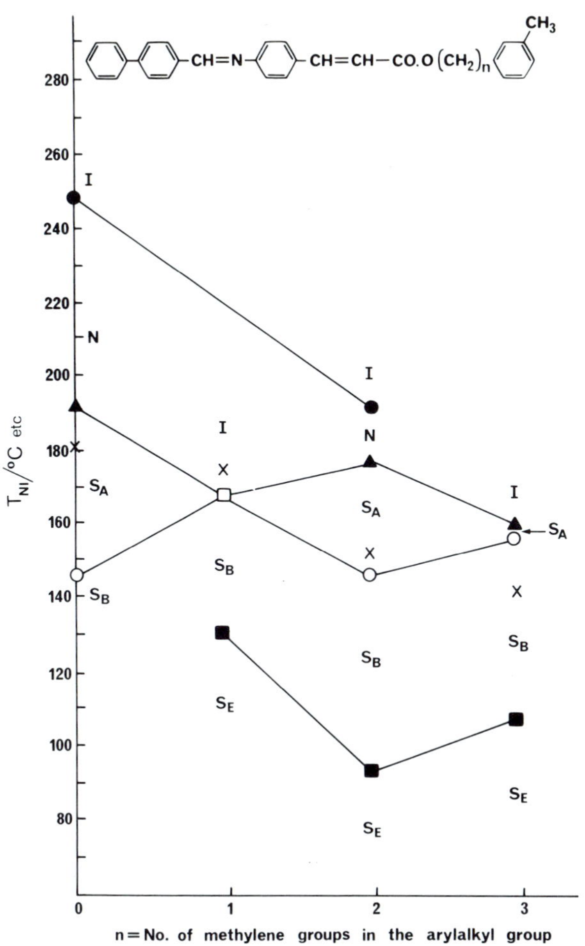

Fig. 6. *Alternation of transition temperatures in an homologous series.*

n = 3, the range of the S_A phase is quite small. The alternation of the S_E-S_B transition temperatures follows the sense of alternation of the S_B-S_A temperatures. Therefore, as we ascend this series we have the interesting set of phase sequences on cooling the isotropic liquid:

n = 0 I, N, S_A, S_B, C;
n = 1 I, S_B, S_E, C;
n = 2 I, N, S_A, S_B, S_E, C;
n = 3 I, S_A, S_B, S_E, C.

That protrusion of a substituent and molecular bending so favour the thermal stabilities of the more ordered S_B and S_E phases which involve some form of co-operative rotation is interesting, and we should certainly not pretend to understand these phases thoroughly until such effects can be explained.

Finally, it is noted that the different effects of a methyl branch in a terminal alkyl chain being moved from C_1 to C_2 to C_3 (see under General Considerations) on the S_A, S_B, and S_E thermal stabilities are a further example of the ability of more ordered S_B and S_E phases to tolerate molecular broadening and bending effects [4,15].

Also, we should not be satisfied until we can explain why the sequence of phases on cooling is not always S_A, S_B, S_E. That is, we do not understand the fact that for compounds [9] such as (5), no S_A phase is formed and a direct N-S_B transition occurs [23] whereas for system (6),

Br–C₆H₄–C₆H₄–C₆H₄–CO·C_6H_{13} (5)

when R = R' = C_2H_5, *only* an S_E phase arises and the S_B and S_A phases do not appear [24]. Yet, when R = C_2H_5 and R' = C_4H_9 (a lower melting com-

RO–C₆H₄–C₆H₄–CO·R' (6)

pound), only S_A and S_E phases are formed, and no S_B phase is produced [24].

Smectic C Phases

This smectic phase is the tilted analogue of the S_A phase, i.e. the smectic layers have a liquid-like, unstructured arrangement of the lath-like molecules which are tilted with respect to the layer planes at angles which vary from compound to compound, and which for a given compound may either vary with temperature over the range of the phase, or stay relatively constant (figure 4c). When it occurs in the company of other phases, the biaxial S_C phase always occurs lower in the temperature scale than either a S_A or N phase, but higher in the scale than a S_B phase. A common enough sequence of phases with falling temperature is therefore $I \rightarrow N \rightarrow S_A \rightarrow S_C \rightarrow S_B$ or S_H.

The S_B phase that forms from the S_C phase on further cooling regains [28] the orthogonal alignment of the molecules relative to the smectic layer planes. In other compounds however, the phase formed from the tilted C phase is also tilted. This would be called by some a tilted S_B phase, but we prefer to designate it as S_H.

X-ray studies of S_C phases reveal a diffraction pattern very similar to that of the S_A phase, i.e. there is an inner ring corresponding to the lamellar thickness, and a diffuse outer ring associated with the unstructured nature of the layers (cf. Chapter 13).

Considerable interest has attached to the reason why the molecules should tilt over when the S_C phase forms from an S_A phase. Several theories have in fact been proposed and notable amongst these are those by McMillan [26] and by Wulf [27]. The Wulf theory associates the driving force to tilt with the molecules having a zig-zag shape, as in the case of TBBA (7):

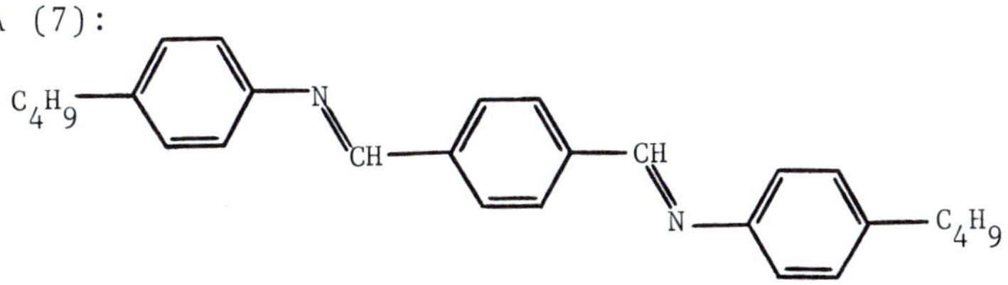

$$C_4H_9-C_6H_4-N=CH-C_6H_4-CH=N-C_6H_4-C_4H_9 \qquad (7)$$

However, we have observed instances in which the S_C character of the system apparently becomes more pronounced as a result of changes in molecular structure which make the systems *deviate* from a zig-zag shape.

The McMillan model depends upon the molecules having two, large, oppositely directed, outboard electric dipoles associated with oxygen or nitrogen atoms. If the molecule possesses an additional dipole also directed across the molecular axis, it is proposed that the outboard dipoles should be equidistant from this central dipole. Thus for the ester molecules which we have been studying recently [6,28], the core structure can be represented as

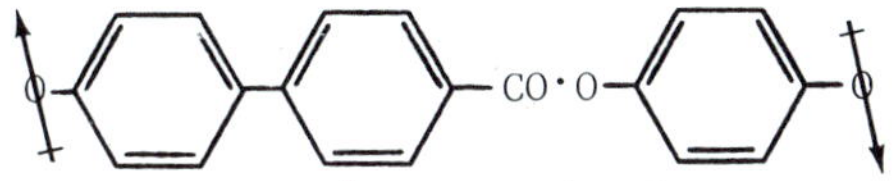

$$O-C_6H_4-C_6H_4-CO\cdot O-C_6H_4-O$$

The outboard dipoles occur at the terminal oxygen atoms of the alkoxy groups, and the ester function constitutes the central dipole. The plot of transition temperatures against alkyl chain length for one series is shown in figure 7; the nematic and S_A behaviour is normal. The S_C-S_A transition temperatures constitute two curves, resulting from an alternation which is of the *same* sense as that for the S_A-N, S_A-I, and N-I temperatures. Note that for these aryl esters, the S_C- S_A curves do not fall away steeply as the chain is extended, as they do for the alkyl ester analogues depicted in figure 3. In this series, the S_B phases are orthogonal and uniaxial. The plot of lamellar spacing, which is related to the tilt angle, against temperature for the 4-n-octyloxyphenyl ester is shown in figure 8. Note the changing tilt angle over the range of the S_C phase, the sudden increase in lamellar spacing on formation of the S_B phase where the spacing is now equivalent to the molecular length, and the smaller lamellar spacing for the S_A phase.

Returning to the McMillan model [26] for the S_C phase, a simple interpretation of the requirement for the outboard dipoles would be in terms of their exerting a torque on the system of orthogonal molecules in the precursor S_A phase, so causing a tilting over of the long axes. However, this would seem to require a freezing out or at least a severe quenching of the molecular rotations known [29] to occur in the S_A phase. It would

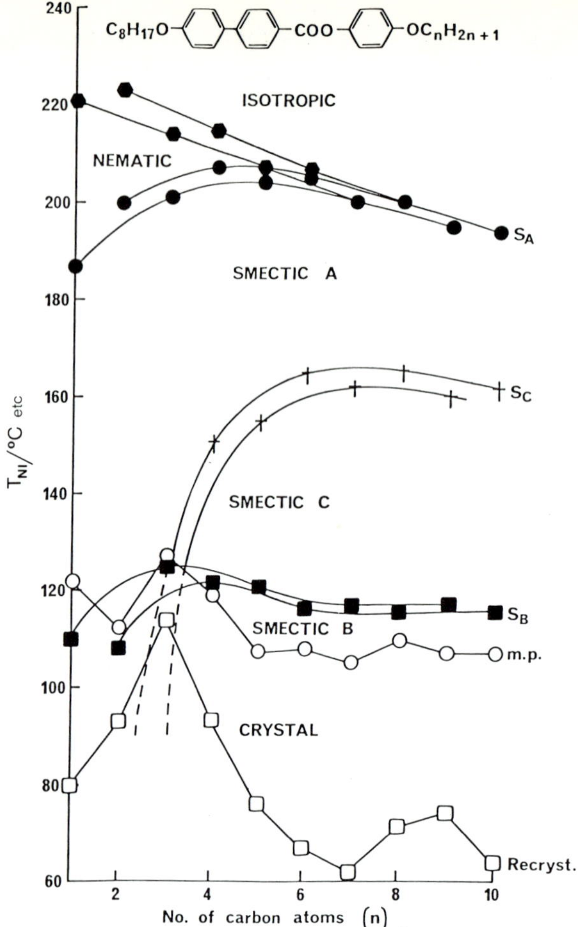

Fig. 7. *Transition temperatures for a series exhibiting N, S_A, S_C, and S_B (orthogonal) phases.*

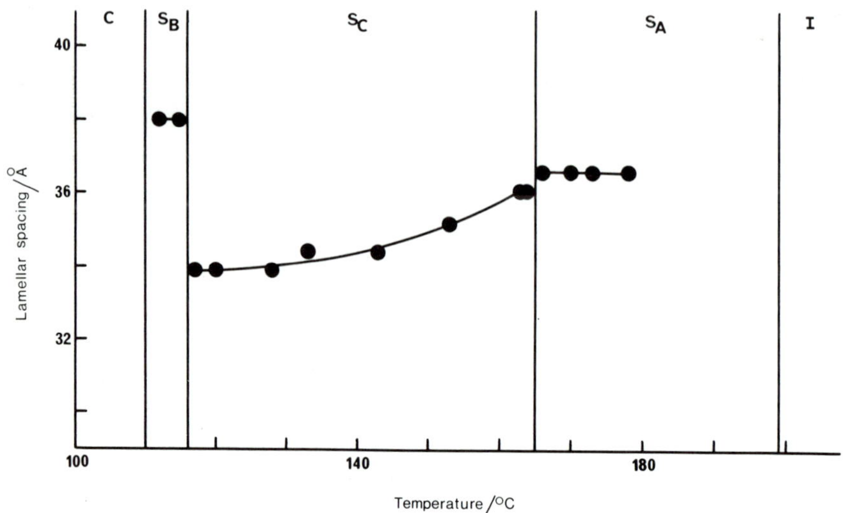

Fig. 8. *Plot of lamellar spacing against temperature for the ester with n=8 from the series shown in figure 7.*

also require an explanation of the regeneration of these molecular rotations which are known [29] to occur in the lower temperature S_B phase which succeeds the S_C phase. Further doubts about the validity of the outboard dipole model are generated by considerations of figure 3 and similar plots for related series, since these show that in a *constant* dipolar situation, the incidence and thermal stability of the S_C phase are dependent strongly on terminal alkyl chain length.

In our test of the Wulf theory [27] we had investigated the effects of alkyl chain branching on the incidence of S_C properties and had shown that in the system

$$C_8H_{17}O-C_6H_4-C_6H_4-CO\cdot OCH(CH_3)C_nH_{2n+1}$$

the incidence of S_C phases is increased [28]. In the light of this, we came to wonder whether it might be possible to design some compounds which do not have terminal, outboard dipoles, and yet exhibit S_C properties. Eventually we synthesised compounds (8) to (11), with the results [30] shown below

$$n\text{-}C_8H_{17}O-C_6H_4-C_6H_4-CO\cdot O-C_6H_4-OC_8H_{17}\text{-}n \qquad (8)$$

$$C \xrightarrow{100^{\circ}} S_B \xrightarrow{116^{\circ}} S_C \xrightarrow{165^{\circ}} S_A \xrightarrow{200^{\circ}} I$$

$$n\text{-}C_8H_{17}O-C_6H_4-C_6H_4-CO\cdot O-C_6H_4-C_8H_{17}\text{-}n \qquad (9)$$

$$C \xrightarrow{110^{\circ}} S_B \xrightarrow{110.5^{\circ}} S_C \xrightarrow{132.5^{\circ}} S_A \xrightarrow{184^{\circ}} I$$

$$C_2H_5\overset{*}{C}H(CH_3)(CH_2)_3-C_6H_4-C_6H_4-CO\cdot O-C_6H_4-CH_2\overset{*}{C}H(CH_3)C_2H_5 \qquad (10)$$

$$C \xrightarrow{83.4^{\circ}} Ch \xrightarrow{114^{\circ}} I \ (Ch \xrightarrow{81.0^{\circ}} S_A \xrightarrow{74.3^{\circ}} S_C) \text{ monotropic}$$

$$C_2H_5\overset{*}{C}H(CH_3)(CH_2)_3-C_6H_4-C_6H_4-CO\cdot O-C_6H_4-C_7H_{15}\text{-}n \qquad (11)$$

$$C \xrightarrow{91.5^{\circ}} S_C \xrightarrow{93^{\circ}} S_A \xrightarrow{112^{\circ}} Ch \xrightarrow{131^{\circ}} I$$

A S_C phase therefore occurs for each of the above four compounds which range from compound (8) with two outboard dipoles and one central dipole, to compounds (10) and (11) with only one dipole associated with a central group through the intermediate case of compound (9) with one central and only one outboard dipole.

McMillan's theory was of course based firmly on the knowledge existing at that time and which suggested that compounds exhibiting the S_C phase should have a rather symmetrical structure and possess terminal outboard dipole moments. Thus, the 4,4'-di-n-alkoxyazoxybenzenes exhibi S_C phases, whereas the related 4,4'-di-n-alkylazoxybenzenes exhibit S_A phases. The synthesis of these new compounds now alters the basis of knowledge (see also ref. 31), and contradicts the idea that two terminal outboard dipoles are necessary. It should however be added that such outboard dipoles do seem to enhance the thermal stability of a S_C phase, cf. the S_C-S_A transition temperatures of compounds (8) to (11).

We now believe that the molecules in the S_C phase do rotate about their long axes and that a combination of the interactions between the resulting rotational dipoles and the shapes of the volumes swept out by the rotating molecules provides the reason for the adoption of a tilted molecular arrangement [32].

Chiral S_C Phases

The chirality of mesogen molecules does not affect the situation in S_A, S_B, and S_E phases where the molecules tend to lie orthogonal to the layer planes. In a tilted situation such as the S_C phase, the chirality of the molecules induces a helical distribution of the tilt directions (the tilt angles being constant at a constant temperature) about an axis drawn at right angles through a stack of layers [33]. Helfrich and Oh [34] first showed that such a phase does occur and that a thin film of the S_C phase having the smectic layers parallel to the microscope slide and cover slip, showed optical activity and rotatory dispersion. However, the pitch (3.2 μm) was not such that selective reflection of light occurred [35].

More recently [36], we have prepared a range of compounds such as (12), in which the chiral 2-methylbutyl group is directly attached to

$$C_2H_5\overset{*}{C}H(CH_3)CH_2-C_6H_4-C_6H_4-CO\cdot O-C_6H_4-OC_6H_{13}\text{-n} \qquad (12)$$

the ring system. Much smaller pitches are obtained, and the materials exhibit S_C and Ch phases, both of which give the selective reflection of specific wavelengths of coloured light. The pitches are in the range 0.2-0.5 μm and change with temperature, but in the opposite sense for the two phases. Thus, for compound (12), on cooling through the cholesteric phase range, the colour of reflected light shifts from blue to red and then into the infra-red at the Ch-S_C transition. As the temperature of the chiral S_C phase is decreased, the colour of reflected light then shifts very quickly back from the infra-red to red and yellow. A mixture of compounds of the type (12) exhibits [36] a complete change from red to blue for the S_C phase.

Because of symmetry considerations, a ferroelectric dipole in the plane of the chiral S_C layers is generated [46,47]. That this dipole is not providing the driving force to tilt is shown by the fact that the racemic modification of (10) exhibits an ordinary, non-ferroelectric S_C phase. Hence, the ferroelectric properties of chiral S_C phases are a consequence and not the cause of tilting.

Smectic D Phases

Only two materials, cf. (13) in which R = n-$C_{16}H_{33}$ or n-$C_{18}H_{37}$, are known [37] to exhibit the S_D phase. The sequence of

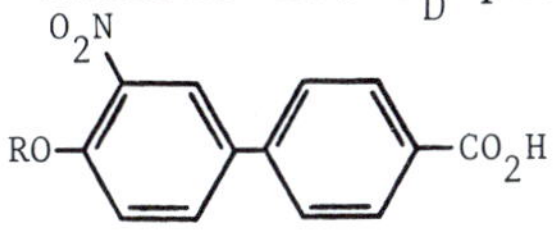

O_2N

RO–(biphenyl)–CO_2H (dimer) (13)

phases on cooling is I → S_D → S_C (C_{18}) and I → S_A → S_D → S_C (C_{16}). The birefringent S_C phase, on heating, suddenly develops optically extinct areas which are often polygonal in shape. These areas extend and eventually fill the entire field of view. The phase is not however homeotropic; it is completely optically isotropic [1].

The molecular structure of this phase is not understood, but it must be of a cubic nature to account for the optical properties, i.e. isotropic, but ordered. One possibility [38] is that the S_C phase must break up and form micelles, perhaps spherical in shape and containing many molecules, and that these then constitute a cubic lattice. In one case, these micelles must break down at a higher temperature to form an isotropic liquid, whereas in the other they must reconstitute a lamellar S_A phase on further heating. Analogies to this possible behaviour are to be found in the field of lyotropic liquid crystals, i.e. in the case of optically isotropic, micellar phases of amphiphiles. From X-ray studies made on a rotating sample of the S_D phase, Tardieu [39] has concluded, however, that there are strong structural similarities to the cubic, amphiphilic phases of certain lipid systems for which Luzzati *et al.* [40] had earlier proposed a jointed rod structure with overall cubic symmetry.

With only two closely related materials forming the phase, no comment about the relation between molecular structure and the occurrence of the phase can be made, except the obvious one that a combination of the long alkyl terminal chain and the strong, lateral ring-NO_2 dipole may be critical.

Smectic F Phases

Until fairly recently, compound (14) was unique [41] in exhibiting the S_F phase. In addition, the rare S_G phase was also formed.

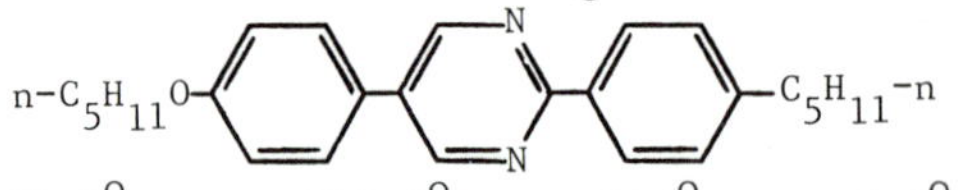

n-C_5H_{11}O–(phenyl)–(pyrimidine)–(phenyl)–C_5H_{11}-n (14)

$$C \xrightarrow{79^o} S_G \xrightarrow{102.7^o} S_F \xrightarrow{113.8^o} S_C \xrightarrow{114^o} S_A \xrightarrow{210^o} I$$

Demus *et al* [41] assigned the new letters F and G to these phases because they were separated by a first order transition and were individually immiscible with all other smectic types.

Recently, however, we have prepared [48] members of more than one series of quite different materials which also form the S_F phase. Compound (14) is therefore no longer unique. X-ray studies [42] confirm the results obtained on the S_F phase of (14). The diffraction pattern has a sharp inner ring which corresponds to a lamellar spacing considerably shorter than the molecular length, and the outer ring is diffuse, perhaps less diffuse than for S_C, but nowhere near as sharp as for a smectic B phase. The molecular arrangement is therefore tilted, and the arrangement within the layers is relatively disordered. A clear distinction between the S_C and the S_F phases based on X-ray diffraction is not therefore obvious, since the tilt angle does not seem to change appreciably at the S_C-S_F transition. However, compared with the very small enthalpy and entropy changes associated with the S_C-S_A transition, the enthalpy and entropy changes at the S_C-S_F transition are relatively large. This implies a considerable degree of change in order at this transition, a change which does not manifest itself clearly in the X-ray diffraction patterns. This remains a puzzling point. Nothing can yet be said about molecular structural correlations for the S_F phase, except that, as a tilted phase, the molecular factors important to its formation may be similar to those for the S_C phase.

Smectic G Phases

At one time, compound (14) was also unique [41] in forming a S_G phase, but the phase is now known [43] to occur in compound (15) and also in TBBA (7). For TBBA, the sequence of phases observed on cooling is

$$I \xrightarrow{235^o} N \xrightarrow{199^o} S_A \xrightarrow{172^o} S_C \xrightarrow{144.5^o} S_H \xrightarrow{89.2^o} S_G \xrightarrow{73\text{-}74^o} S? \rightarrow C$$

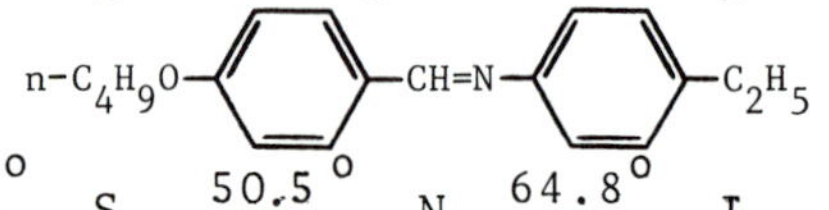

n-C_4H_9O–(phenyl)–CH=N–(phenyl)–C_2H_5 (15)

$$C \xrightarrow{38.7^o} S_G \xrightarrow{50.5^o} N \xrightarrow{64.8^o} I$$

S? has not been classified, and it is not certain whether it is a smectic phase or a polymorphic solid form. Other probable cases of S_G phase formation occur, but full confirmation of the S_G classification by miscibility methods has yet to be obtained. The homologue of compound (14) with C_4H_9 alkyl groups instead of C_5H_{11} also forms a S_G phase [43] in the sequence I, S_A, S_C, S_G, C, on cooling, i.e. no S_F phase is formed.

It is obvious that with such a limited range of S_G materials, no comments can be made about those molecular structural features which favour the occurrence of this type of phase.

The phase has close similarities to the S_E phase. From the X-ray diffraction patterns, the layers are highly structured. For compound (14), the molecular length is about 30 Å and the lamellar spacing [41] of the S_G phase is about 22.1Å. We are therefore dealing with structured layers in which the molecules are inclined with respect to the layer planes. Sometimes indeed, the S_G phase is referred to as a tilted S_E phase, and evidence [44,45] suggests that the planes of the tilted molecules again adopt a chevron arrangement with two-fold disorder about their long axes. The molecules undergo rapid ($\sim 10^{-11}$s) though restricted reorientational motions about their axes. The arrangement in the layers can be described as monoclinic compared with the orthorhombic, orthogonal arrangement of the S_E phase.

Smectic H Phases

Smectic phases having structured layers similar tc those of orthogonal S_B phases, i.e. with a hexagonal arrangement of the molecular centres in the layer planes (see figure 5a), but with a tilted arrangement of the molecular long axes relative to the layer planes, have been recognised for several years. X-ray studies of monodomain samples of these phases were again possible [20]. However, it appeared from earlier studies that these tilted analogues of the orthogonal B phases were always completely miscible with orthogonal S_B phases. Consequently, these orthogonal and tilted phases were not assigned a different code letter; they were considered to be identical in type. As a result the phases are usually referred to as orthogonal S_B phases (S_{B_A} phases) and tilted S_B phases (S_{B_C} phases). However, de Vries [44] had studied such tilted phases in several compounds, including TBBA, and, despite the fact that one of these materials was compound (15) which according to Richter *et al.* [43] is now S_G, he considered that the X-ray results were sufficiently different to those for orthogonal S_B phases to justify distinguishing them clearly from tilted S_B (S_{B_C}) phases.

The matter currently rests in this situation, but we have obtained some new information [32] over the last year. In carrying out miscibility studies of tilted S_B phases *with one another* we found that in an intermediate composition range an orthogonal S_B phase is sometimes produced. This made us look very carefully at the effects of small changes in mixture composition in the vicinity of the change over from an orthogonal to a tilted situation for mixtures of an orthogonal S_B material with a second tilted S_B material. We have established that over narrow composition ranges of 2-3%, the two phases are immiscible. In the narrow range over which immiscibility occurs, we have clear photomicrographs showing heterogeneous textures consisting of areas with S_{B_A} texture and areas with S_{B_C} texture. We are forced to conclude then that the two phases should not be classed together and that S_B should mean an orthogonal S_B phase. The tilted analogue must now be given a different letter, and it would seem that S_H is good enough.

It would appear therefore that a mixture of a S_H and a S_B phase can fairly readily adapt, dependent on composition, to either an overall S_B or an overall S_H arrangement, except for a quite narrow region of composition where some parts assume a S_H and some a S_B texture. Nonetheless, an immiscibility region exists.

As a consequence, we no longer have the dilemma that the unstructured analogues of these phases, namely the S_C phase and the S_A phase, are totally immiscible, whereas their structured analogues, S_B and S_H, appeared to be miscible. They are not, but because of the narrow ranges over which immiscibility occurs, it is not surprising that this has hitherto escaped observation.

Chiral S_F, S_G, and S_H Phases

Optically active materials giving the S_C phase form a chiral S_C with a helical structure and having quite distinctive properties. The same possibility exists for the other tilted phases, S_F, S_G, and S_H. A chiral version of the S_H phase (published [35] as a chiral tilted S_B phase) has already been reported. In our preparation of new S_F materials [48], several of the compounds are chiral, and so the S_F phases observed are chiral S_F phases. The probability also exists that we have made some chiral S_G materials, but we have yet to investigate the optical activity of the phases thoroughly.

Smectic Properties and Cholesteric Materials

In Chapter 1, the effects of molecular structure on the cholesteric properties of non-sterol cholesterogens were mentioned and the reader was referred to a review [14] of this subject in relation to sterol cholesterogens. In this latter context, the review by Elser and Ennulat [49] on the selective reflection of cholesteric liquid crystals is also important.

Both types of cholesterogen are of course of interest in the area of surface thermography, and in both cases, sensitivity of colour response to temperature is closely related to the onset of smectic properties at the lower temperature end of the range in which a sharp response is desired. Therefore some of the sections in this Chapter which deal with molecular structural effects on the incidence of S_A and S_C phases are important to this area of liquid crystal technology, as is the fact that the chiral S_C phases of suitably chosen materials themselves exhibit a thermochromic response which could prove to be useful.

References

1. D. Coates and G.W. Gray, *The Microscope*, 24, 117 (1976).
2. G.W. Gray in *"Liquid Crystals and Plastic Crystals"* (G.W. Gray and P.A. Winsor, eds.), Vol. 1, Chap. 4.1, Ellis Horwood, Chichester, England, (1974).
3. D. Coates and G.W. Gray, *J. Phys. (Paris)*. 36, 365 (1975).
4. G.W. Gray and K.J. Harrison, *Symp. Faraday Soc.*, 5, 54 (1971).
5. A. Biering, D. Demus, G.W. Gray and H. Sackmann, *Mol. Cryst. Liq. Cryst.*, 28, 275 (1974).
6. J.W. Goodby and G.W. Gray, *J. Phys. (Paris)*, 37, 17 (1976).
7. S. Diele, *Phys. Status Solidi*, A25, K183 (1974).
8. G.W. Gray, *"Molecular Structure and the Properties of Liquid Crystals"*, Academic Press Inc., N.Y. (1962).
9. G.W. Gray, *J. Phys. (Paris)*, 36, 337 (1975).
10. G.W. Gray, K.J. Harrison and J.A. Nash, *Pramana, Suppl.*, 1, 381 (1975).
11. J.E. Lydon and C.J. Coakley, *J. Phys. (Paris)*, 36, 45 (1975).
12. G.W. Gray and J.E. Lydon, *Nature (London)*, 252, 221 (1974).
13. A.J. Leadbetter, J.L.A. Durrant and M. Rugman, *Mol. Cryst. Liq. Cryst. Letters*, **34**, 231 (1977).
14. G.W. Gray in *"Advances in Liquid Crystals"* (G.H. Brown, ed.), Vol. 1, p.1, Academic Press Inc., N.Y., (1976).
15. G.W. Gray and K.J. Harrison, *Mol. Cryst. Liq. Cryst.*, 13, 37 (1971).
16. D.J. Byron, G.W. Gray and B.M. Worrall, *J. Chem. Soc.*, 3706, (1965).
17. J.P. Van Meter and B.H. Klanderman, *Mol. Cryst. Liq. Cryst.*, 22, 271 (1973); J.P. Van Meter and A.K. Seidel, *J. Org. Chem.*, 40, 2998 (1975).
18. R. Steinsträsser, *Z. Naturforsch.*, B27, 529 (1972).
19. A. Lösche, S. Grande and K. Eider, First Specialised "Colloque Ampere", Krakow, Poland, p.103, (1973); A. Lösche and S. Grande, 18th Ampere Congress, Nottingham, p.201, (1974).
20. A.M. Levelut and M. Lambert, *C.R. hebd. Séanc. Acad. Sci., Paris*, 272, 1018 (1971); A.M. Levelut, *J. Phys. (Paris)*, 37, 51 (1976).
21. J. Doucet, A.M. Levelut, M. Lambert, L. Liebert and S. Strzelecki, *J. Phys. (Paris)*, 36, 13 (1975).
22. A.J. Leadbetter, unpublished results.
23. J. Nehring and M.A. Osman, *Z. Naturforsch.*, 31A, 786 (1976).
24. D. Demus, L. Richter, C.-E. Rürup, H. Sackmann and H. Schubert, *J. Phys. (Paris)* 36, 349 (1975).
25. S. Diele, P. Brand and H. Sackmann, *Mol. Cryst. Liq. Cryst.*, 17, 163 (1972).
26. W.L. McMillan, *Phys. Rev.*, A8, 1921 (1973).
27. A. Wulf, *Phys. Rev.*, A11, 365 (1975).
28. J.W. Goodby and G.W. Gray, *Mol. Cryst. Liq. Cryst.*, 37, 157 (1976).
29. G.W. Gray and A.J. Leadbetter, *Phys. Bull.*, 28, (1977).

30. J.W. Goodby, G.W. Gray and D.G. McDonnell, *Mol. Cryst. Liq. Cryst. Letters,* 34, 183 (1976).
31. W.H. de Jeu, *J. Phys. (Paris),* **38**, 1265 (1977).
32. J.W. Goodby and G.W. Gray, *J. Phys. (Paris),* **40**, 363 (1979).
33. A. Saupe, *Mol. Cryst. Liq. Cryst.,* 7, 59 (1969).
34. W. Helfrich and Chan S. Oh, *Mol. Cryst. Liq. Cryst.,* 14, 289 (1971).
35. D. Coates and G.W. Gray, *Mol. Cryst. Liq. Cryst. Letters,* 34, 1 (1976).
36. G.W. Gray and D.G. McDonnell, *Mol. Cryst. Liq. Cryst,* 37, 189 (1976).
37. D. Demus, G. Kunicke, J. Neelson and H. Sackmann, *Z. Naturforsch.,* 23a, 84 (1968); S. Diele, P. Brand and H. Sackmann, *Mol. Cryst. Liq. Cryst.,* 17, 1963 (1972).
38. G.W. Gray and P.A. Winsor, *Mol. Cryst. Liq. Cryst.,* 26, 305 (1974).
39. A. Tardieu and J. Billard, *J. Phys. (Paris),* 37, 79 (1976).
40. V. Luzzati and P.A. Spegt, *Nature (London),* 210, 1351 (1966); A. Tardieu and V. Luzzati, *Biochim. biophys. Acta,* 219, 11 (1970); see also K. Fontell in *"Liquid Crystals and Plastic Crystals"* (G.W. Gray and P.A. Winsor, eds.), Vol. 2, Chap. 4, pp. 94-102, Ellis Horwood, Chichester, England, (1974).
41. D. Demus, S. Diele, M. Klapperstück, V. Link and H. Zaschke, *Mol. Cryst. Liq. Cryst.,* 15, 161 (1971).
42. A.J. Leadbetter, unpublished results.
43. L. Richter, D. Demus and H. Sackmann, *J. Phys. (Paris),* 37, 41 (1976).
44. A. de Vries and D.L. Fishel, *Mol. Cryst. Liq. Cryst.,* 16, 311 (1972).
45. J. Doucet, A.M. Levelut and M. Lambert, *Phys. Rev. Lett.,* 32, 301 (1974).
46. R.B. Meyer, L. Liebert, L. Strzelecki and P. Keller, *J. Phys. Lett. (Paris),* 36, L69 (1975).
47. P. Keller, L. Liebert and L. Strzelecki, *J. Phys. (Paris),* 37, 27 (1976).
48. J.W. Goodby and G.W. Gray, *Mol. Cryst. Liq. Cryst. Lett.,* **41**, 145 (1978).
49. W. Elser and R.D. Ennulat in *"Advances in Liquid Crystals"* G.H. Brown, ed), Vol. 2, p.73, Academic Press Inc., N.Y., (1976).

Chapter 13

STRUCTURAL STUDIES OF NEMATIC, SMECTIC A AND SMECTIC C PHASES

A.J. LEADBETTER

Department of Chemistry, University of Exeter
Stocker Road, Exeter EX4 4QD, England

Introduction

A large number of X-ray measurements on the various liquid-crystalline phases, and recently a few neutron diffraction investigations, have been reported. These span the range from the almost crystallographic studies of the highly ordered S_E phases, to the nematic phase in which the formalism required is that appropriate to liquids. The fascination, and the difficulty, of structural studies of the liquid-crystalline phases is precisely that elements of crystal symmetry exist together with considerable disorder characteristic of liquids. This means that no simple formalism is appropriate for all phases and the difficulties are further compounded because the molecules are generally rather large and non-rigid. Consequently a complete solution cannot be expected from diffraction studies alone (or indeed from any single technique) and most of the diffraction work reported has been aimed first at characterisation or verification of the phase behaviour, where it is extremely valuable and generally unambiguous, and then determination of layer and molecular spacings and correlation lengths, which can give valuable insight into the molecular packings and a measure of the extent of order. These studies require, in essence, only measurement of positions and widths of the diffraction features. A small number of more detailed investigations have also been made to try to determine quantitative aspects of the distribution functions. These studies require the measurement of the intensity of scattering in reciprocal space and are consequently much more difficult and time consuming.

We shall not attempt to review all of this work here but will restrict our consideration to the relatively more disordered phases: smectic C, A and nematic. First, a resumé will be given of some of the background to the interpretation of X-ray and neutron diffraction experiments appropriate to these phases. Second, we give a brief general description of the diffraction patterns from the different phases and selected results concerning molecular packing and correlation lengths. Third, we will consider some of the results which have been obtained on the quantitative determination of distribution functions in the different phases.

General Background

Some basics [1,2]

We shall consider both X-ray and neutron scattering measurements; the basic scattering experiment is

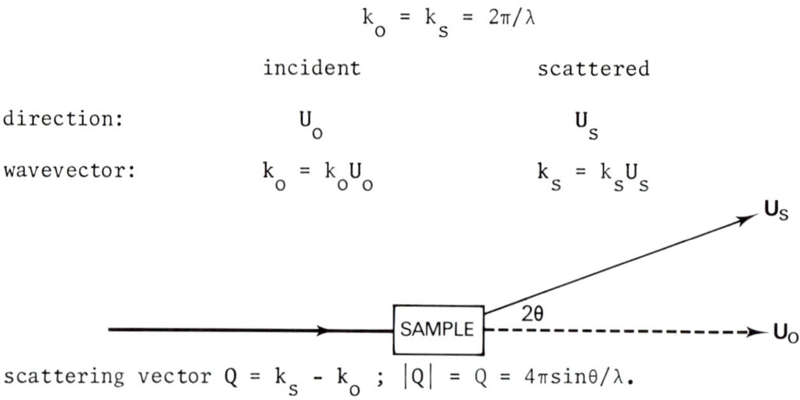

The experiment measures $I_s(Q)/I_o$, the ratio of scattered intensity per unit solid angle to incident intensity, which we will call simply $I(Q)$.

For neutrons the energy exchange with the molecular motions on scattering is significant, i.e. k_o and k_s may differ in magnitude. This is of great importance, of course, to the study of the molecular dynamics which will be discussed in Chapter 20 but for the present purposes is a nuisance! If an energy analysis of the scattered neutrons is made then, in principle, it is possible to determine the purely elastic scattering, for which $k_s = k_o$, which determines the time averaged structure $g(\mathbf{r},\infty)$. More usually, and analagously to X-ray diffraction, the counter sums over neutrons of all energies received at a given scattering angle and the assumption that the result may be treated as if energy exchange were negligible is called the static approximation. For heavy atoms this is quite a good approximation and corrections, the so-called Placzek corrections, are readily made. For very light atoms the recoil effects are more important and the corrections more difficult but no neutron measurements of sufficient sophistication have yet been made on liquid crystals for these effects to be important (but see the section on page 287). For X-rays the static approximation is valid and then (within this approximation) both X-ray and neutron diffraction give the ensemble average of instantaneous structure $g(\mathbf{r},0)$. There is indeed energy exchange of the X-rays with the electrons, the Compton modified radiation, but we assume that this (incoherent) contribution to the total scattered intensity may either be calculated or eliminated experimentally.

The basic equations for scattering intensity may then be written, for neutrons, as

$$I(Q) = \sum_{m,n} \langle b_m b_n \exp(iQ.\mathbf{r}_{mn})\rangle, \tag{1}$$

where b is the scattering length for a nucleus and $\mathbf{r}_{mn} = \mathbf{r}_m - \mathbf{r}_n$. The angular brackets denote an ensemble average, and for X-rays:

$$I(Q) = Pr_e^2 \sum_{m,n} \langle f_m(Q) f_n(Q) \exp(iQ.\mathbf{r}_{mn})\rangle, \tag{2}$$

where P is the polarisation factor which arises because X-rays are electromagnetic radiation; for unpolarised incident radiation $P=\frac{1}{2}(1+2\cos^2 2\theta)$. The classical scattering length of an electron is denoted by r_e. f(Q) is the atomic form factor which describes the effective scattering power of an atom, considered as a point in space, relative to an electron.

We may generalise eqs. (1) and (2) to write for both X-ray and neutron scattering

$$I(Q) = \sum_{m,n} \langle a_m a_n \exp(iQ.\mathbf{r}_{mn})\rangle, \tag{3}$$

where, for X-rays I(Q) now denotes the polarisation-corrected scattering relative to that for a classical electron and $a \hat{=} f(Q)$ while for neutrons I(Q) is the differential cross section $d\sigma/d\Omega$, and $a \hat{=} b$. This equation is quite general.

Coherent and Incoherent Scattering

For simplicity, consider first a system with only one type of atom; the summation in eq. (3) can be taken in two parts, the self term for which m = n and the distinct or interference term where $m \neq n$:

$$I(Q) = N\langle a^2\rangle + \sum_{m,n} \langle a_m a_n \exp(iQ.\mathbf{r}_{mn})\rangle. \tag{4}$$

Now if the atoms may have different scattering lengths, and for monatomic systems this is only possible for neutron scattering because of different isotopes and different nuclear spin states, then providing the scattering amplitude of the atom is independent of its position (the atoms are statistically equivalent) then

$$I(Q) = N\langle a^2\rangle + \sum_{m\neq n} \langle a\rangle^2 \langle \exp(iQ.\mathbf{r}_{mn})\rangle, \tag{5}$$

$$= N(\langle a^2\rangle - \langle a\rangle^2) + \sum_{m,n} \langle a\rangle^2 \langle \exp(iQ.\mathbf{r}_{mn})\rangle, \tag{6}$$

$$= I_{inc}(Q) + I_{coh}(Q) \tag{7}$$

and generalising to the case where several different types of atom (n) are present then

$$I(Q) = \sum_n N(\langle a_n^2 \rangle - \langle a_n \rangle^2) + \sum_{n,n'} \langle a_n \rangle \langle a_{n'} \rangle \langle \exp(iQ.r_{nn'}) \rangle, \tag{8}$$

$$= I_{inc}(Q) + I_{coh}(Q).$$

This is the same as eq. (3) except that the average scattering amplitudes of the appropriate species appear in the coherent term and there is, in addition, a structure-independent incoherent term. This latter is important for neutrons, especially when the system contains protons since here $4\pi\langle b^2 \rangle = 81 \times 10^{-28}\ m^2$ while $4\pi\langle b \rangle^2 = 1.8 \times 10^{-28}\ m^2$ so that $I_{inc}(Q)$ is very large and since b is independent of Q this appears as a flat background. For structural work on hydrogenous materials (like liquid crystals) it is therefore usually necessary to employ deuteriated specimens.

Although eq. (8) is most commonly applied to neutron scattering it has been left in its general form to emphasise its generality since the same form of equation must be found for any system comprising statistically equivalent units of different scattering amplitudes (e.g. the atom types in a disordered alloy or molecules of different orientation in a crystal). In what follows we shall concentrate almost entirely on $I_{coh}(Q)$ since it is this which tells us about the structure. We assume that the incoherent scattering from whatever source has been separated in the experiment or in the preliminary treatment of the data.

Crystals

For a perfect crystal the structure may be described as a convolution of the lattice with the unit cell. The summations in eq. (3) or (8) may be expressed as Fourier transforms [3] over the δ-functions describing the structure, i.e.,

$$\sum_n \exp(iQ.\mathbf{r}_n) = \int d\mathbf{r} \sum_n \delta(\mathbf{r}-\mathbf{r}_n) \exp(iQ.\mathbf{r})$$

and since the transform of a convolution is the product of the transforms of the individual components the intensity from the crystal is

$$I(Q) = F^2(Q) \left| \int d\mathbf{r} \sum_{pmn} \delta(\mathbf{r}-\mathbf{r}_{pmn}) \exp(iQ.\mathbf{r}) \right|^2 \tag{9}$$

and since the transform of a three-dimensional array of δ-functions (the lattice) is also a three-dimensional array of δ-functions (the reciprocal lattice) then eq. (9) is only finite at the nodes of the reciprocal lattice designated by the numbers hkℓ. These are, of course, the

Bragg reflections and the structure factor $F_{hk\ell}(Q)$ determines the relative intensities of these reflections,

$$F_{hk\ell}(Q) = \sum_n a_n \exp\{2\pi i(hx_n + ky_n + \ell z_n)\}, \tag{10}$$

where x_n, y_n, z_n are the fractional coordinates of the atoms in the cell.

If the crystal is disordered so that the unit cells are not all identical, but not sufficiently so as to destroy the long-range order, then by arguments similar to those leading to eq. (8) it may be shown that

$$I_{coh}(Q)/N = \langle F(Q)^2\rangle - \langle F(Q)\rangle^2 + \langle F(Q)\rangle^2 \left|\int dr \sum_{pmn} \delta(r-\mathbf{r}_{pmn}) \exp(iQ.\mathbf{r})\right|^2 + D(Q), \tag{11}$$

where $\{\langle F^2(Q)\rangle - \langle F(Q)\rangle^2\}$ is the first-order diffuse scattering and D(Q) is the diffuse scattering resulting from correlations in the disorder between different unit cells. If these correlations are zero then clearly D(Q) = 0. If the disorder is represented simply by some distribution, h(r), of the atoms about their equilibrium sites r_n (due, for example, to thermal motion) then the structure is a convolution of these distributions with the perfect structure, and the structure factor $\langle F\rangle$ becomes a product of the perfect crystal structure factor with the transforms, H(Q), of the distributions, h(r), for each atom. For example, for an isotropic gaussian distribution

$$h(u) = A \exp(-3u^2/2\langle u^2\rangle),$$

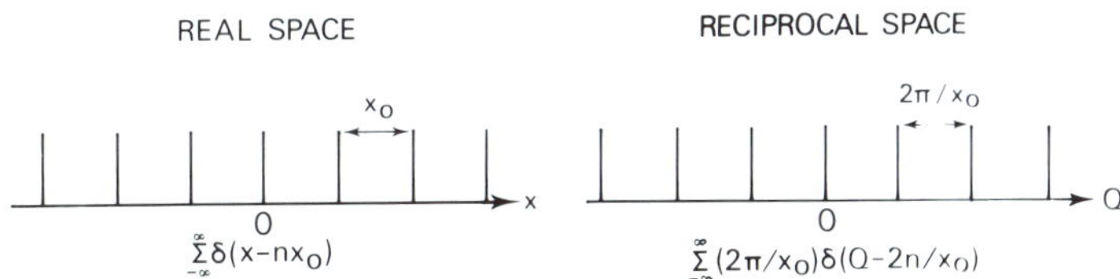

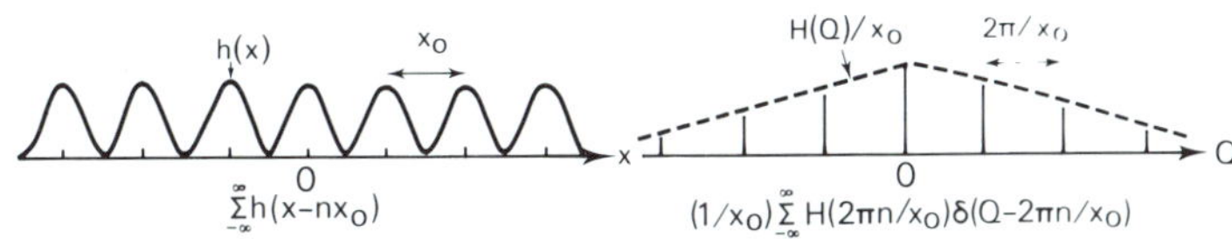

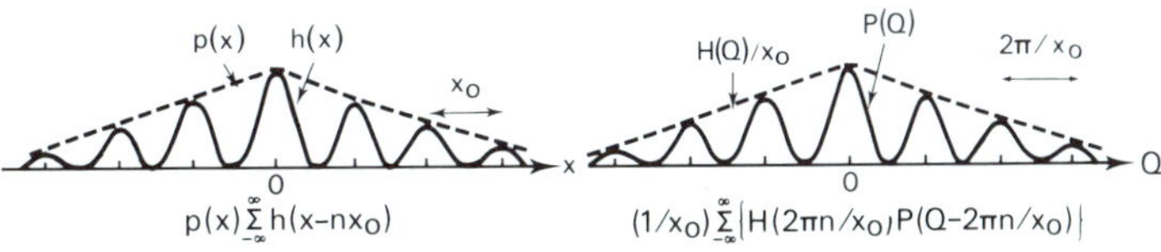

Fig.1 *The spatial distribution functions and their Fourier transforms for a one dimensional array of atoms.*

each term in eq. (10) is multiplied by a Debye-Waller factor

$$H(Q) = \exp(-Q^2 \langle u_n^2 \rangle / 6).$$

If the disorder is such as to destroy long-range order (which is a case clearly relevant to liquid crystals) then the structure may be described by multiplication of the lattice $\delta(\mathbf{r}-\mathbf{r}_{pmn})$ by some damping function $p(\mathbf{r})$ (e.g. a top hat, exponential or gaussian) which reduces the long-range order to zero beyond some finite value of $\mathbf{r}$. The intensity is then a convolution of the lattice transform in eq. (9) with the transform, $P(Q)$, of $p(r)$. This has the effect of broadening the Bragg reflections by an amount inversely dependent on the extent of long-range order (i.e. some correlation length L) but note that the *positions* Q of the reflections remain unchanged provided that some periodicity remains ($Q_p \sim 2\pi/d$). These effects of disorder are illustrated for the one dimensional case in figures 1 and 2. The ideas outlined here have been carried much further

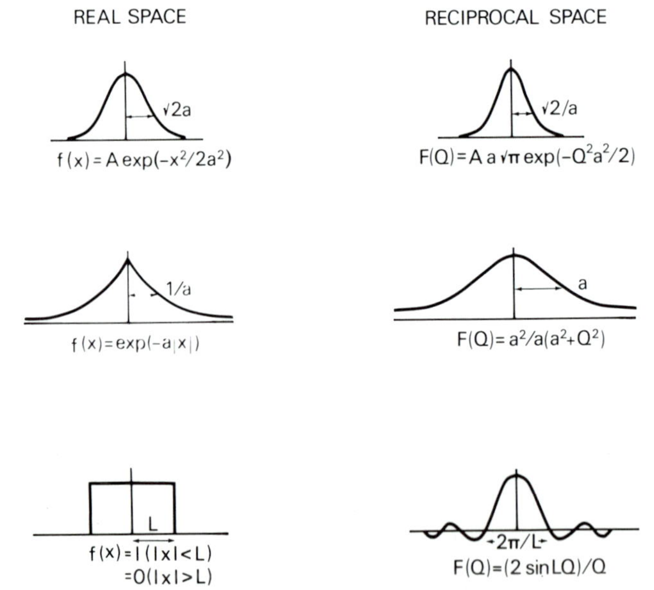

Fig. 2 *Damping functions and their Fourier transforms.*

by Hosemann [4] in his development of the theory of the paracrystal but the simple relations we have developed relating displacement disorder to intensity and extent of correlation to the shape of diffraction peaks will suffice for our later discussions. In general one may relate the width of the diffraction peaks ΔQ to a correlation length L by $2\pi/\Delta Q = cL$ where c is a number of order unity which depends on the model used to define L.

Liquids

Monatomic Consider first a liquid with only one atom type

$$I_{coh}(Q) = \sum_{m,n} a \langle \exp(iQ.\mathbf{r}_{mn})\rangle,$$

$$= Na^2\{1+N/V\int d\mathbf{r}\exp(iQ.\mathbf{r})g(r)\}, \tag{12}$$

where g(r) is the normalized radial distribution function described in Chapter 3. This includes the unobservable forward scattering component from the average sample density which may be removed to give

$$I_{coh}(Q) = Na^2[1+(N/V)\int d\mathbf{r}\ \exp(iQ.\mathbf{r})\{g(r)-1\}]. \tag{13}$$

The liquid structure factor is defined as,

$$S_{coh}(Q) = I_{coh}(Q)/Na^2,$$

$$= 1 + (N/V)\int d\mathbf{r}\ \exp(iQ.\mathbf{r})\{g(r)-1\}, \tag{14}$$

$$= 1 + (N/V)\int_0^\infty d\mathbf{r}\{\sin(Qr)/Qr\}\{g(r)-1\} \tag{15}$$

and note that $S_{coh}(Q) \to 1$ as $Q \to \infty$. In the case of neutron scattering $a^2 \hat{=} b^2$ and there may also be an incoherent scattering component when

$$S_{inc}(Q) = (\langle b^2\rangle - \langle b\rangle^2)/\langle b\rangle^2.$$

Fourier transformation of eq. (15) then gives

$$g(r) = 1 + (V/N2\pi r^2)\int_0^\infty dQ\ Q\{S_{coh}(Q)-1\}\sin(Qr). \tag{16}$$

The problems of carrying out this transform in practice are quite severe because of various corrections required to the data (e.g. multiple scattering, Compton scattering, static approximation), uncertainties in a and the fact that data are obtained only for a finite range of Q. For a discussion see references [5] and [6].

Polyatomic For liquids containing several different species (n) the previous discussion may readily be generalised and the result is that n(n+1)/2 different distribution functions $g_{nn'}(r)$ are required to specify the structure. These are clearly unobtainable from a single diffraction experiment. In a number of cases isotopic substitution methods have been used to provide different scattering powers for the same chemical species and so determine explicitly some of the separate $g_{nn'}(r)$ functions [7] but liquid crystals are much too complex for this approach to be fruitful. A simple Fourier transformation of the intensity data may still be performed but the resulting $g_{tot}(r)$ is a linear combination of the $g_{nn'}(r)$ functions weighted by concentration terms (C_n) and by the atom scattering powers a_n i.e.

$$g_{tot}(r) = \sum_{n\neq n'} C_n C_{n'} a_n a_{n'} g_{nn'}(r).$$

For X-rays a further complication ensues because the a_n's{f(Q)} are Q dependent and different for the different atoms. Hence it is not strictly possible to remove, as in eq. (14), all the Q-dependent terms from the integral over r and this has the effect of modifying the peak shapes of $g_{nn'}(r)$ after transformation [2,8]. This effect can be calculated but we will not consider it further here.

Molecular For molecular liquids it is convenient to separate out the known molecular structure in the scattering equations. We may first define a structure factor for the molecular liquid as a whole [9]

$$N\left(\sum_n^{mol} a_n\right)^2 S_m^{coh}(Q) = I_{coh}(Q),$$

$$= \sum_{m,n} \langle a_m a_n \exp(iQ.\mathbf{r}_{mn})\rangle, \qquad (17)$$

where $\left(\sum_n^{mol} a_n\right)^2$ is a convenient normalisation factor to give a dimensionless $S_m^{coh}(Q)$. In fact for X-rays this is not too useful because of the Q-dependence of the atom form factor, f(Q), and it is more common to carry forward the discussion in terms of the intensity I(Q). Then we take out intramolecular terms and refer the vectors to the molecular centres when

$$S_m^{coh}(Q) = \left(\sum_n^{mol} a_n\right)^{-2} \langle|\Sigma a_n \exp(iQ.R_n)|^2\rangle$$

$$+ N^{-1}\left(\sum_n a_n\right)^{-2} \langle \sum_{i\neq j} \exp(iQ.r_{ij}) \sum_{ni,nj} a_{ni} a_{nj} \exp\{iQ.(R_n^i - R_n^j)\}\rangle. \qquad (18)$$

Here n labels an atom in the molecule, i and j label different molecules so ni labels the nth atom in the ith molecule. r_{ij} is the distance between the centres of molecules i and j.

We define the molecular structure factor

$$f_1(Q) \equiv \left(\sum_n a_n\right)^{-2} \langle|\sum_n a_n \exp(iQ.R_n)|^2\rangle, \qquad (19)$$

which is the (normalised) scattering intensity which would be observed from molecules in a dilute gas. Note that for X-rays the normalisation factor is usually omitted. Now the second term in eq. (18) is an interference function and will contain no components for r less than the molecular diameter d so at high Q ($Q \gtrsim 2\pi/d$), where the small distance contributions are dominant, this term will tend to zero. This decrease with Q is greatly enhanced for a system without long-range order since for *zero* structural correlation the interference term must go to zero (cf. figure 3). $f_1(Q)$ is a perfectly tractable quantity to calculate, as we shall see, so if the scattering intensity can be measured in a Q range ($Q \gtrsim 2\pi/d$) where $f_1(Q)$ is dominant one should be able to infer the distribution of molecular orientation. This has been done in one case for a liquid crystal which will be discussed later.

The interference term in eq. (18) is very much more difficult to handle because of the interdependence of molecular orientations R_n^i and R_n^j and their separation r_{ij}. It is useful to consider two extreme cases.

(a) Uncorrelated orientation We assume that molecular orientations are completely uncorrelated with each other or with molecular separation. That is the R_n^i, R_n^j and r_{ij} are all statistically independent. Consequently the averages in eq. (18) may all be separated and if we define a structure factor for the molecular centres by

$$S_c(Q) = 1+N^{-1} \langle \sum_{i \neq j} \exp(iQ.\mathbf{r}_{ij}) \rangle, \tag{20}$$

then (for a rigid molecule)

$$S_m^{coh}(Q) = f_1(Q)+\{S_c(Q)-1\}(\sum_n a_n)^{-2}\{\sum a_n \sin(QR_n)/QR_n\}^2, \tag{21}$$

$$= (\sum a_n)^{-2}\{\langle F^2\rangle - \langle F\rangle^2 + \langle F\rangle^2 S_c(Q)\}, \tag{22}$$

where

$$\langle F^2\rangle = \sum_{n,n'} a_n a_{n'} \sin(QR_{nn'})/R_{nn'} \tag{22a}$$

and F is called a molecular scattering amplitude. Molecular vibrations are easily incorporated by introducing appropriate Debye-Waller factors.

This is not a valid approximation even for simple molecular liquids although it has often been used, and it will be quite inadequate for liquid crystals.

(b) Complete correlation of orientation We assume that the relative orientation of molecules i and j is statistically independent of their relative separation. The statistical averages in eq. (18) for the factor r_{ij} may then be separated from that in $(R_n^i - R_n^j)$ and the latter becomes

$$F(Q) = (\sum_n a_n)^{-2} \langle \{\sum_{ni} a_{ni} \exp(iQ.R_n^i)\}\{(\sum_{nj} \exp(-iQ.R_n^j)\} \rangle,$$

$$= (\sum_n a_n)^{-2} \langle |\sum_n a_n \exp(iQ.R_n|^2 \rangle,$$

$$= f_1(Q), \tag{23}$$

hence

$$S_m(Q) = f_1(Q)+f_1(Q)\{S_c(Q)-1\},$$
$$= f_1(Q)S_c(Q). \quad (24)$$

Note that $f_1(Q)$ here is not identical with that in eq. (22) because the molecular orientations are not now random.

The result is obtained that the total structure factor for the liquid is simply the product of those for the molecule and the molecular centres. This is precisely analogous to the crystal situation where $S_c(Q)$ becomes the δ-functions of the reciprocal lattice. Eq. (24) ought to be a reasonable first approximation for liquid crystals provided due care is taken in its application. In a formal way it is always possible to write the structure factor for a liquid in the general form

$$S_m(Q) = f_1(Q)+D_m(Q),$$
$$= f_1(Q)+f_2(Q)\{S_c(Q)-1\}, \quad (25)$$

although in the general case $f_2(Q)$ will be complicated and will contain information about the molecular centres.

Some simple examples Before considering the results for liquid crystals themselves it is useful to illustrate this discussion by considering some results for simple molecules. Figure 3 shows structure factor data obtained by neutron diffraction from bromine [10]. From $Q \gtrsim 6\text{Å}^{-1}$ up to the maximum values measured ($\sim 26\text{Å}^{-1}$), $S_m(Q)$ is essentially identical to

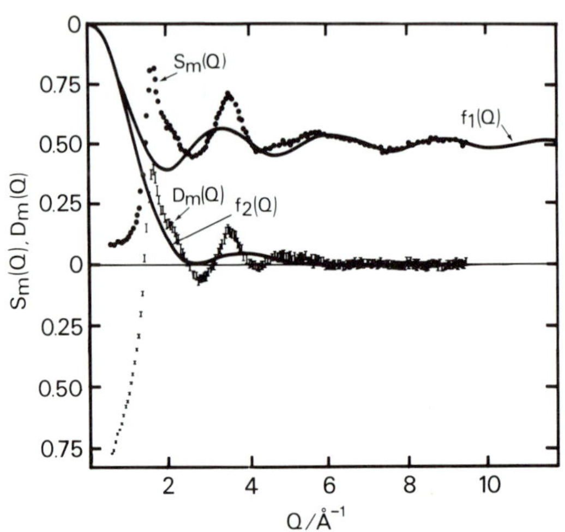

Fig. 3 *The structure factors for liquid bromine.*

$f_1(Q)$. The $D_m(Q)$ is shown also, together with a curve calculated for $f_2(Q)$ assuming no orientational correlations, cf. eq. (21). If this were a good approximation $D_m(Q)$ must be zero when $f_2(Q)$ is zero and this is not so, which demonstrates directly the importance of orientational correlations even for this simple case.

The next example concerns the molecule As_4S_6 which has a rather symmetrical shape (cf. figure 4). Figure 4a shows the calculated X-ray

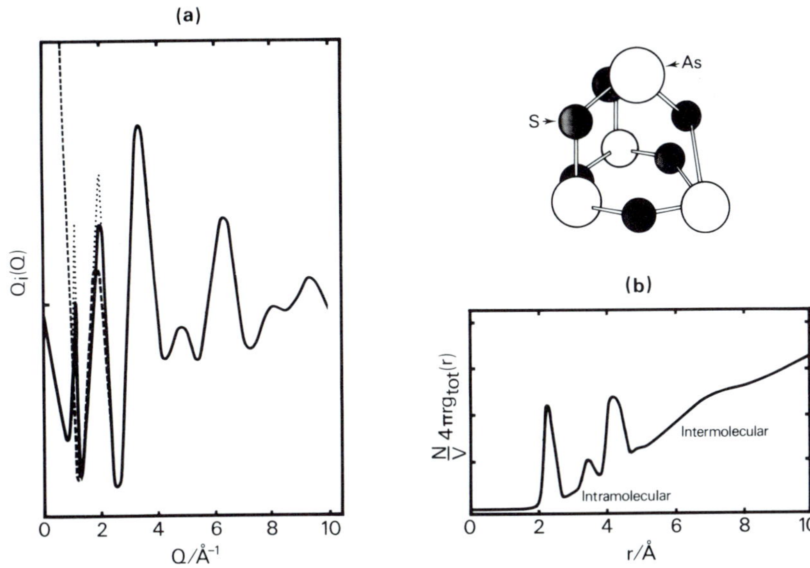

Fig. 4 *(a) The X-ray intensity function $Q_i(Q)$ for random packing of As_4S_6 molecules $Q_i(Q) = Q[I_{coh}(Q) - \Sigma f^2(Q)]$. The solid and dotted lines refer to different packing fractions; the dashed line is the intramolecular term $f_1(Q)$. (b) Distribution function obtained by Fourier transform of data in (a).*

intensity functions for models of random packings of these molecules, cf. eq. (21). Only the first two peaks in the diffraction pattern contain *any* information about the *liquid* structure, the remainder of the pattern is simply $f_1(Q)$ (i.e.$\langle F^2 \rangle$.) the molecular structure factor. Most of the information about the random molecular packing is in the first peak at $Q_p \sim 1.1\text{Å}^{-1}$ and Fourier transformation of $Q_i(Q)$ gives the distribution function shown in figure 4b. The sharp peaks at low r are purely intramolecular, the intermolecular packing is seen in the broad peak at 6.5 - 7.0Å. The relationship between the position of the strong first diffraction peak (1.1Å^{-1}) and the molecular diameter ($\sim$ 6.5Å) (or the nearest intermolecular distance) is $d \sim 1.2(2\pi/Q_p)$. This result is typical of sphere packings and has often been used to obtain intermolecular distances for more complex systems with local orientational correlations (like liquid crystals) where its use must be viewed with caution. The exact relation between molecular spacings and Q_p must depend on the details of the molecular packing. For diffuse reflections there is no unique simple relation which will give the intermolecular spacings, although a Fourier transform may help to reveal the essentially intermolecular correlations; but remember that the experiment really sees only atom-atom correlations, cf. eq. (18).

Application to Liquid Crystals. Descriptive Outline

Some typical X-ray diffraction patterns are shown in figure 5b. These show the kind of X-ray photograph obtained with very simple apparatus employing a monochromated beam, transmission geometry and a flat film as illustrated in figure 5a.

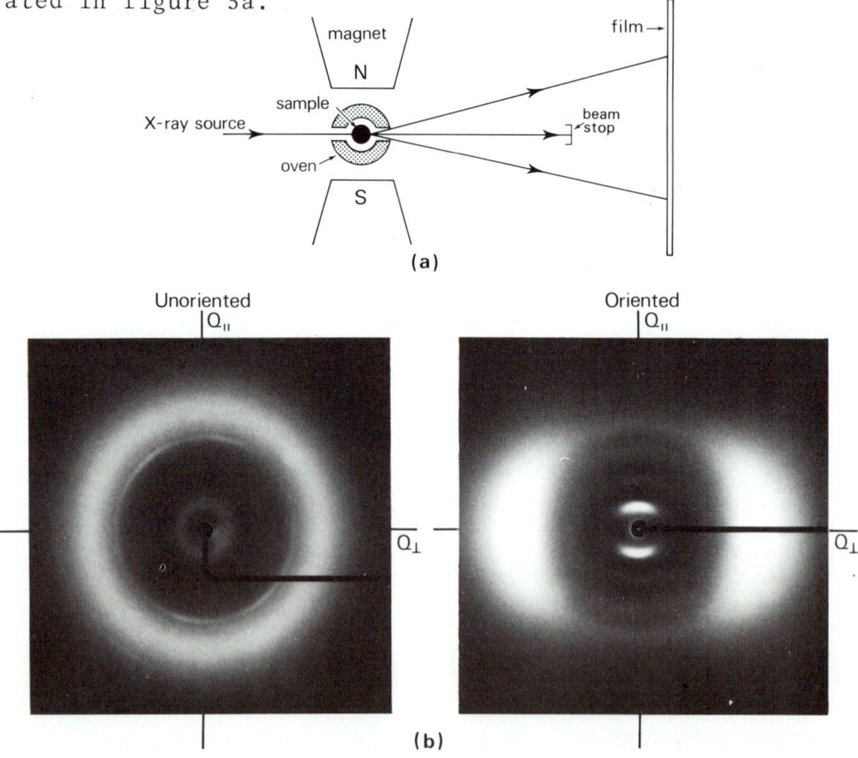

Fig. 5. *(a) A simple X-ray diffraction apparatus and (b) some typical diffraction patterns for nematics.*

Data on unoriented specimens can be very useful for initial characterisation of samples where it is usually possible to distinguish unambiguously between a number of phases (e.g. S_E and S_B, S_A and N) but this is not always so (e.g. S_A and S_C may be confused). Powder data are also very useful for obtaining intermolecular spacings and for determining relative intensities where it may be preferable to study a good powder specimen rather than a poor monodomain. In general, however, for more quantitative work it is necessary to study oriented specimens. For nematics this is readily achieved by a modest magnetic field ($\sim$ 0.1 T) and aligned S_A and S_C may then be prepared by careful cooling from the aligned nematic phase. This may in fact be very difficult and the quality of the result depends markedly on the specimen and the conditions (e.g. surface treatment). When the S_A or S_C phase is formed directly from the isotropic liquid, aligned specimens may still sometimes be prepared by cooling in a somewhat greater field ($\sim$ 2 T). An alternative procedure, useful for preparing monodomains of the more ordered (and more rigid) smectics, is by careful melting of a single crystal.

Nematics

Comparison of powder patterns for nematic and isotropic phases shows that, with regard to the local molecular packing, these phases are little different. This is illustrated in figure 6 for a neutron scattering result on 4,4'-dimethoxyazoxybenzene (PAA) [14] and figure 7 for an

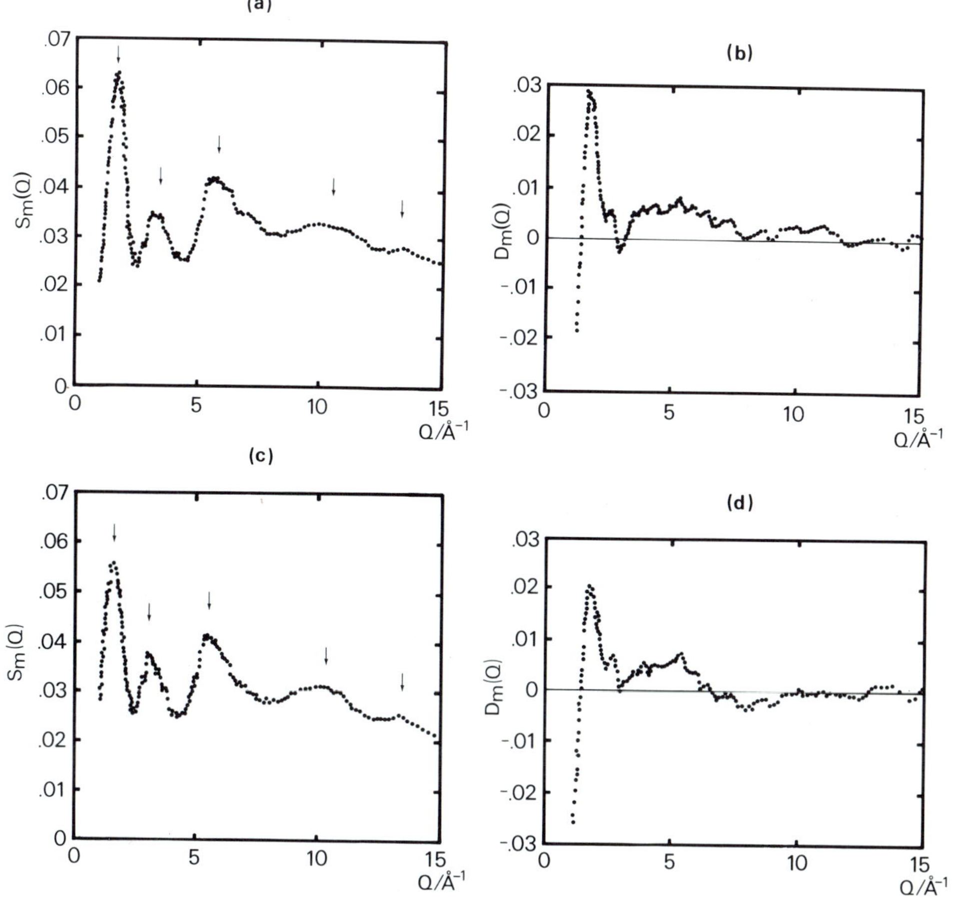

Fig. 6. *The structure factors for perdeuteriated PAA in the nematic mesophase at 122°± 2°C (a and b) and the isotropic phase at 195° ± 3°C (c and d).*

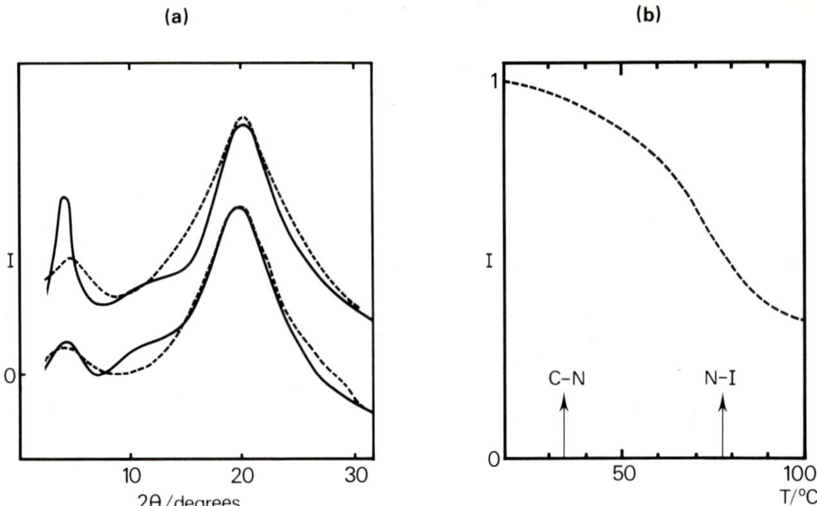

Fig. 7. *(a) Diffraction patterns for 7 CB (solid lines) and 5 CB (dotted lines) taken with CuKα radiation (λ = 1.54Å). Upper curves are for the unoriented nematics and lower curves for the isotropic liquids. Temperatures were: 5 CB 23.5° and 38°C and 7 CB, 27° and 46°C. Intensities are in arbitrary units but have been normalized to the highest angle (1 E) peaks for the isotropic phase. (b) The peak height of the first (1 M) peak for 7 CB as a function of temperature.*

X-ray diffraction study on 4-n-pentyl-and 4-n-heptyl-4'cyanobiphenyl (5 CB and 7 CB) [14]. For the former case the intramolecular structure factor $f_1(Q)$ was also obtained as previously described and the result of subtracting this from $S_m(Q)$ to give $D_m(Q)$ is also shown. Beyond about 7Å^{-1} essentially all the scattering is intramolecular. The intermolecular components at low Q are little different in the two phases; some of the peaks in the nematic phase are slightly stronger and sharper. This means that to a first approximation the molecular packing (within the coherence volume where $g(r) \neq 1$) is similar with respect both to spacings and correlation length so that the nematic phase differs from the isotropic basically in the extent of long-range order of the locally ordered molecules. This result is rather obvious and might be inferred from the small density change (< 1% at T_{NI}) but it is important to eradicate the idea of molecules on average free to tumble relative to their near neighbours, even in the isotropic liquid, much less the nematic phase.

For oriented nematics typical diffraction patterns are shown in figure 5b. The anisotropy is clearly shown and there are correlations of two distinct periods perpendicular and parallel to the director **n** which correspond approximately to an average molecule width and length respectively. For **Q⊥n** the diffraction pattern is very liquid-like in N, S_A and S_C phases and for a wide variety of compounds the first strong peak (see figures 4 and 5) is found near $Q \sim 1.5\text{Å}^{-1}$ corresponding to a spacing, $2\pi/Q_p$, of $\sim 4.5\text{Å}$. An approximate molecular average spacing (d_m) may be inferred from this by considering the geometry of close packed cylinders (e.g. S_B phase) where the relation is $d_m \simeq 1.15\ 2\pi/Q_p$ or from the position of the first peak in the Bessel function $J_o(Qd_m)$ which is the appropriate function to describe the scattering from a random packing of cylinders, which gives $d_m = 1.12(2\pi/Q_p)$. For the reasons discussed earlier it must be recognised that the experimental spacing for disordered structures is inherently difficult to define, and use of some factor like 1.12 to relate the position of the first diffraction peak to an average d_m must be considered a first approximation. Nevertheless, an average molecular spacing within a few per cent of 5Å is found for a wide variety of compounds, the value increasing slightly with temperature and from the more ordered to the less ordered phases. The value of about 5Å lies midway between maximum (6.5Å) and minimum ($\sim$ 3.5Å) lateral dimensions of a typical liquid crystal molecule which shows clearly that there must be strong local correlations in orientation about the long axes even in the *isotropic* phase, and although molecular rotation about the long axes is known to be very rapid even in the S_B phase this must in fact be strongly cooperative in nature. This result also shows that an assumption of average cylindrical symmetry cannot be valid for nearest neighbour interactions. Beyond the first peak for $Q \perp n$ the scattering pattern is increasingly dominated by the intramolecular component $f_1(Q)$ (cf. figure 6).

The dominant feature of the scattering with $Q || n$ is usually a peak in the shape of a short bar, of intensity comparable to, or smaller than, that of the strong equatorial peak. Significant structure in the scattering is observed at higher Q and again this becomes increasingly dominated by $f_1(Q)$. The detailed intensity profile of the first meridional peak is strongly sample and temperature dependent. In the majority of cases studied its position (Q_p) corresponds to a repeat distance d ($d = 2\pi/Q_p$) of the order of the molecular length (ℓ). In some cases $d \approx \ell$ [14,16,17] but in many cases the value of d is significantly shorter than that of ℓ estimated for the most extended molecular configuration (typically $\ell\text{-}d \simeq 2\text{Å}$ and $\ell \sim 17$ to 25Å) [14, 18-20]. In a general way this diffraction peak must arise from correlations in molecular arrangement along **n** so the most obvious explanation of the ℓ-d values is in terms of a tilt effect. The distribution function for the number of molecules with their long axes (**m**) at an angle β to **n** is $g(\beta)$(i.e. $f(\beta)\sin\beta$) and for $\overline{P_2(\cos\beta)} \sim 0.5$, $g(\beta)$ has a broad maximum near $\sim 30^o$. Hence if correlations along **n** are for molecules with an average tilt $\langle\beta\rangle$ 30° the correlation length $d \sim \ell \cos 30^o$ which is approximately of the correct magnitude. Alternatively, one could consider simply the mean projection of the molecular length along the director $d = \ell|\overline{\cos\beta}|$, and use this, identifying d with the correlation length, to define an average excursion angle of the molecule from its preferred direction at $\beta = 0$, by $\bar{\beta} = \cos^{-1}d/\ell$. For $d/\ell = 0.9$ this gives $\bar{\beta} \sim 26^{\circ}$ which is reasonable for a nematic. Of course molecular tilt here must be interpreted widely so as to include conformational disorder, for example of long-end chains, in such a way as to reduce the average molecular length but this alone is usually insufficient to account for the whole of $(\ell\text{-}d)$.

A number of instances is also known where $d > \ell$ [14, 21] and apart from obvious cases where end-to-end dimers may form, the maximum d/ℓ is ~ 1.4 for the cyanobiphenyls. This will be discussed later under S_A materials since some of these compounds also form S_A phases [22] but it must imply some form of modified bilayer structure arising from strong short-range interactions.

More detailed information about the molecular packings is contained in the intensity profiles of the diffraction pattern. These may be qualitatively understood in terms of the following simple picture (cf. figure 8) [2, 23-25]. The diffraction pattern for a system of independ-

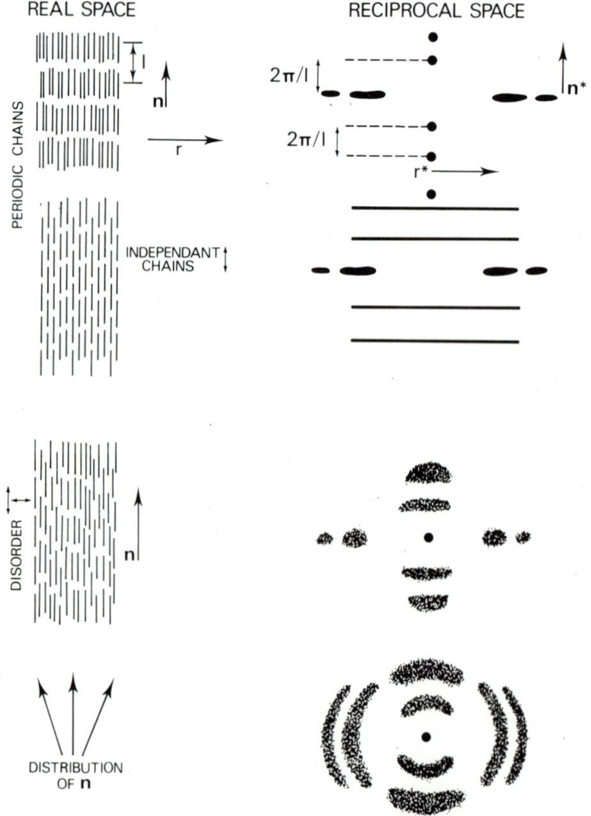

Fig. 8. *Illustration of structure and diffraction patterns for systems of rods.*

ent chains of rods (molecules) exactly parallel and periodic (periodicity) ℓ along the chains (**n**) but with a disordered net of chain centres is a series of sheets of scattering, normal to n^*($n.n^* = 1$) with periodicity $2\pi/\ell$. In the equatorial plane (r^*) the intensity profile is not uniform but has a liquid-like distribution from the disordered net.

Introduction of disorder along **n** and lateral displacement of the rods perpendicular to **n** results in the scattering sheets broadening along n^* and shortening in extent in the radial direction (r^*). If the reduction in periodicity along **n** is not too great but there is complete lateral disorder this implies (for a dense structure) the formation of layers, a S_A structure. Then the diffraction sheets away from the equatorial plane become layer spots (but with a width along n^* depending on the extent of order), whilst in the equatorial plane the diffraction pattern remains liquid-like. Finally, if the coherently scattering regions are given a distribution of angular orientations (β) relative to the average (**n**) the diffraction pattern will be spread into arcs. The relative importance of the various kinds of disorder will depend both on the material and on temperature but we may make a few general observations.

Measurement of widths of diffraction peaks shows that correlation lengths L (beyond which no significant correlations of molecular positions are found) are generally less than about five molecular dimensions. The shape of the equatorial reflection is often a pronounced arc which suggests that the dominant component results from the singlet orientational distribution, $f(\beta)$, of the long axes. The determination of $f(\beta)$ from analysis of the equatorial reflections will be discussed in more detail later.

The meridional reflections sometimes also show a distinct arc-shape but this is not usually the dominant component of the shape of this part of the diffraction pattern. Detailed analysis has been attempted only for a few materials [16,17]. These are materials in which pronounced smectic order parameter fluctuations have been observed as the N-S transition is approached. As we have seen, the strength of correlations along **n** shows a marked temperature dependence in all cases studied but this is especially true of certain substances which transform to S_A or S_C phases at lower temperatures, where very strong pretransition effects are found which may extend over a wide temperature range. The diffraction pattern of the nematic then shows the development of enhanced order characteristic of the smectic phase. These strong short-range order effects have been studied quantitatively by MacMillan [16,17] and were called cybotactic groups by de Vries [26,27]. The effect is especially striking for S_C pretransition phenomena where the diffraction patterns (figure 9a) clearly show the effect of local order, characteristic of the S_C phase (figure 9b). At the transition these short-range fluctua-

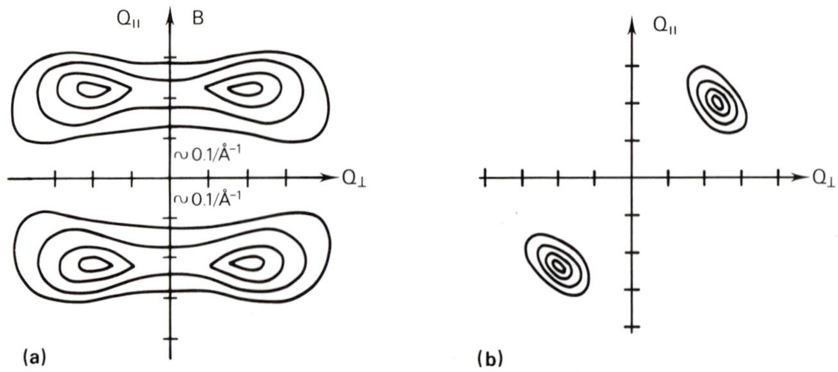

Fig. 9. *Schematic diffraction profiles of the 1st meridional reflection (a) nematic phase, showing S_C order near N-S_C transition, (b) S_C phase.*

tions condense to give the Bragg reflections characteristic of the long-range order of the true S_C phase. Similar effects are found above S_A phases but here, of course, only a quantitative intensity increase of the diffraction peak is observed as the extent of order increases as the transition is approached [16,32], rather than a qualitative change in the appearance of the pattern as in the S_C case.

Smectic A

Typical diffraction patterns for an oriented and an unoriented sample of a S_A phase are shown in figure 10. The pattern in the equatorial plane (outer rings) is qualitatively like that for a nematic but rather more intense and with a less pronounced arc. Nevertheless, the structure within the layers is undoubtedly disordered. The very strong sharp re-

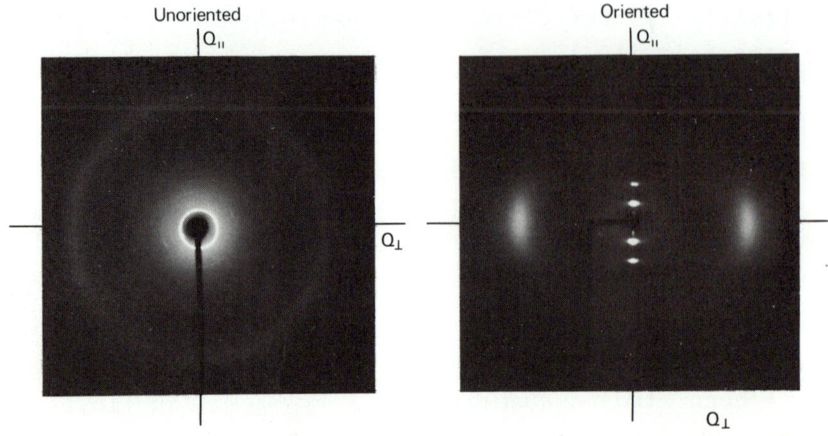

Fig. 10 *Schematic diffraction pattern for a S_A phase.*

flections for $Q||n$ (usually two orders are observed) show the existence of extensive layer-like correlations. The quantitative analysis of the intensity of these spots to characterise the S_A order will be discussed later.

In a S_A phase the director is normal to the layers, which are about one molecule thick. Conventional wisdom has it that the singlet orientational distribution $f(\beta)$ also has a maximum in this direction, although there have been suggestions that this could be tilted relative to n and disordered around the cone.

The d spacing of the layers $(2\pi/Q_p)$ in a few cases has been found to be almost identical with the molecular length [16,28]. In many more cases, however, it is significantly less [18,19,21,29,30]. The amount is dependent on the compound but is typically about 10%. If this is interpreted as arising from the molecular distribution $f(\beta)$ as discussed earlier for nematics then $\bar{\beta} \sim 25^o$ is implied. The value of $\bar{\beta}$ determined in this very simple manner thus appears to vary from 0° to $\sim 25^o$ depending on the substance and ought to be correlated with $\bar{P}_2$. However, much more systematic work needs to be done in this area before more detailed structural conclusions can be drawn. Interpenetration of molecules in different layers is sometimes mentioned as a possible explanation of $d-\ell>0$ but this is an ill-defined concept and raises considerable steric problems.

We have recently been investigating a number of compounds which show the phase sequence $C \rightarrow S_C \rightarrow S_A \rightarrow I$. For example, for

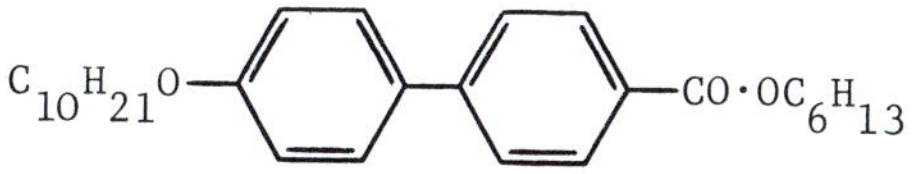

$$C_{10}H_{21}O-\langle\bigcirc\rangle-\langle\bigcirc\rangle-CO\cdot OC_6H_{13}$$

(which has C 57^oC S_C 64^oC S_A 81^oC I) the layer spacing d of 29.8 ± 0.3Å is independent of temperature through the C, S_C and S_A phases. The fully extended molecular length is 33.8Å implying an average tilt angle of $\sim 28^o$. In view of the similarity from crystal to S_A phase it seems unlikely that intramolecular effects alone could account for this but a crystal structure analysis would be useful here. On the other hand a number of examples are now known where $d > \ell$ [21,22,31,32]. Compounds containing a cyano end group are particularly susceptible to this behaviour and here $(d - \ell)$ is of the order of a chain length leading to the schematic model [33] for their structure illustrated in figure 11. Similar, if less pronounced effects are also found in other materials [21,31] where presumably some modified bilayer structure arises because of particular intermolecular dipoles or steric effects (or both).

This brief discussion shows that while the main features of the S_A

phase are indeed understood much yet remains to be established about the details of the molecular arrangement in this phase.

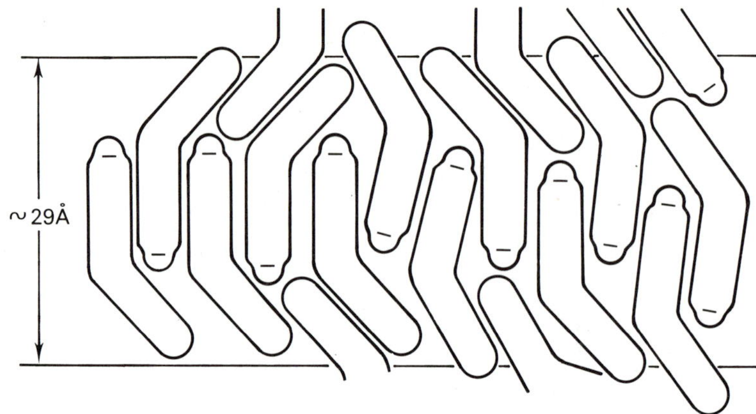

Fig. 11. *Schematic 2-dimensional representation of proposed local structure in an 8CB layer in the S_A phase; approximately to scale, with the width of molecular cores and tails proportional to the square root of their areas. Cyano groups are indicated by negative signs.*

Smectic C

Typical diffraction patterns of ordered S_C phases are shown in figure 12. Diffraction patterns of powder samples are essentially the same as those for S_A except that the layer spacings are usually relatively smaller. The accepted structure for the S_C is that the director is tilted with respect to the layers. By cooling from the nematic phase in a magnetic field different types of ordering may be obtained depending on the material and the conditions (e.g. surface treatment and rate of cooling). These are illustrated with their accompanying diffraction patterns in figure 9. It thus appears that the transition from the nematic phase can occur with either an ordering of the (vertical) molecules into tilted layers which (a) may or (b) may not be rotationally disordered about the vertical (n) axis or (c) a tilting of the molecules to form horizontal layers. All kinds of behaviour may be observed for the same substance [37] and some of these are described in Chapters 12 and 14.

The tilt angle may be determined directly from the diffraction patterns as illustrated in figure 12. In cases (a) and (b) if the average direction of the long axes lies along the field, the maximum intensity of the equatorial reflections is on a plane perpendicular to this and the positions of the layer spots give the tilt angle directly. A tilt angle may also be inferred from the layer spacings although there is the possibility of some ambiguity here because of unknown molecular conformations. The most extensively studied example is probably TBBA, the re-

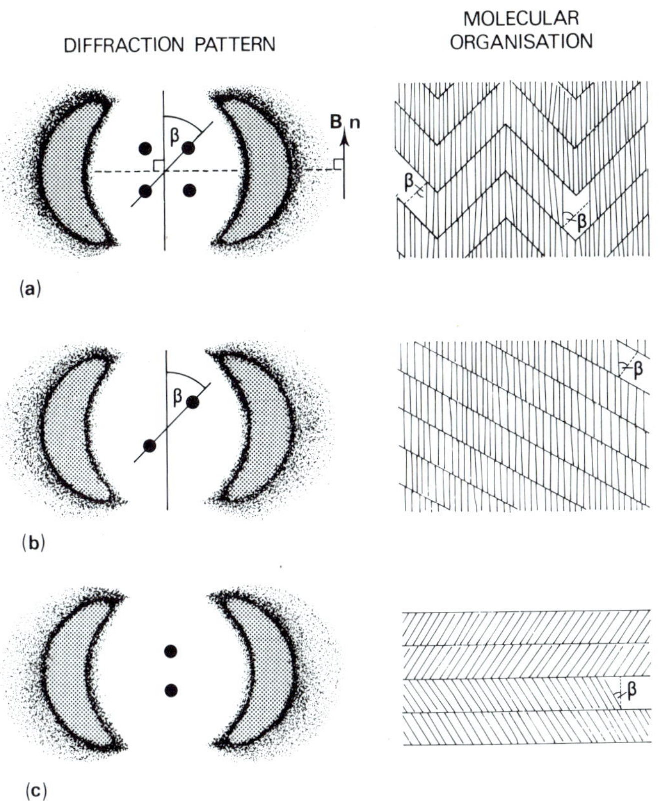

Fig. 12. *Schematic diffraction patterns and structures for S_C phases.*

sults for which are shown in figure 13 [28,34]. Agreement was obtained between the direct measurement [34] and the values inferred from layer spacings [28] (as well as with diffraction methods other than X-ray) and the classic behaviour is shown where the tilt angle increases from zero at the S_A-S_C transition to $\sim 25^o$ at the S_B-S_C transition. In contrast, for the 4,4'-di-n-heptyloxy- and octyloxy-azoxybenzenes, which have a S_C-N transition, the cybotactic groups in the nematic phase show tilt

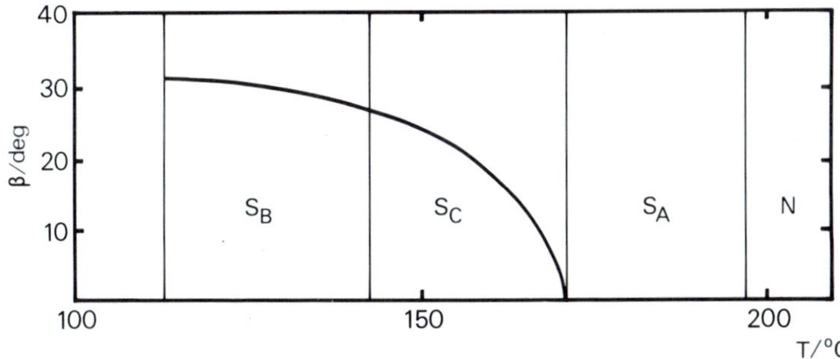

Fig. 13. *The tilt angle for TBBA.*

angles of $\sim 30^{\circ}$ and in the S_C phase the tilt angle increases only very slightly.

The behaviour of the 4'-n-decyloxybiphenyl-4'-carboxylates mentioned under S_A systems is worth referring to again here explicitly, since to a good approximation the layer spacings are independent of temperature through S_C and S_A phases which suggests that in some materials the tilt angle certainly varies only slightly through these phases and that there may be a non-zero tilt in a S_A phase.

Again, as for the S_A case much more systematic work needs to be done to characterise, for example, the molecular correlations between layers, the nature of the azimuthal ordering both locally and on a macroscale and whether different kinds of S_C-S_A transitions do exist.

Determination of the Distribution Function

Cylindrical distribution functions

In principle, scattering data, {cf. eq. (18)} contain all the information about the general distribution function $G(\mathbf{r},\Omega)$ needed to specify the system fully but at this stage at least we must be satisfied with only partial solutions.

Nematic and smectic A phases have cylindrical symmetry and the general equations derived previously can be reformulated in cylindrical coordinates [24]. We will not discuss the general formulation but merely some particular aspects. Consider first an assembly comprising one atom species in which the distribution of atom positions on the equatorial (xy) plane is given by $2\pi Np(r)/A$ where N is the total number of atoms and A the total area in the xy plane. The scattering intensity for $Q\perp z$ is

$$I_{coh}(Q_{\perp}) = Na^2[1 + (2\pi N/A) \int_0^{\infty} drr\{p(r)-1\ J_o(Qr)\}], \qquad (26)$$

with $J_o(x)$ appearing due to the circular averaging, in contrast with $j_o(x)$ for the spherical case. As before this equation may be transformed to give a distribution function according to:

$$p(r) = 1 + (A/2\pi N)(Na^2)^{-2}\int_0^{\infty} dQ\ Q\{I_{coh}(Q\ )-Na^2\}J_o(Qr). \qquad (27)$$

Again, these results may readily be generalised to the situation where there are several atomic species but then the distribution function $p_{tot}(r)$ is a linear combination of those for the different possible pairs, weighted according to concentration and scattering powers.

Accurate data and much care are needed in order to obtain reliable distribution functions and although a number of investigations have been reported few give results which inspire great confidence, although it must be said that the positions of the peaks (as opposed to their areas) obtained are probably meaningful. The results of some recent work by Delord and Malet [35,36] on 4,4'-dimethoxyazoxybenzene are shown in figure 14. The peaks at low r are clearly of intramolecular origin while those

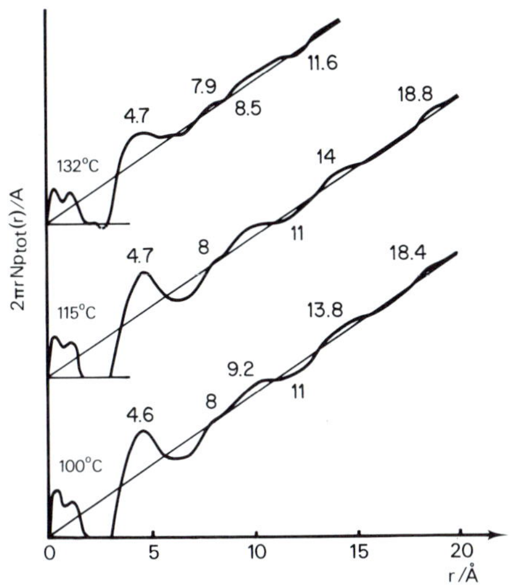

Fig. 14 *Atomic cylindrical distribution functions for the nematic phase of PAA at various temperatures.*

beyond about 4Å must arise from intermolecular atom pairs and the strong peak around 4.6Å must come from pairs of atoms in adjacent molecules. Note that the distribution function for molecular centres is not given by this experiment. Attempts have been made to determine this directly by making the assumption that the molecular orientations about the z axis are random and also quite uncorrelated with the molecular separation, when {cf. eqs. (20-22)}

$$I(Q_\perp) = N\langle F^2\rangle + \langle F\rangle^2 (2\pi N/A)\int dr r\{P_m(r)-1\}J_o(Qr), \tag{28}$$

where
$$\langle F^2\rangle = \sum_{i,j} a_i a_j J_o(Qr_{ij}) \tag{29}$$

and
$$\langle F\rangle^2 = \left|\sum_i a_i J_o(Qr_{ci})\right|^2, \tag{30}$$

with all distances being projections on the xy plane. However, this simple model cannot be valid any more than the equivalent assumption in 3-dimensions considered earlier, and results derived by its use must be viewed with caution. In any case there are problems in evaluating eqs. (29) and (30) because the r_{ij} depend also on the molecular tilt [39].

The intensity of scattering for $Q \parallel z$ is simply related to the projection along Z, thus

$$I_{coh}(Z) = \sum_{m,n} \langle a_m a_n \exp(iQ_z Z_{mn}) \rangle. \quad (31)$$

A 1-d Fourier transformation of the data for $Q \parallel z$ has sometimes been carried out [44] but, of course, the distribution function is entirely dominated by the intramolecular structure. The intermolecular arrangement appears as the long wavelength ($\lambda \sim \ell$) Fourier components associated with the intermolecular scattering peaks at ($2\pi n/\ell$). However, eq. (31) leads to a method for determining smectic order parameters which will be outlined later.

The distribution function for the orientation of the long molecular axes

Nematics The singlet distribution describing the orientation of the molecules in a mesophase is, in principle, readily determined from the single molecule component of the scattering $f_1(Q)$

$$f_1(Q) = \sum_{m,n}^{mol} a_m a_n \langle \exp(iQ.r_{mn}) \rangle. \quad (32)$$

The average involves an integral over the distribution function $f(\alpha\beta\gamma)$ which denotes the probability density for finding a molecule whose orientation is defined by ($\alpha\beta\gamma$). For a uniaxial system the distribution is independent of α and assuming molecular cylindrical symmetry eliminates the dependence on γ. It may then be shown that [40,42]

$$f_1(Q) = \sum_{m,n} (a_m a_n/2\pi) \int_0^{2\pi} d\alpha \int_0^{\pi} d\beta f(\beta) \sin\beta$$

$$\times \exp\{-iQ(\sin\beta\sin\theta_Q\cos\alpha + \cos\theta_Q\cos\beta) r_{mn}^{\parallel}\}$$

$$\times [J_o(Qr_{mn}^{\perp}\{1-(\sin\beta\cos\alpha\sin\theta_Q+\cos\beta\cos\theta_Q)^2\}^{\frac{1}{2}}] \quad (33)$$

or

$$f_1(Q) = \sum_{L(even)} (-1)^{1/2}(2L+1) I_L(Q) P_L(\cos\theta_Q) \bar{P}_L, \quad (34)$$

where

$$I_L(Q) = \sum_{m,n} a_m a_n j_L(Qr_{mn}) P_L(\cos\theta_{mn})$$

and

$$\cos\theta_{mn} = r_{mn}^{\parallel}/\mathbf{r}_{mn}^{\perp}.$$

$r_{mn}^{\parallel}$ and $r_{mn}^{\perp}$ are components of $\mathbf{r}_{mn}$ parallel and perpendicular to the molecular long axis, $j_L(x)$ is a spherical Bessel function, $P_L(x)$ a Legendre polynomial and θ_Q is the angle between Q and n. Eq. (34) is basically a very simple result in which the molecular structure ($I_L(Q)$), the scattering geometry (θ_Q) and the order parameters $\bar{P}_L$ are clearly separated, but it is an infinite series. Eq. (33) was used [42] with the mean field result for the singlet distribution f(β) to investigate the convergence of eq. (34) and it was concluded that for nematics terms up to L = 4 are probably adequate for most purposes.

Neutron data for a fully deuteriated specimen of PAA have been analysed [42] using eq. (34) and the results are shown in figures 15 and 16.

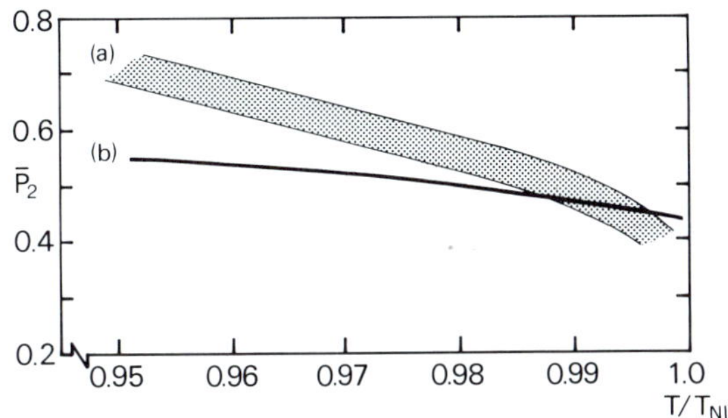

Fig. 15 *The temperature dependence of $\bar{P}_2$ for deuteriated PAA (a) determined from neutron diffraction experiments and (b) calculated from the Maier-Saupe theory.*

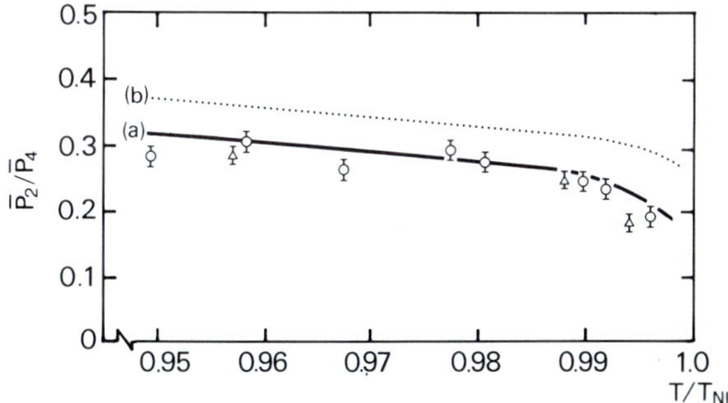

Fig. 16 *The results for $\bar{P}_4/\bar{P}_2$ obtained from an analysis of neutron diffraction data for deuteriated PAA. The dotted line is based on the simple Maier-Saupe theory.*

The analysis concentrated on two broad peaks in the diffraction pattern centred on Q of about 1.8Å^{-1} and 3.0Å^{-1}. The former is dominated by interference terms and the latter by the intramolecular (self) term but analysis in terms of eq. (34) gives identical results for both. This result, at first sight surprising, can be understood if the molecular long axes are locally well aligned when eq. (24) for the case of perfect orientational correlation becomes at least approximately applicable and $f_1(Q)$ is the structure factor also for the interference term. The results of this analysis are therefore consistent with the picture of locally very well aligned regions.

The $\bar{P}_2$ data in figure 15 are liable to considerable error owing to uncertainties in the background intensity, but differ appreciably in their temperature dependence from the results of the Maier-Saupe mean field theory described in Chapter 4. (However they also differ from a variety of other data which are in fair agreement with the simple theory in this respect). The $\bar{P}_4/\bar{P}_2$ ratio from experiment has the expected temperature dependence but the absolute values are low relative to the predictions of the simple theory. The reasons for these discrepancies are not yet understood.

It is possible also to use the strong first diffraction peak centred on the equatorial plane and dominated by the interference term to determine the angular distribution function [43,44,39]. In the simplest approach the basic assumption is made that for a perfectly ordered situation the equatorial scattering would be confined strictly to the equatorial (xy) plane. This in turn requires the structure to consist of parallel chains of colinear molecules, the chain net in the xy plane, however, being disordered (cf. figure 8). The parallel and colinear arrangement must be maintained throughout the coherence volume of the structure which is the volume within which $g_{tot}(r)$ shows significant oscillations (cf. figure 14). The appearance of a pronounced arc in the scattering, centred on the equatorial plane, is then assumed to arise from differently aligned *regions* with the local director m making an angle β with the average director n. In this approximation the evaluation of the intensity profile around the arc I(θ) where θ is the angle relative to the x axis (figure 17) becomes a straightforward geometrical problem and the result

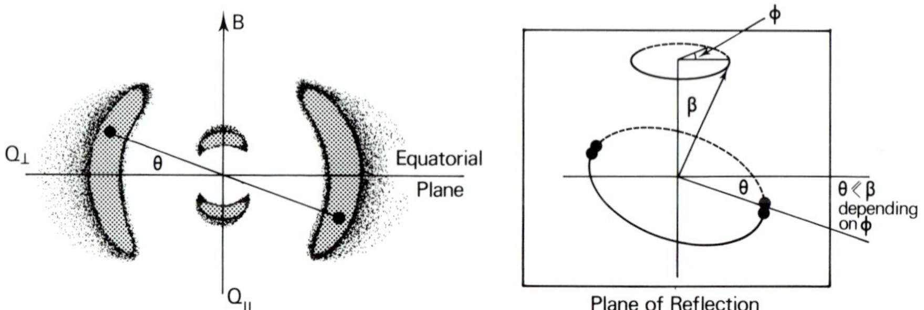

Fig. 17 *The intensity distribution around the arc.*

is

$$I(\theta) = \int_{\beta}^{\pi/2} d\beta \sin\beta f_d(\beta) \sec^2\theta (\tan^2\theta - \tan^2\beta)^{\frac{1}{2}}. \qquad (35)$$

This equation (or its variants) can be inverted to obtain $f_d(\beta)$ and hence the order parameters $\bar{P}_2$, $\bar{P}_4$ etc. Values of $\bar{P}_2$ have been obtained in this way by Delord for PAA [43,44] and the results are in good agreement with other measurements as shown in figure 18. Note that the derivation

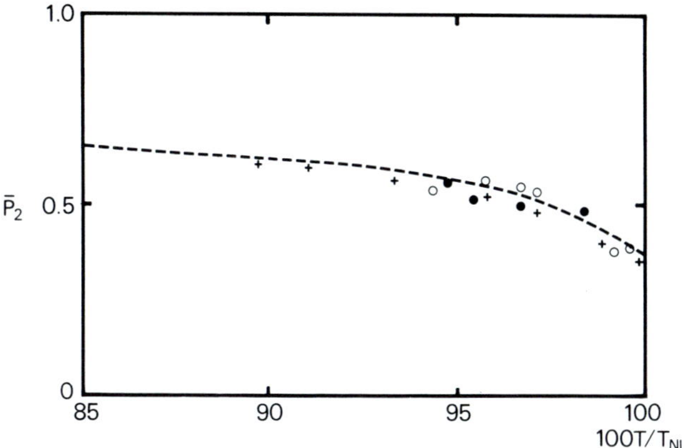

Fig. 18 *Effect of temperature on the order parameter $\bar{P}_2$ determined by different methods, for the nematic phase of PAA. Experimental points were determined as follows: (+) optical studies (●) X-ray diffraction (O) nuclear magnetic resonance spectroscopy.*

of eq. (35) involves an average over the azimuthal angle ϕ shown in figure 17 and sometimes this is erroneously neglected by assuming $I(\theta) \simeq f(\beta)$ which only leads to a rough approximation to an order parameter.

The distribution function $f_d(\beta)$ given by this method is, of course, that for a local cluster of molecules which are assumed to have a mutually perfect alignment. This orientational distribution is expected therefore to be the same as, or more ordered than, the true singlet distribution $f(\beta)$. On the other hand, the effect of the basic assumption of perfect local order will be to give an over estimate of the width of $f_d(\beta)$. This is because the effect of a finite correlation length L along the chains in the z direction will be to broaden the first equatorial reflection in the z direction by an amount of order $2\pi/L$. Since the minimum value of L must be about a molecular length ℓ then the contribution from this source to the angular width of the equatorial arc at Q_E is given by $\Delta\theta/\text{deg} \sim 180/\ell Q_E$ which is about 6 for a molecular length of 20Å and this will have an almost negligible effect unless the order is very high, $\bar{P}_2 \gtrsim 0.8$ [37,39]. The experimental fact that the equatorial reflections do indeed comprise a clearly defined arc support this conclusion.

Some typical recent results [37] for 4-4'di-n-octyloxyazoxybenzene (OAB) for which $f_d(\beta)$ has been derived are shown in figure 19. The distribution function is in very good agreement with the simple Maier-Saupe theory while the absolute values of $\bar{P}_2$ are a little higher than the predictions of the theory. This work shows that the picture of locally very well ordered regions with fluctuations of orientation about the average director is indeed essentially correct.

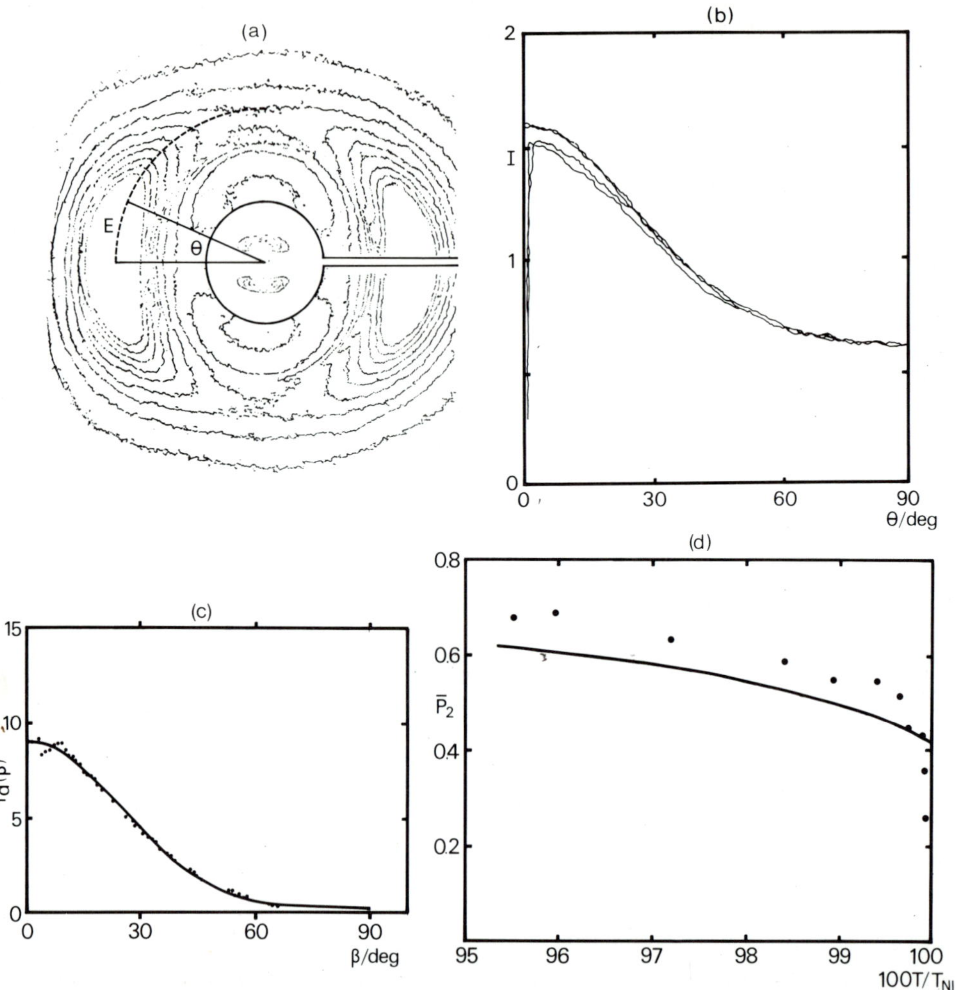

Fig. 19 *X-ray diffraction results for OAB in the nematic phase. (a) Contour map of intensity data at a reduced temperature of 0.997. (b) Intensity versus θ along equatorial arc E for the same temperature. (c) $f_d(\beta)$ at the T/T_{NI} of 0.997. The line through data points is a fit to the simple Maier-Saupe theory. (d) The temperature dependence of $\bar{P}_2$; the line is predicted by the Maier-Saupe theory.*

Smectic A and C The analysis of the equatorial arcs in the diffraction pattern of S_A and S_C phases to determine properties of $f(\beta)$ should follow the same lines as discussed in the previous section. Little work appears to have been done, however, in this direction, but some unpublished work from our laboratory [37] on the S_A phase of EABAC:

$$CH_3CO{\cdot}O-\langle\bigcirc\rangle-CH{=}N-\langle\bigcirc\rangle-CH{=}CHCO{\cdot}OC_2H_5$$

(which has C 105°C; S_A 125°C; N 150°C I) and the S_A phase of OAB give $\bar{P}_2$ values from $f_d(\beta)$ near the middle of the smectic ranges of $\gtrsim 0.65$ and $\gtrsim 0.70$ respectively. These are perhaps a little lower than the predictions of simple theory but further discussion would be premature.

The smectic A order parameters

Simple theory The distribution function of molecular positions normal to the layers in a S_A phase (the density wave describing the S_A layers) may be written in terms of a Fourier series [45] as we saw in Chapter 3

$$f(z) = \sum_{M=0} \alpha_M \cos(2\pi Mz/d), \tag{36}$$

where

$$\alpha_M = (2/d)\langle\cos(2\pi Mz/d)\rangle, \text{ for } M > 0;$$

$\alpha_o = 1/d$ and d is the layer spacing; the distribution function is normalised so that

$$\int_o^d dz\, f(z) = \int_{-d/2}^{d/2} dz\, f(z), \tag{37}$$
$$= 1.$$

The smectic order parameters are defined by

$$\tau_M = \langle\cos(2\pi Mz/d)\rangle. \tag{38}$$

The scattering amplitude for Q along the z direction, in crystallographic terms the OOM reflections, may be written

$$F_{OOM} = \sum_m \langle a_m \exp 2\pi i M z_m' /d\rangle, \tag{39}$$

where m denotes the atoms in a molecule and the average is taken over all molecular configurations. If we now denote the z coordinates for a perfectly ordered smectic (τ_M=1) by z_m^p then $z_m' = z_m^p + z_m$ when

$$F_{OOM} = \sum_m \langle\exp 2\pi i M\, z_m/d\rangle\, a_m \exp 2\pi i M z_m^p/d \tag{40}$$

and if we assume that the z_m are the same for all atoms in the molecule

$$I_{OOM} = |F_{OOM}|^2 = \langle\cos 2\pi Mz/d\rangle^2 |F_p|^2, \tag{41}$$

so that

$$I_{OOM}/I_{OOM}^P = \langle\cos 2\pi Mz/d\rangle^2,$$
$$= \tau_M^2. \tag{42}$$

For a gaussian distribution

$$f(z) = [2\pi<z^2>]^{-\frac{1}{2}}\exp(-z^2/2<z^2>), \tag{43}$$

$$\begin{aligned}\tau_M &= <\cos(2\pi Mz/d)>,\\ &= \exp(-2\pi^2 M^2<z^2>/d^2),\\ &= \exp(-W_{OOM}),\end{aligned} \tag{44}$$

$$\tau_M = \tau_1 M_2 \tag{45}$$

and W_{OOM} is simply the Debye-Waller factor for the Mth order reflection.

Applications Eq. (42) shows that the intensity of the **first order** layer reflection is proportional to the smectic order parameter τ_1^2. The proportionality constant is the intensity for the perfect smectic. McMillan [46] used this result to compare the experimental data with simple theory for two substances, using an empirical normalisation (I^p_{OOM}). The result for cholesteryl nonanoate is shown in figure 20 which demonstrates fair agreement with theory.

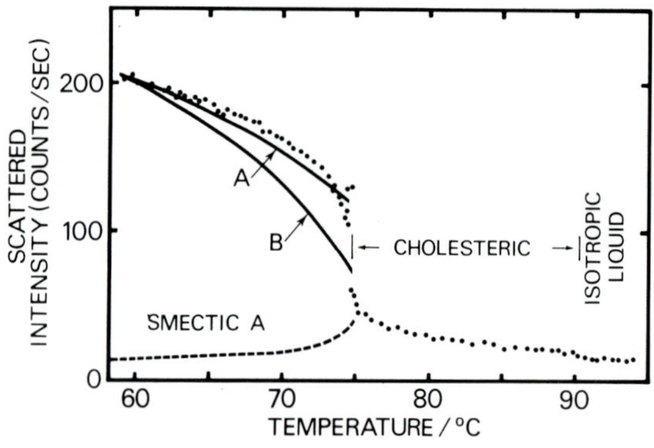

Fig. 20 *X-ray scattered intensity of the 1st order layer spacing (001 reflection) for cholesteryl nonanoate. The lines A and B are the results of calculations based on microscopic models for the S_A phase; they have been fitted to the experimental curve at the lowest temperature.*

To obtain an absolute value for τ_1 from the first order reflection alone requires an absolute intensity I_{OOM} together with a knowledge of I^p_{OOM} which will be quite difficult. On the other hand if two or more orders are observed then assumption of a simple model for I^p_{OOM} and a form for f(z), enables τ_1 and $<z^2>^{\frac{1}{2}}$ to be obtained. This has been done [37] for EABAC in the middle of its S_A range ($\sim$ 114°C) where τ_1 = 0.56 and $<z^2>^{\frac{1}{2}}$ = 3.8Å. These results were based on the assumption of a gaussian distribution and a model in which the molecules are on average tilted with respect to the layer normal by 15-20°. The results are not very sensitive to the details of these assumptions and are in good agreement with simple theory [47]. They must be considered typical for S_A materials where usually (as in this case) only two orders of reflection are observed, which immediately implies a considerable disorder of the layers. The sharpness of the reflections on the other hand implies that the layer correlation itself is essentially long range.

References

1. For general accounts of neutron scattering see, W. Marshall and S. Lovesey, *Theory of Thermal Neutron Scattering;* B.T.M. Willis (Ed.) *Chemical Applications of Neutron Scattering, (O.U.P.)* (1973).
2. For general accounts of X-ray diffraction see, A. Guinier, *X-ray Diffraction* (W.H. Freeman) (1973); B.E. Warren, *X-ray Diffraction* (Addison Wesley) (1969).
3. See D.C. Champeney, *Fourier Transforms and their Physical Applications* (Academic Press) (1973).
4. R. Hosemann and S.N. Bagchi, *Direct Analysis of Diffraction by Matter* (North Holland) (1962).
5. J.G. Powles, *Adv. Phys.,* **22**, 1 (1973).
6. A.J. Leadbetter and A.C. Wright, *J. Non. Cryst. Solids,* **7**, 141 (1972).
7. G.W. Neilson, R.A. Howe and J.E. Enderby, *Chem. Phys. Lett.,* **33**, 284 (1975).
8. A.J. Leadbetter and A.C. Wright, *J. Non. Cryst. Solids,* **7**, 23 (1972).
9. P.A. Egelstaff, D.I. Page and J.G. Powles, *Mol. Phys.* **20**, 881 (1971).
10. J.H. Clarke, J.C. Dore, A. Walford and R.N. Sinclair, *Mol. Phys.,* **31**, 883 (1976).
11. A.J. Apling, A.J. Leadbetter and A.C. Wright, *J. Non. Cryst. Solids,* **23**, 369 (1977).
12. A.J. Leadbetter, A.J. Apling, M.F. Daniel, A.C. Wright and R.N. Sinclair, *The Structure of Non Crystalline Materials* (Ed. P.H. Gaskell, Society of Glass Technology 1977) p.23.
13. N. Nimura, *Mol. Cryst. Liq. Cryst.,* **31**, 123 (1975).
14. A.J. Leadbetter, R.M. Richardson and C.N. Colling, *J. Phys.* (Paris), **36**, C1-37 (1975).
15. A. de Vries, *Liquid Crystals* (Pramana Suppl. No.1 Ed. S. Chandrasekhar 1975) p.93.
16. W.L. McMillan, *Phys. Rev.* **A7**, 1673 (1973).
17. W.L. McMillan, *Phys. Rev.* **A8**, 328 (1973).
18. A. de Vries, *Mol. Cryst. Liq. Cryst.,* **11**, 361 (1970).
19. A. de Vries, *Mol. Cryst. Liq. Cryst.,* **20**, 119 (1973).
20. B.K. Vainshtein, I.G. Chistyakov, E.A. Kosterin and U.M. Chaikovskii, *Mol. Cryst. Liq. Cryst.,* **8**, 457 (1969).
21. A.J. Leadbetter, J.C. Frost, J.P. Gaughan,
22. J.E. Lydon and C.J. Coakley, *J. de Phys.,* **36**, C1-44 (1975).
23. B.K. Vainshtein and I.G. Chistyakov, *Liquid Crystals* (Pramana Suppl. No.1. Ed. S. Chandrasekhar 1975) p.79.
24. B.K. Vainshtein, *Diffraction of X-rays by Chain Molecules* (Elsevier 1966).
25. E.A. Kosterin, *Soviet Phys. Crystallography,* **17**, 549 (1972).
26. A. de Vries, *Mol. Cryst. Liq. Cryst.,* **10**, 31 (1970).
27. A. de Vries, *Mol. Cryst. Liq. Cryst.,* **10**, 219 (1970).
28. J. Doucet, A.M. Levelut and M. Lambert, *Mol. Cryst. Liq. Cryst.* **24**, 317 (1973)
29. S. Diele, P. Brand and H. Sackmann, *Mol. Cryst. Liq. Cryst.* **16**, 105 (1972)
30. D. Demus, S. Diele, M. Klapperstück, V. Link and H. Zaschke, *Mol. Cryst. Liq. Cryst.,* **15**, 161 (1971).
31. H. Terauchi, T. Takenchi and S. Kusabayashi, *Japan J. Appl. Phys.* **11**, 1862 (1972).
32. W.L. McMillan, *Phys. Rev.,* **A7**, 1419 (1973).
33. A.J. Leadbetter, J.L.A. Durrant and M. Rugman, *Mol. Cryst. Liq. Cryst. Letters,* **34**, 231 (1977).
34. F. Rusticelli, private communication
35. P. Delord and G. Malet, *Mol. Cryst. & Liq. Cryst.,* **27**, 231 (1974).
36. P. Delord and G. Malet, *Mol. Cryst. Liq. Cryst.* **28**, 223 (1974).
37. A.J. Leadbetter and E.K. Norris, unpublished work.
38. I.G. Chistyakov and W.M. Chaikovsky, *Mol. Cryst. Liq. Cryst.,* **7**, 269 (1969).
39. A. de Vries, *J. Chem. Phys.,* **56**, 4489 (1972).
40. P.G. de Gennes, *Comp. Rend.,* **274B**, 142 (1972).

41. R. Pynn, *J. Phys. Chem. Solids*, 34, 735 (1973).
42. M. Kohli, K. Otnes, R. Pynn and T. Riste, *Z. Phys.*, B24, 147 (1976).
43. P. Delord and J. Falgueirettes, *Comp. Rend.*, 260, 2468 (1965).
44. J. Falgueirettes and P. Delord, *Liquid Crystals & Plastic Crystals* (Ed. G.W. Gray and P.A. Winsor, Ellis Horwood, 1974) Vol. 2. Ch. 3.
45. P.J. Wojtowicz, *Introduction to Liquid Crystals* (Eds. E.B. Priestley, P.J. Wojtowicz, Ping Sheng, Plenum 1974). Ch.7.
46. W.L. McMillan, *Phys. Rev.* A6, 936 (1972).
47. R.L. Humphries and G.R. Luckhurst, *Mol. Phys.*, 35, 1201 (1978).

Chapter 14

X-RAY STUDIES OF ORDERED SMECTIC PHASES

J. DOUCET

Laboratoire de Physique des Solides,
Université Paris-Sud, Bâtiment 510,
91405 Orsay, France.

Introduction

The X-ray study of liquid crystals (mesophases) is not new; the first experiments were performed as early as 1913 by van Lingen [1] on the nematic phases of *p*-azoxyanisole and *p*-azoxyphenetole. The first X-ray patterns of smectic phases were obtained ten years later by de Broglie and Friedel [2]. In 1932, Hermann and Krummacher [3] found, again through X-ray investigations, a new kind of smectic phase, whose diffraction pattern indicates the presence of order within the smectic layer. Unfortunately, this study was not continued and we had to wait until 1960 to see a new, growing interest in X-ray diffraction by liquid crystals. Much work, published during the succeeding years, has eventually allowed us to arrive at a better understanding of mesophase structures.

Our aim, in this Chapter, is to explain what kind of information can be gleaned from the study of the X-ray diagrams of ordered smectic phases and to describe how this information is obtained. We shall examine successively each of the ordered phases and discuss their common structural features and their differences, together with the importance, the scope and the utility of X-ray experiments on liquid crystals. We begin by recalling the main results concerning X-ray diffraction.

Summary of X-ray Diffraction by Crystals

The wavelength of the X-rays used in diffraction is of the order of magnitude of the distances between atoms in condensed matter. This means that X-ray interference phenomena can occur, thus providing us with information about the structure of this matter. Let us assume that an X-ray beam passes through matter. The interaction between the radiation and the electrons of that matter is complex, and the photons, which do not appear in the transmitted beam, may have been subjected to various transformations:

(i) some are deviated without change of wavelength; this is the scattered radiation (coherent scattering);

(ii) some are scattered with a wavelength change; this is the Compton effect (incoherent scattering);

(iii) some are absorbed by atoms; this is the photoelectric effect. The excited atom can return to its ground state, emitting either another electron (Auger effect) or X-photons (X-ray fluorescence).

In the following sections we shall be concerned only with scattered radiation; a few books on X-ray diffraction and crystallography are listed in the Bibliography prior to the References.

General formula for X-ray diffraction

Consider an X-ray beam with wavevector k_o ($|k_o| = 1/\lambda$) hitting a volume v of matter; the X-photons will interact with electrons, and some of the photons will be scattered. Let us look at the photons with wave vector k_s, where $(k_o, k_s) = 2\theta$ scattered by an elementary volume centred on the origin O and by another volume centred on M, as in figure 1.

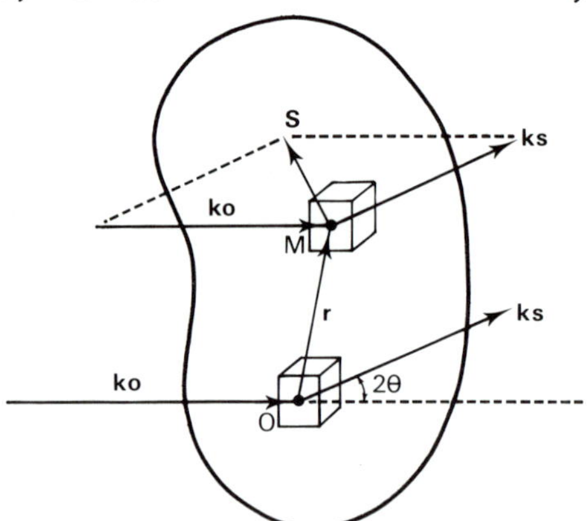

Fig. 1. *Schematic representation of X-ray diffraction.*

Taking OM as r, the phase difference between the two scattered beams is

$$\psi = 2\pi r.S,$$

where

$$S = k_s - k_o;$$

S is called the scattering vector. If $\rho(r)$ is the electron density, the amplitude scattered by the elementary volume dv centred on M is $\rho(r)dv$ times the amplitude scattered by one electron located at M. This implies that the contribution of dv to the total diffracted amplitude is equal (in complex notation) to $\rho(r)dv \exp(2i\pi r.S)$, and so the total amplitude diffracted by volume v is

$$A(S) = \int dv\ \rho(r)\ \exp(2i\pi r.S).$$

From this formula, it can be seen that A(S) is the Fourier transform of the electron density.

This formula may be written in a system of coordinates defined by three basis vectors (a_1, a_2, a_3) in which

$$r = xa_1 + ya_2 + za_3;$$

for S, whose dimension is L^{-1}, let us choose the three basis vectors a_1^*, a_2^*, a_3^* such that

$$a_i . a_j^* = \delta_{ij},\ i,j = 1,3.$$

The three vectors a^* are the basis vectors of reciprocal space. If

$$S = Xa_1^* + Ya_2^* + Za_3^*,$$

the expression for the diffracted amplitude becomes

$$A(X,Y,Z) = v_o \iiint dxdydx\ \rho(x,y,z)\exp\{2i\pi(xX + yY + zZ)\},$$

where v_o is the elementary volume (a_1, a_2, a_3).

Diffraction by a crystal

If we assume now that v is a crystal, (x,y,z) is a triperiodic function which can be described by an elementary cell and three basis translational vectors (a_1, a_2, a_3). With these conditions, it can be demonstrated that the diffracted amplitude A(X,Y,Z) is zero except at the nodes of the reciprocal lattice (a_1^*, a_2^*, a_3^*). This means that the X-rays are scattered in discrete directions which depend only on the geometry of the elementary cell; in contrast, the intensity of the diffracted beam depends on the electron density inside the elementary cell. At a node ($hk\ell$) (where h,k,ℓ are integers), the intensity is proportional to $F^2(hk\ell)$; $F(hk\ell)$ is called the structure factor.

The diffraction conditions can be interpreted in a geometrical way by means of the reciprocal lattice, the scattering vector S (which is a vector in the reciprocal space) and a sphere of radius $1/\lambda$ centred in the crystal at O (Ewald's sphere). In figure 2, it can easily be seen that there is a diffracted beam each time a node of the reciprocal lattice (centred on I) is located on the Ewald sphere. It can also be demonstrated that a family of reticular planes ($hk\ell$) of the crystalline lattice is orthogonal to the reciprocal row, [$hk\ell$] and that the distance between the ($hk\ell$) node and the origin I of the reciprocal lattice is equal to $1/d_{hk\ell}$ where $d_{hk\ell}$ is the reticular spacing between [$hk\ell$] planes. These remarks lead to the well-known Bragg relation

$$\lambda = 2d_{hk\ell} \sin\theta.$$

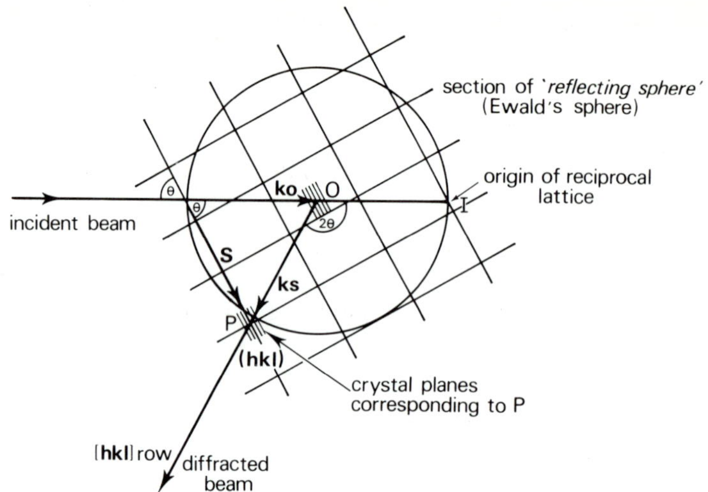

Fig. 2. *Geometrical representation of the conditions for X-ray diffraction*.

Diffraction by an imperfect crystal

In the previous section, we have been concerned only with perfect and infinite crystals. However, the diffraction patterns are affected by the finite size of the crystal and its imperfections, as we saw in Chapter 13. Let us examine some of these modifications which we shall meet again when studying liquid crystals.

Effect of crystal size The effect of the finite size of the crystal is to distribute the intensity in a small volume around each node of the reciprocal lattice, where the intensity was concentrated in a delta function when the crystal was infinite. This volume is inversely proportional to the crystal size L, as we can see from the Scherrer formula which gives the integral width $\Delta(2\theta)$ of the Bragg spot as

$$\Delta(2\theta) = (\lambda/L)\cos\theta.$$

Thermal agitation In a crystal, the atoms are not motionless; they move about an average position, through the influence of thermal agitation. Let us assume that the atoms vibrate about their average position in a random way. This motion has two effects on the X-ray diffraction: it produces a general background of scattering in all directions and, as a result, the intensity of the reflections decreases by a factor, $\exp(-2B\sin^2\theta/\lambda^2)$, where B is related to the mean quadratic amplitude of the atomic displacements u,

$$B = (8/3)\pi^2\langle u^2\rangle.$$

However in most cases, the motions of neighbouring atoms are correlated, and the atomic displacements can be described as a superposition of phonons.

Diffusion by phonons In the case of phonon diffusion, the intensity subtracted from the Bragg reflections is not distributed evenly in reciprocal space, but is located around each node in a very definite way. The intensity scattered by one phonon of wavevector k is concentrated at the extremities of the factors ± k originating at the nodes of the reciprocal lattice. The intensity of these two spots or satellites is

$$I \propto F^2_{(s)} (u.S)^2,$$

where **S** is the scattering vector, u is the polarization of the phonon (atomic displacement) and F is the structure factor of the reciprocal node. Experimentally, the superpositions of phonons produce a diffuse area around each node which is caused by the juxtaposition of such satellites. To a rough approximation, the profile of the intensity of the diffuse scattering around a node is

$$I \propto S^2/q^2,$$

where q is the distance between the point in reciprocal space and the studied node, as shown in figure 3. We note that theoretically, the phonons produce a slight shift in the frequency of the scattered X-rays,

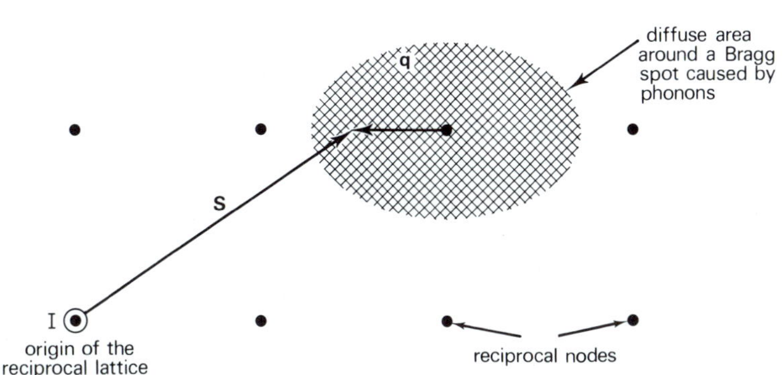

Fig. 3. *Diffusion by phonons*.

but this effect may be neglected since the X-ray frequencies are about 10^7 times greater than the frequencies of the thermal vibrations i.e. 10^{18} for X-rays and 10^{11} Hz for phonons. This is the reason why an energy analysis of the diffuse area around each node is impossible by X-ray experiments.

In contrast, this study is possible with neutron scattering experiments because both the wavelength and the energy of the thermal neutrons are of the same order of magnitude as the interatomic distances and the vibrational energy, as we can see from figure 4. Thus it is possible,

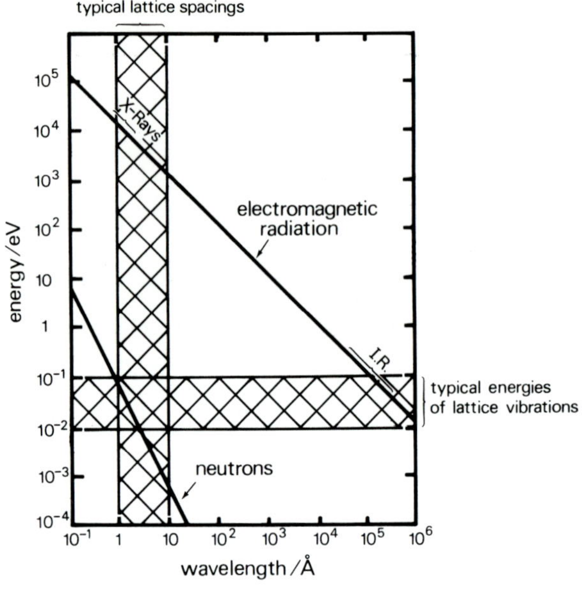

Fig. 4. *Diagram of energy and wavelength.*

by neutron scattering experiments performed with a three-axis spectrometer, to determine the dispersion curves $\omega(q)$; in other words, the study of collective modes is possible by coherent neutron scattering but not by X-ray scattering. With X-ray experiments, it is impossible to differentiate between dynamic and static phenomena.

Linear disorder Let us consider the linear disorder which often occurs in liquid crystals. By linear disorder, we mean that which occurs when periodic parallel rows are not arranged regularly with respect to one-another, as seen in figure 5. It is possible to show that the scatter- is then concentrated on a family of parallel planes, normal to the rows, and separated by a distance 1/a, where a is the period of the linear

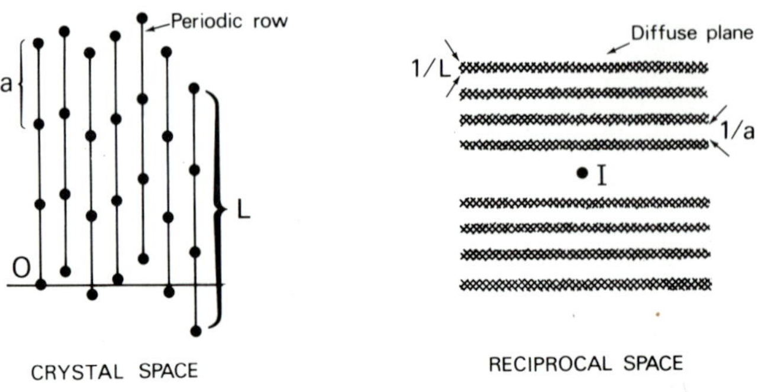

Fig. 5. *Linear disorder with longitudinal motion.*

lattice. This result appears obvious, especially when one realises that the Fourier transform of a linear periodic row is formed by a family of parallel planes perpendicular to the row. We note that the layer thickness is equal to the inverse of the length of the rows and that when the row displacements are longitudinal, the intensity of the plane passing through the origin of the reciprocal lattice vanishes. This behaviour is shown in figure 5.

X-ray diffraction techniques Diffraction techniques can be divided into two groups namely the techniques using monodomain samples and those employing powder samples. It is obvious that monodomain samples provide much more information than powder samples, but they are not always available. Monodomains of liquid crystals can sometimes be obtained, either by melting a single crystal or by orientating a powder sample with a magnetic field.

The usual diffraction device for the study of monodomains is represented schematically in figure 6. A monochromatic X-ray beam is diff-

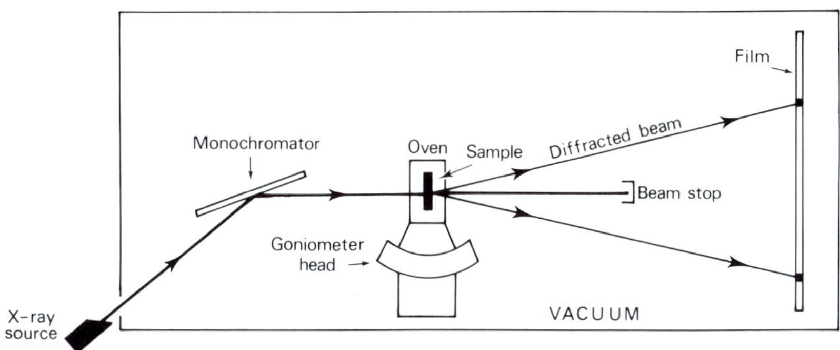

Fig. 6. *X-ray equipment for the study of monodomain samples.*

racted by the sample which is held at a constant temperature in an oven and orientated on a goniometer head. The diffracted beams are collected at a film placed behind the sample; the sample is sometimes oscillating, in order to collect more information. Instead of a film, it is also possible to use an X-ray counter.

For powder samples, the well-known Debye-Scherrer technique is used. The equipment is similar to that for monodomains, except that the goniometer head is not needed. Because all possible orientations of the crystallites occur in the powder, each Bragg reflection produces many diffracted beams, with the same value of 2θ, located on a cone whose axis corresponds to the direct beam. The intersections of such cones, with a plane film perpendicular to the direct beam form concentric circles; each circle corresponds to a Bragg reflection and so this method gives all the possible reticular spacings, but no information about the spatial orientation of these planes. X-ray counters are often used instead of films because intensity measurement in only one-dimension is required.

Liquid Crystal Phases

In this Chapter, we shall use the classification of the various kinds of liquid crystal phase proposed by Sackmann and his coworkers [4]. This classification is based on miscibility criteria, and these authors have identified one nematic phase and seven smectic phases, namely S_A, S_B, S_C, S_D, S_E, S_F and S_G. These phases can be divided into two groups according to the characteristics of their X-ray patterns at large diffraction angles. By large diffraction angles, we mean angles of the order of 20° which correspond to distances of approximately 4.5 Å for the usual wavelengths (CuKα = 1.54 Å and CoKα = 1.79 Å); small diffraction angles are taken to be of the order of 3°, corresponding to distances of about 30 Å. In the first group (nematic, S_A, S_C and S_D) or the disordered phases, we find those which give only one diffuse, broad diffraction maximum at large diffraction angles. The second group is composed of phases whose diffraction patterns present one or a few sharp Bragg reflections instead of a diffuse ring. In this group we find three smectic phases, S_B, S_E and S_G. The S_F phase seems to be a phase intermediate between the two groups.

Here, we shall deal only with the phases belonging to the second group, that is to say with phases whose constituent molecules are regularly arranged within the layers i.e. the ordered smectic phases. The X-ray study of disordered phases is discussed in the previous Chapter but let us just summarise the results concerning these phases. At large diffraction angles they all give a diffuse, broad ring which corresponds to the lateral distances between molecules (by lateral distances, we mean intermolecular distances perpendicular to the molecular long axes). This ring, which is quite similar to the ring given by isotropic liquid phases, indicates a lack of periodic lateral order, that is to say the distribution of the molecular centres of mass is random,

The X-ray patterns of nematics and smectics differ mainly in their appearance at small diffraction angles. Nematic patterns present a diffuse ring, corresponding to distances equal to the molecular length which indicates that there is no order in the direction of the molecular long axes; the pattern for a nematic phase is quite similar to that given by its isotropic liquid phase. In contrast, X-ray patterns obtained from smectics present one sharp ring at small angles which is indicative of a periodic layer structure, corresponding to the smectic layers. The difference between S_A and S_C lies in the fact that in S_A

phases, the director is orthogonal to the layers whereas it is inclined in the S_C; in the two phases, the order within the layers is liquid-like.

Study of the Ordered Smectic Phases of TBBA

The structures of ordered smectic phases are much more complicated than the structures of disordered mesophases, but X-rays can provide much *more* information for ordered phases. To clarify the situation, we shall examine the ordered smectic phases exhibited by three compounds, TBBA, PBAPC and PPP; then, we shall discuss a few problems, such as the relations between the structures of these phases, the miscibility identifications and their crystalline character.

First we study the ordered smectic phases of TBBA. This compound is especially interesting because it has recently been the subject of much published work and because it exhibits two ordered smectic phases which are of great concern for our study. The phase sequence for TBBA (terephthalylidene-bis-4-n-butylaniline):

$$C_4H_9-C_6H_4-N{=}CH-C_6H_4-CH{=}N-C_6H_4-C_4H_9$$

is

$$\underset{\text{Crystal}}{\text{(IX)}} \underset{}{\overset{-33^{\circ}}{\rightleftarrows}} \underset{\text{Crystal}}{\text{(VIII)}} \xrightarrow{113^{\circ}C} \underset{S_B}{\text{(V)}} \overset{144^{\circ}C}{\rightleftarrows} \underset{S_C}{\text{(IV)}} \overset{172^{\circ}C}{\rightleftarrows} \underset{S_A}{\text{(III)}} \overset{200^{\circ}C}{\rightleftarrows} \underset{N}{\text{(II)}} \overset{236^{\circ}C}{\rightleftarrows} \underset{I}{\text{(I)}}$$

$$\text{(VIII)} \xleftarrow{52^{\circ}C} \text{(VII)} \overset{68^{\circ}C}{\rightleftarrows} \text{(VI)} \overset{84^{\circ}C}{\rightleftarrows} \text{(V)}$$

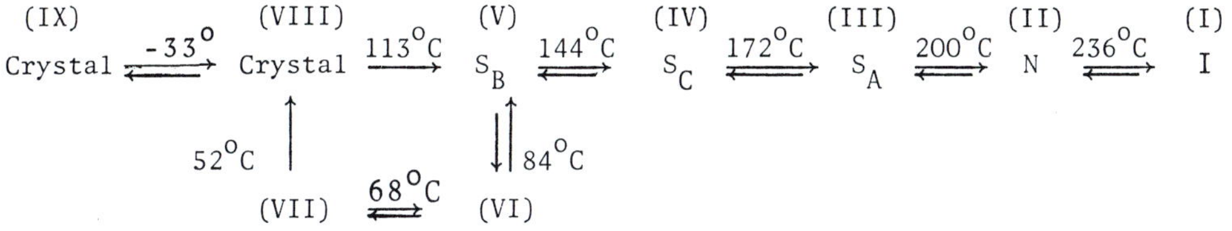

of all these mesophases, two {phases V(S_B) and VI} are ordered smectic phases.

The smectic B phase of TBBA

We shall examine the S_B phase of TBBA in some detail because its structure is a good starting point for a study of all other ordered smectic phases.

X-ray patterns Monodomains of the S_B phase can be obtained by melting a single crystal, without an external field; the structure can then be determined from monodomain X-ray photographs [5] and two of these are shown in figure 7.

In figure 7(a), the X-ray beam is perpendicular to the long molecular axes and a set of parallel, diffuse sheets appears on the photograph; these sheets correspond to reciprocal diffuse planes perpendicular to the molecules and their spacing is the inverse of the molecular length

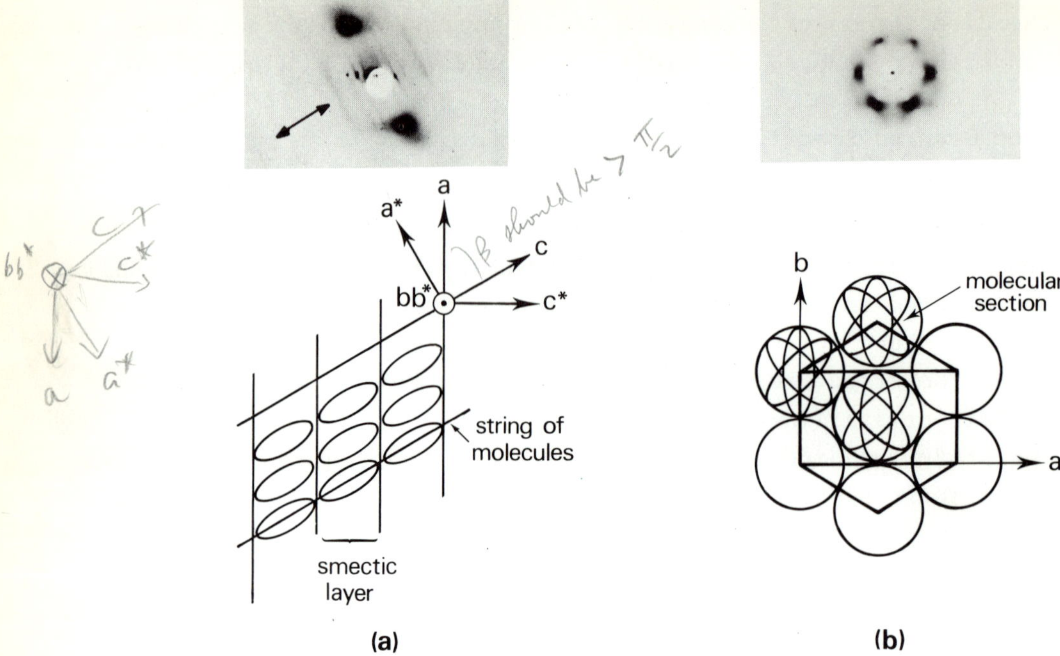

Fig. 7. *(a) X-ray pattern obtained for the S_B phase of TBBA with the direct beam parallel to b (i.e. to the smectic layers); the arrow indicates the direction of the long molecular axes. (b) X-ray pattern obtained with the direct beam parallel to c (i.e. to the long molecular axes). Schematic representations of the corresponding structures are given below the X-ray patterns.*

($\sim$ 1/29 Å^{-1}). As we have seen such a feature in the sample indicates the existence of longitudinal displacements of parallel strings of molecules which are uncorrelated from one string to another. The correlation length along the molecular direction (c) can be measured from the thickness of the diffuse sheets and is of the order of 200 Å (i.e. 7 molecules) near the solid to S_B transition. From the extent of the diffuse scattering pattern, we can deduce [6] the fluctuation angle (2°) and the amplitude of the longitudinal motion (2.5 Å). We note that contrary to the behaviour in some nematics, the diffuse sheets are not curved [7]; this implies that all the strings are orientated in the same direction i.e. there are almost no fluctuations in the direction of the string. The intense spots appearing on the diffuse sheets are the Bragg diffraction spots originating from the smectic layers. The tilt angle of the molecules with respect to the normal to the smectic layers can be directly measured from the photograph and is found to be 32° at 115^oC.

The photograph in figure 7(b) represents the equatorial plane (a^*, b^*) of the reciprocal lattice (perpendicular to the molecular direction). The six Bragg spots forming a pseudo-hexagon are indicative of a periodic order within the layers (the photograph is over-exposed in order to show the diffuse scattering, and so the sharpness of the spots is concealed).

At large diffraction angles, the powder pattern of the S_B phase of TBBA, given in figure 8, shows the presence of several sharp, well-defined rings, instead of the diffuse, broad ring observed for disordered phases. Nevertheless, at very large angles, no rings are visible whereas some can be seen on the pattern for the solid phase (VIII in figure 8). This indicates a very high degree of thermal agitation, in fact the Debye-Waller factor $B \simeq 10\ \text{Å}^2$ for S_B but only $\sim 3\ \text{Å}^2$ for the solid phase).

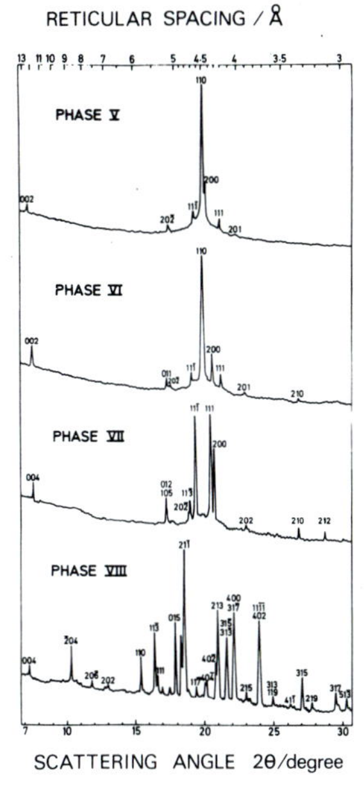

Fig. 8. *Debye-Scherrer patterns for phases V, VI, VII and VIII of TBBA (λ = CuKα_1) analysed using a microdensitometer.*

Structure of the smectic B phase Using both the single domain patterns shown in figure 7 and the powder pattern, it is relatively easy to determine a local elementary cell describing the S_B phase. This elementary cell has monoclinic symmetry but, in a plane perpendicular to the molecules, the packing is hexagonal to within 1%. The parameters for the cell at 125°C are:

$$a = 10.10 \pm 0.05\ \text{Å},$$
$$b = 5.18 \pm 0.02\ \text{Å},$$
$$c = 28.3 \pm 0.2\ \text{Å},$$
$$\beta = 119.0 \pm 0.5.$$

From these experimental results, the structure of the S_B phase has the following features which are shown in figure 9. Firstly the molecules are arranged in parallel layers with their long axes tilted by 30° to the normal. The molecules are lined up from one layer to another, forming strings of molecules. Secondly within each layer, the molecular packing is pseudo-hexagonal and all the molecular positions are equivalent. This order is three-dimensional and one layer cannot rotate with respect to a neighbouring layer.

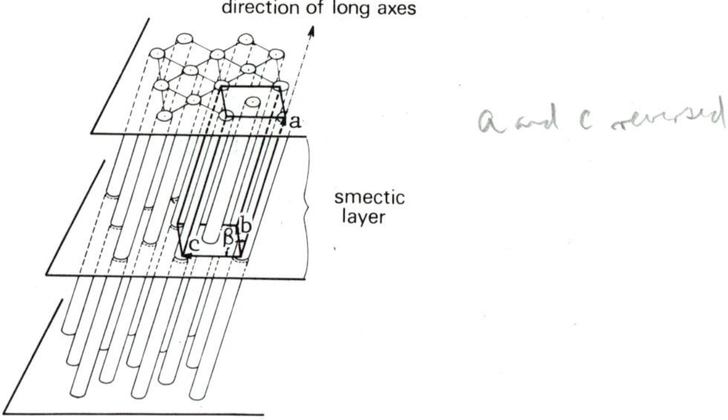

Fig. 9 *Schematic representation of a tilted S_B phase.*

Description of the disorder In fact, such a description of the structure of this S_B phase is incomplete, since it does not take into account additional features observed on the equatorial plane, shown in figure 7(b). These consist of twelve diffuse spots which are represented schematically in figure 10. The presence of these spots indicates that the

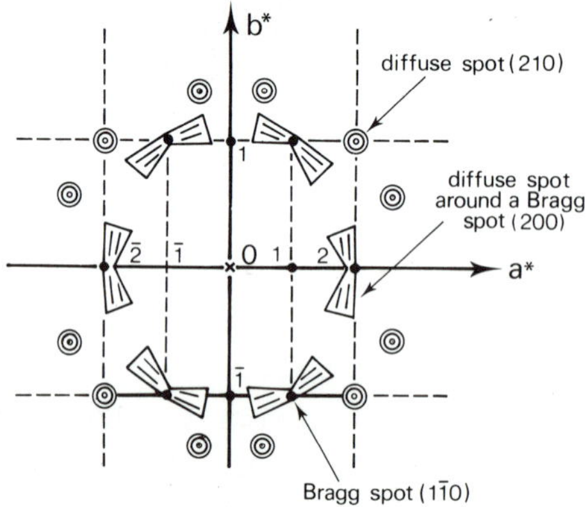

Fig. 10. *Schematic representation of the X-ray pattern of the equatorial plane (a*,b*) of the S_B phase of TBBA.*

local symmetry is lower than the global symmetry. Let us discuss this point more precisely. If a and b are the two axes of the monoclinic lattice lying in the smectic layers, the diffuse spots have the following indices: (±2±1 0), (±5/2±1/2 0), (±1/2±3/2 0). Only the spots of the first group are located on reciprocal nodes; the others, with semi-integer indices, which can be deduced from the first group by rotations of $2\pi/3$ do not belong to the reciprocal lattice.

What is the significance of the four spots (±2±1 0)? We have said previously that all the molecular positions were equivalent because the

Bragg spots whose indices h and k are such that $(h + k) = 2n + 1$ are not present in the pattern (the corresponding structure factors $F(hk\ell)$ are equal to zero when the C face of the elementary cell is centred). But, in fact, we observe diffuse spots at (±2±1 0); this implies that the C face is not centred. In other words, the molecular positions at the centre and the apices of the rectangular bidimensional cell are not equivalent. However this conclusion is only valid for a small domain because of the width of the diffuse spots; the size L of the domains is given by the Scherrer formula, and is about 20 Å to 30 Å. Within this small domain the extinction conditions of the structure factors ($h0\ell$; $h = 2n + 1$) correspond to the existence of a glide mirror a (a mirror perpendicular to b, plus the translation a/2). This allows us to relate the two molecular orientations; the resulting molecular packing parallel to the smectic layers, inside one domain, is shown in figure 11(a) where each molecular section is represented by an ellipse.

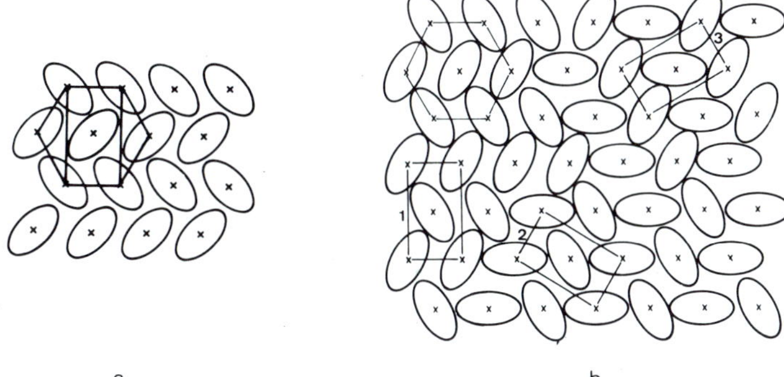

Fig. 11. *Lateral order in (a) one domain and (b) a S_B phase.*

The existence of the two other families of four diffuse spots is explained by assuming that there are three possible orientations for such domains which are related to one another by rotations of ± (2π/3) (cf. figure 11b), corresponding to three orientations of the rectangular cell (a,b). Thus, on the whole, all the molecular positions become equivalent, assuming that each molecule can also adopt two symmetrical positions with respect to b; at each node, there are, on average, six identical molecular orientations [8]. This conclusion is valid only if one is dealing with a true hexagonal lattice. In our case, the three types of domain are not quite equivalent because of the pseudo-hexagonal lattice and also the tilt angle of the molecules which distinguish the domains of type 1 from domains 2 and 3 (the c direction of the molecular long axis is the same for the three domains); the spots (±2±1 0) are more intense and less broad that the other eight.

One major problem is to know whether this rotational disorder of the molecules around their long axes is static or dynamic. X-rays cannot differentiate between these two possibilities, but other techniques, such as NMR [9] and above all neutron scattering [10] can make this distinction. The commonly accepted answer is that the disorder is dynamic, with a relaxation time of the order of 10^{-11}s. It is worth noting that neutron experiments cannot distinguish easily between a disorder involving six molecular orientations around their long axis and an isotropic rotation.

Let us conclude this discussion of the S_B phase of TBBA with a final remark. The rotational disorder, which is characterized by the existence of three small domains does not destroy the long-range periodic order of the distribution of the molecular centres of mass. This is clearly proved by the sharpness of the diffraction rings in the powder pattern which is similar to that given by crystalline samples. The disappearance of the rings at very large angles is due to a very large thermal agitation factor and not to the absence of long-range order. The disorder type is first order (not destroying the long-range order) and not second order (destroying the long-range order) as observed in paracrystals [11].

The smectic VI phase of TBBA The X-ray patterns of the S_{VI} phase of TBBA are very similar to those of the S_B (V) phase, except for one major difference; four of the twelve diffuse spots (±2±1 0) are now condensed into Bragg spots and the other eight have disappeared [12]. This indicates that domains of type 1 have grown at the expense of domains 2 and 3 shown in figure 11(b). Such a structure implies that the molecular positions are no longer equivalent, but are connected by a glide mirror a; reflections with $h + k = 2n + 1$ are now visible, but not the reflections $h0\ell$ with $h = 2n + 1$ as we can see in figure 8. Therefore, the rotational disorder around the molecular long axes seems to be frozen in the S_{VI} phase; the only possible disorder is a rotation of π around the long axis. This phenomenon has probably been observed by neutron scattering [13]. All the other main features of the S_B phase, such as the tilt angle and the strings of molecules are still present in the S_{VI} phase. The problem concerning the classification of this phase will be discussed later.

Study of the S_B and S_E Phases of PBAPC

The chemical formula of PBAPC (*p*-phenylbenzylidene-*p*-amino n-pentylcinnamate) is:

The phase sequence is: $C_6H_5-C_6H_4-CH{=}N-C_6H_4-CH{=}CH-CO{\cdot}OC_5H_{11}$

$$\text{Crystal} \underset{}{\overset{92^{\circ}C}{\rightleftarrows}} S_E \overset{101.5^{\circ}C}{\rightleftarrows} S_B \overset{168^{\circ}C}{\rightleftarrows} S_A \overset{204^{\circ}C}{\rightleftarrows} I$$

The structures of the S_B and S_E phases of PBAPC can easily be deduced from the structure of the ordered smectic phases of TBBA. The X-ray patterns obtained from both monodomain and powder samples, given in figure 12, show that the structure of the S_B and S_E phases of PBAPC

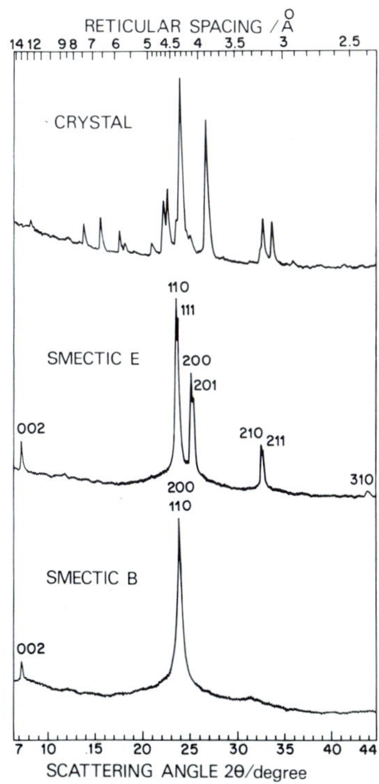

Fig. 12 *Debye-Scherrer patterns of the phases of PBAPC (λ=CoKα_1) analysed using a microdensitometer.*

are respectively analogous to the structures of the S_B and S_{VI} phases of TBBA [14]. The only major difference arises from the fact that in PBAPC the molecules are perpendicular to the layers and not tilted as in TBBA.

Under these conditions, the local cell of the S_B phase of PBAPC is a true hexagonal cell and this explains why only one diffraction ring is visible in the powder pattern (the rings 200 and 110 are superimposed). In the S_E phase, the hexagon is distorted and the diffraction rings are separated. We note also that the ring with $\ell = 1$ is not visible in the S_B phase; this corresponds to a decrease of correlation between layers at the S_E to S_B transition. However this is not a characteristic of S_B phases, since for other S_B phases with the molecules normal to the layers, the ring 111 (201) is visible. Similarly, the cell of the S_{VI} phase which is monoclinic becomes orthorhombic in the S_E phase of PBAPC (the c axis, parallel to the molecular long axes, is perpendicular to a and b) and the extinction conditions are due to the presence of two perpendicular glide mirrors. This arrangement is shown in figure 13.

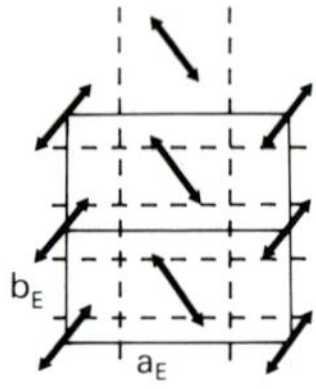

Fig. 13. *Lateral order in a S_E phase.*

However, the conclusions concerning the molecular disorder are the same for PBAPC and TBBA. The S_B phase is characterized by rotational disorder of the molecules around their long axes in such a way that on average, all the molecular positions are equivalent and that locally the lateral molecular arrangement is a herring bone packing with three orientations of the local domains. In the S_E phase, the herring bone packing extends over long distances; the presence of two perpendicular glide mirrors implies a disorder involving a two-fold axis of reorientation through an angle π. From neutron experiments, this disorder seems to be dynamic [15].

The study of the ordered phases of TBBA and PBAPC has provided two examples of S_B phases differing mainly in the tilt angle of the molecules. In the two phases, the order within the layers is hexagonal or pseudo-hexagonal and this corresponds to reorientational motions of the molecules about their long axes. In order to distinguish between the tilted and the non-tilted S_B phases, we shall name them SB_C and SB_A respectively; but we note that this distinction is purely structural and is not based on miscibility criteria. In the same way, the S_{VI} phase of TBBA and the S_E phase of PBAPC have a similar order within the layers; the difference between the two phases lies only in the tilt angle. Thus, as for the S_B type, we can distinguish SE_A (type PBAPC) from SE_C (type TBBA).

SB_A, SB_C and SE_A phases are exhibited by many compounds; in contrast the only example known to date of SE_C is the S_{VI} phase of TBBA.

Study of Smectic F and Smectic G Phases

In the literature on liquid crystals, it is common to find references to smectic phases of the type F, G and H. In this section, we will discuss the nature of these phases and some of the classification problems which are, at the present, rather confusing.

Sackmann *et al.* [16] have designated two of the smectic phases of 2-(4-n-pentylphenyl)-5-(4-n-pentyloxyphenyl)pyrimidine (PPP):

$$C_5H_{11}O-\langle\bigcirc\rangle-\langle\bigcirc_{N}^{N}\rangle-\langle\bigcirc\rangle-C_5H_{11}$$

by new symbols (F and G) because the phases are not miscible in all proportions with other smectic phases (A, B, C, D, E). The phase sequence for PPP is

$$\text{Crystal} \underset{}{\overset{79^\circ C}{\rightleftharpoons}} S_G \overset{102.7^\circ C}{\rightleftharpoons} S_F \overset{113.8^\circ C}{\rightleftharpoons} S_C \overset{144^\circ C}{\rightleftharpoons} S_A \overset{210^\circ C}{\rightleftharpoons} I$$

The S_F phase

At small angles the X-ray patterns of the S_F phase display a sharp Bragg reflection from the smectic layer; the layer thickness indicates that the molecules are tilted. At large diffraction angles, the only feature is a diffuse ring which is about half as broad as in the S_A or S_C phases as we can see in Figure 14. This means that the order inside the layers is neither liquid-like as in the S_C phases, nor hexagonal as in S_B phases. The order is probably intermediate in character between these two phases, i.e. it is locally hexagonal, but with no long-range

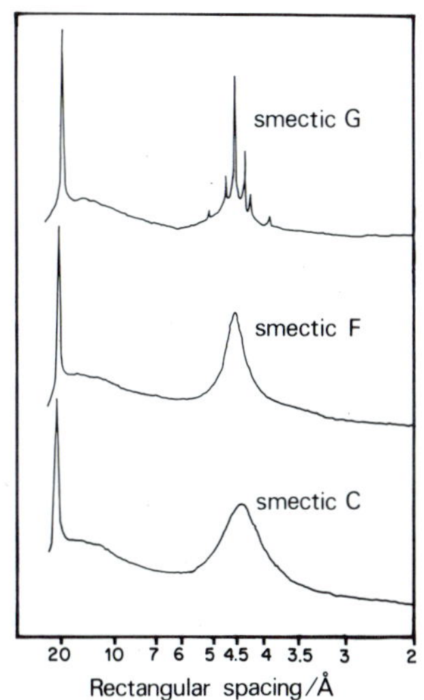

Fig. 14 *Debye-Scherrer patterns of the S_G, S_F and S_C phases of PPP ($\lambda = CoK\alpha_1$) analysed using a microdensitometer.*

order; if this hypothesis is correct, the order inside the layers could be described as a paracrystalline order.

Further X-ray experiments are necessary to determine the exact structure of the S_F phase.

The S_G phase

The powder pattern of the S_G phase of PPP can be interpreted in exactly the same way as that for the SB_C phase of TBBA. More generally, from X-ray investigations, the S_G phases appear the same as SB_C phases; in particular, the structure of the ordered smectic phase of 4-n-butyloxybenzylidene-4'-n-butylaniline, which has recently been identified as a S_G phase [17], is analogous to the structure of the SB_C phase of TBBA. This phase of the Schiff's base had previously been studied and labelled as S_H by de Vries [18].

Miscibility phase identifications and X-ray structures are not therefore in agreement as regards the S_G and SB_C phases. These two phases appear with the same structure from X-ray investigations. The problem is to understand why structures which appear as SB_C from X-ray photographs are not always miscible, and why SB_A and SB_C phases are sometimes miscible [19] whereas in some cases they are not. Moreover some compounds exhibit, with increasing temperature, a SB_C phase and a SB_A phase [20,21] (in these compounds, the SB_C phase is, by miscibility, a S_G phase). A tentative explanation is that miscibility investigations are probably a good criterion when they are applied to non-ordered liquid crystals, but are insufficient to determine the nature of the highly ordered smectic phases which differ only in the extent of the order within the layers.

The structures of the various kinds of ordered smectic phases and the possible transitions observed may be summarised as:

	←	—	→		
X-ray structural studies	SE_C	SE_A	SB_C	SB_A	S_F (molecules tilted)
		←	—	→	
			←	—	→
identification by miscibility	only one phase (S_{VI} of TBBA) (S_G?)	SE	S_B or S_G	S_B	S_F
			←	→	

The phases are classified from the more ordered, SE_C, to the less ordered, S_F; the arrows indicate the transitions.

X-ray and Neutron Analysis of the Diffuse Spots Located around the Bragg Spots (hk0)

Until now we have only considered the positions of the Bragg spots which provide us with information on the geometrical parameters of the structure. However X-rays can also yield information about the dynamic parameters as we shall now see.

X-ray analysis

The study of the diffuse spots located around the Bragg spots in the equatorial reciprocal plane is important, since it gives a measure of the degree of crystallinity of the ordered smectic phases [22]. Let us recall how this parameter can be determined.

As we have seen the thermal motion of the atoms or molecules about their average positions produces a diffuse scattering volume around each Bragg spot. The intensity profile I(q) of this scattering can be represented by the function $1/q^{\alpha}$ where q is the distance to the Bragg spot. In a crystal, where thermal motions can be described in terms of phonons, we have $\alpha = 2$. However de Gennes and Sarma [23], choosing for the S_B phase a model with almost uncorrelated layers, predict a coefficient $\alpha < 2$ and this different behaviour is illustrated in figure 15. Thus the magnitude of α is a good criterion for the degree of rigidity or crystallinity.

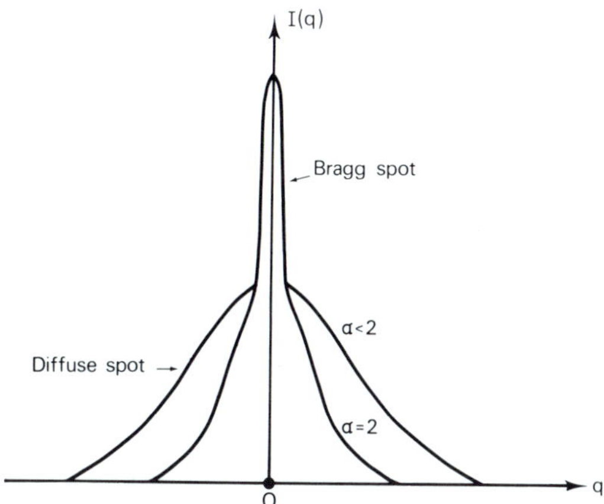

Fig. 15 *Diffuse scattering around a Bragg spot.*

The interesting result which emerges is that the behaviour of S_B phases is different depending on the compound. For instance, for the S_B phase of TBBA, we have $\alpha \simeq 2$ (near the S_B to crystal transition which means that the smectic layers are well correlated, whereas we have $\alpha \simeq 1.4$ for 4-n-butyloxybenzylidene-4'-n-octylaniline (BOBOA), which corresponds better to de Genne's model with the smectic layers gliding over each other.

It must be pointed out that the phase rigidity is related to the longitudinal order between molecules; this longitudinal order in the direction of the long axes can be measured from the sharpness of the diffuse plane. It is significant that the larger the correlation length, the more rigid is the phase and the better defined is the hexagonal order within the layers. The longitudinal correlation seems to be a major factor in relation to the structure of ordered smectic phases. The longitudinal correlation itself seems to be related to the molecular structure, and more precisely, as observed with TBBA and BOBOA, the shorter the aliphatic tails, the larger is the correlation.

In this case, we see how X-ray patterns can provide very valuable information about the relations between smectic structure and molecular structure.

Neutron analysis

Let us now examine the same diffuse spots obtained by coherent neutron scattering. In contrast to what occurs with X-rays, it is possible, with neutron scattering, to perform a frequency analysis of the diffuse spots by measuring the energy shift of the scattered beams because as we have seen thermal neutron frequencies are of the same order of magnitude as lattice vibrations. Thus, it is possible to determine dispersion curves $\omega(q)$ and from these to establish a dynamic model for the molecular and lattice vibrations.

It is important to note that here we deal with coherent inelastic neutron scattering which is sensitive to collective modes and not with incoherent quasielastic neutron scattering which is sensitive to noncollective molecular motions. The experimental arrangement needed to observe the phonons is a three-axis spectrometer shown in figure 16. Its mode of operation is identical to that of an X-ray diffractometer with the additional feature that an analyser crystal is located between the sample and the counter and selects the wavelength of the scattered neutrons that are directed towards the counter. At each point of the diffuse volume around a Bragg spot, it is possible to measure the frequency shift resulting from the phonon-neutron interaction.

Only one such experiment has been performed up to now [24]; this was carried out on the S_B, S_{VI} and crystalline phases of TBBA. Figure 17 shows an example of the phonons observed in these three phases. We can see that phonons similar to those found in the crystalline phase, can also propagate in the ordered smectic phases. Longitudinal and also

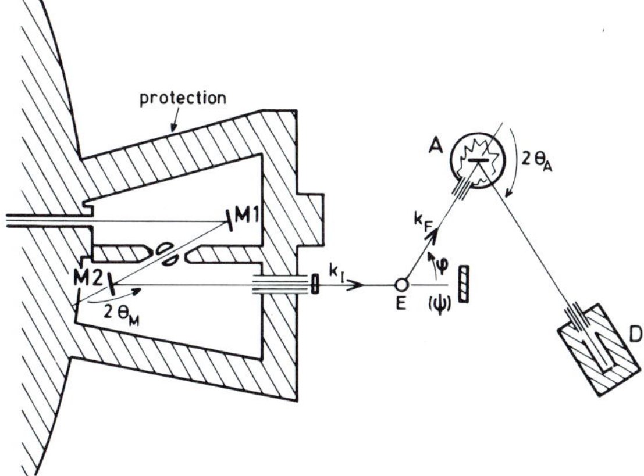

Fig. 16. *Schematic representation of a three-axis spectrometer (IN2 at ILL, Grenoble). 1st axis-monochromator, $M(2\theta_M)$; 2nd axis-sample, $E(\psi)$; 3rd axis-analyser, $A(2\theta_A)$.*

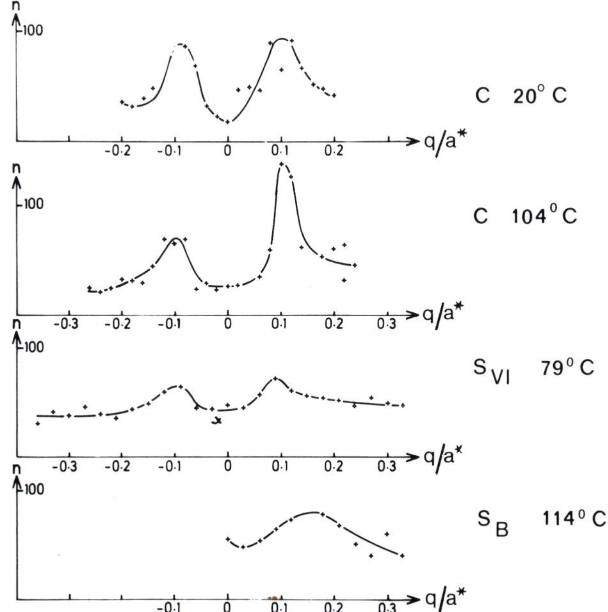

Fig. 17. *Examples of phonons observed in TBBA around the Bragg spot (20C). These correspond to longitudinal acoustic modes measured at constant energy (E=0.15 THz)*

transverse acoustic modes are observed in S_B and S_{VI}; the existence of the transverse phonons is indicative of a three-dimensional order.

Examples of dispersion curves are given in figure 18. We shall not discuss these in detail but simply note that the slope, at the origin of a curve, is equal to the velocity of sound for the mode studied. The velocities of sound measured for TBBA have values analogous (500 $m^{-1}s$ to 1500 $m^{-1}s$) to those of lead; this gives some idea of the rigidity of ordered smectic phases. We also note that the velocity of sound is related to the elastic coefficients and so some of these coefficients can be determined.

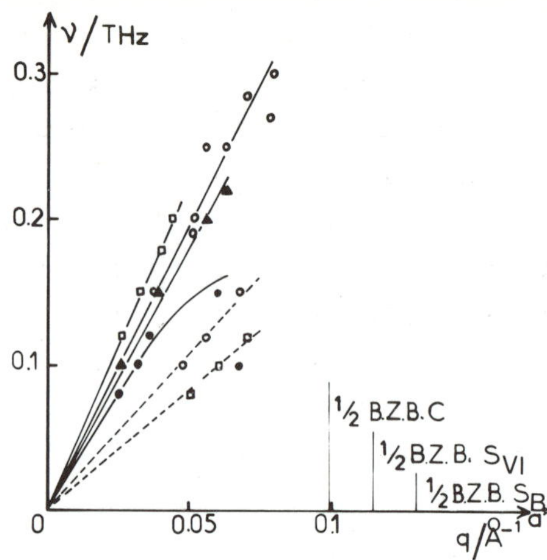

Fig. 18 *Examples of dispersion curves observed for TBBA. The solid line denotes the longitudinal acoustic modes while the transverse are indicated by the dashed lines; crystal at 20°C (○), crystal at 104° (▲), smectic VI (□), smectic B (●).*

Through this example, we have shown what information can be obtained from coherent neutron scattering. It is clear that many dynamic parameters, which are inaccessible by X-ray scattering, can be determined by this method. Unfortunately, contrary to incoherent neutron scattering which requires only powder samples, this method needs large single domains (about 0.5 cm^3) and this is one of the reasons why coherent inelastic neutron scattering is not extensively used. A great deal of very interesting work can, however, be done in this area.

Diffraction by Mesogenic Crystals

In this last section, we shall examine what information, related to mesomorphic behaviour, can be obtained from X-ray diffraction studies on mesogenic crystals.

X-ray diffraction experiments performed on crystals allow us to determine the structure at the atomic level whereas in ordered smectic phases only information at the molecular level is available. The basic data obtained from X-ray diffraction investigations of crystals are measurements of the intensities of the Bragg spots $F^2(hk\ell)$ at a great number of reciprocal nodes $hk\ell$; about 2000 to 5000 intensity measurements are necessary. We have seen that the electron density in crystals is the Fourier transform of $F(hk\ell)$; the major problem in the determination of crystal structures is the phase determination of the complex $F(hk\ell)$, but it is not our purpose here to explain what methods are used for its solution.

Only a few structures of mesogenic compounds have been determined and so it is difficult to establish rules concerning the relations between solid phases and the mesomorphic phases which appear on melting. Nevertheless, it seems that smectogenic crystals already have a structure characterized by layers [25, 26] and that in nematogenic crystals, all the molecular long axes are often approximately parallel [27]. It can thus be assumed that the transition to the mesophase involves mainly the lateral order between molecules. A pretransitional phenomenon involving the lateral order between molecules has been observed in p-azoxyanisole [28] by neutron investigations. In a few cases, a lateral disorder already exists in the crystal.

The structure of the crystalline phase VIII of TBBA at room temperature [27] allows us to propose a hypothesis about the nature of the longitudinal correlation along the molecular strings which are also found in phase VIII. From the projections of this structure, shown in figure 19,

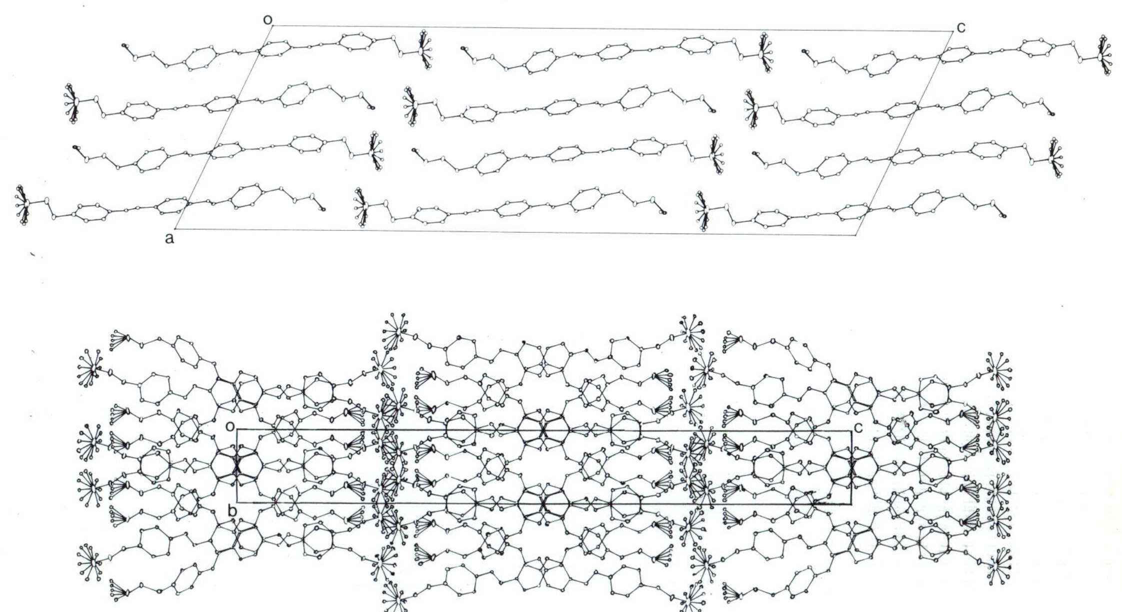

Fig. 19 *Projections (ac) and (bc) of the structure for the crystalline phase VIII of TBBA.*

it can be seen that the two atoms located at the extremities of each butyl chain are highly disordered, one being nearly delocalized on a complete circle. Calculations of interatomic distances between these disordered atoms belonging to neighbouring molecules in the same string prove that their displacements are correlated. These atomic displacements probably involve slight longitudinal correlated displacements of a molecule along a string. Thus the longitudinal correlation would proceed from correlated atomic motions of the extremities of the butyl chains, but this result needs to be confirmed by other precise structure determinations and analyses of disorder for mesogenic crystals. It is worth noting that linear correlations along the molecular long axes are a rather general feature of liquid crystal phases and sometimes occur in crystalline phases (e.g. TBBA and PAA). This is an unexpected result which has been demonstrated for liquid crystal phases by X-ray investigations and for which an explanation has been provided by the study of a mesogenic crystal. We can see therefore that the study of mesogenic crystals is of direct interest for an understanding of the mesomorphic state.

Conclusion

We have tried in this Chapter to describe the information that is available from the X-ray diffraction patterns obtained from ordered liquid crystals. It is quite obvious that X-ray diffraction is a very relevant technique for structural studies of liquid crystals. Moreover, X-ray techniques have a very practical use in relation to the classification of smectic phases; in this connection, the experiments should be performed on monodomains. The phase classification obtained from X-rays is not quite in accord with the usual miscibility classification, especially as regards ordered smectic phases; but perhaps this is not too surprising since one technique is based on structural parameters and the other on thermodynamic parameters.

In addition we have seen that X-ray diffraction is an excellent technique for the study of order in ordered smectic phases, of the degree of crystallinity and the linear correlation of molecules in mesophases. Much work remains to be done on this topic.

However it must be remembered that X-ray frequencies are about 10^7 times larger than the usual molecular vibrational frequencies. Therefore X-ray diffraction cannot differentiate between static and dynamic phenomena; for this reason interesting results can be obtained by correlating X-ray data with other techniques, which are sensitive to dynamic phenomena, such as NMR, EPR, Raman, etc. However the most fruitful interaction is that with neutron scattering which permits us to construct a dynamic model of the mesophase.

Bibliography

A.Guinier, *X-ray diffraction*, Freeman and Company (1963).

M.M. Woolfson, *An introduction to X-ray crystallography*, Cambridge University Press, (1970).

M. van Meerssche, and J. Feneau-Dupont, *Introduction à la cristallographie et à la chimie structurale, Oyez* (1976).

References

1. J.S. van der Lingen, *Ber. Deutsch. Phys. Ges,* **15**, 913 (1913).
2. M. de Broglie and E. Friedel, *C.R. Acad. Sci.* **176**, 475 (1923).
3. K. Hermann, and A.H. Krummacher, *Z. Krist,* **81**, 317 (1932).
4. H. Sackmann and D. Demus, *Fors. der. Chimie Forschung,* **12**, 317 (1974).
5. J. Doucet, A.M. Levelut, and M. Lambert, *Mol. Cryst. Liq. Cryst.* **24**, 317 (1974).
6. A.M. Levelut, and M. Lambert, *C.R. Acad. Sci.,* **272B**, 1018 (1971).
7. A.J. Leadbetter, R.M. Richardson, and C.N. Colling, *J. Phys*. (Paris), **36**, C-1-37 (1975).
8. A.M. Levelut, *J. Phys*. (Paris), **37**, C-3-51 (1976).
9. J. Charvolin, and B. Deloche, *J. Phys*. (Paris), **37**, C-3-69 (1976).
10. H. Hervet, F. Volino, A.J. Dianoux and R.E. Lechner, *J. Phys. Lett.,* **35**, L 151 (1974).
11. R. Hosemann, K. Lemon and W. Wilke, *Mol. Cryst.* **1**, 333 (1967).
12. J. Doucet, A.M. Levelut and M. Lambert, *Phys. Rev. Lett.* **32**, (6), 301 (1974).
13. F. Volino, A.J. Dianoux and H. Hervet, *J. Phys*. (Paris), **37**, C3-55 (1976).
14. J. Doucet, A.M. Levelut, M. Lambert, L. Liebert and L. Strzelecki, *J. Phys*. (Paris), **36**, C1-13 (1975).
15. A.J. Leadbetter, R.M. Richardson and C.J. Carlile, *J. Phys.* (Paris), **37**, C3-65 (1976).
16. D. Demus, S. Diele, M. Klapperstuck, V. Link and H. Zaschke, *Mol. Cryst. Liq. Cryst.* **15**, 161 (1971).
17. L. Richter, D. Demus and H. Sackmann, *J. Phys.* (Paris), **37**, C3-41 (1976).
18. A. de Vries and D.L. Fishel, *Mol. Cryst. Liq. Cryst.* **16**, 311 (1972).
19. W. Urbach, Thése Université Paris VI (1973)
20. J. Doucet and A.M. Levelut, *J. Phys.* (Paris), **38**, 1161 (1977).
21. W.M. de Jeu and J.A. de Poorter, *Phys. Letters,* **61A**, No.2, 114 (1977).
22. A.M. Levelut, J. Doucet and M. Lambert, *J. Phys.* (Paris), **35**, 773 (1974).
23. P.G. de Gennes and G. Sarma, *Phys. Lett.* **38A**, 219 (1972).
24. P. Porquet, Thése Université Paris-Sud (1977).
25. W.R. Krigbaum, and T. Taga, *Mol. Cryst. Liq. Cryst.* **28**, 85 (1974).
26. J. Doucet, J.P. Mornon, R. Chevalier and A. Lifchitz, *Act. Cryst.,* **B33**, 1701 (1977).
    J. Doucet, A.M. Levelut and M. Lanbert, *Act. Cryst.,* **B33**, 1710 (1977).
27. D.P. Lesser, A. de Vries, J.W. Reed and G.H. Brown, *Act. Cryst.* **B31**, 653 (1975).
28. R. Pynn, and T. Riste, NATO Institute, *Anharmonic lattice, structural transitions and melting,* 361 (1974).

Chapter 15

NUCLEAR MAGNETIC RESONANCE STUDIES OF MOLECULAR BEHAVIOUR

J. CHARVOLIN and B. DELOCHE

Laboratoire de Physique des Solides
(Laboratoire associé au CNRS)
Université Paris-Sud, 91405 Orsay, France.

Introduction

A common feature of mesogenic molecules is their elongated shape. This explains qualitatively the orientational order of thermotropic mesophases. As a matter of fact studies of static properties, such as elasticity, the dynamical properties, particularly those of the collective fluctuations of order, have been interpreted using models in which the molecule is likened to a rigid cylindrical rod. However, the actual molecule is much more complicated; it can be divided approximately in two parts, an elongated aromatic core with alkyl or alkoxy chains attached at both ends. The different chemical affinities of those two parts, the possibility of their deformations by isomeric rotations around some bonds might play a role in the thermodynamics of the mesophases as suggested by theoretical studies of smectic A[1] and nematic [2,3] phases. Moreover, going further into the details of the chemical structures, some molecules contain specific groups which might also intervene, as shown by systematic studies of phase diagrams in relation to the structures [4,5] and suggested by models for the biaxial smectic C and H phases [6-8]. A first step towards the understanding of the relation between the molecule and the macroscopic structure of the mesophase consists in collecting information about the local behaviour of the molecule in the ordering potential. Besides the analysis of the orientational order of the long molecular axis, the following questions arise: does the molecule keep the low symmetry of its ground state conformation and does the symmetry of the "cage" of molecules surrounding it correspond to that of the bulk phase?

Several techniques sensitive to local fluctuations have been applied to answer such questions. They are neutron scattering (cf. Chapter 13) electron paramagnetic resonance (EPR) (cf. Chapter 16), nuclear magnetic resonance (NMR), infra-red and Raman spectroscopies (cf. Chapter 17).

Each of them provides specific information, coarsely, gyration radii for neutron experiments, populations of conformational states for infra-red and Raman spectroscopies, orientations of molecular tensors or vectors in external fields for Raman spectroscopy, EPR and NMR. This Chapter will try to provide an account of those NMR studies which we find representative and it will be limited to those undertaken to study the local molecular behaviour. Thus the investigations of the dynamics of the collective orientational modes are not considered here.

Relevant Aspects of Magnetic Resonance

The magnetic fields used in NMR experiments are approximately 1T. In such fields the sample is macroscopically oriented with the director parallel on average to the magnetic field. As we shall see, NMR is sensitive to the orientation of some molecular tensor with respect to the field and hence to the director. The protons, naturally present in the molecule, are the more direct NMR probes to use but, in some cases, the lack of specificity of the method has led to the development of more sophisticated investigations through the use of other nuclei.

Proton NMR

The value of this type of NMR experiment can be illustrated by first considering a simple representation of the molecule.

The rod molecule The molecule is idealised as a rigid rod bearing only two protons H_1 and H_2 with a separation d.

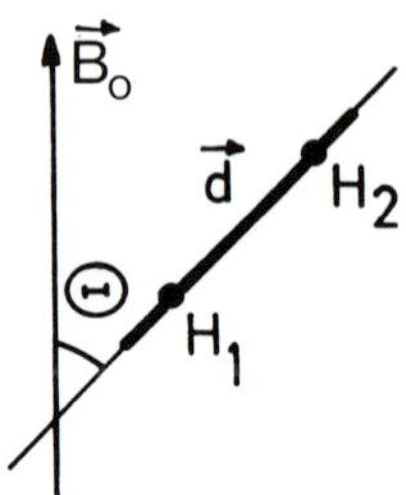

In an external magnetic field B_o the energy levels of the proton spins, I_1 and I_2 are determined by their total spin hamiltonian (cf. [9] Chapters III and IV),

$$H = H_Z + H_D .$$

The Zeeman hamiltonian,

$$H_Z = -\gamma_H \hbar B_o . (I_1 + I_2) ,$$

represents the coupling of the magnetic moment of each proton with the external field and the dipolar hamiltonian,

$$H_D = I_1.D.I_2,$$

represents the spin-spin dipolar coupling between the nuclear magnetic moments. The coupling tensor D depends on the length, d, and its orientation relative to B_o. In general the matrix elements of H_D are less than those of H_Z and the system of two spins 1/2 has a structure of energy levels equivalent to that for a system of a single nucleus with spin 1. At thermal equilibrium the nuclear spins occupy those levels according to the Maxwell-Boltzmann law and a nuclear paramagnetism ensues. Irradiation with an electromagnetic field of appropriate frequency induces transitions between the levels and a NMR signal can be observed. The energy levels, transition frequencies and lineshapes observed under irradiation with a radiofrequency field, in the absence and presence of H_D, are (cf. [9] Chapter VII):

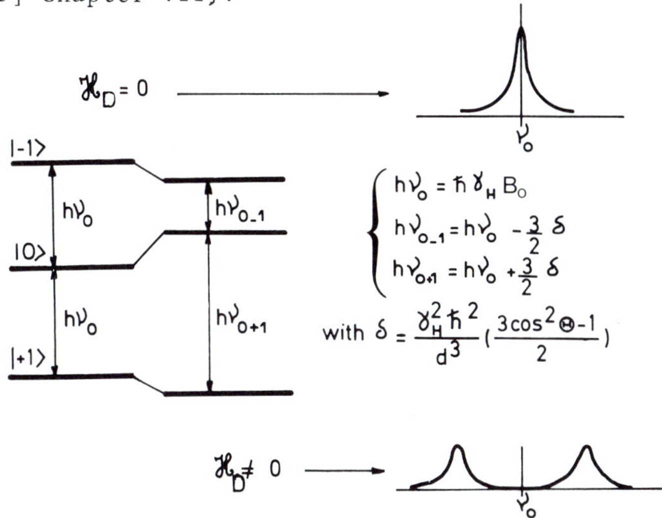

where γ_H is the proton gyromagnetic ratio and $\nu_o = \gamma_H B_o/2\pi$ the proton Larmor frequency (ν_o is 42.5 MHz for a field of 1T). In the presence of H_D the splitting of the lines is

$$\Delta\nu_D = (3\gamma_H^2\hbar/2\pi d^3)P_2(\cos\theta),$$

where $P_2(\cos\theta)$ is the second Legendre polynomial and the ratio $\gamma_H^2\hbar/d^3$ is typically 5kHz. This is the splitting measured for a static rod, but in the liquid crystal the orientation of the rod fluctuates. If this motion is rapid enough, i.e. if its correlation time, τ, is such that $2\pi\Delta\nu_D\tau<1$ (cf. [9] Chapter 16), it averages the splittings. The measured splitting is reduced to zero in the case of isotropic reorientation, to something between zero and the static value in the case of anisotropic reorientation. The residual dipolar splitting is therefore a measure of the degree of orientational disorder and, for the rod in the liquid crystal, the observed splitting can be written

$$\Delta\nu_D = (3\gamma_H^2\hbar/2\pi d^3)\overline{P_2(\cos\theta)},$$

$$= (3\gamma_H^2\hbar/2\pi d^3)\bar{P}_2,$$

where $\bar{P}_2$ is the second rank order parameter, also denoted by S.

Real molecules As the number of protons increases, so does the number of dipolar couplings and the complexity of the PMR spectrum. Two experiments on 4-methoxybenzylidene-4'-n-butylaniline (MBBA) illustrate this [10]. First, the natural MBBA molecule is studied in the nematic phase and the following PMR spectrum is obtained as we see in figure 1.

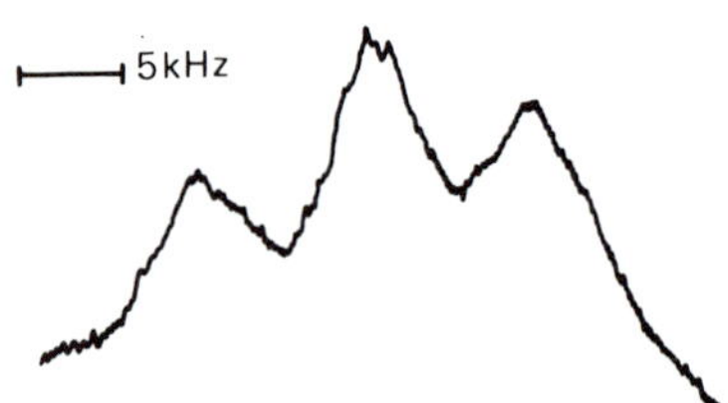

Fig. 1 *PMR spectrum of MBBA in the nematic phase at 30°C.*

The spectrum can be analyzed as consisting of a central peak plus a doublet. The central peak comes from the methoxy, butyl and methine protons while the doublet originates from those in the phenyl groups. The signals are broadened mostly by intramolecular dipolar couplings. Consequently when protons are replaced by deuterons, except those of the phenyl ring of the butylaniline, a much simpler PMR spectrum, shown in figure 2, is observed. The simplification comes from the fact that the magnetic moment of the deuteron is smaller than that of the proton, thus their NMR signals are distinct and the dipolar couplings between them are small in comparison with protons. The spectrum looks like that expected from the rod-like molecule but the situation is not quite analogous, because in the real molecule, the HH vectors are not aligned parallel to

Fig. 2 *PMR spectrum of partially deuteriated MBBA in the nematic phase at 30°C.*

the long molecular inertial axis (m). The internuclear vector for the *ortho* protons is almost parallel to (m) whereas the vector for the *meta* protons is almost perpendicular and that of the *para* protons is at an angle of about 60°. Corrections may be introduced by assuming that the molecule is effectively axially symmetric about m. It is then easy to write the average $P_2(\cos\Theta)$, through the spherical harmonic addition theorem, as

$$\overline{P_2(\cos\Theta)} = \overline{P_2(\cos\theta)}P_2(\cos\gamma)$$

and so

$$\Delta\nu_D = (3\gamma_H^2 h/2\pi d^3)\bar{P}_2 P_2(\cos\gamma),$$

where the angles are related by

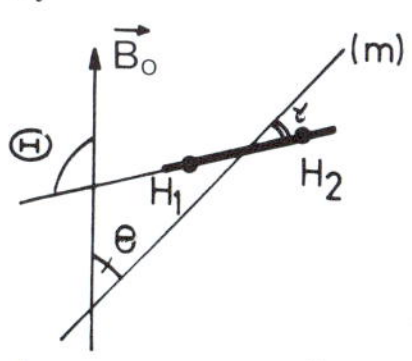

The determination of $\bar{P}_2$ depends upon a knowledge of d and γ. Good estimates of d can be obtained from molecular models but that of γ relies upon the orientation of the long axis which, although certainly close to the major inertial axis, is not known accurately. However the shape of the $P_2(\cos\gamma)$ curve is flat around 0° and 90° but steep in between; consequently an uncertainty in γ has an important effect on $\bar{P}_2$ when γ is close to 60°. It is then necessary to exercise considerable care in the choice of the magnetic interaction used to determine the order parameter and it is important to rely on the interaction between *ortho* protons because their γ is rather close to 0°.

Exact knowledge of the angles γ cannot rely on the assumption of a rigid molecular model, for two reasons. Firstly the bond angles can be distorted by the intermolecular potential [11], and secondly the molecule can change its shape by isomeric rotations around some bonds, thus moving its inertial axis with respect to its structural groups. The study of the intramolecular behaviour might provide information about the occurrence of such effects; it involves the comparison of one chemical group of the molecule with another. For this it is necessary, in a first step, to obtain information specific for each group. This is not easy with PMR studies, because of the coupling between the spins, and their interpretation in such terms requires the use of certain complex aspects of spin dynamics [12,13] which are outside the scope of this Chapter. This is not always essential because of the development of more selective magnetic resonance studies.

Carbon-13 NMR

This nucleus is present naturally in the skeleton of the molecule and, because of its low isotopic abundance ($\sim$ 1%), there is no dipolar coupling between like spins. There remains a dipolar coupling with unlike spins, such as those of the neighbouring protons, but this can be removed by the so-called ^{13}CH decoupling. Here an intense irradiation at the proton frequency equalizes the proton distribution between the energy levels so that their resulting magnetic moment, and hence their dipolar field at the ^{13}C sites vanish. Under these conditions each ^{13}C behaves as an isolated nucleus and the total hamiltonian can be written

$$H = H_Z + H_{CS},$$

where

$$H_Z = -\gamma_c \hbar B_o . I$$

and

$$H_{CS} = -\gamma_c \hbar B_o . \sigma . I,$$

which represents the coupling of the nucleus with the orbital diamagnetism of the bonding electrons around it. This term is responsible for the so-called "chemical" shift of the resonance; it exists also for other nuclei, particularly the protons, but its contribution is much smaller than for ^{13}C and negligible when compared to the PMR linewidth. The coupling tensor $\boldsymbol{\sigma}$ depends very sensitively on the electronic structure of the chemical bond so that chemically different carbons have distinct resonances. An example of the spectrum obtained with 4,4'-di-n-butyloxyazoxybenzene [14] is given in figure 3. This NMR experiment contains

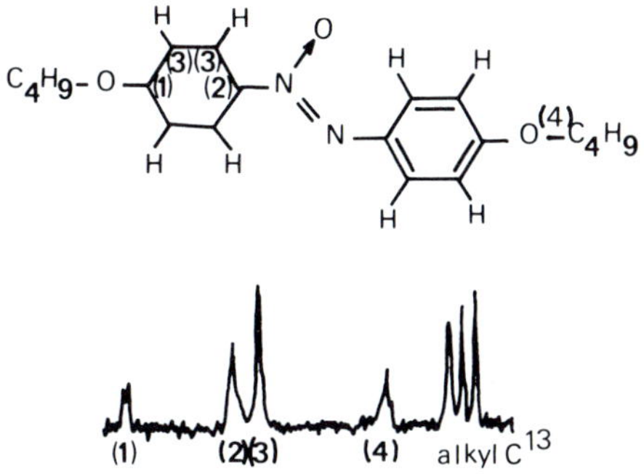

Fig.3 ^{13}C *MR spectrum, with proton decoupling, of BOAB in the nematic phase; NMR lines and* ^{13}C *are numbered accordingly.*

very specific information. In principle, it could be exploited in terms of motions and order parameters because H_{CS} is orientation dependent but, unfortunately, the static tensors in the alkyl chain are not known yet.

EPR of Spin Probes

This technique is described in detail in Chapter 16 and elsewhere [15], nevertheless a few essential aspects will be recalled here in order to prepare for a future discussion concerning the nature of the motions responsible for the measured order parameters. The hamiltonian for an unpaired electron of spin S in an external magnetic field can be written

$$H = H_Z + H_{HF},$$

where

$$H_Z = B_o . g . S$$

represents the Zeeman coupling with the field and

$$H_{HF} = S.A.I$$

is the hyperfine coupling with the closest nuclei: both contributions are orientation dependent. The static value of the latter is rather large, 10^8 Hz, and only motions with correlation times shorter than 10^{-8}s can have an averaging effect. EPR and PMR have therefore different time scales, around 10^{-8}s and 10^{-4}s respectively and they appear complementary in this respect. The unpaired electron, which is not naturally present in mesogen molecules, is introduced with a radical either dissolved in the phase [29] or attached at variable places on the molecule [16].

Quadrupolar Effects in NMR

As stated earlier, deuterons have small magnetic moments and hence negligible dipolar couplings, consequently their NMR spectra can then be a way to obtain specific information about molecular groups after isotopic substitution of the protons [17]. The deuteron NMR spectrum is dominated by the electrostatic interaction between the quadrupole of the charge distribution of the nucleus with the electric field gradients of the electronic or ionic distributions around it. This is also the case for ^{14}N which has been used in liquid crystal studies [17,18]. (The quadrupolar interaction does not exist for nuclei with spin $I < 1$; this is why it was not considered for proton and ^{13}C which have $I = 1/2$). This technique is now widely used and so we shall consider the underlying theory in some detail.

As we have already said dipolar couplings are generally weak and will be neglected in a first approximation (examples of their influence can be found in references [11, 20, 21]). The spin hamiltonian can then be approximated by

$$\mathcal{H} = \mathcal{H}_Z + \mathcal{H}_Q,$$

where
$$\mathcal{H}_Z = -\gamma_D \hbar B_0 . I$$

and the quadrupolar hamiltonian

$$\mathcal{H}_Q = I.Q.I,$$

which is orientation dependent. The coupling tensor **Q** is proportional to the electric quadrupolar moment eQ of the nucleus and to the electric field gradient tensor **V**. This is a second rank symmetric tensor with zero trace. It is diagonal in a local frame $(x_o y_o z_o)$; in the case of a CD bond the z_o axis of the **V** tensor is along the bond, the principal value $V_{z_o z_o}$ is written as eq and the symmetry around the bond is defined by the asymmetry parameter $\eta = (V_{xoxo} - V_{yoyo})/V_{zozo}$. To simplify the calculation of the energy levels **V** has to be expressed in the laboratory frame (XYZ), the Z axis of which is along B_o. This transformation of coordinates makes use of the Euler angles $(0,\theta,\alpha)$ defined as

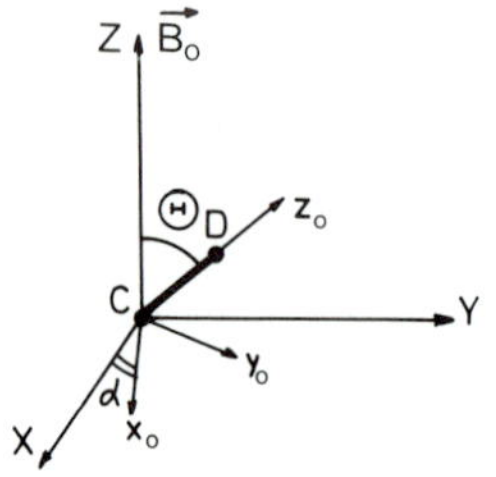

With the matrix elements of $\mathcal{H}_Q$ less than those of $\mathcal{H}_Z$ a first order calculation leads to the following structure of levels, NMR transition frequencies and lineshapes under irradiation with a radiofrequency field (cf. [9] Chapter VII, [22]):

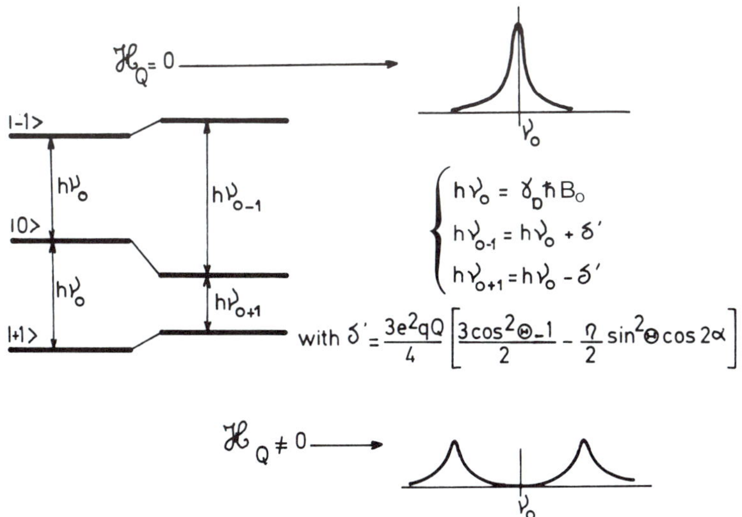

where γ_D is the deuteron gyromagnetic ratio and $\nu_o = \gamma_D B_o/2\pi$, the deuteron Larmor frequency (ν_o is 6.53 MHz for a field of 10 kG). In the presence of H_Q the splitting of the lines is

$$\Delta\nu_Q = \frac{3}{2}\frac{e^2qQ}{h}\left(\frac{3\cos^2\theta - 1}{2} + \frac{1}{2}\eta\sin^2\theta\cos 2\alpha\right).$$

Typical values of the static quadrupolar coupling constant, $e^2qQ/h = \nu_Q$, are around 10^5 Hz for deuterium and 10^6 Hz for nitrogen; the time scales of the experiments are then 10^{-5} and 10^{-6} s. If there is a rapid uniaxial motion around the long molecular axis, m, then as before the splitting can be written as

$$\Delta\nu_Q = (3e^2qQ/2h)\ \bar{P}_2 P_2(\cos\gamma).$$

The determination of $\bar{P}_2$ is not generally accurate with deuterons because ν_Q is not known precisely and because γ is usually far from 0° or 90°. This drawback is counter-balanced by the very interesting resolution of the spectra which is illustrated in figure 4 by the DMR study [23] of deuteriated 4-n-butyloxybenzylidene-4'n-octylaniline (BBOA). Each deuteriated group is perfectly distinguishable; the methylene groups of the alkyl chain are numbered from the phenyl group towards the methyl end. This spectrum will be analyzed later, in terms of the conformational states of the alkyl chain.

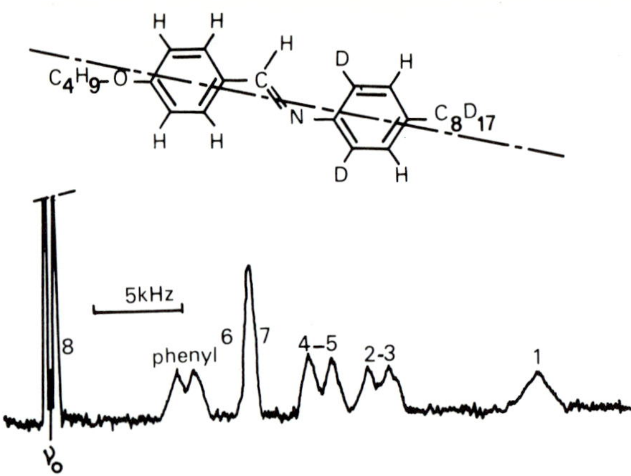

Fig. 4 *DMR spectrum of partially deuteriated BBOA in the nematic phase at 74°C (only half of the spectrum, which is symmetric around the frequency* ν_0 *is shown).*

The Ordering Matrix

The rapid reorientations of a molecule modulate the orientation of the spin interaction tensors with respect to the external magnetic field $\mathbf{B}_0$. The techniques of magnetic resonance, which study the averaged interactions, give access to the characteristics of the corresponding averaged tensors. When $\mathbf{B}_0$ is parallel to the director **n**, as is the case in most NMR and EPR studies, these average tensors can be analysed in terms of the orientational order tensor **S** describing the way the molecular axes fluctuate about **n**. In particular the order parameter value of the long molecular axis provides a test for the theories of the intermolecular potential. Moreover, from some experiments, it should be possible to extract the orientational order of the molecular axes with respect to the laboratory axes normal to the director and so to determine the symmetry of the molecular behaviour around **n**; this point has to be discussed in relation to the macroscopic symmetry of the phase.

Uniaxial phase ; nematics

In the nematic phase no departure from uniaxiality has been detected on the characteristic times of NMR and EPR experiments, *i.e.* 10^{-8}s for the shortest. This means that the long molecular axis reorients in every direction with respect to the director. In the case of the ideal experiment with the rod-like molecule the orientational order was described by the single order parameter $\bar{P}_2$, which corresponds to the component of **S** for the long axis. The other two principal elements of the ordering matrix are equivalent because of the assumed cylindrical symmetry of the molecule. This assumption was adopted even in the case of real molecules, but is not strictly valid if the molecule keeps the lath-like shape implied by its chemical formula. In such a case the molecule, rather than have a uniaxial behaviour in agreement with the symmetry of the phase, might prefer to reorient in its plane rather than out of it and the two coordinate axes normal to the long axis would have different order parameters with respect to the director [24]. The orientational order of the molecule would have to be described by a tensor with three principal values S_{xx}, S_{yy} and S_{zz} defined by

$$S_{\alpha\alpha} = \overline{P_2(\cos\Theta_\alpha)} \,.$$

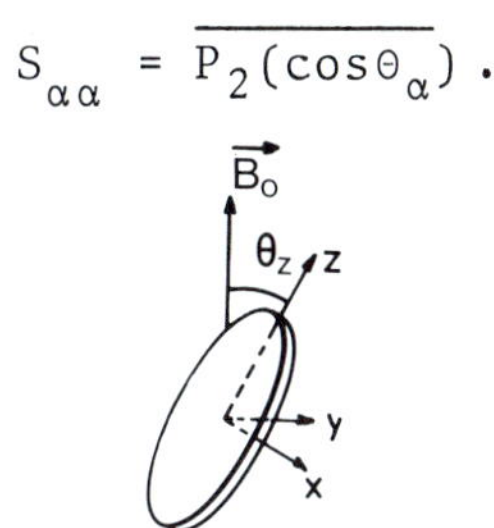

As $\sum_{\alpha\alpha} S_{\alpha\alpha}=0$ the order and its symmetry are well described by only S_{zz} and $S_{xx}-S_{yy}$. A simple way to determine these two quantities is to measure the dipolar splittings associated with HH vectors respectively parallel and perpendicular to the z axis. Considering any molecule it would appear that the vectors joining *ortho* and *meta* protons are only about 10° from the ideal geometry and, as we have seen, such a deviation should not introduce a large error. The required information is therefore contained in a spectrum such as the one shown in figure 2, but it is hidden in the large linewidth. Quite recently it has been shown that the major cause of broadening comes from the HD residual dipolar coupling [25]. Applying a HD decoupling experiment, quite analogous to that described previously for ^{13}C, to a partially deuteriated 4-cyano-4'-n-pentyl-biphenyl (5-CB) the PMR lines were narrowed in such a way that structure typical of the dipolar interactions in a four proton system appears, as we see in figure 5. From the values of the dipolar couplings the ordering matrix of one phenyl group for 5-CB in the nematic phase at 31^oC is characterized by $S_{zz} = 0.66$ and $S_{xx}-S_{yy} = 0.012$. The asymmetry parameter is so low, that the ordering matrix may be considered as cylindrical. At the moment it is not clear if this is a property of the total molecule or of only the group investigated and if this property is common to all nematogenic molecules or only to a few representatives of them. For instance a larger value of $S_{xx}-S_{yy} \sim 0.06$ for the end methyl of di-n-heptyloxyazoxybenzene (HOAB) has been deduced from another DMR study; however the basis of the analysis (the independence of the average conform-

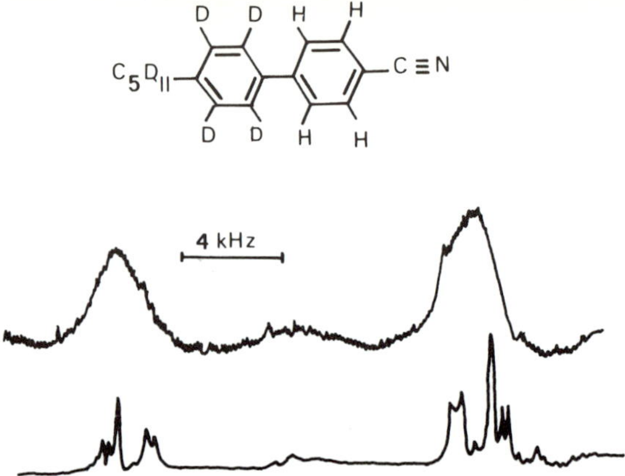

Fig. 5 *PMR spectra without and with deuteron decoupling, of partially deuteriated 5-CB in the nematic phase at 31°C.*

ational state of the molecule on temperature) is still a matter of discussion. In any case the ordering matrix of the phenyl group in 5-CB suggests that the molecule has not the low symmetry suggested by its chemical formula or has lost it prior to the orientational fluctuations. In this second case the higher symmetry observed for the phenyl group might result from a fast rotation around this axis and/or internal deformations.

The evolution of S_{zz} with temperature determined for MBBA [10] is shown in figure 6. The order parameter curves for different systems,

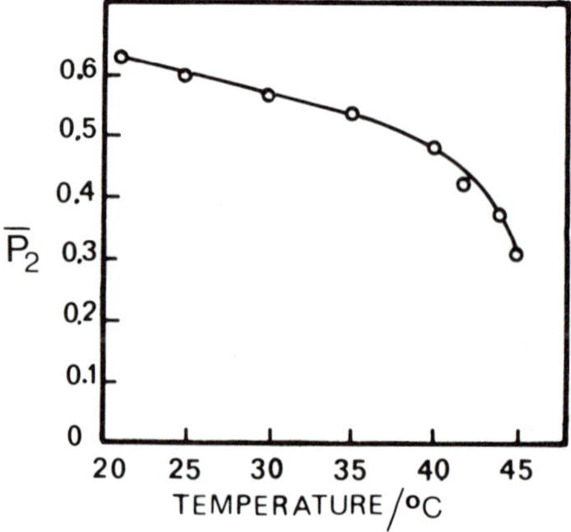

Fig. 6 *PMR determination of the order parameter curve for MBBA in the nematic phase.*

plotted against a reduced temperature scale, are rather similar. Their comparisons with the results of the Maier-Saupe mean field theory of the intermolecular potential has led to significant improvements of the latter [27,28]. The comparison of these order parameter curves with those obtained from EPR is also worthwhile. They are very close as can be seen from the curve in figure 7, obtained from a nitroxide spin probe in the nematic phase of 4-4'di-ethoxyazoxybenzene [27]. This shows first

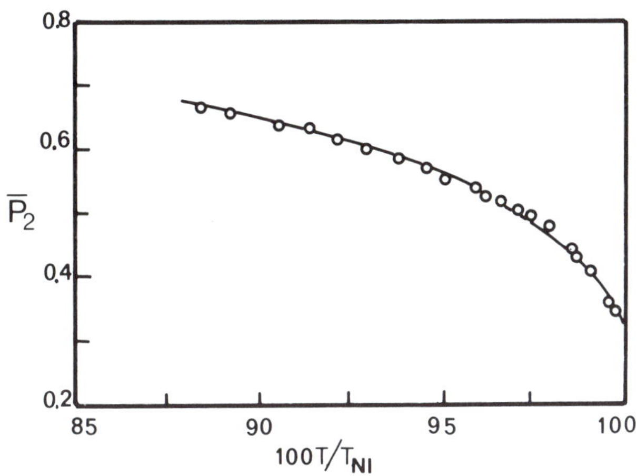

Fig. 7 *EPR determination, through the use of a nitroxide probe, of the order parameter curve in the nematic phase of PEAB. The points are compared with the extended Maier-Saupe theory.*

that, although it is a dissolved probe which is investigated and not the material itself, the EPR method appears to give faithful information about the anisotropic intermolecular potential. Second, the similarity between the results of EPR and NMR also suggests that most of the orientations responsible for the order parameter have correlation times shorter than the EPR time scale, *i.e.* 10^{-8}s. If slower motions were important they should introduce a supplementary averaging effect which should be detected by NMR due to its longer time scale so that the NMR order parameter should be smaller than that determined by EPR. In fact motions slower than 10^{-8}s do exist; they are the thermal collective modes of fluctuations, but those in that range of time have a very low amplitude which is insufficient to influence S. This is confirmed by the analysis of the small difference between the NMR and EPR results which attributes to the reorientations, with times in the range 10^{-8} to 10^{-5}s, small amplitudes whose root mean square fluctuation is of the order of a few degrees [29] whereas the observed values of S require angular deviations of a few 10°. Therefore most of the motions responsible for S are to be considered as rapid ($\tau<10^{-8}$s) large reorientations of the long axis of one molecule in the cage formed by its neighbours. Owing to their large amplitude these local reorientations are certainly comparable with the large distributions of molecular orientations determined by X-ray [30], neutrons [49] and EPR studies of frozen nematics [31].

The results discussed in this section suggest the following scheme for the local behaviour in a uniaxial nematic phase: owing to its own conformation or to rapid internal deformations and/or rotations around its long axis the molecule can be likened to an object of high symmetry, certainly not far from cylindrical and the orientation of its long axis then fluctuates in an environment of uniaxial symmetry. Both events, the motions around the axis and its reorientation might occur simultaneously or sequentially as described. A clue in favour of the second possibility will be presented in a later section. In any case the effect of these motions is to determine an average molecule which is the object to be considered on longer times at the hydrodynamic scale, *i.e.* in the collective phenomena whose wavelengths are larger than the dimensions of the volume occupied by one molecule and its neighbours.

Biaxial phases

The determination of the local symmetry can be achieved, for example, through the measurement of the asymmetry parameter η of a molecular quadrupolar tensor averaged by the rapid motions of the molecule. The expression for the quadrupolar splitting, given previously, is also valid for a partially averaged tensor; it contains η which can be obtained by varying the angle Θ, *i.e.* the orientation of a monodomain sample, in the external magnetic field. No systematic investigation of this type has been made yet but the expected effect might be comparable to the linewidth and hardly detectable [32]. In fact none of the attempts presented so far is really convincing. NMR lacks the necessary sensitivity because, schematically, the effect of η is not the principal term in the total spin hamiltonian; it is a correction to H_Q which is itself a perturbation on H_Z. An original solution would be to work in conditions such that H_Q is the principal term, *i.e.* to perform a pure nuclear quadrupole resonance (NQR) experiment. Such an experiment has been developed recently [33]. Some nuclei have large quadrupole moments eQ and their couplings with the electric field gradients are strong enough to split the energy levels considerably so that resonances can be observed in the radiofrequency range, without the application of an external magnetic field. This is the case for ^{14}N but not for deuterons. The energy levels, transition frequencies and lineshapes for a spin I = 1 are [34]:

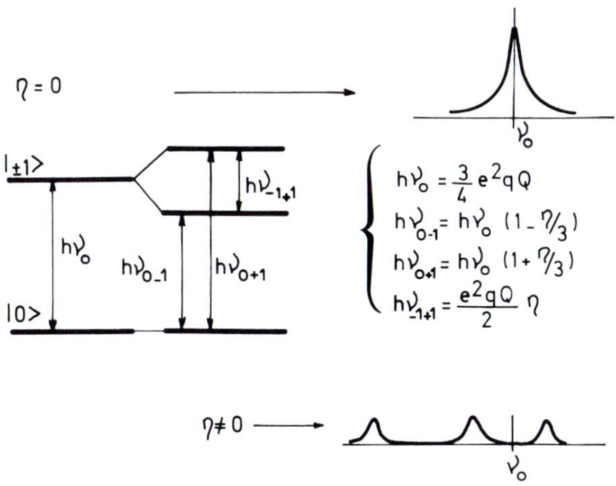

The values of $e^2q\,Q/h$ and η of the average tensor in different phases of terephthalylidene-bis-4-butylaniline, TBBA, are listed in the table.

Table *The ^{14}N quadrupole coupling constant and asymmetry parameter in TBBA*

Phase Temperature/^oC	Solid 20	S_{VI} 91	S_H 134	S_C 150	S_A 185
$\frac{e^2qQ}{h}$ /kHz	4220	1170	1020	1010	810
η	0.26	0.7	0.24	0.08	0

The measured values of η *a posteriori* explain why the asymmetry was detected with certainty, by an ordinary rotation pattern DMR experiment, only in the smectic VI phase which is very close to the solid [21].

A correspondence then appears between the microscopic and macroscopic symmetries. The averaged tensor is uniaxial in the smectic A, but not in the smectic C, H, VI phases. The weakness of the biaxiality in the smectic C phase implies that it is the tilt which induces the local biaxiality and not the opposite, as assumed previously [6]. For the moment there is still the choice between two possible local descriptions; the molecule may, or may not, be likened to a cylindrical object moving in a biaxial medium. Thus, a biased rotation around the long axis has been proposed [33] to interpret the results. As neutron quasielastic scattering experiments have failed to detect any biasing in the smectic C phase [35] another explanation might be thought of. For example, the anisotropy of the orientational fluctuations in the smectic C, easy only at constant tilt angle, would lead qualitatively to a similar result.

The Molecular Geometry

The NMR study of terephthalylidene-bis-4-n-butylaniline is a good example because of the complexity of the molecule, the possibility of varied deformations, and its rich polymorphism which has been the subject of detailed structural investigations [36,37].

C_4H_9–C_6H_4–N=CH–C_6H_4–CH=N–C_6H_4–C_4H_9 (m)

Cr(VIII) $\xrightarrow{113^\circ C}$ S_H(V) $\rightleftharpoons^{142^\circ}$ S_C $\rightleftharpoons^{172^\circ}$ S_A $\rightleftharpoons^{195^\circ}$ Nem $\rightleftharpoons^{236^\circ}$ Iso

S_H(V) $\xrightarrow{84^\circ}$ S_{VI} $\xrightarrow{68^\circ}$ S_{VII} $\xrightarrow{52^\circ}$ Cr(VIII)

The alkyl chain [21]

The DMR spectrum of a molecule with deuteriated butyl chains is well resolved as we can see in figure 8. As with the DMR spectrum of deuteriated BBOA shown in figure 4 each methylene and methyl group is perfectly

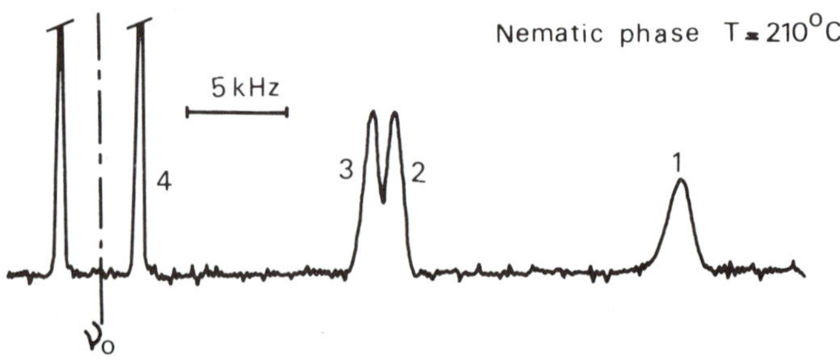

Fig. 8 *DMR spectrum of TBBA with deuteriated alkyl chains, in the nematic phase (only half the spectrum, which is symmetric around the frequency ν_o, is shown).*

distinguishable; the proposed numbering goes from the phenyl group towards the methyl end. The spectra, from either BBOA or TBBA can be analysed in the same way. As e^2qQ/h is constant along the chain (cf. note 1 in [21]) the chain deuterons differ by their value of $(3\cos^2\gamma-1)$ as we saw earlier. As there are as many lines as CD_2 the chain first admits a symmetry plane and the molecular axis lies in this plane (if not the deuterons of the same CD_2 would be distinct). Secondly the chain is not rigid and is deformed by isomeric rotations around CC bonds most probably (if not, there would be only one signal for all the CD_2

because they would be equivalent with respect to the molecular axis; this case has been found in a lyotropic liquid crystal [38]). This shows that the local molecular motions are not only reorientations of the long axis or around it but also deformations of the molecule. The chain links exchange between trans and gauche conformations as shown by Raman spectroscopy experiments [39]. The angle γ is difficult to estimate and in addition the treatment of $P_2(\cos\gamma)$ in $\Delta\nu$ could differ according to the relative rates of the motions; for example $P_2(\cos\bar{\gamma})$ should be considered if the deformations are the fastest and $\overline{P_2(\cos\gamma)}$ if they are the slowest. In any case it is plausible to assume that the motional averaging is more and more effective going from the phenyl group towards the methyl end, hence the assignment of the spectral lines in figures 4 and 8. The evolution of the splittings along the chains in the different phases of TBBA is shown in figure 9, with the first methylene as reference. These

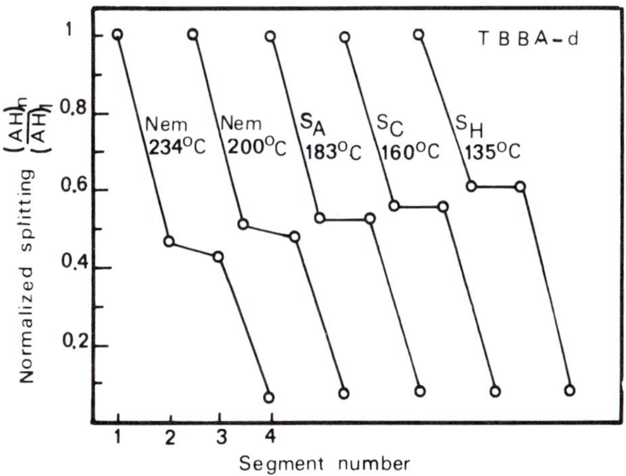

Fig. 9 *Evolution of the normalized splittings of the methylene groups for the deuteriated chains of TBBA in different mesophases (the horizontal scales relative to each curve have been shifted by constant amounts with respect to each other).*

curves are representative of the order of the different links of the chain with respect to the first one. The methylene groups are paired (as observed also on figure 4 (2-3, 4-5, 6-7)); this may be an even-odd effect, the mean orientation of the corresponding CC bonds being preferentially parallel to the long molecular axis. On the other hand, the order decreases strongly for the methyl end group and this may be due to the fact that its splitting depends on the orientation of the last CC bond, the CD_3 symmetry axis, which is at an angle with the long axis not far from the so-called magic angle of 54°44' which annuls $P_2(\cos\phi)$. The latter decrease of order can lead to complete cancellation for longer chains with an even number of links [23,26], but is much less pronounced for chains with an odd number of links [23,40], thus confirming the previous interpretation for the low order for the CH_3 group.

The decrease of the order along the chain varies slightly with the temperature. The general trend is an increase of order from nematic to smectic H and a levelling of the order of the two intermediate CH_2 groups. Similar deformations of the order curves with temperature have been observed also for longer chains (OC_8 and C_8). This shows that the average conformation of the chain changes with temperature. Unfortunately, this is a weak effect and the chains appear hardly more mobile in the nematic than in the smectic H phase. This may be understood considering first that the free space available to the chain does not vary very much throughout the mesophases. (The mean area per chain [37] varies by 2 $Å^2$ while it is known, from DMR studies of lyotropic mesophases [41], that this area must vary by about 10 $Å^2$ to observe appreciable effects). The second point, suggested by the behaviour of the CD_3 in BBOA [23], might be that DMR is not very sensitive for the study of CD_2 order because their orientation relative to the long axis does not vary in the steepest region of the function $P_2(\cos\phi)$.

In the crystalline phase the spectrum is no longer resolved; the lines are broadened by the slowing down of the motions. This indicates that most of the chain disordering takes place at the melting point (see also [16]).

The aromatic core [42]

The DMR spectra of the central or lateral phenyl groups have similar splittings while that of the methine group is about twice as large [20,21]. This difference is illustrated in figure 10. Due to uncer-

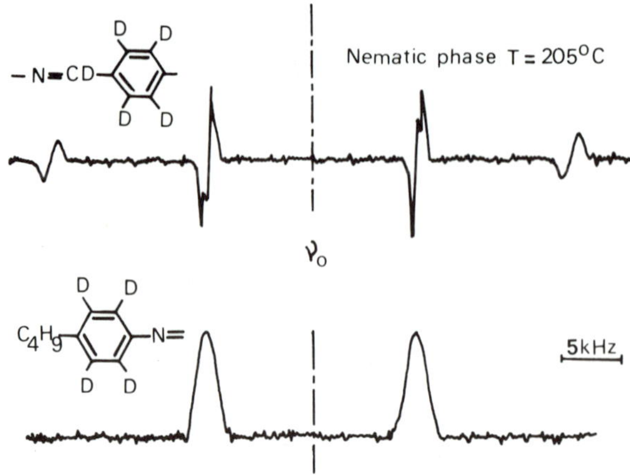

Fig. 10 *DMR spectra of the phenyl rings of TBBA in the nematic phase. The upper spectrum is a derivative.*

tainties in our knowledge of bond angles and static quadrupolar coupling constants we cannot rely too heavily upon a direct exploitation of the DMR results to decide about the relative orientations of the phenyl rings and the methine groups [11]. In any case, the geometry of the core is not rigid but changes with the temperature. This conclusion results from considering the ratio of the methine splitting to that of the phenyl rings shown in figure 11. (The splitting of a group depends on the geometry of this group with respect to the long axis and on the reorientations of the latter. The ratio, R, of the splittings for the two groups contains only information about the internal geometry provided the ordering matrix is cylindrical). If the molecule was rigid, its geometry

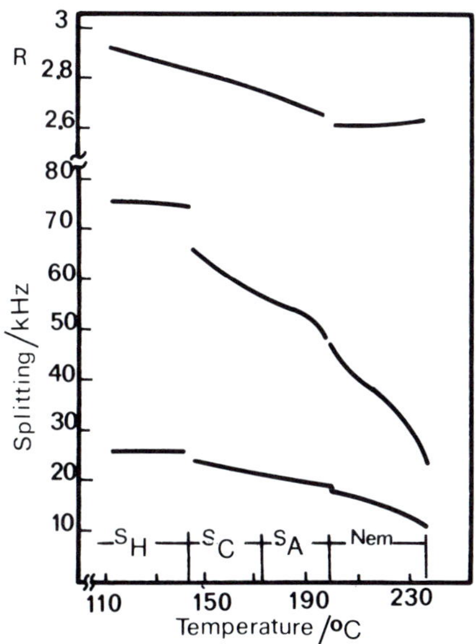

Fig. 11 *Quadrupolar splittings of the aromatic core of TBBA as a function of the temperature [20]. The lower curve applies to the central ring deuterons, the middle one to the methine deuterons, and the top curve plots their ratio R.*

around the long axis would be fixed and the ratio R would be constant to a first approximation. In fact it is not constant and so the average geometry of the molecule around the long axis must change. The chain deformations might intervene modifying the mass distribution of the molecule but the DMR signals of the molecule without the chain (TBMA) behave in the same way [43]. This evolution of R has therefore its origin in the core. A plausible model relies on core deformations through rapid exchanges between several conformations the extremes of which are represented by structures 1 and 2. This motion modulates the orientation of the CD bonds relative to the long axis from $60^\circ + \alpha$ to 60° for the methine deuteron and from $60^\circ + \alpha$ to $60^\circ - \alpha$ for the phenyl deuterons. The averaged value of the quadrupolar interaction, i.e. the splitting, is then larger for the methine than for the phenyl and R > 1. In this model, the temperature variation of R arises from the variations of the statistical weight of the conformations. When the temperature increases, the intermolecular constraints become less stringent, conformations of type 2 become more probable and R decreases because the methine and phenyl splittings tend to be similar in such states. However, different values of $S_{xx}-S_{yy}$ for the two groups could contribute to the temperature dependence of R, but the sign and the magnitude of this effect is not yet determined.

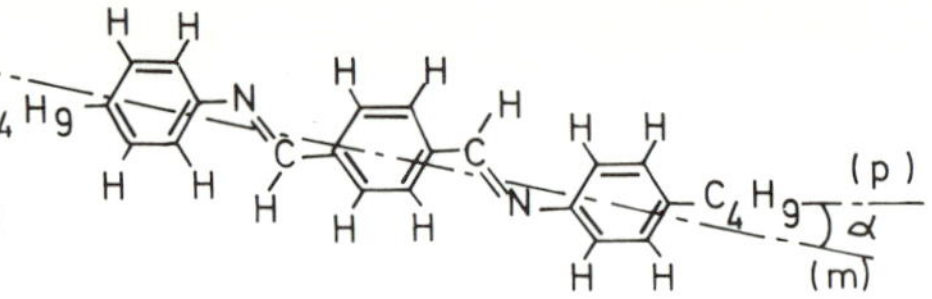

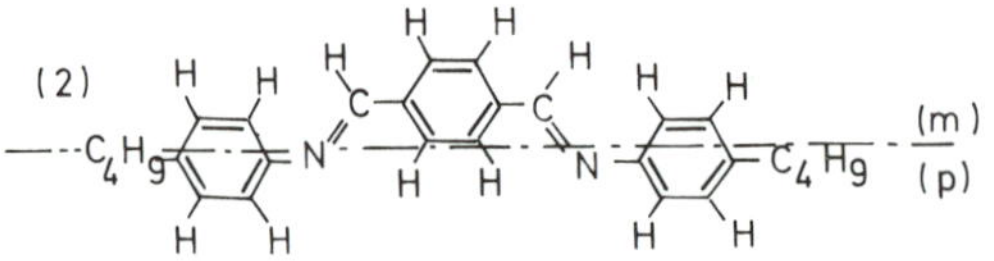

The test point of the deformation model is the relative orientation of the two C=N bonds of the molecule. It is not directly accessible in the liquid crystalline states but various considerations favour the idea of such conformational changes. First, the barriers to rotations around the phenyl *para* axis have been calculated [42,44] and are rather low, less than 4kJ/mol. Secondly, the dipole moment of the molecule in solution is intermediate between that expected from conformers 1 and 2 [45] demonstrating the reorientation of one methine group relative to the other in an isolated molecule.

Motions and Analysis of the Results

In the preceding sections, it has been shown that the analysis of the magnetic resonance results involves the description of the motions of a molecular tensor in a laboratory frame containing the magnetic field. This is rather complex because several motions are involved: deformations, rotational diffusion and orientational fluctuations. Decompositions can be proposed in which each particular motion is described as the rotation of one frame of reference with respect to another through Euler angles and rotation matrices [15]; the motional averages are then taken step by step. For instance, the first frame may be one in which the intramolecular motions are described; it can be attached to a phenyl group and the motions of the other groups are referred to it. As the internal deformations of the molecule change its geometry around the long axis the first frame has to be referred to one related to this axis. The long axis moves with respect to a frame attached to the director which, finally, has to be described in the laboratory frame. There are not less than four steps; indeed there may be more, and it is not unthinkable that different sequences give account of the same result. On the other hand, the dynamical characteristics of the motions have to be known in order to perform the averaging procedure from the fastest to the slowest motion; a slow motion can average only the residual part of the interaction modulated by faster anisotropic motions. Estimates can be attempted.

As already stated in an earlier section the thermal collective fluctuations of the director hardly affect the lineshapes. In a more dettailed way these fluctuations can be analysed in modes of wavevector q and lifetime $\tau_q = Kq^2/\eta$ (the elastic constant K is typically 10^{-7} dyn, the viscosity η of the order of 10^{-1} cP) [46]; the fastest mode, which has $q = 2\pi/a$, where a is a distance of about the molecular length, has then a time $>10^{-9}$s. As the motions responsible for the lineshapes have times shorter than 10^{-8}s the fastest mode might intervene but their small amplitudes, proportional to kT/Kq^2, make them of low efficiency. The largest contribution certainly comes from the local behaviour of one molecule in the cage of its neighbours. The cage fluctuates as the local density does and this momentarily releases the confined molecule which can change its shape and its orientation. The value or the upper limit of the times of the local motions is then the lifetime of the cage which is directly related to the time for a translational diffusive jump: as soon as the cage opens the molecule can move forward. This time can be estimated from the diffusion formula $\ell^2 \simeq 2D\tau$ giving the mean quadratic distance covered in a time τ; with $\ell^2 \sim 25\text{Å}^2$ and $D \sim 10^{-5}$- 10^{-6}cm^{2}s^{-1} [47], the lifetime of the cage should be 10^{-10} or 10^{-9}s. This is just the time scale which can be expected for the other local motions. Thus NMR relaxation studies [48] suggest the characteristic times for the long axis reorientation, with large amplitudes, to be about 10^{-9}s and this is partially confirmed by a recent neutron scattering experiment [49] which indicates that their lowest limit is the characteristic time of this experiment, namely 10^{-10}s. From NMR relaxation studies in related systems, such as lyotropic liquid crystals [50] or polymers [51] the local or segmental deformations can be estimated to have times in the 10^{-10}s range. At least, the rotation around the long axis, which may be a meaningless concept if the internal deformations are more rapid, appears in the same range of times as those from NMR relaxation studies[48] or even faster than those from neutron quasielastic scattering studies [35]. In the light of those estimates the model proposed, in which the molecule achieves a high symmetry about its long axis before the fluctuations of this axis, would appear reasonable.

Conclusion

The number of NMR studies of the molecular behaviour in liquid-crystalline phases is relatively small when compared to the wide variety of molecules and phases. Also, because of certain difficulties in chemical preparation, the choice of the systems under investigation was most often dictated solely by the availability of the mesogen. In spite of these restrictions we think it is possible to sketch a general picture for the studied molecules already described in the preceding sections. This picture is obviously not to be considered as a definitive model but rather as a starting point for more extended and systematic investigations.

The hypothesis of a rigid molecule, appears insufficient to provide an account of the experimental results whatever its conformation might be. The molecule is deformed by bond distortions and isomeric rotations around some bonds of the aromatic core or of the end chains. In the case of TBBA it is proposed that the molecule rapidly exchanges between several conformations the statistical weights of which are temperature dependent. The possible conformations have different overall dimensions so that the exchange between them can only occur if the surrounding molecules move accordingly. Those deformations are thus coupled to cage fluctuations, the same may also be true for rotational diffusion and more particularly for the rapid reorientations of the long molecular axis which are mostly responsible for the order parameter, S_{zz}. The determination of S_{zz}, under well defined experimental conditions with an interaction axis nearly parallel to the long axis, has provided a reliable test for the theoretical models of the intermolecular potential. The few complete characterizations of the ordering matrix indicate that its two other components, S_{xx} and S_{yy}, are very close. This suggests that the molecules achieve a rather high symmetry around their long axes prior to the reorientations of this axis. As a matter of fact disparate dynamical data from thermotropic liquid crystals but also from related fields, such as lyotropic liquid crystals and polymers, suggest that the long-axis reorientation is somewhat slower (10^{-9}s) than the rotation and internal deformations ($\sim 10^{-10}$s) but the ranges are most likely very close and it is difficult to go far in any detailed, quantitative interpretation at the moment. In any case those local motions determine an average molecule and the reciprocal adjustment of the molecule and the intermolecular potential induces the macroscopic structure. In fact recent work shows that a relation exists between the symmetry of the structure and that of this average molecule: the local symmetry, on the time of an NMR experiment ($\sim 10^{-5}$s), is uniaxial or biaxial when the macroscopic one is so. Clearly the understanding of the above points needs a detailed analysis which is not possible for molecules studied so far which several motions are simultaneously possible. It is necessary to isolate each local motion and this makes it necessary to work with molecules specifically synthesised with well characterized mesophases.

Acknowledgements

Professor G.R. Luckhurst and Drs. J.M. Emsley and J.W. Doane are thanked for communication of their work prior to publication.

References

1. W.L. MacMillan, *Phys. Rev.* A4, 1238 (1971).
2. S. Marcelja, *J. Chem. Phys.* **60**, 3599 (1974).
3. D. E. Martire, Chapter 10.
4. See the related papers in *Les smectiques thermotropes et leurs applications*, *J. Phys.* (Paris), C-3 (1976).
5. G.W. Gray, Chapters 1 and 12.
6. R.J. Meyer and W.L. MacMillan *Phys. Rev.* **A9**, 899 (1974).
7. D. Cabib and L. Benguigui, *J. Phys.* (Paris), **38**, 419 (1977).
8. A. Wulf, *Phys. Rev.* **A11**, 365 (1975).
9. A. Abragam, *The Principles of Nuclear Magnetism*, Oxford University Press (1961).
10. Y.S. Lee, Y.Y. Hsu and D. Dolphin, *Liquid Crystals and Ordered Fluids*, vol.2, edited by J.F. Johnson and R.S. Porter, Plenum Press, New York (1974), see also a similar more recent work by J.J. Visintainer, E. Bock, R.Y. Dong and E. Tomchuck, *Can. J. Phys.* **53**, 1483 (1975).
11. P. Diehl and A.S. Tracey, *Mol. Phys.* 30, 1917 (1975).
12. N. Boden, Y.K. Levine, D. Lightowlers and R.T. Squires, *Chem. Phys. Lett.* 34, 63 (1975).
13. M. Bloom, E. Burnell, S.B.W. Roeder and M.I. Valic, *J. Chem. Phys.* 66, 3012 (1977).
14. A. Pines, D.J. Ruben and S. Allison, *Phys. Rev. Lett.* 33, 1002 (1974).
15. L.J. Berliner (ed)., *Spin Labelling Theory and Applications*, Academic Press N.Y.(1976).
16. F. Poldy, M. Dvolaitsky and C. Taupin, *J. Phys.* (Paris), **36**, C1-27 (1975).
17. J.C. Rowell, W.D. Philips, L.R. Melby and M.J. Panar, *J. Chem. Phys.* 43, 3442, (1965).
18. B. Cabane and W.G. Clark, *Phys. Rev. Lett.* 25, 91 (1970).
19. B. Cabane and W.G. Clark, *Sol. St. Comm.* **13**, 129 (1973).
20. Z. Luz, R.C. Hewitt and S.L. Meiboom, *J. Chem. Phys.* 61, 1758 (1974).
21. B. Deloche, J. Charvolin, L. Liébert and L. Strzelecki, *J. Phys.* (Paris), 36, C1-21 (1975).
22. M.H. Cohen and F. Reiff, *Solid State Physics*, **5**, 321 (1957), edited by F. Seitz and D. Turnbull.
23. B. Deloche and J. Charvolin, *J. Phys.* (Paris), **37**, 1497 (1976).
24. R. Alben, J.R. McColl and C.S. Shih, *Solid State Comm* **11**, 1081 (1972).
25. J.W. Emsley, G.R. Luckhurst, G.W. Gray and A. Mosley, *Mol. Phys.* **35**, 1499 (1978).
26. P.J. Bos, J. Pirs, P. Uklega, J.W. Doane and M.E. Neubert, *Mol. Cryst. Liq. Cryst.* 40, 59 (1977).
27. R.L. Humpries, P.G. James and G.R. Luckhurst, *J. Chem. Soc. Faraday Trans. II*, 68, 1031 (1972).

28. G.R. Luckhurst, Chapter 4.
29. G.R. Luckhurst, *Mol. Cryst. Liq. Cryst.* 21, 125 (1973).
30. I. Chistyakov, *Adv. Liq. Cryst.* **1**, 143, Acad. Press (1975).
31. P.G. James and G.R. Luckhurst, *Mol. Phys.* 20, 761 (1971).
32. A. Wulf, *J. Chem. Phys.* **63**, 1654 (1975).
33. J. Seliger, R. Osredkar, V. Zagar and R. Blinc, *Phys. Rev. Lett.* 38, 411 (1977).
34. T.P. Das and E.L. Hahn, *NQR Spectroscopy, Solid State Physics, Suppl. 1,* Edited by F. Seitz, D. Turnbull and H. Ehrenreich, Acad. Press (New York, 1958).
35. F. Volino, A.J. Dianoux and J. Hervet, *J. Phys.* (Paris), **37**, C3-55 (1976).
36. J. Doucet, Chapter 14.
37. D. Guillon and A. Skoulios, *J. Phys.* (Paris), **38**, 79 (1977).
38. B. Mely and J. Charvolin, *Chem. Phys. Lip.* **19**, 43 (1977).
39. C. Destrade and H. Gasparoux, *J. Phys.* (Paris), **36**, L-105 (1975).
40. J.W. Emsley, J.C. Lindon and G.R. Luckhurst, *Mol. Phys.* 32, 1187 (1976).
41. B. Mely, J. Charvolin and P. Keller, *Chem. Phys. Lip.* **15**, 161 (1975).
42. J. Charvolin and B. Deloche, *J. Phys.* (Paris), **37**, C3-69 (1976).
43. B. Deloche and J. Charvolin, to be published.
44. J. Berger, H. Perrin and M. Jaffrain, *C.R. Hebd. Acad. Sci.* 281, C-441 (1975).
45. Obtained by Prof. H. Lumbroso (Université Paris VI).
46. Orsay Liquid Crystal Group, *J. Chem. Phys.* **51**, 816 (1969).
47. G.J. Kruger, H. Spiesecke and R. Van Steenwinkel, *J. Phys.* (Paris), **37**, C3-123 (1976); G.J. Kruger and R. Weiss, *J. de Phys.* **38**, 353 (1977).
48. R.D. Orwoll, C.G. Wade and B.M. Feug, *J. Chem. Phys.* 63, 986 (1975).
49. A.J. Leadbetter and R.M. Richardson, Chapter 20.
50. J. Charvolin and P. Rigny, *J. Chem. Phys.* **58**, 3999 (1973).
51. G. Hermann and G. Weill, *Macromolecules,* **8**, 171 (1975).

Chapter 16

MAGNETIC RESONANCE SPECTROSCOPY. STATIC BEHAVIOUR

PIER LUIGI NORDIO and ULDERICO SEGRE

Istituto di Chimica Fisica dell'Universita' di Padova, Italy.

Introduction

Magnetic resonance spectroscopy is a powerful tool currently used to investigate macroscopic ordering and molecular motions in liquid crystals. The line positions and the spacings in the multiplet structures of the spectra give direct information on the degree of orientational order of the molecules. This is because most of the magnetic interactions which are responsible for the spectral structures are anisotropic, *i.e.* their magnitude is dependent upon the molecular orientation with respect to the direction of the applied magnetic field. In partially oriented liquids, the anisotropies are averaged by the molecular tumbling to a value which is determined by the mean molecular orientation. In addition, the spectral profiles or the dynamical behaviour of the macroscopic magnetization when the equilibrium conditions are abruptly altered in relaxation experiments, are extremely sensitive to the nature of the motions responsible for averaging the anisotropic interactions, and therefore provide information on the characteristics of the molecular dynamics.

The success of the magnetic resonance techniques resides in the fact that the problem of calculating the spin energy levels is solved exactly, when the magnetic parameters corresponding to the interactions operating in a molecular system are known, and in turn these parameters can be either extracted from experiments, for example from single crystal measurements, or evaluated theoretically with sufficient accuracy. The solutions of the problem for the dynamics of the spin systems also rest on a well-established theoretical background, and so the interpretation of the relaxation effects can, in general, be accepted with confidence.

Since the theoretical description of the nuclear spin and the electron spin resonance techniques is formally the same, the general aspects of their applications in the study of the liquid-crystalline systems will be treated as much as possible in a unitary way. Obviously, since mesogenic molecules are diamagnetic, NMR spectroscopy can be performed directly on these systems as we saw in the previous Chapter whereas doping with paramagnetic probes is required for ESR investigations.

The application of this branch of spectroscopy is confined to the stat ic properties in this Chapter and the dynamic aspects will be discussed in Chapter 19.

Magnetic Interactions

Magnetic resonance spectra can be interpreted as plots of radio-freque power absorbed as a function of frequency at constant magnetic field. In practice, commercial spectrometers operate by sweeping the field at constant frequency to satisfy the resonance conditions, and in ESR instruments the first derivative of the absorption curve is recorded, rather than the absorption itself. It is always possible, however, to express the field in frequency units, and this makes the theoretical interpretation more direct, by analogy with other spectroscopic techniques.

The positions of the absorption lines in the spectra can be computed from the quantum mechanical spin hamiltonian which describes the interactions of the magnetic moments ascribed to the spins with the external fields and among themselves. The interaction of the spins with the static magnetic field is called the Zeeman interaction, and, very often, it constitutes the major contribution to the total magnetic energy of the system. Typical interactions between different spins are the exchange and the dipolar interactions. In addition, interactions of the nuclear quadrupole charge distribution with the gradient of the electric field generated at the nuclear position by the electronic distribution may occur. All of these interactions are, in general, anisotropic and therefore they need to be represented by tensorial quantities. Almost invariably, the inter- and intramolecular interactions which give rise to the energy levels in magnetic resonance spectroscopy are described by scalars and second rank tensors. The relevant interactions encountered in NMR and ESR experiments are as follows.

a) Zeeman interactions

(i) electronic $\mu_B \mathbf{B}.\mathbf{g}.\mathbf{S}$,

(ii) nuclear $-\mu_N \mathbf{B}.\boldsymbol{\sigma}.\mathbf{I}$,

where μ_N and μ_B are the nuclear and Bohr magnetons respectively.

Here **g** and σ are called the g-tensor and chemical shielding tensor respectively, B is the magnetic flux density, S is the electron spin operator and I is the nuclear spin operator.

b) Anisotropic spin-spin couplings

(i) electron-electron, or zero-field splitting term

$$\mathbf{S}.\mathbf{D}.\mathbf{S},$$

and the zero-field splitting tensor, **D**, is traceless.

(ii) electron-nucleus, or hyperfine interaction

$$\mathbf{I}.\mathbf{A}.\mathbf{S},$$

where A is the nuclear hyperfine tensor.

(iii) nucleus-nucleus

$$I_1.D.I_2,$$

where **D** is the spin-spin coupling tensor. The scalar parts of interactions (ii) and (iii) give rise to the isotropic Fermi contact term a**I**.**S** and the isotropic spin-spin coupling $JI_1.I_2$ respectively.

c) Nuclear quadrupole interaction

$$I.Q.I.$$

This term is present whenever the nuclear spin I $\geqslant$ 1 (*e.g.* deuterium and nitrogen). The quadrupole tensor **Q** always has zero trace. Consequently the spin hamiltonian is expressed as a sum of terms of the form **U**.**T**.**V**, where U and V are spin operators (or the magnetic field) and the interaction tensor **T** is a molecular property. The scalar product **U**.**T**.**V** is obviously invariant under rotation of the reference axis system, and we shall find it convenient to choose this axis system as the fixed laboratory frame with the Z axis parallel to the static field. In this way the spin operators are well defined, but it is necessary to express the tensor T in the laboratory system in terms of its components in the molecular frame. As we saw in Chapter 17 this is achieved by means of the Euler matrix R($\alpha\beta\gamma$) which performs the transformation from the fixed (laboratory) system to the mobile (molecular) system [1]. Note that with this definition the angle α refers to rotation about the static field direction. Actually we are interested in the inverse transformation, *i.e.*

$$\mathbf{T}_L = \mathbf{R}^{-1}\mathbf{T}_M\mathbf{R},$$

$$= \tilde{\mathbf{R}}\mathbf{T}_M\mathbf{R}, \qquad (1)$$

where $\tilde{\mathbf{R}}$ is the transpose of **R**.

Instead of using cartesian components, we will find it more useful to express the tensors in terms of their irreducible components in a spherical basis [1,2]. This is because the irreducible spherical tensors transform under rotation according to definite quantum numbers. In the spherical notation, the hamiltonian becomes

$$H = \sum_{\mu}\sum_{L,m} A_{\mu}^{(L,m)} T_{L,\mu}^{(L,m)^*},$$

$$= \sum_{\mu}\sum_{L,m,n} A_{\mu}^{(L,m)} D_{m,n}^{L}(\alpha\beta\gamma) T_{M,\mu}^{(L,n)^*}, \qquad (2)$$

where μ denotes the different interactions, $A_\mu^{(L,m)}$ the corresponding spin operators of rank L in the laboratory frame. The second line of eq. (2) expresses the same transformation given in eq. (1), and $D^L_{m,n}(\alpha\beta\gamma)$ are Wigner rotation functions [1,2]. The general form of the Wigner functions (or generalized spherical harmonics) is given in the Appendix of Chapter 3. The general expressions are given in table 1, for the irreducible spherical components corresponding to a second rank cartesian tensor of components T_{ij} or U_iV_j; in the latter case the spherical vector components $U_\pm = U_x \pm iU_y$ are introduced. This notation is particularly useful when dealing with spin operators. If the magnetic interaction tensors are symmetrical, the first rank components are zero, but this is not necessarily the case in molecules of symmetry C_{nh} or lower. However, we shall see that there are no contributions of the first rank components in determining the resonance frequencies or line splittings in a uniaxial liquid crystalline phase, provided this has no polar character.

Table 1. *Irreducible spherical components of the interaction tensor T and of the tensor operator A = U⊗V.*

(L,m)	$T^{(L,m)}$	$A^{(L,m)}$
(0,0)	$-(T_{xx}+T_{yy}+T_{zz})/\sqrt{3}$	$-\{U_zV_z+\frac{1}{2}(U_+V_-+U_-V_+)\}/\sqrt{3}$
(1,1)	$-\{T_{xz}-T_{zx}+i(T_{yz}-T_{zy})\}/2$	$-(U_+V_z-U_zV_+)/2$
(1,0)	$-i(T_{yx}-T_{xy})/\sqrt{2}$	$-(U_+V_--U_-V_+)/\sqrt{8}$
(1,-1)	$-\{T_{xz}-T_{zx}-i(T_{yz}-T_{zy})\}/2$	$-(U_-V_z-U_zV_-)/2$
(2,2)	$\{T_{xx}-T_{yy}+i(T_{xy}+T_{yx})\}/2$	$U_+V_+/2$
(2,1)	$-\{T_{xz}+T_{zx}+i(T_{yz}+T_{zy})\}/2$	$-(U_+V_z+U_zV_+)/2$
(2,0)	$\{2T_{zz}-(T_{xx}+T_{yy})\}/\sqrt{6}$	$\{2U_zV_z-(1/2)(U_+V_-+U_-V_+)\}/\sqrt{6}$
(2,-1)	$\{T_{xz}+T_{zx}-i(T_{yz}+T_{zy})\}/2$	$(U_-V_z+U_zV_-)/2$
(2,-2)	$\{T_{xx}-T_{yy}-i(T_{xy}+T_{yx})\}/2$	$U_-V_-/2$

Order Parameters

Relations (1) and (2) are appropriate for molecules incorporated in crystal lattices, so that the magnetic field direction (laboratory Z axis) has a definite orientation with respect to the molecular frame. In the liquid state, because of brownian motions, the interactions must be properly averaged over all the molecular orientations. If the motions are sufficiently fast, in the sense that the frequency of molecular tumbling is much higher than the magnetic anisotropies expressed in frequency units, the positions of the spectral lines are obtained by averaging the spin hamiltonian with a proper orientational distribution function. In the opposite case of very slow motions (rigid limit), a spectrum will appear which could be interpreted as the superposition of all spectra corresponding to any possible orientation, again weighted by a suitable distribution function. The first condition applies to isotropic and nematic liquids, and we shall concentrate our attention on this case.

An aligned nematic phase is optically uniaxial and the orientations of the optic axis is represented by a unit vector called the director. In most situations the director coincides with the preferred molecular orientation. The orientational distribution function must be therefore independent of the angle α. The same conclusion applies to the smectic A phase, but not necessarily to the smectic C phase which we shall discuss later. In addition, the nematic or smectic phases do not show polar character; the directions Z and -Z are indistinguishable, and therefore the orientational distribution function must be an even function of $\cos\beta$. We shall call it $f(\beta,\gamma)$, and we shall assume it to be normalized,

$$\int_0^{2\pi} d\gamma \int_0^{\pi} d\beta f(\beta,\gamma)\sin\beta = 1. \tag{3}$$

The angles β,γ are the polar angles which describe the orientation of the director in the molecular frame. In an isotropic phase $f(\beta,\gamma)$ is independent of all angles and equal to $1/4\pi$. The procedure of averaging the hamiltonian by means of the distribution $f(\beta,\gamma)$ relies on the so-called ergodic hypothesis, which states that the time and ensemble averages are equivalent.

The experimental determination of the partially averaged anisotropic terms of the hamiltonian gives information on the ordering characteristic of the mesophase and its temperature dependence. To derive useful expressions for the order parameters, we shall consider the spin hamiltonian both in cartesian and in spherical notation.

Hamiltonian in cartesian form

From eq. (1) the tensor components T^D_{IJ} in the laboratory axis system (XYZ) containing the director as Z axis are

$$T^D_{IJ} = \sum_{i,j} R_{iI} R_{jJ} T^M_{ij}, \qquad (i,j = x,y,z), \tag{4}$$

where (xyz) denote a molecular axis set. From the previous arguments, any partially averaged tensorial property of a liquid-crystalline phase is described in the (XYZ) system by an axially symmetric tensor in diagonal form: $\overline{T^D_{IJ}} = 0$ for I≠J, and $\overline{T^D_{XX}} = \overline{T^D_{YY}}$. Upper bars are used to denote the equilibrium averages over the orientational distribution specified by f(β,γ). The following relations must then hold,

$$\overline{R_{iJ}R_{jJ}} = 0 \quad \text{for } I \neq J, \tag{5}$$

$$\overline{R_{iX}R_{jX}} = \overline{R_{iY}R_{jY}}$$

and

$$\sum \overline{R_{iI}R_{jI}} = \delta_{ij}. \tag{6}$$

The last equality results from the orthogonality properties of the direction cosines. Consequently

$$\overline{T^D_{XX}} = \overline{T^D_{YY}},$$

$$= \tfrac{1}{2}\sum_{i,j} (\delta_{ij} - \overline{R_{iZ}R_{jZ}})T^M_{ij} \tag{7}$$

and

$$\overline{T^D_{ZZ}} = \sum_{i,j} \overline{R_{iZ}R_{jZ}}T^M_{ij}. \tag{8}$$

We observe that R_{iZ} represents the cosine of the angle between the optic axis (Z) and the molecular axis (i). By introducing the symmetric and traceless matrix **S** with components

$$S_{ij} = (3\overline{R_{iZ}R_{jZ}} - \delta_{ij})/2, \tag{9}$$

we find, after some manipulation,

$$\overline{T^D_{XX}} = \overline{T^D_{YY}},$$

$$= \frac{1}{3}\mathrm{Tr}\mathbf{T}^M - \frac{1}{3}\mathrm{Tr}(\mathbf{S}.\mathbf{T}^M), \tag{10}$$

$$\overline{T^D_{ZZ}} = (1/3)\mathrm{Tr}\mathbf{T}^M + (2/3)\mathrm{Tr}(\mathbf{S}.\mathbf{T}^M). \tag{11}$$

For the case of a magnetic interaction with the general form **U.T.V** we have

$$\overline{\mathbf{U.T.V}} = (1/3)(\mathrm{Tr}\mathbf{T}^M)\mathbf{U.V} + (2/3)(\mathrm{Tr}\mathbf{S}.\mathbf{T}^M)[U_ZV_Z - \frac{1}{2}(U_XV_X + U_YV_Y)]. \tag{12}$$

The S matrix is commonly called the ordering matrix, and it can be calculated from experimentally determined anisotropic coupling components, if the coupling tensor T^M is known from theoretical studies or independent measurements.

Hamiltonian in Spherical Notation

If the average of $T^{(L,m)}$ in eq. (2) is taken with an orientational distribution function independent of α, it follows immediately from the functional form of the Wigner matrix elements that

$$\overline{T_D^{(L,m)}} = \delta_{mo}\overline{T_D^{(L,o)}},$$

$$\overline{T_D^{(L,o)}} = \sum_m \overline{D_{o,m}^{L*}}\, T_M^{(L,m)},$$

$$= \sum_m \overline{D_{o,m}^{L}}\, T_M^{(L,m)^*}. \qquad (13)$$

If the mesophase is assumed to have no polar character, this means that $f(\beta,\gamma)$ is invariant under space inversion, *i.e.* $f(\beta,\gamma) = f(\pi-\beta, \gamma+\pi)$. From the symmetry relation [2]

$$D_{o,m}^{L}(\pi-\beta, \pi+\gamma) = (-)^L D_{o,m}^{L}(\beta,\gamma), \qquad (14)$$

it follows that for uniaxial liquid crystals L must be even. Therefore first-rank components of the interaction vanish under these conditions. For second-rank interactions there are five independent parameters $\overline{D_{o,m}^2}$, as in the case of the symmetric, traceless ordering matrix **S**. The relations between the two sets of parameters are easily found and are given in Chapter 3. If the molecule has some symmetry, one can choose a favourable molecule-fixed coordinate system to reduce the number of independent orientational parameters to a minimum. In general, the non-vanishing order parameters $\overline{D_{o,m}^2}$ must transform under symmetry operations according to the totally symmetric irreducible representation of the molecule point group. The formal symmetry arguments for reducing the number of unknown order parameters are given in Chapter 3. These show that certain simplifying conditions are as follows:

a) three-fold (or higher) symmetry axis. The choice of this as z axis causes only S_{zz} or $\overline{D_{o,o}^2}$ to be non-zero;
b) two perpendicular planes of symmetry. If the z axis is taken parallel to the intersection of the planes, with the x axis in one plane and the y-axis in the other, only S_{zz} and $S_{xx}-S_{yy}$ are non-zero (or alternatively, $\overline{D_{o,o}^2}$ and $\overline{D_{o,2}^2} = \overline{D_{o,-2}^2}$).
c) single plane of symmetry. By taking this as the xy plane, only S_{zz}, $S_{xx}-S_{yy}$ and S_{xy} will have non-vanishing values (or alternatively $\overline{D_{o,o}^2}$, $\overline{D_{o,2}^2}$, and $\overline{D_{o-2}^2}$).

In the last case, if all non-zero parameters are available from the experiment, a new molecular axis system (x'y'z') can be found, in which the ordering matrix S' is put in diagonal form.

Orientational Distribution Function

Even if all five independent order parameters, $\overline{D_{o,m}^2}$ or S_{ij}, are known it is important to realise that the knowledge we obtain from magnetic resonance data about the orientational distribution function is still incomplete. If we denote by Ω the set of three Euler angles necessary to specify the molecular orientation then $f(\Omega)$ can be conveniently expanded in terms of the orthogonal Wigner functions $D^L_{m,n}(\Omega)$

$$f(\Omega) = \sum f_{Lmn} D^L_{m,n}(\Omega), \tag{15}$$

where, as we saw in Chapter 3, the expansion coefficients are given by

$$f_{Lmn} = (2L+1)\overline{D^{L*}_{m,n}}/8\pi^2. \tag{16}$$

It follows that from second rank interaction tensors alone one can obtain only the L=2 coefficients of the expansion for $f(\Omega)$. If we think of $f(\Omega)$ as originating from an orientational pseudo-potential $U(\Omega)$ then

$$f(\Omega) = \exp\{-\beta U(\Omega)\}/Z, \tag{17}$$

where

$$Z = \int d\Omega \, \exp\{-\beta U(\Omega)\},$$

we shall find no unique relationship between $U(\Omega)$ and the experimental set of $\overline{D^2_{m,n}}$ values. We shall see that, in principle, magnetic relaxation studies can partially overcome this limitation, for the calculated effects are sensitive to the complete expansion of $f(\Omega)$.

Angular Dependence of the Magnetic Interactions

When magnetic resonance measurements are performed on nematic phases with positive diamagnetic anisotropy, the direction of preferential alignment will always be, on the average, parallel to the static field direction. However, the use of a thin film oriented by surface effects, or the application of electric fields, allows the director to be oriented at any angle θ to the direction of **B**. In this way, the angular dependence of the magnetic interactions can be measured.

In the case of most smectic phases, it is impossible for the director to change its orientation when the sample is rotated in a magnetic field. This feature makes the smectic phase particularly suitable for the investigation of the angular dependence of the magnetic parameters, but it is essential to obtain a monodomain sample. In fact, in a multidomain sample severe broadening of the spectral lines results from the superposition of spectra corresponding to different director orientations. A single crystal of a smectic A phase can be obtained by cooling the sample in a magnetic field from a nematic phase or an isotropic melt. The phase is aligned with the director parallel to the field direction. Once aligned, the sample can be rotated with the director following the rotation.

Smectic C mesophases differ from smectic A in having the director tilted away from the normal to the smectic planes. If the smectic layer is taken as the xy plane, and z is the normal to the layer, then the polar coordinates of the director are ν and ϕ where ν is the tilt angle; these angles are shown in figure 1. In contrast with the previous case, the

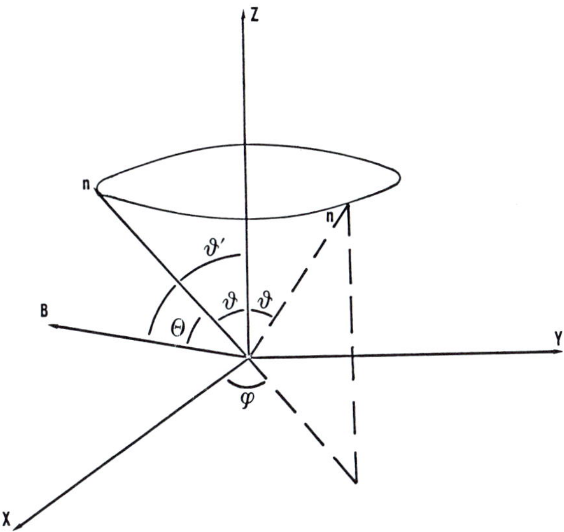

Fig 1. *Orienting effects of surface and magnetic forces on a smectic C phase. The xy plane defines the smectic layer, parallel to the glass plate, and* **B** *gives the direction of the magnetic field in the xz plane. The director* **n** *initially defined by the polar coordinates* ν, ϕ *is forced to be aligned in the same plane as* **B** *and the normal to the smectic layers.*

phase is expected to be biaxial and the action of the magnetic field alone is insufficient to provide a monodomain sample, with uniform orientation of both the smectic layers and of the azimuthal angle ϕ. A monodomain smectic C phase can be obtained if glass walls are used to orient the smectic planes, and a static field, applied at an angle ν' to the normal to the glass plates, is employed to give a uniform orientation to the director [4]. According to the traditional model of the smectic C structure, the tilt angle ν depends uniquely upon the temperature, but in the presence of a magnetic field the director can move on the surface of a cone of aperture 2ν about the normal to the smectic planes to minimize the magnetic energy. This condition is met when the director lies in the plane defined by the normal to the smectic layers and the direction of the field. For each orientation of the field there is only one orientation of the director, **n**, which satisfies this condition of minimum energy, except when the field is applied perpendicular or parallel to the smectic planes. In the first case, **n** can assume all the possible orientations on the surface of the cone, which are magnetically equivalent if cylindrical symmetry about the director can be assumed. In the second case, two energetically equivalent orientations of the director are obtained, with the same tilt angle ν but with ϕ = o and π respectively.

To interpret the measured angular dependence of the spectral parameters, we start from the spin hamiltonian given in eq. (2) and then perform two successive transformations, first from the magnetic field axis system to the director axis system through the angles Ω', and secondly from the latter to the molecular frame through the angles Ω. We obtain for the anisotropic part of the spin hamiltonian

$$H = \sum_{m,n,p} A^{(2,m)} D^2_{m,n}(\Omega') D^2_{n,p}(\Omega) T^{(2,p)*}. \tag{18}$$

We now take the average for the rotations about the director, again making the assumption of axial summetry. This assumption is strictly not valid for a smectic C phase, where in principle one should consider all the 25 order parameters $\overline{D^2_{n,p}}$ (cf. Chapter 3). However the available ESR [4] and NMR [5] experimental studies do not provide any evidence in conflict with this assumption, which therefore is justified *a posteriori*. Even so, in contrast with the case Ω' = 0 (director aligned along the field) all the spin operator components $A^{(2,m)}$ remain in the expression for the averaged hamiltonian, which therefore must be completely diagonalized in a suitable basis of spin functions for the calculation of the energy levels and of the resonance frequencies. Some approximations can be reasonably introduced, but usually they are different in NMR and ESR.

For the calculation of the nuclear transition frequencies we can, within the framework of first-order perturbation theory, neglect those spin operators which do not conserve the total component of the spin angular momentum in the direction of the magnetic field (non-secular terms). This is equivalent to retaining only the m=o component of the operators $A^{(2,m)}$ (cf. table 1). The spin hamiltonian truncated in this way becomes

$$H = A^{(2,o)} P_2(\cos\theta) \sum_p \overline{D^2_{o,p}} T^{(2,p)*}, \tag{19}$$

where θ is the angle between the director and the magnetic field. Thus the angular dependence of the transition frequencies is given by $P_2(\cos\theta)$.

In ESR, when the angular dependence of the hyperfine splittings is considered, the use of the truncated hamiltonian is a rather poor approximation, because in the presence of a hyperfine interaction, which is normally larger than the nuclear Zeeman term, the nuclear spins are not quantized along the magnetic field. For this reason, the so-called pseudosecular terms $I_{\pm}S_Z$ of the anisotropic hyperfine hamiltonian should be retained, and a diagonalization performed in the basis of the nuclear spin manifold corresponding to the same electron spin Zeeman level. The same problem is encountered when calculating the angular dependence of the hyperfine splittings for free radicals aligned in single crystals. In this case, if the hyperfine tensor has axial symmetry, the frequency separation of the hyperfine lines is given by

$$a^2(\theta) = A_{\parallel}^2 \cos^2\theta + A_{\perp}^2 \sin^2\theta,$$

$$= (1/3)(A_{\parallel}^2 + 2A_{\perp}^2) + (2/3)(A_{\parallel}^2 - A_{\perp}^2)P_2(\cos\theta). \quad (20)$$

This expression is rather accurate if the g tensor anisotropy is small and if the hyperfine terms are unimportant in comparison with the Zeeman splitting, conditions which are generally satisfied, for example by nitroxide free radicals, but not necessarily for paramagnetic transition metal complexes, such as vanadyl acetylacetonate.

We can then use an expression analogous to eq. (20) in angular studies of a mesophase by recalling that the axial symmetry of the mesophase leads for any magnetic interaction **T** to an axially symmetric averaged tensor with component $\bar{T}_{\parallel}$ parallel to the director and $\bar{T}_{\perp}$ perpendicular to it, given by

$$\bar{T}_{\parallel} = T_o + (2/3)^{\frac{1}{2}} \sum_m \overline{D^2_{o,m}} T^{(2,m)*} \quad (21a)$$

and

$$\bar{T}_{\perp} = T_o - (1/6)^{\frac{1}{2}} \sum_m \overline{D^2_{o,m}} T^{(2,m)*}, \quad (21b)$$

where T_o is the isotropic part of the interaction, which is usually assumed to equal the scalar coupling and can be measured in the isotropic phase of the liquid crystal.

Spin Probes for Molecular Ordering

In NMR investigations, the best probes to be used to gain information on the molecular ordering in a mesophase are obviously the mesogen molecules themeselves. However the spectra are often complex and unresolved because of the very large number of lines arising from the partially averaged dipolar couplings between the numerous magnetic nuclei present in the molecules. Selective isotopic substitution may then be necessary to reduce this source of broadening (cf. Chapter 15). On the other hand, small solute molecules give well-resolved spectra, but in this case it is questionable, for reasons that we shall discuss very shortly, if information on the solvent alignment can be inferred from the measured solute ordering.

In contrast to NMR, the ESR spectra of free radicals or more specifically of paramagnetic systems with only one unpaired electron, have the same number of lines in monodomain liquid crystalline phases as in the isotropic phase, and so the spectra are usually well-resolved and readily interpreted. Thus, spin-labelled molecular systems of convenient structure and dimensions can be conveniently used to obtain the ordering parameters, and also information on the molecular dynamics. It is important to realize, however, that mesomorphic molecules are made up of a rigid aromatic core and flexible alkyl end chains, so that it is reasonable to expect that a spin probe can be accommodated in at least two different sites, depending upon its molecular structure and shape. Such a situation is very likely to take place particularly in the highly organized smectic phases, where aromatic layers are separated by less ordered aliphatic regions. Thus, an elongated rigid probe will be intercalated between the aromatic parts of the mesogen molecules, assuming their average orientation, while probes with the same flexible structure as the alkyl chains may be excluded from the aromatic regions of the liquid-crystalline phase by specific intermolecular interactions, and be confined in the more fluid aliphatic chain region. Small molecules might not have a unique ordering even in the same liquid crystalline phase, and complications arise in this case in the interpretation of the experimental data if exchange occurs between the different sites. In fact, if the exchange frequency is low compared with the difference in frequency of the splittings corresponding to the different locations of the probe, a simple superposition of spectra is observed and the order parameters corresponding to the various local environments experienced by the probe can be obtained. On the other hand, if the exchange frequency is high, a single spectrum is observed where the magnetic parameters are the average of the values corresponding to the different sites, weighted by the fractional population of the probe in each site. In this respect, an important difference in the results of the NMR and ESR experiments exists, because of the characteristic time scales for these two branches of spectroscopy. Since the difference in the line splittings (or shifts due to different alignment may be typically of the order of kHz for NMR, and MHz for ESR, an exchange frequency of say 10^4Hz between different sites is *fast* for NMR and *slow* on the ESR time scale.

Paramagnetic transition metal-ion complexes have occasionally been used to probe molecular ordering in nematic and smectic phases, but the most widely employed spin probes are certainly nitroxide free radicals. The nitroxide group is very stable, it can be linked to a variety of molecular substrates, and it has a very simple three line spectrum resulting from the hyperfine coupling with the nitrogen nucleus. Typical nitroxide spin probes are shown in figure 2. By choosing a molecular

Fig. 2. *The molecular structure of typical nitroxide spin probes.*

axis system as shown

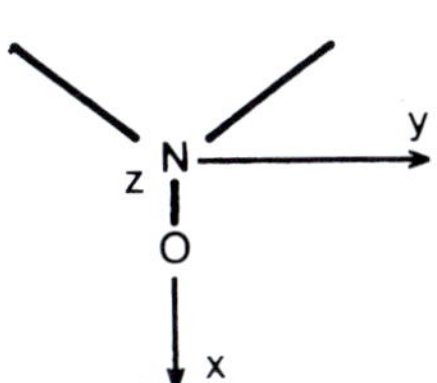

typical values for the hyperfine and g-tensor components are

$$A_{xx} = A_{yy} = 17 \text{ MHz}, \qquad A_{zz} = 90 \text{ MHz},$$

$$g_{xx} = 2.0089, \qquad g_{yy} = 2.0058, \qquad g_{zz} = 2.0021.$$

For this principal axis system, which does not necessarily coincide with that for the ordering matrix, the averaged g and A components parallel to the director in a uniaxial liquid crystal are

$$\bar{g} = g_{iso} + (2/3)\{g_{zz} - (1/2)(g_{xx}+g_{yy})\}\overline{D^2_{o,o}} + (g_{xx}-g_{yy})(\overline{D^2_{o,2}}+\overline{D^2_{o,-2}})/\sqrt{6} \tag{22a}$$

and

$$\bar{A} = a_{iso} + (2/3)(A_{\parallel} - A_{\perp})\overline{D^2_{o,o}}, \tag{22b}$$

where $A_{\perp} = A_{xx}(=A_{yy})$, $A_{\parallel} = A_{zz}$, and g_{iso} and a_{iso} are the isotropic values given by

$$g_{iso} = (1/3)\mathrm{Tr}\mathbf{g} = (g_{xx}+g_{yy}+g_{zz})/3$$

and

$$a_{iso} = (1/3)\mathrm{Tr}\mathbf{A} = (A_{xx}+A_{yy}+A_{zz})/3. \tag{23}$$

Since the argument β of $\overline{D^2_{o,o}}$ refers to the angle between the molecular z axis and the director, $\overline{D^2_{o,o}}$ will be positive ($0<\overline{D^2_{o,o}}<1$) if the z axis tends to be aligned parallel to the director, and negative ($-\frac{1}{2}<\overline{D^2_{o,o}}<0$) if it is aligned perpendicularly.

Probes of type I, spin-labelled derivatives of steroids often referred to as cholestane spin probes, are long and rigid molecules which tend generally to align parallel to the director of the mesophase. If this is the case, it is the y axis which on average is aligned along the director, and it may be convenient therefore to define β as the angle between the director and the long molecular axis. In fact, if the probe molecule is assumed to be cylindically symmetric about the long axis, this choice makes $\overline{D^2_{o,2}}$ and $\overline{D^2_{o,-2}}$ vanish. The y axis actually makes an angle of about 20° with the long molecular axis, and the corrections on the order parameters are small but not negligible, although they will be neglected here. If the principal axis system of the magnetic interactions on the nitroxide moiety is considered to be practically coincident with that of the ordering matrix, one has

$$\bar{g}-g_{iso} = (g_{yy}-g_{iso})\overline{D^{2'}_{o,o}}+(g_{zz}-g_{xx})(\overline{D^{2'}_{o,2}}+\overline{D^{2'}_{o,-2}})/\sqrt{6},$$

$$= (g_{yy}-g_{iso})\overline{(3\cos^2\beta-1)}/2+(g_{zz}-g_{xx})\overline{\sin^2\beta\cos 2\gamma}/2, \quad (24)$$

since

$$D^2_{o,\pm 2}(\alpha\beta\gamma) = \sqrt{3/8}\ \sin^2\beta\exp(\mp 2i\gamma),$$

$$g_{pp}-(g_{qq}+g_{rr})/2 = 3(g_{pp}-\mathrm{Tr}g)/2$$

and

$$\bar{A}-a_{iso} = -(A_{\parallel}-A_{\perp})\{\overline{D^{2'}_{o,o}}/3-(\overline{D^{2'}_{o,2}}+\overline{D^{2'}_{o,-2}})/\sqrt{6}\}. \quad (25)$$

The primed order parameters are related to those given in eq. (22) by the transformation

$$\overline{D^{2'}_{o,m}} = \sum_n \overline{D^2_{o,n}} D^2_{n,m}(\pi/2,\pi/2,0). \quad (26)$$

Now $\overline{D^2_{o,o}}$ varies between 0 and 1 and $\overline{D^2_{o\pm 2}}$ should be small, given the rod-like shape of the probe. If the assumption of cylindrical symmetry is correct, then the plot of $\bar{g}-g_{iso}$ against $\bar{A}-a_{iso}$ should give a straight line passing through the origin, of slope

$$(g_{yy}-g_{iso})/(A_{yy}-a_{iso}).$$

In any case, going from the isotropic melt to the oriented phase one expects the hyperfine splitting to decrease. If the deviation from cylindrical symmetry can be detected, a negative value of $\overline{\sin^2\beta\cos 2\gamma}$

means the director lies preferentially parallel to the pentatomic ring containing the nitroxide group. Typical ESR spectra in isotropic and anisotropic phases are shown in figure 3.

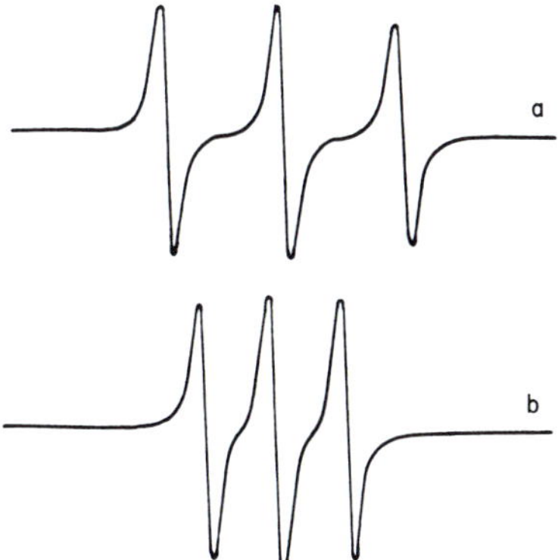

Fig. 3. *ESR spectra of a steroidal spin probe (a) in the isotropic phase and (b) in the mesophase of a nematogen.*

Spin probes of type II, that is nitroxide derivatives of fatty acids, are very flexible and tend to be accommodated in the aliphatic chain region of smectic mesophases, totally or partially depending upon the molecular length.

Type III spin probes are short and disc-like, and in nematic phases appear to be oriented preferentially with the plane of the ring parallel to the director. Perdeuteriated 2,2,6,6-tetramethyl-4-piperidone N-oxide (PD-Tempone) has been used for very accurate determinations of the temperature dependence of the ordering parameters, and for spectral line-shape analysis in connection with studies of molecular dynamics [8]. In smectic phases, however, the probe is partially excluded from the aromatic ring site and it tends to be oriented in the chain region with the ring parallel to the smectic planes (perpendicular to the director)[7]. Rapid exchange between these two sites is observed in some cases, and this results in an unusual temperature dependence of the hyperfine splitting because the relative populations of the sites change.

Experimental Applications

Numerous experimental determinations of temperature dependence of the order parameters in nematic and smectic phases have been performed in recent years, by measuring the ring *ortho*-proton dipolar couplings or ring deuteron quadrupole splittings of mesogen molecules by NMR, and hyperfine splittings of paramagnetic probes by ESR. The results have been discussed in exhaustive review articles [9] and demonstrate how magnetic resonance techniques can be used as a routine tool to obtain accurate information on the molecular organization and phase changes in liquid-crystalline systems. Of particular interest are the investigations on the angular dependence of the magnetic parameters, which give, in principle, more detailed information and provide a critical test of the models proposed for the structures of the various types of mesophases. Some of the more significant results will be mentioned here.

The splitting resulting from the dipolar interaction between the two equivalent proton spins in the probe molecule CH_2Cl_2 dissolved in the smectic A phase of TBBA has been measured as a function of the director orientation with respect to the field [5a]. If the probe molecules are assumed to experience a cylindrically symmetric environment, the spacing between the two resonance lines is

$$\Delta = \overline{P}_2(3h\gamma_1\gamma_2/2\pi d^3)P_2(\cos\theta), \tag{27}$$

where the order parameter $\overline{P}_2 = \overline{(3\cos^2\beta-1)}/2$ is expressed in terms of the angle β between the director and the interproton vector, and d is the internuclear separation. The experimental points fit the predicted angular law exactly.

Two other experiments confirm the validity of the proposed structure of smectic A phases. The first refers to measurements of the doublet spacing resulting from the deuteron quadrupole interaction in partially deuteriated TBBA [5b] (the intramolecular proton dipolar splitting in TBBA itself are impossible to resolve because of the intermolecular interactions with the other solvent molecules). In this case the spacing of the quadrupole doublets is given by

$$\Delta = \overline{P}_2(3e^2qQ/2h)P_2(\cos\theta)P_2(\cos\delta), \tag{28}$$

where $\overline{P}_2$ is expressed now in terms of the angle β between the director and the long molecular axis, and δ is the angle between this axis and the principal axis of the quadrupole tensor, which is believed to have axial symmetry. If this is the case, the principal axis of the quadrupole tensor is expected to coincide with the C-D bond direction. It should be noted that the presence of only one quadrupole splitting for the ring deuterons implies relatively fast 180° flips of the central benzene ring, which interchange pairs of *ortho*-deuterons making them equivalent on the average.

In another experiment the angular dependence of the hyperfine splitting of a cholestane spin probe has been measured by electron spin resonance [10]. The system investigated was the smectic A phase of 4-n-butyloxybenzylidene-4'-acetoaniline which exists in a nematic state from 110° to 98°C and then forms a smectic A phase. The angular dependence of the nitrogen hyperfine splitting of the probe, calculated from eq. (20), fits exactly the experimental points, confirming that the director remains aligned in the direction of the magnetic field when the smectic A phase is reached by cooling the system from the nematic state, and that it follows rigidly the rotation of the sample.

Another striking success of magnetic resonance spectroscopy has been the determination of the tilt angle in smectic C phases. We have previously seen that the combined effect of surface and magnetic field aligning forces is to produce a monodomain sample. When a liquid crystal sample contained between two parallel glass plates is oriented in a magnetic field, it is convenient to define the orientation of the sample in terms of the angle ν' between the normal to the walls and the magnetic field direction. In the case of the smectic A sample, the director is normal to the plates so that ν' is the same as θ, but this is not the case for a smectic C phase (cf. figure 1). In general, one might resort to the addition theorem for spherical harmonics [1], to express $P_2(\cos\theta)$ in terms of the polar angles (ν,ϕ) and (ν',ϕ') which define the orientations of the director and of the field in a reference system with the (xy) plane parallel to the glass plates, that is to the smectic layers,

$$P_2(\cos\theta) = \sum_q e^{iq(\phi-\phi')} d^2_{q,p}(\nu) d^2_{q,o}(\nu').$$

According to the proposed model for smectic C phases, in equilibrium conditions the director lies in the plane defined by the direction of the field and the normal to the plates, and so $\phi-\phi'$ is zero. Therefore, the angle θ is simply $|\nu-\nu'|$, and the angular dependence of the line spacings in magnetic resonance experiments, when performed on smectic C phases, is the same as that given in eqs. (20), (27) and (28) with $\cos^2\theta$ replaced by $\cos^2(\nu-\nu')$. In the ESR experiment performed on the smectic C phase of 4,4'-di-n-heptyloxyazoxybenzene using the cholestane spin probe [4], monodomain samples are obtained by cooling the nematic phase contained between glass plates in the presence of the magnetic field oriented at an angle near to the tilt angle. The direction of the field and the normal to the plates define the xz plane as in figure 1, and when the field is varied in this plane the angular dependence of the hyperfine splitting fits accurately eq. (20) modified as we have just seen, and a value of $38.5^o \pm 1^o$ is obtained for the tilt angle. The monodomain structure of the sample remains unaltered if the field is moved off the xz plane, showing that the intensity of the field used in normal ESR experiments (*ca.* 0.3T) is below the critical value necessary to reorient the xy component of the director [11]. In contrast, NMR experiments on a multidomain sample of the smectic C phase of TBBA at a field intensity of 1.4T [5b], confirm that the director is free to change its orientation in order to minimize the magnetic energy under the constraint of keeping the tilt angle constant.

References

1. M.E. Rose, *"Elementary Theory of Angular Momentum"* (J. Wiley and Sons, New York, 1957).
2. D.M. Brink and G.R. Satchler, *"Angular Momentum"* (Clarendon Press, 2nd Ed., Oxford, 1968).
3. A. Carrington and A.D. McLachlan, *"Introduction to Magnetic Resonance"* (Harper and Row, New York, 1967).
4. G.R. Luckhurst, M. Ptak and A. Sanson, *J.C.S. Faraday II*, 69, 1752 (1973).
5. a) Z. Luz and S. Meiboom, *J. Chem. Phys.*, 59, 275 (1973);
   b) Z. Luz, R.C. Hewitt and S. Meiboom, *J. Chem. Phys.*, 61, 1758 (1974).
6. J. Seelig *"Spin Labelling"*, ed. L.J. Berliner (Academic Press, New York, 1976).
7. D. Sy and M. Ptak, *Mol. Cryst. Liq. Cryst.*, 39, 53 (1977).
8. C.F. Polnaszek and J.H. Freed, *J. Phys. Chem.*, 79, 2283 (1975). (It should be noted that these authors do not use the Rose definition of $D^{L}_{m,n}$)
9. G.R. Luckhurst in *"Liquid Crystals and Plastic Crystals"*, G.W. Gray and P.A. Winsor eds. (Ellis Horwood Ltd., Chichester, 1974).
10. G.R. Luckhurst and M. Setaka, *Mol. Cryst. Liq. Cryst.*, 19, 179 (1972).
11. P.G. de Gennes, *"The Physics of Liquid Crystals"* (Clarendon Press, Oxford, 1974) page 310.

Chapter 17

RAMAN STUDIES OF ORIENTATIONAL ORDER IN LIQUID CRYSTALS

P.S. PERSHAN

Division of Applied Sciences, Harvard University, Cambridge, MA 02138, U.S.A.

Introduction

In this Chapter we shall discuss the use of polarized Raman scattering to measure the orientational statistics in liquid crystals. The number of publications from different laboratories on this technique is still somewhat limited and we shall not attempt to review the literature. Instead we shall concentrate on the basic ideas associated with this application of Raman scattering and also some of the practical problems encountered in the experiments. Since this is most effectively done in the context of real examples, I shall discuss the results from our own laboratory in depth. This work has now been published and interested parties can read about those aspects of the work that are not covered here in the paper by Shen Jen, Noel Clark, Eldon Priestley and myself [1]. The earlier publications on one or other aspects of this work are also cited in that paper.

In principle, the Raman technique is capable of measuring the average of the square of a molecular polarizability [2]. We shall show that this quantity depends on spherical harmonics of order 0, 2 and 4. Thus, in addition to determining the usual nematic order parameter, the Raman technique is capable of measuring higher rank order parameters that have not otherwise been measured.

In the simplest cases, both the concept and the execution of these experiments is straightforward. On the other hand, there are a number of complicating factors and, in the absence of some type of empirical demonstration, it is not obvious that the technique is practical. We shall discuss both the simple idea and the complications that might arise. Finally, we shall present some of the empirical evidence in support of the argument that the technique is practical.

Molecular Polarizability

Orientational Dependence

For our purpose the optical properties of liquid crystals are adequately described through the introduction of a molecular polarizability tensor $\boldsymbol{\alpha}$ that is both real and symmetric [2]. The presence of an electric field E induces a change in the electric dipole moment $\delta p = \boldsymbol{\alpha}.E$. For a given electric field both the orientation and magnitude of δp will depend on the orientation of the molecule relative to **E**. Consider some reference frame fixed with regard to the nuclear coordinates of the molecule and let $(\alpha_M)_{ij}$ designate the components of $\boldsymbol{\alpha}$ relative to that frame. Then to obtain the components of $\boldsymbol{\alpha}$ in the laboratory frame, where E and δp are measured, we must perform a transformation. The transformation $R(\alpha\beta\gamma)$ that brings the laboratory frame into coincidence with the molecular frame can be described in terms of the series of counter clockwise rotations shown in figure 1; rotation by α about the z axis, β about

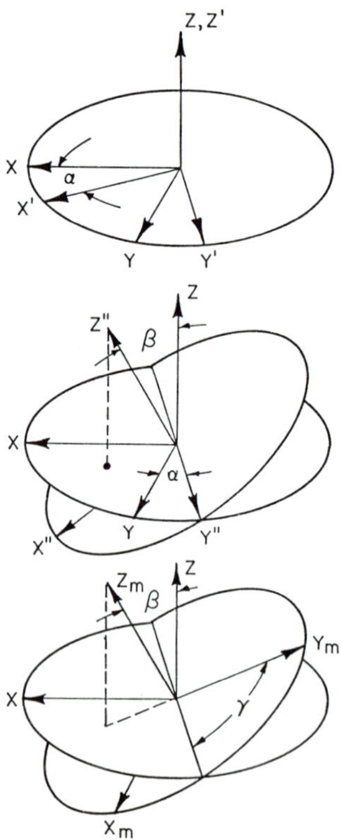

Fig. 1 *Illustration of the transformation $R(\alpha\beta\gamma)$ that rotates the laboratory frame into coincidence with the molecular frame.*

the new y axis, and γ about the new z axis. Alternatively one can show that an equivalent series consists of counter-clockwise rotations by γ about z, β about y [*n.b.* this is about the original y axis and not the new one that results from the γ rotation], and α about the original z axis. For example, if the unit vector along z is held fixed the components of $\hat{z}$ along the $\hat{x}_m$, $\hat{y}_m$, $\hat{z}_m$ are $(-\sin\beta\cos\gamma, \sin\beta\sin\gamma, \cos\beta)$. Alternatively one can consider a vector that is transformed such that it remains fixed in the rotating reference frame. Then the components of the rotated vector $R(\alpha\beta\gamma)\hat{z}$ along the original $(\hat{x},\hat{y},\hat{z})$ axis are $(\sin\beta\cos\alpha, \sin\beta\sin\alpha, \cos\beta)$.

The easiest way to proceed is to adopt the notation that has been developed in Chapter 3, to deal with rotations of general tensorial quantities. The general form for the polarizability tensor in the laboratory frame is

$$\boldsymbol{\alpha} = \boldsymbol{\beta}^{o}\alpha^{(0,0)} + \sum \boldsymbol{\beta}_{m}^{2}\alpha^{(2,m)}, \tag{1}$$

where the components of $\boldsymbol{\alpha}$ are given by

$$\alpha^{(0,0)} = \mathrm{Tr}\boldsymbol{\alpha}/\sqrt{3}, \tag{2}$$

$$\alpha^{(2,0)} = (2\alpha_{zz}-\alpha_{xx}-\alpha_{yy})/\sqrt{6},$$

$$\alpha^{(2,\pm1)} = \mp(\alpha_{xz}\pm i\alpha_{yz}),$$

$$\alpha^{(2,\pm2)} = (\alpha_{xx}-\alpha_{yy}\pm 2i\alpha_{xy})/2,$$

and the basis vectors $\boldsymbol{\beta}^{o}$ and $\boldsymbol{\beta}_{m}^{2}$ are listed in table 1. A similar ex-

Table 1. *Basis vectors for symmetric second rank tensors.*

$$\boldsymbol{\beta}^{o} = (3)^{-1/2}\begin{bmatrix} 1 & 0 & 0 \\ 0 & 1 & 0 \\ 0 & 0 & 1 \end{bmatrix}$$

$$\boldsymbol{\beta}_{o}^{2} = (6)^{-1/2}\begin{bmatrix} -1 & 0 & 0 \\ 0 & -1 & 0 \\ 0 & 0 & 2 \end{bmatrix}$$

$$\boldsymbol{\beta}_{\pm1}^{2} = (2)^{-1}\begin{bmatrix} 0 & 0 & \mp1 \\ 0 & 0 & i \\ \mp1 & i & 0 \end{bmatrix}$$

$$\boldsymbol{\beta}_{\pm2}^{2} = (2)^{-1}\begin{bmatrix} 1 & \mp i & 0 \\ \mp i & -1 & 0 \\ 0 & 0 & 0 \end{bmatrix}$$

pression relates the cartesian components of $(\alpha_{M})_{ij}$ in the molecular frame to the spherical components $\alpha_{M}^{(0,0)}$ and $\alpha_{M}^{(2,m)}$. The formal theory relates the $\alpha^{(2,m)}$ to the $\alpha_{M}^{(2,m)}$ by [3]

$$\alpha^{(2,m)} = \sum D^2_{m,n}(\Omega)^* \alpha_M^{(2,n)} \qquad (3)$$

and $\alpha^{(0,0)} = \alpha_M^{(0,0)}$. The Wigner rotation matrices $D^2_{m,n}(\Omega)$, can be found in the Appendix of Chapter 3.

At this point it is tempting to reflect on the fact that the optical dielectric constant is related to the induced macroscopic polarization of a material. Thus we might hope that a statistical average $\rho\langle\delta p\rangle = \mathbf{P}$, $= (\boldsymbol{\varepsilon}-1).\mathbf{E}/4\pi$ would give the dielectric tensor where ρ is the number density. For example for a cylindrically symmetric mesophase composed of particles with the same symmetry, the only non-vanishing term obtained from eq.(3) would be $D^2_{0,0}(\Omega) = \overline{(3\cos^2\beta-1)}/2$, so that

$$\bar{\alpha}_{zz} = \mathrm{Tr}\bar{\boldsymbol{\alpha}}/3 + \overline{(3\cos^2\beta-1)}\alpha_M^{(2,0)}/\sqrt{6} \qquad (4)$$

and since $\mathrm{Tr}\,\bar{\boldsymbol{\alpha}} = \mathrm{Tr}\boldsymbol{\alpha}$,

$$\varepsilon_{zz}-\varepsilon_{xx} = 2\pi\rho(3/2)^{\frac{1}{2}}\overline{(3\cos^2\beta-1)}\alpha_M^{(2,0)}; \qquad (5)$$

eq. 5 implies a direct relation between the anisotropy in the optical dielectric constant and the nematic order parameter. Unfortunately, the derivation of eq. 5 ignored the local field corrections and at first glance one might fear that this would render the equation useless for quantitative purposes. In fact we shall argue that the only effect of the local fields is to alter the meaning of $\boldsymbol{\alpha}_M$ and that eq. (5) can be used [4,5]. This is basic to the Raman studies since if local field corrections were necessary for the dielectric constant they would also be necessary for the Raman effects.

Local field effects

If $\boldsymbol{\alpha}$ is considered to be the electronic polarizability of a bare molecule in free space, the electric field in the relation $\delta p = \boldsymbol{\alpha}.\mathbf{E}$ is the sum of the macroscopic field of the propagating wave and the local fields resulting from the dipole moments on all the neighbouring molecules. Since a part of these dipole moments on neighbouring molecules are themselves induced by the macroscopic field, δp contains two terms proportional to the macroscopic field and one of them involves a summation over many neighbouring molecules. Furthermore since the system is anisotropic, δp will not be parallel, in general, to the macroscopic field. Many authors have attempted quantitative calculations of the relation between δp and the macroscopic field. However, in our opinion there is no way to solve the problem for molecular systems without introducing *ad hoc* assumptions that are hard to justify.

The principal difficulty is that the dipole-dipole interaction falls off as $(r)^{-3}$ with a numerator, whose angular dependence is such that the average over a spherical surface is zero. Since volume elements vary as $r^2 dr d\Omega$ the summation will be conditionally convergent in that one can get different answers depending on whether one first sums over a spherical shell and then over shells at different radii or the reverse [6,7]. Finally the answer depends on the detailed way one treats the limits at large and small r. The large r limit is treated automatically on satisfying the boundary conditions at the material surface. The small r limit, on the other hand is intractable since it depends ultimately on both near neighbour correlations and the shape of the molecule. Furthermore, the entire value of the summation will derive from a few near neighbouring molecules. Thus, different treatments of the near neighbours does not simply produce different small corrections to the effects of the neighbours on the local field of an individual molecule. Rather they represent the entire answer. A simple illustration of this fact is the local electric field inside a cavity in an otherwise uniform dielectric. Different cavity shapes give E(internal)-E(macroscopic) that vary from zero for a long thin cavity whose axis is parallel to E(macroscopic) to $(\varepsilon^{-1}-1)$ for a disc shape cavity with surfaces normal to E(macroscopic).

On the other hand we are fortunate that the near neighbour correlations in most condensed phases are not significantly temperature dependent. For example, the local environment of an individual molecule is probably the same in the isotropic liquid phase, the nematic and smectic liquid-crystalline phases, and even the crystalline phases immediately below the smectic phase. Thus, the local fields are automatically included in the treatment leading to eq. (5) if the $\boldsymbol{\alpha}$ introduced above is considered to be the dressed molecular polarizability tensor that includes the local field due to the surrounding molecules. If α is regarded as a constant microscopic property of a liquid crystal, the temperature dependence of $(\varepsilon_{zz}-\varepsilon_{xx})$ will yield the orientational order parameter of a liquid crystal [4,5]. We should not forget however that eq. (5) followed from eqs. (1) and (3) by assuming $\overline{D^2_{m,n}} = 0$ if $m,n \neq 0$; this need not be true generally.

Raman intensities

Raman cross sections can be obtained from the dressed polarizability if we imagine that for one molecule

$$\alpha = \alpha^o + \sum(\partial\boldsymbol{\alpha}/\partial Q_\mu)Q_\mu + \cdots\cdots , \qquad (6)$$

where $\{Q_\mu\}$ are the normal modes of vibration for the molecule [2]. Although there are a number of approximations involved here the most serious for our purposes is the implicit neglect of interactions with either different normal modes of the same molecule or with other degrees of freedom such as the modes on neighbouring molecules. In all probability the dominant effects of these interactions will be a small shift in the resonance frequency and an increased linewidth. On the other hand if the vibration is accompanied by a strong oscillating electric dipole moment intermolecular coupling will cause correlations between molecules that cannot be neglected.

In an isotropic material the Raman cross section for an individual molecule can be obtained by assuming an induced dipole moment $\delta p = \boldsymbol{\alpha}.E^o$ proportional to the incident field E^o. If the incident field is at frequency ω_o the cross section for Raman scattering by the mode Q_μ at frequency ω_μ is obtained by calculating the radiated power from a dipole $\boldsymbol{\delta p} = (\partial\boldsymbol{\alpha}/\partial Q_\mu)Q_\mu.\mathbf{E}^o$. Since Q_μ oscillates at ω_μ there are terms at $\omega_o+\omega_\mu$ (anti-Stokes) and $\omega_o-\omega_\mu$ (Stokes). If $\hbar\omega_\mu \gtrsim kT$ a quantum mechanical treatment of $|Q_\mu|$ is required but for our purposes the classical result is sufficient. For an incident field with polarization components $(\mathbf{e}^o)_i$ and scattered field with components $(\mathbf{e}^s)_j$ the Raman cross section is proportional to

$$\left(\sum_{i,j}(\mathbf{e}^s)_j(\partial\boldsymbol{\alpha}/\partial Q_\mu)_{ji}(\mathbf{e}^o)_i\right)^2. \tag{7}$$

Note that $\partial\boldsymbol{\alpha}/\partial Q_\mu$ is also a second rank tensor and everything we have said concerning the dependence of $\boldsymbol{\alpha}$ on the orientation of the molecule also holds for $(\partial\boldsymbol{\alpha}/\partial Q_\mu)$. The notation may be condensed by defining $\boldsymbol{\alpha}_\mu \equiv (\partial\boldsymbol{\alpha}/\partial Q_\mu)$ and $\boldsymbol{\alpha}_{M,\mu} \equiv \partial\boldsymbol{\alpha}_M/\partial Q_\mu$ so that in the laboratory frame

$$\alpha_\mu = \beta^o\alpha_\mu^{(o,o)} + \sum \boldsymbol{\beta}_m^2\alpha_{M,\mu}^{(2,n)}D_{m,n}^2(\Omega)^* \tag{8}$$

and

$$I_{ij}^\mu \equiv \overline{(\alpha_\mu)_{ij}(\alpha_\mu)_{ij}}. \tag{9}$$

The relative intensity of a Raman line due to mode μ when observed with incident light polarized along j and scattered light along i to the same mode observed with polarizations j' and i' is just the ratio $I_{ij}^\mu/I_{i'j'}^\mu$. Eq. (9) *cannot* be used to compare intensities of different modes, nor can it be used to compare Stokes to Anti-Stokes spectra without introducing other factors that we shall not go into here. Further, one should note that the derivation of eqs. (7) and (9) neglected, implicitly, correlations between different modes μ. This would be questionable if there were degenerate modes. Also we have assumed, implicitly, that the Raman line shape of the mode is identical when observed in two different polarizations. If this were not the case it would indicate that either correlation effects are important or the mode is degenerate so that the single variable Q_μ does not specify an eigenmode of the many body system. Alternatively, if rotational motions are so rapid that they must be included, explicitly, different polarizations will yield different line shapes and again Q_μ will not specify an eigenmode. For the large molecules that form liquid crystals, rotational motions are generally slow in comparison with vibrational life-times and can be neglected. If the spectral lineshapes are identical in different polarizations the I^μ_{ij} refer equally well to peak or integrated intensities.

At this point we have also neglected two factors that will be necessary for the study of Raman intensities on ordered phases. Firstly the radiation formula for different polarizations will vary since different dielectric constants are involved. Secondly there are geometrical optical effects that cause the radiation with different polarizations to be detected with different efficiency. These will be discussed in the next section.

Eq. (9) specifies the relative intensity for mode μ due to a single molecule. The experimentally observed intensity is obtained by averaging I^μ_{ij} over the appropriate angular distribution. Assuming α_μ is of the form described by eqs. (1) and (3),

$$I^\mu_{ij} = \overline{(\alpha_\mu)_{ij}(\alpha_\mu)_{ij}}, \tag{10}$$

where

$$\begin{aligned}\overline{\boldsymbol{\alpha}_\mu \boldsymbol{\alpha}_\mu} &= \beta^o \beta^o (\alpha_\mu^{(o,o)})^2 + \sum (\beta^2_m \beta^o + \boldsymbol{\beta}^o \boldsymbol{\beta}^2_m) \alpha_\mu^{(o,o)} \alpha_{M,\mu}^{(2,n)} \overline{D^{2*}_{m,n}} \\ &+ \tfrac{1}{2} \sum (\beta^2_m \beta^2_n \;\; \beta^2_n \beta^2_m) \alpha_{M,\mu}^{(2,n)} \alpha_M^{(2,n')} \overline{D^{2*}_{m,n} D^{2*}_{m',n'}}\end{aligned} \tag{11}$$

and

$$\overline{D^{2*}_{m,n} D^{2*}_{m',n'}} = \sum (2L+1) \begin{pmatrix} 2 & 2 & L \\ m & m' & m'' \end{pmatrix} \begin{pmatrix} 2 & 2 & L \\ n & n' & n'' \end{pmatrix} \overline{D^L_{m'',n''}}, \tag{12}$$

where the symbols are those defined by Edmonds [3]. If we restrict ourselves to systems that are macroscopically uniaxial the averages must be independent of α. Since $D^L_{m,n}(\alpha\beta\gamma) = \exp(im\alpha) d^L_{m,n}(\beta) \exp(in\gamma)$ where $d^L_{m,n}(\beta)$, is a function of β only, this requires m=0 in the second term and m+m'=0 in the third term of eq. (11). Also the identity

$$\begin{pmatrix} 2 & 2 & L \\ m & m' & 0 \end{pmatrix} = (-1)^L \begin{pmatrix} 2 & 2 & L \\ m' & m & 0 \end{pmatrix} \tag{13}$$

is sufficient to prove that the only non-vanishing terms in eq. (12) are those for L = 0,2,4. Further the properties of these 3-j symbols, which are related to the Clebsch-Gordan coefficients, are such that m = -m' in eq. (13)

$$\overline{\alpha_\mu \alpha_\mu} = \beta^o \beta^o (\alpha_\mu^{(o,o)})^2 + (\beta_o^2 \beta^o + \beta^o \beta_o^2) \alpha_\mu^{(o,o)} \alpha_{M,\mu}^{(2,n)} \overline{D_{o,n}^{2*}}$$

$$+ \tfrac{1}{2} \sum (\beta_m^2 \beta_{-m}^2 + \beta_m^2 \beta_{-m}^2) \alpha_{M,\mu}^{(2,n)} \alpha_{M,\mu}^{(2,n')} (2L+1) \begin{pmatrix} 2 & 2 & L \\ m & -m & 0 \end{pmatrix} \begin{pmatrix} 2 & 2 & L \\ n & n' & n'' \end{pmatrix} \overline{D_{o,n''}^{L}} \quad (14)$$

and

$$D_{o,n}^{L}(\alpha\beta\gamma) = (4\pi/2L+1)^{1/2} Y_{L,n}(\beta\gamma).$$

In the absence of any further simplification, eq. (14) indicates that the different Raman intensities are dependent upon five orientational averages of the form $\overline{Y_{2,n}}$, n=0, ±1, ±2 and nine of the form $Y_{4,n}$; n=0, ±1, ±2, ±3, ±4 for a total of 14 temperature dependent parameters. If the orientational statistics of the individual molecule have any symmetries this number can be reduced. In fact the only cases that are practical to study are where either there is some molecular axis such that $\overline{e^{in\gamma} f(\beta)} = 0$ for all n ≠ 0 or that there is some particular molecular axis where the only non-vanishing components of $\alpha_{M,\mu}$ are $\alpha_\mu^{(0,0)}$ and $\alpha_{M,\mu}^{(2,0)}$.

The first of these two cases corresponds to cylindrically symmetric statistics. The average $\overline{\alpha_\mu \alpha_\mu}$ is thus independent of γ implying the only non-vanishing terms in eq. (14) are those with n' and n''=0;

$$\overline{\alpha_\mu \alpha_\mu} = \beta^o \beta^o (\alpha_\mu^{(o,o)})^2 + (\beta_o^2 \beta^o + \beta^o \beta_o^2) \alpha_\mu^{(o,o)} \alpha_{M,}^{2,0)} \overline{D_{o,o}^{2}}$$

$$+ \tfrac{1}{2} \sum_{L(\text{even})} \sum (\beta_m^2 \beta_{-m}^2 + \beta_{-m}^2 \beta_m^2) \alpha_{M,\mu}^{(2,n)} \alpha_{M,\mu}^{(2,-n)} (2L+1) \begin{pmatrix} 2 & 2 & L \\ m & -m & 0 \end{pmatrix} \begin{pmatrix} 2 & 2 & L \\ n & -n & 0 \end{pmatrix} \overline{D_{o,o}^{L}} \quad (15)$$

and the only two temperature dependent quantities are

$$\overline{D_{o,o}^{2}} = \overline{P_2}, = \overline{(3\cos^2\beta - 1)/2}$$

and

$$\overline{D_{o,o}^{4}} = \overline{P_4}, = \overline{(35\cos^4\beta - 30\cos^2\beta + 3)/8}. \quad (16)$$

Although the coefficients of these quantities depend on the six independent components of $\alpha_{M,\mu}$ there are only five independent combinations in eq. (15); $\{\alpha^{(0,0)}\}^2$; $\alpha^{(0,0)}\alpha^{(2,0)}$ and $\alpha^{(2,n)}\alpha^{(2,-n)}$ for n = 0,1,2. In addition, as we are only interested in relative intensities, the number of unknowns is reduced to four relative values of the $(\alpha)^2$ terms plus the two temperature dependent quantities in eq. (16) to give a total of 6 unknowns. Since a second rank tensor can always be diagonalized, these unknowns can also be expressed in terms of the principal values of α_M, and three eulerian angles. In the molecular reference frame for which α is diagonal we take it to have the form

$$\left(\alpha_{M,\mu}\right)^P = \begin{pmatrix} a_\mu & \cdot & \cdot \\ \cdot & b_\mu & \cdot \\ \cdot & \cdot & 1 \end{pmatrix} . \tag{17}$$

The eulerian angles that transform the reference frame in which the molecular statistics have cylindrical symmetry into the frame in which α has the form given by eq. (17) are denoted by $(\alpha^M_\mu \beta^M_\mu \gamma^M_\mu)$. This would correspond to seven unknowns however, since we have assumed cylindrically symmetric statistics about the axis z_m, the Raman intensity will not depend on α^M_μ. This can be seen by considering figure 1. If the z_m axis in the figure represents the major axis for $(\alpha_{M,\mu})^P$ and the x,y,z axes represent the x_m, y_m, z_m axes, the cylindrical symmetry we have assumed implies the results for $\alpha_\mu \alpha_\mu$ must be independent of the angle α^M_μ. Thus the problem is reduced to six unknowns, that correspond to a_μ, b_μ, β^M_μ, γ^M_μ, $\bar{P}_2$ and $\bar{P}_4$.

The other special case, where eq. (14) simplifies, would be if the only non-vanishing components of $\alpha_{M,\mu}$ were $\alpha^{(0,0)}_\mu$ and $\alpha^{(2,0)}_{M,\mu}$. This corresponds to the case that $a_\mu = b_\mu$ and from eq. (2), $\alpha^{(0,0)} = (1+2a_\mu)/\sqrt{3}$, $\alpha^{(2,0)}_{M,\mu} = 2(1-\alpha_\mu)/\sqrt{6}$

$$\begin{aligned} \overline{\beta^o \beta^o \alpha_\mu \alpha_\mu} &= (a^{(0,0)}_\mu)^2 + \\ &(\beta^o \beta^o + \beta^o \beta^2_o) \alpha^{(0,0)}_\mu \alpha^{(2,0)}_M \bar{P}_2 \\ &+ \tfrac{1}{2} \sum_{L(\text{even})} \sum (\beta^2_m \beta^2_{-m} + \beta^2_{-m} \beta^2_m)(\alpha^{(2,0)}_{M,\mu})^2 \\ &\times (2L+1) \begin{pmatrix} 2 & 2 & L \\ m & -m & 0 \end{pmatrix} \begin{pmatrix} 2 & 2 & L \\ 0 & 0 & 0 \end{pmatrix} \bar{P}_L . \end{aligned} \tag{18}$$

The only unknowns in eq. (18) are $\bar{P}_2$, and $\bar{P}_4$ and a_μ. Unfortunately the $\alpha_{M,\mu'}$ for mode $Q_{\mu'}$, will not necessarily be diagonal in the same reference frame as $\alpha_{M,\mu}$ for $\mu \neq \mu'$. The transformation that rotates the laboratory axis into the molecular frame that diagnolizes $\alpha_{M,\mu}$ and $\alpha_{M,\mu}$ are respectively $R(\Omega_\mu)$ and $R(\Omega_{\mu'})$. In general there is no reason to expect $\overline{D^L_{m,n}(\Omega_\mu)}$ to be equal to $\overline{D^L_{m,n}(\Omega_{\mu'})}$ and the $\bar{P}_2$, $\bar{P}_4$ derived from studying the mode Q_μ will not necessarily be the same as the order parameters by studying $Q_{\mu'}$. If, however, there is some special axis that is naturally defined as a *molecular axis* one can define the $\bar{P}_2$ and $\bar{P}_4$ relative to that.

For example in the previous approximation we assumed one special molecular axis for which the statistics were cylindrically symmetric and concluded that there were six unknowns. If we add the condition that $a_\mu = b_\mu$ to this approximation the six unknowns are immediately reduced to five. However, this also eliminates from the final result the dependence on the eulerian angle γ_μ. The relative Raman intensities in this approximation only depend upon $a_\mu = b_\mu$, $\bar{P}_2$, $\bar{P}_4$ and one eulerian angle β_μ^M that rotates the principle axis of $\alpha_{M,\mu}$ into coincidence with the molecular axis about which the statistics are cylindrical; *i.e.* the molecular axis for which $\overline{e^{im\gamma} f(\beta)} = 0$. We will argue below that for the Schiff's base liquid crystals studied here the experimental evidence proves a unique axis does exist. The evidence suggests this axis approximately diagonalizes $\alpha_{M,\mu}$ and because of that the terms in eq. (14) which depend on γ_μ are small. Thus, eq. (15) is a reasonable approximation and the $\{\alpha_{M,\mu}\}$ are obtained from eq. (17) by rotations $R(\Omega_\mu^M)$ with small β_μ^M.

The general result embodied in eq. (15) has been written in terms of the various cartesian components of $\alpha_{M,\mu}$ and the statistical averages of trigonometric functions of α, β and γ [1]. These are rather cumbersome and will not be reproduced here. For a system with cylindrical symmetry there are only four independent intensities, I_{xx}, I_{zz}, I_{xz} and I_{xy}. Since only relative intensities enter our discussion there are three independent measurements that can be made

$$r_1 = I_{xz}/I_{zz},$$

$$r_2 = I_{xy}/I_{xx}$$

and

$$r_3 = I_{xy}/I_{yy}. \tag{19}$$

Although there as many as 19 unknowns in the most general case the simplifying approximations that we have discussed are practical and in the systems studied to date the various unknowns are reduced to a tractable number.

Examples

The physical conclusions that can be drawn from experimental results for $\overline{\alpha_\mu \alpha_\mu}$ will be better understood if we treat an example. Consider the case when incident light is polarized along y and scattered along **x**. Then

$$I_{xy}^\mu = \overline{(\alpha_\mu)_{xy}(\alpha_\mu)_{xy}} \tag{20}$$

and from table 1 we see that the only non-vanishing terms result from

$$(\beta_2^2)_{xy}(\beta_{-2}^2)_{xy} + (\beta_{-2}^2)_{xy}(\beta_2^2)_{xy} = 1/2.$$

In the special case when $a_\mu = b_\mu$ eq. (18) reduces to

$$I^\mu_{xy} = (\alpha^{(2,0)}_{M,\mu})^2/2$$

$$x\left\{\begin{pmatrix}2&2&0\\2&-2&0\end{pmatrix}\begin{pmatrix}2&2&0\\0&0&0\end{pmatrix}+5\begin{pmatrix}2&2&2\\2&-2&0\end{pmatrix}\begin{pmatrix}2&2&2\\0&0&0\end{pmatrix}\bar{P}_2+9\begin{pmatrix}2&2&4\\2&-2&0\end{pmatrix}\begin{pmatrix}2&2&4\\0&0&0\end{pmatrix}\bar{P}_4\right\} \tag{21}$$

so that evaluation of the 3-j symbols gives

$$I^\mu_{xy} = (105)^{-1}(1-a_\mu)^2[7-10\bar{P}_2+3\bar{P}_4] \tag{22}$$

or

$$I^\mu_{xy} = 8^{-1}(1-a\ )^2[1-2\cos^2\beta+\cos^4\beta], \tag{23}$$

since $\alpha^{(2,\ )}_{M,\mu} = (2/3)^{\frac{1}{2}}(1-a_\mu)$. Similarly, if we write $A=(1+2a_\mu)$, $D=2(1-a_\mu)$ then

$$I^\mu_{xx} = \frac{A}{9}+\frac{DA}{18}+\frac{11D}{228}\left\{\left(\frac{DA}{6}\right)+\frac{5D^2}{48}\right\}\overline{\cos^2\beta}+\left(\frac{3D^2}{32}\right)\overline{\cos^4\beta}. \tag{24}$$

An experimental determination of the ratio $r^\mu_3 \equiv I^\mu_{xy}/I^\mu_{xx}$ would fix one relation between $\overline{\cos^2\beta}$, $\overline{\cos^4\beta}$ and a_μ.

Eqs. (23) and (24) should also hold in the isotropic phase where $\overline{\cos^2\beta} = 1/3$ and $\overline{\cos^4\beta} = 1/5$, consequently $I^\mu_{xx} = \{9+12a_\mu+24(a_\mu)^2\}$ and $I^\mu_{xy} = (1-a_\mu)^2/15$, so that

$$(r_\mu)_{iso} \equiv \left(\frac{I^\mu_{xy}}{I^\mu_{xx}}\right)_{iso} = \frac{(1-a_\mu)^2}{3+4a_\mu+12(a_\mu)^2}. \tag{25}$$

If we know $\overline{\cos^2\beta}$ these two measurements are sufficient to determine $\overline{\cos^4\beta}$. Alternatively for a system with cylindrical symmetry there are two other intensity ratios that can be measured, cf. eq. (19), to obtain two further relationships between $\overline{\cos^2\beta}$, $\overline{\cos^4\beta}$ and a_μ. In summary there are three independent measurements that can be made at any temperature where the system is uniaxial. Together with the measurement in the isotropic phase there are four results and in the present approximation there are only three unknowns. The problem is over-determined and the simplifying assumptions can be tested for self-consistency.

Experimental Details

The measurement

All of the experiments to be described here are done on suitably oriented monodomain samples. For practical reasons, all of the samples were thin slabs contained between glass slides whose surfaces we pretreated to obtain either the homogeneous orientation (figure 2a) in which the director is parallel to the glass surface, or the homeotropic orientation (figure 2b) in which the director is normal to the glass surface. The incident laser beam, k_o was normal to the glass surface and

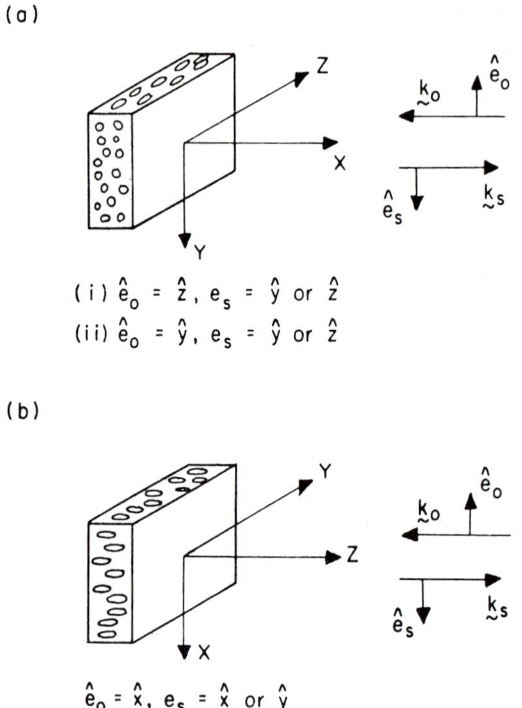

Fig.2 *Experimental sample geometries.*

Raman intensities were measured in the back scattering geometry, $\mathbf{k}_s$. The homogeneous orientation permits measurements of $r_1 = I_{xz}/I_{zz}$ (note $I_{xz} = I_{yz}$ for uniaxial systems) and $r_2 = I_{xz}/I_{xx}$ while the homeotropic orientation allows one to measure $r_3 = I_{xy}/I_{xx}$.

The principal features of the experimental geometry are illustrated in figure 3. The laser was an Argon-ion type and most of the data was recorded using the line at 5145 Å. Laser power was continuously monitored by the power meter as shown. The half-wave plate and polarizer P allowed selection of the incident polarization $\mathbf{e}_o$. The small mirror M_2 permitted the beam to be directed onto the sample. Back scattered light was collected by lens L_2. The aperture A_3 was varied to insure artefacts did not occur because of the finite solid angle of the collection optics. The polarizer P_2 selects the desired scattered polarization $\mathbf{e}_s$ and the light is then passed through the SPEX 1400 II double monochromator. The

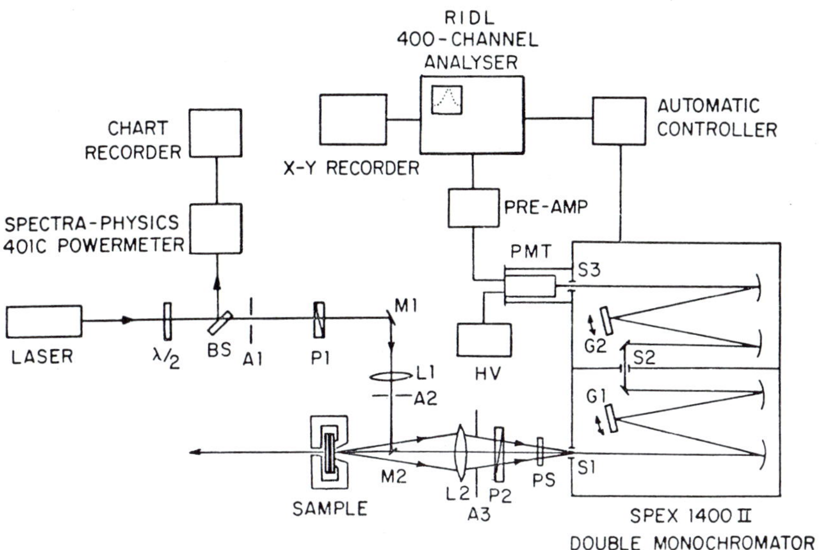

Fig.3 *Experimental details.*

monochromator itself partially polarizes unpolarized light and to avoid an artefact from this the oriented quartz wedge (PS) completely depolarizes the light before it enters the monochromator. The remaining details are fully described elsewhere [1].

Sources of error

Multiple scattering The principal source of error in Raman polarization measurements in the nematic phase is the quasielastic depolarized light scattering due to fluctuations in the director. The phenomenon was first studied by Chatelain in 1948 [8] and explained by de Gennes in 1968 [9]. The *Orsay Liquid Crystal Group* was the first to study the phenomenon using lasers and other modern optical techniques in 1969 [10]. From many points of view this experiment is the prototype of some of the most exciting light scattering studies that can be done on liquid crystals. On the other hand, the information obtained by those measurements relates primarily to the collective or macroscopic properties and since our purpose here is the discussion of microscopic or molecular properties of liquid crystals we limit ourselves to a brief discussion of the effect.

Eq. (4) illustrates the fact that as a result of the nematic ordering the macroscopic optical frequency dielectric constant is anisotropic. What is not illustrated there is that the principal axes of ε are not everywhere parallel to the same direction at a given instant of time.

Thermal fluctuations in the system cause some variation from point to point. Of course the system is macroscopically uniaxial and on *average* the principal axes are everywhere parallel but local fluctuations do exist. If the instantaneous local symmetry axis is defined by a director **n** such that <**n**> = **z** the local fluctuations induce components $\delta\varepsilon_{xz}$ and $\delta\varepsilon_{yz}$ De Gennes explained that if

$$\delta\varepsilon_{xz}(q) = \int dr\ \delta\varepsilon_{xz}(r)\exp(-i\mathbf{q}.\mathbf{r}), \tag{26}$$

the thermal average is given, for small q, by

$$\overline{|\delta\varepsilon_{xz}(q)|^2} \sim kT/Kq^2, \tag{27}$$

where the three Frank elastic constants K have been equated. As a result of the $1/q^2$ dependence there is very intense depolarized light scattering in the near forward direction. Indeed the effect is quite apparent on visual observation of an oriented liquid crystal with polarized light. For our purposes one principal point is that polarized incident light will be nearly elastically scattered and in the process will become partially depolarized on propagating through the sample. Equally important is that Raman scattered light with specific polarizations will be elastically scattered before leaving the sample and the polarization intensities measured outside the sample will not be a true measure of the Raman selection rules. This effect can be accounted for my measuring the Raman intensities at several different thicknesses and extrapolating to zero thickness. Figure 4 is a plot of the observed ratios $R_1 = I_{xz}/_{zz}$ and $R_2 = I_{xz}/I_{xx}$ for one sample as a function of sample thickness. Note that for the thinnest sample that yielded an acceptable signal to noise ratio the depolarization effect is still significant. Note also that the observed R_1 and R_2 differ from the theoretical r_1 and r_2 discussed previously {eq. (19)} because of the effects that are caused by the optical anisotropy of the liquid crystal.

Optical anisotropy The expressions for power radiated by an oscillating electric dipole in an isotropic media depend on the value of the optical dielectric constant. Similarly in anisotropic media the ratio of radiated power to the square of the magnitude of the oscillating dipole depends on the anisotropic dielectric tensor. Lax and Nelson treated this problem and we quote their results [11] for the geometries of interest here (cf. figure 2).

Assuming an incident electro-magnetic field with E = 1 inside the sample the radiated power densities normal to the sample surface are related to the $(\alpha_\mu)_{ij}$ by

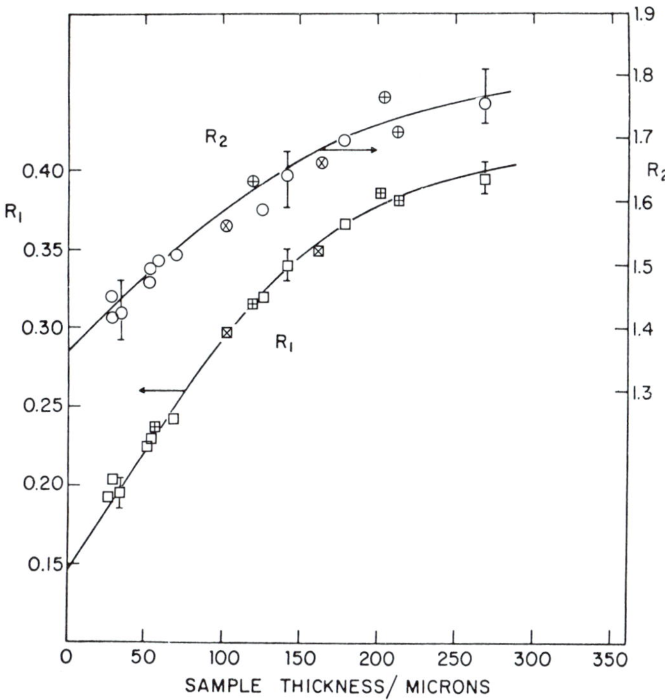

Fig. 4 *Typical Raman intensity ratios as a function of thickness {see eq. (30)}*.

$$(I^{\mu}_{zi}) = A\varepsilon_{xx}\varepsilon_{zz}^{-1/2}\overline{(\alpha_{\mu})_{zi}(\alpha_{\mu})_{zi}}, (I^{\mu}_{xi})=A\varepsilon_{xx}^{1/2}\overline{(\alpha_{\mu})_{xi}(\alpha_{\mu})_{xi}}, \qquad (28)$$

where i = x, y or z, and A is the same numerical factor for both geometries. According to eq. (28) $I^{\mu}_{zx} \neq I^{\mu}_{xz}$ if $\varepsilon_{xx} \neq \varepsilon_{zz}$. This is spurious and points out that we have omitted other corrections. One factor which might be considered is that the spectrometer outside the sample only collects the fraction of radiation that falls within the solid angle defined by the instrument optics. For an optically isotropic system of dielectric constant ε a given small solid angle Ω outside of a flat surface corresponds to the smaller solid angle Ω/ε inside. For both the geometries and all polarizations illustrated in figure 2 this factor is given for a uniaxial medium by $\Omega_{inside}=\Omega_{outside}/\varepsilon_{xx}$. For our purposes this phenomenon has no effect. On the other hand the Fresnel factors for transmission and reflection through the liquid crystal surface are important. The ratio of the intensity outside to inside for light leaving the liquid crystal is proportional to $4\varepsilon_{ii}^{1/2}/(n+\varepsilon_{ii}^{1/2})^2$ for light polarized along i where n is the index of refraction outside. A similar expression without the $\varepsilon_{ii}^{1/2}$ in the numerator, describes the ratio of intensities inside to outside for light entering the liquid crystal. These factors can be combined with eq. (28) to obtain the formulae appropriate to the measured Raman intensities

$$I^{\mu}_{ij} = I^{\mu}_{ji} = A(n+\varepsilon^{1/2}_{ii})^{-2}(n+\varepsilon^{1/2}_{jj})^{-2}\overline{(\alpha_{\mu})_{ij}(\alpha_{\mu})_{ij}}. \tag{29}$$

The I^{μ}_{ij} are now in the form appropriate to cross sections, in that they specify the ratio of measured intensities to incident intensities. In place of eq. (19) the measured ratios of Raman intensities are given by

$$R_1 = (n+\varepsilon^{1/2}_{zz})^2(n+\varepsilon^{1/2}_{xx})^{-2}\ \overline{\alpha^2_{xz}}/\overline{\alpha^2_{zz}},$$

$$R_2 = (n+\varepsilon^{1/2}_{xx})^2(n+\varepsilon^{1/2}_{zz})^{-2}\ \overline{\alpha^2_{xz}}/\overline{\alpha^2_{xx}} \tag{30}$$

and

$$R_2 = r_3 = \overline{\alpha^2_{xy}}\Big/\overline{\alpha^2_{yy}}.$$

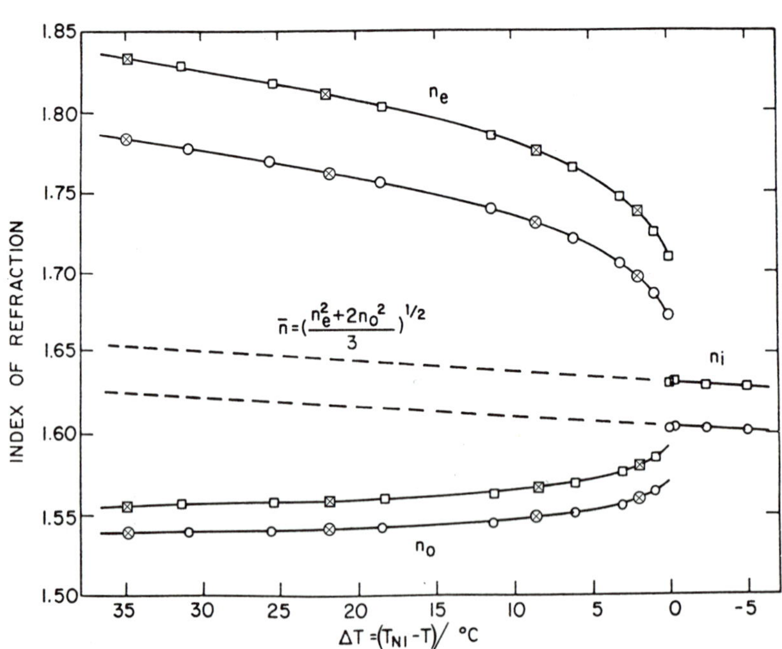

Fig. 5 *Indices of refraction of 20% BBCA in MBBA. Squares for λ = 5145Å, circles for λ = 6328Å. T_{NI} = 55.25°C for crossed points and 53.78°C for open points.*

Figure 5 illustrates values of $n_z = \sqrt{\varepsilon_{zz}}$ and $n_o = \sqrt{\varepsilon_{xx}}$ for a mixture of 4-n-butyloxybenzylidene-4'-cyanoaniline (BBCA) and 4-methoxybenzylidene-4'-n-butylaniline (MBBA); these are typical of most thermotropic liquid crystals. If the index of refraction for the glass substrates containing the liquid crystal is taken to be 1.5 the correction factor R_1/r_1 is $(1.8 + 1.5)^2\ (1.5 + 1.5)^{-2}$ *i.e.* ∼ 1.21; this would produce a 20% error in the experimental value for $\overline{\alpha^2_{xz}}/\overline{\alpha^2_{zz}}$ if it were neglected.

Measurements and Interpretation

Cyano vibration of BBCA in MBBA

The physical case that allows the simplest analysis is a molecular vibration for which $a_\mu = b_\mu << 1$ see eq. (17). The stretching vibration of the cyano group of 4-n-butyloxybenzylidene-4'-cyanoaniline is responsible for a Raman line at 2225 cm^{-1} that has this property. The structural similarity of the two molecules, together with the observed solubility of BBCA in MBBA lead us to hope that the orientational order of BBCA in MBBA would parallel that of pure MBBA. Since the order parameter of MBBA is well known we felt that an ideal system in which to evaluate the theoretical ideas sketched above would consist of trace amounts of BBCA in MBBA. Unfortunately, below 20% concentrations the signal to noise we were able to achieve, with samples that were sufficiently thin to avoid multiple scattering effects was rather poor for practical measuring times. We therefore studied the 20% mixture in depth. Since most of the details of this experiments have been published [1], we will restrict the present remarks to a few of the principal points.

First, experiments in the isotropic phase {see eq. (25)} determined a_μ to be 0.05 ± 0.02. Second, a series of experiments in the nematic phase determined both R_1 and R_2 as a function of $\Delta T = T_{NI} - T$. The errors in R_3 were too large to allow us to make use of this measurement in the 20% mixture. We analyzed the data for R_1 and R_2 by assuming cylindrical statistics about the principal long axis of BBCA. Using a space filling molecular model of BBCA we estimate the CN axis makes an angle β^M_μ of $10^o \pm 2^o$ with respect to that axis. The equations derived for the example (cf. page 395) apply with only slight modifications. For example, in place of $D = 2(1-a_\mu)$ we write $D = (1-a_\mu)(3\cos^2\beta^M_\mu-1)$; for $\beta^M_\mu = 10^o \pm 2^o$ this corresponds to $D = 1.9(1-a_\mu) \pm .03(1-a_\mu)$; the entire analysis has been published [1]. The principal Raman results are shown as open circles in figure 6. This can be compared with the temperature dependence of $\bar{P}_2$ inferred from the anisotropy in the optical index of refraction as measured on the mixture of 20% BBCA in MBBA. The data was scaled such that this determination of $\bar{P}_2$ agreed with the Raman value at $T_{NI}-T = 2^oC$ and is shown as + signs in figure 6.

Figure 6 also contains data from a number of different experiments on pure MBBA in other laboratories. Within experimental error one can see that the relative values of $\bar{P}_2$ determined by the Raman measurements on the BBCA dopant are essentially the same as the results for both pure MBBA

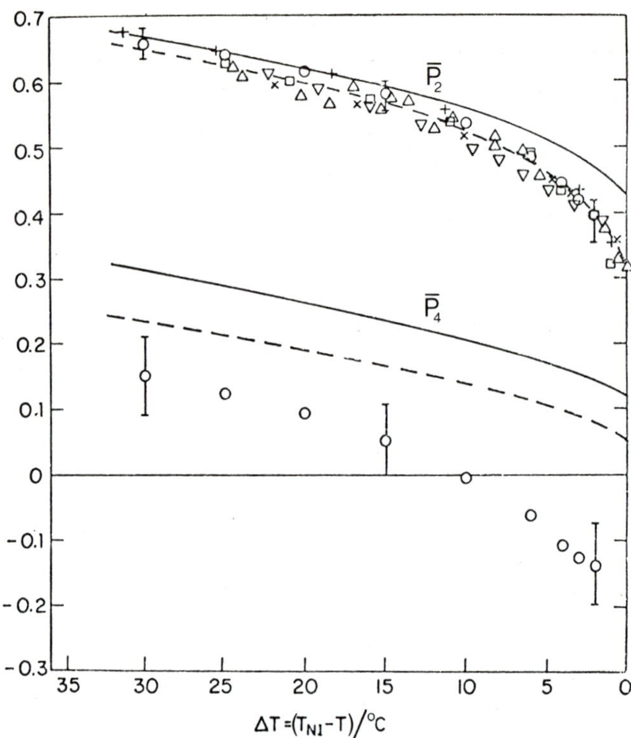

Fig. 6 *Theoretical and experimental values of the nematic order parameters* $\bar{P}_2$ *and* $\bar{P}_4$ *of 20% BBCA in MBBA. Solid lines: Maier-Saupe theory; dashed lines: Humphries-James-Luckhurst theory with* $\lambda = -0.55$ *and* $\gamma = 10 \pm 2$ [12]. *O: Raman measurements;* □*: NMR*[13]*; crosses: relative measurements from optical dielectric anisotropy (+ : 20% mixture; x: neat MBBA); triangles: relative measurements from diamagnetic anisotropy (Δ: from Ref. 14, Δ: from I. Haller,* [15] *∇: from P. Rose* [26]*).*

using a variety of techniques and also for the same mixture using the optical anisotropy. In addition the absolute values of $\bar{P}_2$ determined by deuteron NMR experiments on pure MBBA are identical to the Raman results for the 20% mixture (cf. Chapter 15). The conclusion we draw is that the Raman technique, and the assumptions necessary for analysis of the data are empirically shown to yield acceptable values for $\bar{P}_2$. Unfortunately there have not been other experimental measurements of $\bar{P}_4$ with which to compare the Raman results. The solid and broken lines represent different theoretical models for $\bar{P}_2$ and $\bar{P}_4$; however, since the theory of nematic order is described in Chapters 4, 7, and 8 we will not elaborate on the comparison beyond pointing out the significance of the discrepancy.

The quantity,

$$\sigma^2 = \{\cos^4\beta_\mu - (\cos^2\beta_\mu)^2\}/(\cos^2\beta_\mu)^2 \tag{31}$$

is a different measure of the degree of disorder in a nematic liquid crystal to $\bar{P}_2$. Conceivably there might be a system in which $\bar{P}_2 \neq 1$ but $\sigma^2 = 0$. This would correspond to all molecules making the identical angle $\beta_\mu \neq 0$ with respect to the z axis but with uniform probability for angles α_μ and γ_μ. The order parameter $\bar{P}_4$ can be expressed in terms of $\bar{P}_2$ and σ^2 as

$$\bar{P}_4 = [140(1+\sigma^2)\bar{P}_2)^2-20(2-7\sigma^2)\bar{P}_2-74(4-5\sigma^2)]/72. \quad (32)$$

Figure 7 is a plot of experimental values $\bar{P}_4$ *vs.* $\bar{P}_2$. For reference eq.

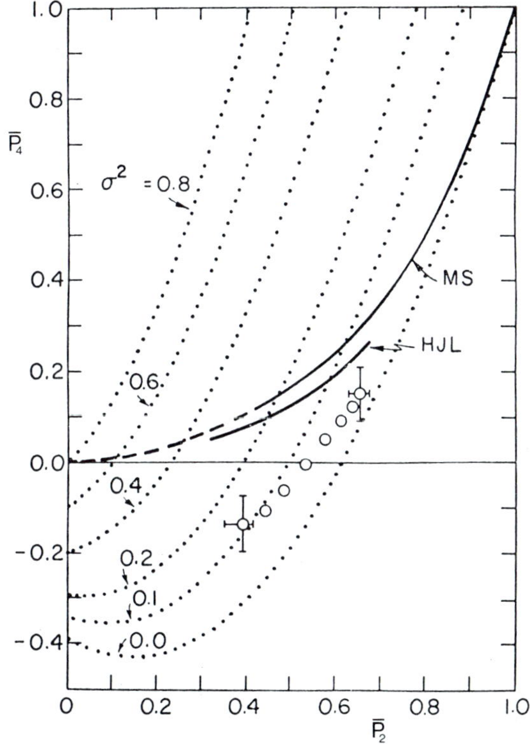

Fig.7 *Variation of $\bar{P}_4$ versus $\bar{P}_2$. Circles are data points from Raman measurements. Dotted lines are contours of constant σ^2. Solid lines are predictions of mean field theories.*

(32) is plotted on the same axis for different values of σ^2. One can immediately see that the experimental data imply values of σ^2 that are smaller than predicted by the theoretical models even though they yield the same value of $\bar{P}_2$ as experiment. This would be consistent with a molecular model in which the most probable molecular orientation was *not* parallel to the macroscopic director [17].

Intrinsic Vibrations of MBBA

Although the symmetry and electronic properties of the cyano group in BBCA satisfied the necessary approximations for analysis of the Raman data the technique of using probe molecules has obvious disadvantages. Most important of these is the need to use concentrations of BBCA far in excess of what might be termed trace amounts. In this section we discuss experiments to determine empirically whether or not any of the normal modes of MBBA can be utilized in the same way as the stretching vibration of the cyano group of BBCA. A comparative study of the Raman spectra of benzene, aniline, benzaldehyde, MBBA, BBCA, etc. suggests that Raman lines at 1576, 1597 and 1626 cm^{-1} are associated with normal modes of the rigid central portion of Schiff's base liquid crystals. Table 2

Table 2. *Raman depolarization ratios of vibrational bands at 1576 and 1625 cm^{-1} for 20% BBCA in MBBA. Ratios other than R_{iso} are extrapolated values at zero thickness.*

$\Delta\omega/cm^{-1}$	$\Delta T/^{o}C$	R_1	R_2	R_3
1576	25	0.160 ± .010	1.916 ± .110	0.496 ± .020
	15	0.182	1.530	0.464
	4	0.227 ± .015	1.105 ± .060	0.419 ± .020
	R_{iso} =	0.318 ± .012		
1625	25	0.145 ± .011	1.320 ± .050	0.362 ± .020
	15	0.165	1.145	0.353
	4	0.205 ± .015	.892 ± .070	0.334 ± .020
	R_{iso} =	0.281 ± .010		

lists selected experimental results for the 1576 and 1625 cm^{-1} lines of the 20% BBCA - MBBA mixture. If $a_\mu = b_\mu = 0$ eq. (25) yields $(r^\mu)_{iso}$ = 0.333. Although the experiment shows that this is not the case it does indicate that Raman polarizabilities $(\alpha_{M,\mu})$ are sufficiently anisotropic that information on the orientational statistics should be available.

Assuming the existence of a molecular axis about which there is cylindrical symmetry we analyzed the data for R_1, R_2, R_3 and R_{iso} using explicit expressions derived from eqs. (15) and (17). At any temperature in one of the ordered phases we have three measurements which, together with R_{iso}, gives four relations between the six unknowns a_μ, b_μ, β^M_μ, γ^M_μ, $\bar{P}_2$ and $\bar{P}_4$. We then assume $\gamma^M_\mu = 0$ and solve for a_μ, b_μ, and $\bar{P}_2$ and $\bar{P}_4$

for different values of β_μ^M. Since we obtain three new measurements for each temperature and there are only two new unknowns for each temperature, with the measurement of R_{iso} and three temperatures there are 10 relations amongst 10 unknowns. In the present case we simplified the procedure by choosing the value of β_μ^M that gave the previously determined value of $\bar{P}_2$ for each temperature. Table 3 shows the results of this exercise for three Raman lines and the following points should be noted.

Firstly, the values of $\bar{P}_2$ and $\bar{P}_4$ quoted in Table 3 are shown in figure 8 superimposed on the previously quoted results from the cyano vibrations of BBCA. The agreement is excellent.

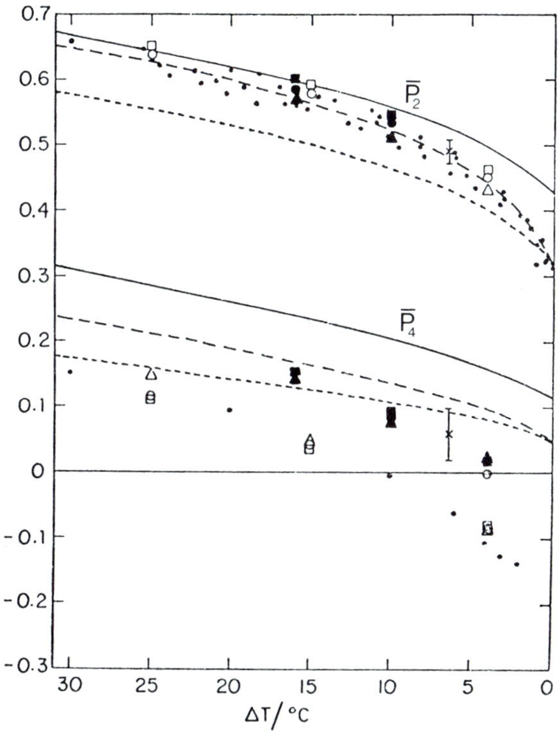

Fig. 8 *Order parameters $\bar{P}_2$ and $\bar{P}_4$ deduced from the measurements of the 1576 (squares), 1597 (triangles) and the 1625 (circles) cm^{-1} bands. The open data points for the 20% mixture, and the closed data points for pure MBBA. The cross is for the 10% mixture. Dots are data from figure 6. Solid lines: MS theory; dashed lines: HJL theory (- - -: γ = 10; ---: γ = 4; λ = -0.55)* [12].

Secondly, the values of a_μ and b_μ are considerably less than unity. This is rather fortuitous since the combination of small values for β_μ and small values for a_μ and b_μ means that our results for $\bar{P}_2$ and $\bar{P}_4$ are very insensitive to the assumptions we have made concerning cylindrical statistics. In fact we repeated the analysis with various choices for γ_μ^M without obtaining any significant variation in the determined values

Table 3. *Calculated values of $\bar{P}_2$, $\bar{P}_4$, a, b, the trace, and the anisotropy for the three bands of the 20% mixture. Uncertainties include all the statistical uncertainties listed in table 2 together with $\beta_\mu = 1^o \pm 1^o$ for the 1576 and 1625 cm^{-1} bands, and $5^o \pm 2^o$ for the 1597 cm^{-1} band.*

$\Delta T/^oC$	$\bar{P}_2$	$\bar{P}_4$	(a.b) (1)		(a.b) (2)		a + b + 1	$\frac{2a - b - 1}{a + b + 1}$	$\frac{2b - a - 1}{a + b + 1}$	$\frac{2 - a - b}{a + b + 1}$
				$\Delta\omega = 1576\ cm^{-1}$						
25	.65±.01	.11±.03	-.14±.01	.20±.13			1.06±.04	-1.39±.08	-.44±.08	1.84±.12
					.20	-.14	1.06	-.43	-1.40	1.84
15	.59	.04	-.15	.21			1.06	-1.43	-.39	1.82
					.22	-.15	1.06	-.39	-1.43	1.82
4	.46±.02	-.08±.03	-.16±.02	.24±.05			1.07±.05	-1.45±.07	-.34±.11	1.80±.14
					.24	-.16	1.07	-.34	-1.45	1.80
				$\Delta\omega = 1597\ cm^{-1}$						
25	.64±.01	.15±.03	-.12±.02	.16±.03			1.04±.04	-1.35±.09	-.52±.08	1.88±.11
	.64±.01	.15±.03			.17	-.13	1.04±.04	-.50± .08	-1.37±.09	1.88±.11
15	.58	.04	-.13	.17			1.04	-1.38	-.50	1.88
	.58	.04			.17	-.13	1.04	-.49	-1.39	1.88
4	.43±.02	-.08±.03	-.14±.02	-1°±.04			1.04±.05	-1.40±.10	-.48±.12	1.89±.16
	.43±.02	-.87±.03			.18	-.14	1.04±.05	-.46±.12	-1.42±.10	1.88±.14
				$\Delta\omega = 1625$ cm-1						
25	.63±.01	.11±.03	-.09±.02	.22±.03			1.13±.04	-1.26±.04	-.40±.08	1.66±.10
					.22	-.09	1.13	-.40	-1.26	1.66
15	.58	.04	-.10	.24			1.13	-1.28	-.37	1.65
					.24	-.10	1.13	-.37	-1.28	1.65
4	.45±.02	-.07±.04	-.11±.03	.25±.04			1.40±.05	-1.30±.09	-.33±.13	1.63±.04
					.25	-.11	1.14	-.33	-1.30	1.63

for $\bar{P}_2$ and $\bar{P}_4$. We also introduced statistical distributions with specific *ad hoc* deviations from cylindrical symmetry and repeated the analysis to find $\bar{P}_2$ and $\bar{P}_4$ were essentially unchanged. Other tests of the self-consistency of this analysis are described in the paper by Jen *et al.* [1].

Figure 8 also contains the results of this same analysis for pure MBBA. It is interesting to note that the values for $\bar{P}_4$ in the 20% mixture are different from the values for pure MBBA. The one data point that is included for 10% BBCA in MBBA is essentially identical to the result for MBBA.

In summary, we have made the following points:

1. the Raman polarizability of the 225 cm^{-1} cyano vibration of BBCA has the simple form that allows straightforward analysis of the Raman intensities;
2. although multiple scattering effects forced us to use thin samples with accompanying small scattering cross sections, it was possible to study the cyano vibration in a mixture of 20% BBCA in MBBA;
3. the value of $\bar{P}_2$ deduced from these measurements agreed with values obtained from the optical anisotropy on the 20% mixture;
4. the value of $\bar{P}_2$ obtained from the 2225 cm^{-1} line in the 20% mixture was identical to the values of pure MBBA that were measured at the same T_{NI}-T by a number of different techniques;
5. the value of $\bar{P}_4$ obtained from the 2225 cm^{-1} line in the 20% mixture was considerably smaller than the values predicted by mean field theories. This corresponds to values of $\overline{(\cos^2\beta_\mu - \overline{\cos^2\beta_\mu})^2}$ that are smaller than the values obtained from mean field theories;
6. the Raman polarizability of some of the intrinsic vibrations of the rigid central portion of MBBA and BBCA have the same simple form as the 2225 cm^{-1} cyano vibration;
7. polarized Raman studies using the modes just mentioned can be analysed to obtain $\bar{P}_2$ and $\bar{P}_4$ for pure MBBA and also on the 20% BBCA-MBBA mixture;
8. the result of such a study on both pure MBBA and on the 20% mixture gave values of $\bar{P}_2$ and $\bar{P}_4$ that were independent of the mode being studied;
9. for the 20% mixture the values of $\bar{P}_2$ and $\bar{P}_4$ deduced from studying the cyano vibration of BBCA were identical to the values deduced from study of the intrinsic MBBA lines in the 20% mixture;
10. the value of $\bar{P}_4$ deduced from studying the intrinsic MBBA lines was different for the 20% mixture and pure MBBA. For one temperature point a 10% mixture obtained results identical to pure MBBA.

BBOA

A series of measurements was made on the compound 4-n-butyloxybenzylidene-4'-n-octylaniline (BBOA). We shall not review the analysis of these results since it would involve almost no new basic ideas. Suffice to say we studied the 1597 cm^{-1} line of BBOA in detail. Values of $\bar{P}_2$ and $\bar{P}_4$ are shown in figure 9 as a function of temperature; values of

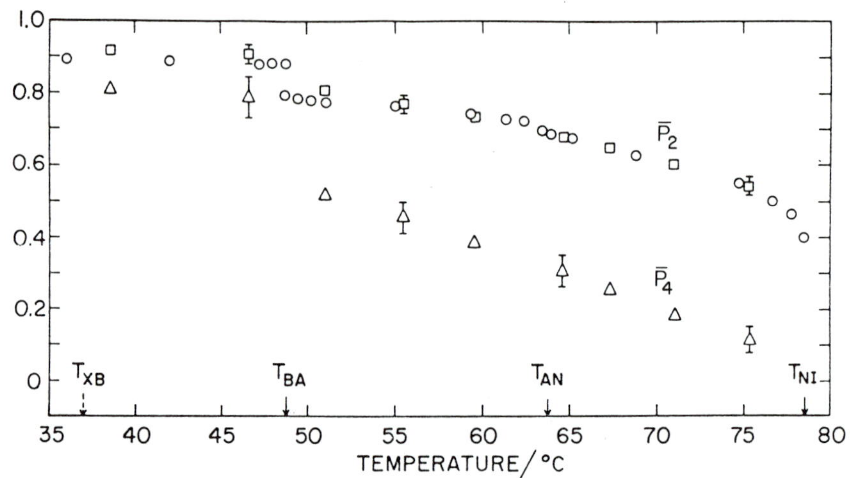

Fig. 9 *Order parameters $\bar{P}_2$ (squares) and $\bar{P}_4$ (triangles) calculated from Raman measurements of the 1597 cm^{-1} band. Circles are relative values obtained from the optical dielectric anisotropy measurements.*

$\bar{P}_2$ deduced from the optical anisotropy are also shown. Firstly, we can see that both the Raman measurement and the optical anisotropy determination of $\bar{P}_2$ are identical in all three liquid crystalline phases. Secondly, in contrast to the 20% BBCA-MBBA mixture, $\bar{P}_4$ is positive.

Figure 10 is a plot of $\bar{P}_4$ *vs.* $\bar{P}_4$ for BBOA. There are two very significant aspects of this plot. Firstly, note that the data never give a pair of values $\bar{P}_2$, $\bar{P}_4$ for which $\sigma^2 < 0$. This is an important check on the analysis used in the Raman technique since positive values of σ^2 are not built into the analysis and if the approximations discussed earlier were not valid negative values of σ^2 could have obtained. Secondly, note that in contrast to the 20% BBCA-MBBA mixture the result for BBOA is consierably closer to the result of the Maier-Saupe mean field theory. Unfortunately, the best agreement only occurs in the smectic phases since the results for the three highest temperatures in the nematic phase all yield σ^2 less than the values predicted by the Maier-Saupe theory. Thirdly, although we did not plot the results for pure MBBA in this same way

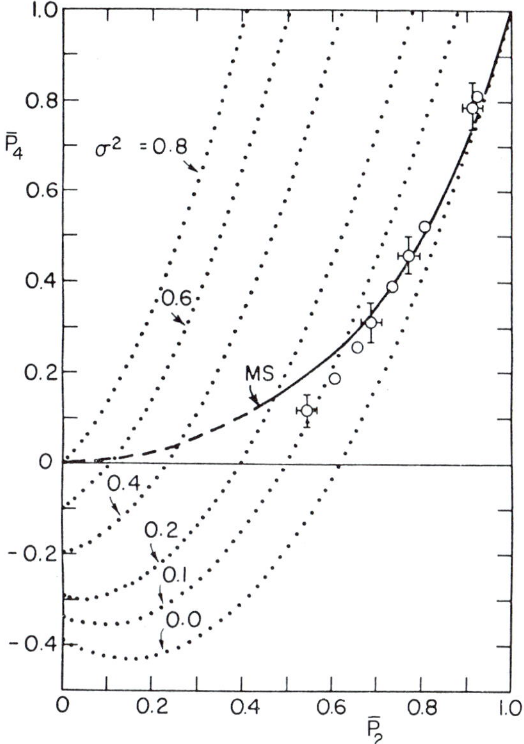

Fig. 10 *$\bar{P}_4$ versus $\bar{P}_2$ for BBOA. The solid line is from the Maier-Saupe theory. The circles are Raman data points taken from figure 9.*

the values of $\bar{P}_2$ and $\bar{P}_4$ are such that they fall on a smooth extrapolation of the BBOA results to lower values of $\bar{P}_2$.

Final Remarks

In conclusion, we would argue that we have empirically demonstrated that polarized Raman studies are capable of determining both the usual nematic order parameter $\bar{P}_2$ and the higher order parameter $\bar{P}_4$ in Schiff's base mesogens. The order parameter $\bar{P}_4$ can be related to the mean square fluctuations in the quantity $\bar{P}_2$. For the mixture of 20% BBCA in MBBA the fluctuations are considerably less than those predicted by mean field theory. In the case of pure MBBA and BBOA the fluctuations in the nematic phase are less than those predicted by mean field theory, but not nearly as much so as in the case of the 20% mixture.

Although we believe the Raman technique should generally be applicable to a wide class of liquid-crystalline compounds, one should not lose sight of the considerable number of plausible approximations that have to be made in order to analyze the results. For each new class of material studied, it is important to examine the results for self-consistency. Wherever possible, multiple experiments should be done to see that identical results are obtained in separate ways.

Acknowledgement

I would like to express my appreciation to Shen Jen who carried out the bulk of the work described in this Chapter as part of his Ph.D. thesis. Both Noel Clark and Eldon Priestley were major contributors to this research and I am grateful to them and Shen Jen.

All of us express our appreciation for the support we received from the Joint Services Electronics Program (U. S. Army, Navy, and Air Force) under Contract No. N00014-75-0-0648 and by the National Science Foundation under Grants No. DMR-72-02088 and DMR-72-03020-A05.

References

1. Shen Jen, Noel A. Clark, P.S. Pershan and E.B. Priestly, *J. Chem. Phys.* 66, 4635, (1977).
2. G. Placzek, *Handbuck der Radiologie* (E. Marx, Ed), *Lpz., Akad, Verlag,* VI 2, 209, (1934).
3. A.R. Edmonds, *Angular Momentum in Quantum Mechanics* (Princeton University Press, Princeton, 1957).
4. S. Chandrasekhar and N.V. Madhusudana, *J. Phys.* (Paris), **30**, C4-24 (1969).
5. M.F. Vuks, *Optics and Spectroscopy,* 20, 361 (1966).
6. B.R.A. Mijboer and F.W. DeWette, *Physica,* 23, 309 (1957); 24, 422 (1958); 24, 1105 (1958).
7. F.W. Dewette, *Physica,* 25, 1225 (1959); *Phys. Rev.* 123, 103 (1961).
8. P. Chatelain, *Acta Cryst.* 1, 315 (1948).
9. P.G. Gennes, *Compt. Rend.* 266B, 15 (1968).
10. Orsay Liquid Crystal Group, *Phys. Rev. Letters,* 22, 1361 (1969).
11. (a) M. Lax and D.F. Nelson, *Phys. Rev.* B4, 3694 (1971).
    (b) M. Lax and D.F. Nelson, *Proc. of the IIIrd Rochester Conf. on Coherent and Quantum Optics,* L. Mandel and E. Wolf, Eds. (Plenum Press, New York, 1972).
12. R.L. Humphries, P.G. James and G.R. Luckhurst, *J. Chem Soc. Faraday Trans II,* **68**, 1031 (1972).
13. J.S. Lee, Y.Y. Hsu and D.H. Dolphin, presentation at the *ACS Symp. on Ordered Fluids and Liquid Cryst.* (Chicago, 1973).
14. H. Gasparoux, B. Regaya, and J. Prost, *C.R. Acad. Sci.* 272B, 1168 (1971); H. Gasparoux and J. Prost, *J. Phys.* (Paris), 32, 953 (1971); G.Sigaud and H. Gasparoux, *J. de Chim. Phys.* **70**, 699 (1973).
15. I. Haller, *J. Chem. Phys.* 57, 1400 (1972).
16. P.I. Rose, presentation at the *IVth International Conference on Liquid Crystals,* (Kent, Ohio, 1972).
17. F. Volino, A.J. Dianoux and H. Hervet, *Mol. Cryst. Liq. Cryst.* 37, 483 (1976).

Chapter 18

ROTATIONAL DYNAMICS

PIER LUIGI NORDIO and ULDERICO SEGRE

Istituto di Chimica Fisica dell'Università di Padova, Italy

Introduction

Molecular orientations are completely specified when the Euler angles $(\alpha\beta\gamma)\equiv\Omega$ which relate the laboratory fixed reference frame to the molecule axis system are known, and any orientation dependent molecular quantity can be expressed in terms of the Wigner rotation matrix elements $D^L_{p,q}(\alpha\beta\gamma)$, which span the space of all functions of Ω [1]. In other Chapters some experimental methods have been described which can probe the dynamics of the orientational motions in liquids. Their common feature is that the relevant spectroscopic quantities are Fourier transforms of the correlation functions

$$G^{LL'}_{pp'qq'}(t) = \overline{D^L_{p,q}\{\Omega(o)\}D^{L'*}_{p',q'}\{\Omega(t)\}} - \overline{D^L_{p,q}}\,\overline{D^{L'*}_{p',q'}}, \qquad (1)$$

where the upper bar denotes an equilibrium average over angular variables.

The exact evaluation of the correlation functions (1) is a formidable task, its mathematical difficulty being comparable to the problem of evaluating the complete partition function of a system, needed to compute any equilibrium property [2]. However the problem of molecular motion in a condensed medium becomes tractable when we relinquish hope of solving it in a deterministic way and resort instead to the stochastic approach. The position and/or the orientation of a molecule is assumed to be a random variable, whose time dependence is not given by a law of motion in a mechanical sense. The well-developed methods of probability theory are then used to derive expressions for the correlation functions. Let us revise briefly the foundations of the theory of the stochastic processes [3], with special reference to molecular reorientations.

Stochastic Processes

Let $X(t)$ be a stochastic variable (or a set of stochastic variables) and $f(x,t)dx$ the probability of finding X between x and $x+dx$ at time t.

The probability distribution $f(x,t)$ is not sufficient to describe completely the time evolution of $X(t)$, and higher order probability distribution functions must be known. The second order distribution function $W_2(x_1t_1;\ x_2t_2)$ is the probability density for finding X between x_1 and $x_1 + dx_1$ at time t_1 *and* between x_2 and $x_2 + dx_2$ at time t_2. The n-th order distribution function is defined in the same way as a joint probability of finding X in x_1, $x_2 \ldots\ldots x_n$ respectively at times t_1, $t_2 \ldots\ldots t_n$. Obviously, when the function $W_n(x_1t_1;\ x_2t_2;\ \ldots\ ;\ x_nt_n)$ is known, all the lower distributions can be found. We have in fact

$$W_{n-1}(x_1t_1;\ldots;x_{i-1}t_{i-1};x_{i+1}t_{i+1};\ldots x_nt_n) = \int dx_i W_n \tag{2}$$

and so on until $W_1(x_1,\ t_1) \equiv f(x_1,\ t_1)$ is obtained. So far no assumption is made concerning the nature of the random process responsible for the time dependence of the variable X. The various types of process may be classified according to the order of the highest distribution function needed to describe the time evolution of X.

First order process or purely random process

Only W_1 is needed, and the higher order distribution functions are given by

$$W_n(x_1t_1;\ldots;\ x_nt_n) = W_{n-1}(x_1t_1;\ldots x_{n-1}t_{n-1})W_1(x_nt_n). \tag{3}$$

For this kind of process no correlation at all exists between the values assumed at different times by the random variable X.

Second order process or Markov process

It is convenient to consider that the time instants are ordered in the sense that $t_1<t_2\ldots<t_n$. Now the conditional probability function $f(x_1t_1|x_2t_2)$ is defined as the probability density that X has the value x_2 at time t_2 *given* that it had the value x_1 at time t_1. For a Markov process the distribution functions are given by $(n > 1)$:

$$W_n(x_1t_1;\ldots x_nt_n) = W_{n-1}(x_1t_1;\ldots x_{n-1}t_{n-1})f(x_{n-1},t_{n-1}|x_n,t_n). \tag{4}$$

All the available information about the time dependence of the stochastic variable $X(t)$ may be extracted from the probability function $f(x,t)$ and the conditional probability $f(x',t'|x,t)$; the values assumed by X at two different times are now correlated. A stochastic process is said to be stationary when the probability functions are invariant to an arbitrary shift of the time scale, so that the probability distribution $f(x,t)$ does not depend upon t and the conditional probability distribution $f(x',t'|x,t)$ depends upon the difference, $t-t' = \tau$, alone; thus

$$f(x,t) \equiv f(x)$$

and

$$f(x',t'|x,t) \equiv f(x'|x,\tau). \tag{5}$$

For this special case eq. (2) becomes

$$f(x) = \int dx' f(x') f(x'|x,t) \tag{6}$$

and $f(x'|x,t)$ must satisfy the normalisation condition,

$$\int dx f(x'|x,t) = 1. \tag{7}$$

Moreover when the system tends to an equilibrium state we have,

$$f(x) = \lim_{t\to\infty} f(x'|x,t). \tag{8}$$

Finally it may be demonstrated [4] that for a stationary Markov process the conditional probability is subject to the Smoluchowski-Chapman-Kolmogorov (SCK) equation,

$$f(x'|x,t) = \int dx'' f(x'|x'',t-\tau) f(x''|x,\tau). \tag{9}$$

When it can be assumed that the conditional probability is a differentiable function of time, the SCK equation may be rewritten in the integro-differential form

$$\partial f(x'|x,t)/\partial t = \int dx'' R(x'',x) f(x'|x'',t), \tag{10}$$

where $R(x'',x) = \{\partial f(x''|x,t)/\partial t\}_{t=0}$ is the transition probability rate from x'' to x.

The molecular reorientation in liquids will be considered, in the next section, as a markovian stationary stochastic process.

Molecular Reorientation in Anisotropic Liquids

A number of different models have been proposed to deal with molecular reorientation in isotropic fluids, and a few of them have been developed to include oriented liquids. In this section a stochastic approach will be followed, assuming that molecular reorientation is a Markov process which obeys a master equation such as eq. (10). The set of random variables $X(t)$ should include in principle *all* the quantities describing the rotational motion, that is the Euler angles Ω and the components of the angular momentum L. In the different stochastic models the transition rate $R(x',x)$ of eq. (10) is constructed in terms of elementary events describing collisions between molecules and the motion between collisions [5]. When the intermolecular torques are strong, practically no free rotation occurs and the molecule reorients through a sequence of collisions. The free inertial motion gives important contributions when rotations of small molecules in gases or liquids of low viscosity are considered, but generally in liquid crystals the viscosity is high and the molecules are large enough to encourage us to neglect these inertial effects. (The presence of inertial effects could be detected by the spin-rotational contribution to NMR relaxation times. This contribution can be extracted from the temperature dependence of proton T_1 or the carbon-13 nuclear Overhauser enhancement [6]. However, measurements [7] of ^{13}C relaxation in the isotropic phase of 4,4'-dimethoxyazoxybenzene show no evidence of spin-rotational effects). Therefore our basic hypothesis is to assume that molecular reorientation takes place via collisions only and that each collision randomizes the molecular angular momentum, so that no correlation exists between the components of **L** at any time. As a consequence of this hypothesis the number of variables needed to describe a Markov process is reduced, and only the Euler angles are considered.

The master equation is

$$\partial f(\Omega_0|\Omega,t)/\partial t = \int d\Omega' R(\Omega',\Omega) f(\Omega_0|\Omega',t), \tag{11}$$

where the conditional probability must obey the initial condition

$$f(\Omega_0|\Omega,0) = \delta(\Omega-\Omega_0) \tag{12}$$

and the transition rate is subject to the conservation law [4],

$$\int d\Omega R(\Omega',\Omega) = 0. \tag{13}$$

In the general case the integral operator R is not symmetric, as we will see in the following examples. However, it is convenient to make a unitary transformation to symmetrize R. Since, on physical grounds, R must obey the condition of detailed balance,

$$f(\Omega')R(\Omega',\Omega) = f(\Omega)R(\Omega,\Omega'), \tag{14}$$

where $f(\Omega)$ is the equilibrium orientational distribution, the required transformation is given by [8,9]

$$R_s(\Omega',\Omega) = f^{-\frac{1}{2}}(\Omega)R(\Omega',\Omega)f^{\frac{1}{2}}(\Omega') \tag{15}$$

and the symmetrized version of eq. (11) is

$$\partial f_s(\Omega_0|\Omega,t)/\partial t=\int d\Omega' R_s(\Omega',\Omega)f_s(\Omega_0|\Omega',t), \tag{16}$$

where

$$f_s(\Omega_0|\Omega,t) = f^{-\frac{1}{2}}(\Omega)f(\Omega_0|\Omega,t)f^{\frac{1}{2}}(\Omega_0). \tag{17}$$

Eq. (16) may be solved by expanding the conditional probability in terms of the eigenvectors of R_s

$$f_s(\Omega_0|\Omega,t) = \sum_n C_n(t)\psi_n(\Omega), \tag{18}$$

where

$$\int d\Omega' R_s(\Omega',\Omega)\psi_n(\Omega') = -\alpha_n\psi_n(\Omega). \tag{19}$$

The α's are non-negative real numbers as a consequence of eqs. (7) and (14) [10]. The existence of an eigenvalue, $\alpha_o = 0$, is ensured by condition (8), which provides the link between the conditional distribution function and the equilibrium distribution function. The corresponding eigenvector is given by

$$\psi_o(\Omega) = f^{\frac{1}{2}}(\Omega). \tag{20}$$

The coefficients $C_n(t)$ in eq. (18) are determined by the initial condition

$$f_s(\Omega_o|\Omega,0) = \delta(\Omega-\Omega_o) = \sum\psi_n{}^*(\Omega_o)\psi_n(\Omega), \tag{21}$$

which gives the final result as

$$f_s(\Omega_o|\Omega,t) = \sum_n \psi_n{}^*(\Omega_o)\psi_n(\Omega)\exp(-\alpha_n t). \tag{22}$$

The problem of calculating the conditional probability function is reduced, therefore, to the solution of the eigenvalue equation (19).

The required correlation functions can be computed, by making use of the definition of a stationary Markov process, as

$$\begin{aligned}
\overline{D^L_{p,q}(\Omega_o)D^{L'*}_{p',q'}(\Omega)} &= \int\int d\Omega_o d\Omega W_2(\Omega_o;\ \Omega,t)\, D^L_{p,q}(\Omega_o)D^{L'*}_{p',q'}(\Omega),\\
&= \int d\Omega_o f(\Omega_o)D^L_{p,q}(\Omega_o)\int d\Omega f(\Omega_o|\Omega,t)D^{L'*}_{p',q'}(\Omega),\\
&= \int d\Omega_o f^{\frac{1}{2}}(\Omega_o)D^L_{p,q}(\Omega_o)\int d\Omega f_s(\Omega_o|\Omega,t)f^{\frac{1}{2}}(\Omega)D^{L'*}_{p',q'}(\Omega)\\
&= \sum_n\int d\Omega_o\psi_o(\Omega_o)D^L_{p,q}(\Omega_o)\psi^*_n(\Omega_o)\exp(-\alpha_n t)\\
&\quad \times \int d\Omega\psi_n(\Omega)D^{L'*}_{p',q'}(\Omega)\psi_o(\Omega).
\end{aligned} \tag{23}$$

In the last equality we have made use of eq. (20). The average values of the Wigner matrix elements must be subtracted from eq. (23) to give

$$G^{LL'}_{pp'qq'}(t) = \sum_{n\neq o} \langle\psi_n|D^L_{p,q}|\psi_o\rangle\langle\psi_o|D^{L'*}_{p',q'}|\psi_n\rangle\exp(-\alpha_n t), \tag{24}$$

where the bracket notation is employed to denote the integral over the angular variables

$$\langle\psi_n|D^L_{p,q}|\psi_o\rangle = \int d\Omega\psi_n(\Omega)^* D^L_{p,q}(\Omega)\psi_o(\Omega). \tag{25}$$

We observe that in an isotropic medium composed of symmetric top molecules the equilibrium distribution f(Ω) is a constant and the $D^L_{p,q}(\Omega)$ themselves are eigenfunctions of the transition operator R, which must depend upon the difference $(\Omega'-\Omega)$ only [9],

$$R(\Omega',\Omega) = R(\Omega'-\Omega),$$
$$= R(\Omega-\Omega')$$

and

$$\int d\Omega' R(\Omega'-\Omega) D^{L}_{p,q}(\Omega') = -\alpha^{L}_{q} D^{L}_{p,q}(\Omega). \quad (26)$$

In this case, eq. (24) reduces to a single exponential because of the orthogonality of the Wigner matrices [1],

$$G^{LL'}_{pp'qq'}(t) = \delta_{LL'}\delta_{pp'}\delta_{qq'}(2L+1)^{-1}\exp(-\alpha^{L}_{q}t), \quad (27)$$

which is independent of q for spherical tops. The eigenvalues α^{L}_{q} are dependent upon the different models of reorientation.

If the anisotropic phase has $D_{\infty h}$ symmetry, as nematic liquid crystals, and if it is composed of rod-like molecules, then the correlation function must be invariant for rotations about both the director (Z in the fixed reference frame) and the long molecular axis (z in the mobile reference frame), so it reduces to

$$G^{LL'}_{pp'qq'}(t) = \delta_{pp'}\delta_{qq'}g^{LL'}_{pq}(t). \quad (28)$$

Moreover, under these conditions the correlation functions must obey the following symmetry requirements [11],

$$g^{LL'}_{pq}(t) = g^{L'L}_{pq}(t) = g^{LL'}_{-p-q}(t) = (-)^{L+L'}g^{LL'}_{p-q}(t). \quad (29)$$

Therefore the correlation functions vanish when (L+L') is odd and p or q is zero.

In most physical problems, only correlation functions with L=L'=1 or 2 have to be considered. Typically, the case L=1 occurs in the description of dielectric relaxation and L=2 in magnetic resonance relaxation effects. Finally it is important to note that we have implicitly assumed the eigenvalue spectrum of the transition operator to be discrete. It has been pointed out that in principle a continuous spectrum may also appear [9], but this possibility is ruled out for the models discussed in the following sections.

The Strong Collision Model

The simplest form for the transition rate $R(\Omega',\Omega)$ is obtained when it is assumed that the orientation after a collision is independent of the orientation before the collision, and that the time taken for the transition is negligible with respect to the residence time τ spent by the molecules at any orientation [12]. The angular variation during a single collision is a random process and since no correlation exists before and after the rotational jump, the probability of the orientation Ω after the jump is determined by the equilibrium distribution. Therefore the transition rate is given by

$$R(\Omega',\Omega) = \{f(\Omega)-\delta(\Omega'-\Omega)\}/\tau \tag{30a}$$

and

$$R_s(\Omega',\Omega) = \{f^{\frac{1}{2}}(\Omega')f^{\frac{1}{2}}(\Omega)-\delta(\Omega'-\Omega)\}/\tau. \tag{30b}$$

It is easily verified that, besides the zero eigenvalue corresponding to $\psi_o(\Omega) = f^{\frac{1}{2}}(\Omega)$, the operator R_s has an infinitely degenerate set of eigenvectors corresponding to the eigenvalue $\alpha_n = \tau^{-1}$

$$\psi_n(\Omega) = f^{-\frac{1}{2}}(\Omega)D^L_{p,q}(\Omega), \tag{31}$$

where the index n is used to denote the three labels Lpq. The autocorrelation functions calculated according to eqs. (24) and (28) are

$$g^L_{pq}(t) = \exp(-t/\tau)\sum_{n\neq o}\langle\psi_o|D^{L*}_{p,q}|\psi_n\rangle\langle\psi_n|D^L_{p,q}|\psi_o\rangle. \tag{32}$$

Adding to and subtracting from the right hand side of eq. (32) the term

$$\langle\psi_o|D^{L*}_{p,q}|\psi_o\rangle\langle\psi_o|D^L_{p,q}|\psi_o\rangle\exp(-t/\tau) = \delta_{po}\delta_{qo}\overline{P}^2_L\exp(-t/\tau) \tag{33}$$

and using the completeness of the eigenfunctions given by eq. (21) we obtain [13],

$$g^L_{pq}(t) = \{\overline{|D^L_{p,q}|^2}-\delta_{po}\delta_{qp}\overline{P}^2_L\}\exp(-t/\tau). \tag{34}$$

This is indeed a very simple result, but the assumption of a single correlation time is a major limitation of the model, especially when experimental data concerning tensorial quantities of different rank are compared. For example, the very large decrease (two orders of magnitude) of the molecular correlation times displayed in dielectric relaxation experiments in nematics conflicts with the results of magnetic relaxation experiments, which do not show evidence of such a dramatic slowing down of the molecular reorientation rate. However, when homogeneous quantities are considered, this simple model may be used satisfactorily, the anisotropic features of the motions being contained in the average values of the Wigner matrix components. The correlation function may be improved in a phenomenological way by allowing the correlation time to depend upon the index q for a symmetric top [13].

The Diffusion Model

In the diffusion model it is supposed that molecules reorient through small angular steps [14], so that a strong correlation exists before and after each collision. Thus the integral in the master equation (11) is evaluated for values of Ω' close to Ω. The master equation reduces to a differential equation in the variables $(\alpha\beta\gamma)$ and it can be expressed by means of the vector operator **L** which generates infinitesimal rotations.

The equilibrium distribution function is determined by the Boltzmann distribution

$$f(\Omega) = \exp[-\beta U(\Omega)]/\int d\Omega \exp[-\beta U(\Omega)], \tag{35}$$

where $U(\Omega)$ is the orientational energy of a molecule. The conditional probability distribution is the solution of the diffusion equation given by [15,16],

$$\partial f(\Omega_0|\Omega,t)/\partial t = -\mathbf{L}.\mathbf{D}.\{\mathbf{L}+[\mathbf{L}\beta U(\Omega)]\}f(\Omega_0|\Omega,t), \tag{36}$$

where **D** is the molecular rotational diffusion tensor and **L** is formally equivalent to the quantum-mechanical angular momentum operator. If **D** is assumed to be axially symmetric with principal components $D_\perp$ and $D_{||} = \kappa D_\perp$, the diffusion equation becomes [17],

$$\partial f/\partial t = D_\perp[\nabla^2+\beta\{(\nabla^2 U)+\nabla^2 U-U\nabla^2\}/2]f(\Omega_0|\Omega,t\},$$

$$= -\Gamma f(\Omega_0|\Omega,t), \tag{37}$$

where we have made use of the definition of the laplacian operator in Euler angle space:

$$\nabla^2 = -\{L^2+(\kappa-1)L_z^2\}. \tag{38}$$

The eigenfunctions of the operator ∇^2 are the Wigner functions $D^L_{p,q}(\Omega)$ with eigenvalues $-\{L(L+1) + (\kappa-1)q^2\}$.

The symmetrized version of the operator Γ is obtained by means of the unitary transformation [18,19],

$$\Gamma_s = \exp\{\beta U(\Omega)/2\}\Gamma\exp\{-\beta U(\Omega)/2\},$$

$$= -D_\perp[\nabla^2-\beta\{2(L^2U)+\beta(L_+U)(L_-U)+\beta\kappa(L_zU)^2\}/4] \ . \tag{39}$$

To solve the eigenvalue equation for the operator Γ (or Γ_s) the probability $f(\Omega_0|\Omega,t)$ is expanded in a basis of Wigner functions. The effect of the operators $L_\pm$ and L_z on these functions is [20],

$$L_\pm D^L_{p,q} = \{L(L+1)-q(q\pm1)\}^{\frac{1}{2}}D^L_{p,q\pm1}$$

and

$$L_z D^L_{p,q} = -qD^L_{p,q}. \tag{40}$$

The matrix elements of the diffusion operator are easily calculated when an appropriate expression for the orienting potential is chosen. To comply with the axial symmetry of the mesophase the potential must be a function of β and γ only, and it can be expanded in the same basis set,

$$U(\beta,\gamma) = \sum_{L,q} \varepsilon_{L,q} D^L_{o,q}(\beta,\gamma). \tag{41}$$

The coefficient $\varepsilon_{L,q}$ could be interpreted, through the mean field theory, in terms of molecular interactions [21], but here they are treated as independent parameters. If the mesophase has $D_{\infty h}$ symmetry, only even values of L appear in eq. (41) and for rod-like molecules q must be zero, in other words the potential depends only upon the angle β. The leading term in the expansion is expected to be $\varepsilon_{2,0}$ and, to avoid the introduction in the model of too many undetermined parameters, the truncated form of the potential will be used:

$$U(\beta) \simeq \varepsilon_{2,0} D^2_{o,o}(\beta),$$

$$= \lambda kT\, P_2(\cos\beta). \tag{42}$$

We note that, when the potential U=U(β) is considered, a simplification of the eigenvalue problem occurs, because, in this special case, the operator Γ does not connect Wigner functions with different p or q indices, and so the eigenfunctions are given by

$$\psi^n_{pq} = \sum_L D^L_{p,q}(\Omega) X^{(pq)}_{Ln}$$

and

$$\Gamma\psi^n_{pq} = D_\perp \alpha^n_{pq} \psi^n_{pq}. \tag{43}$$

The eigenvalues are the same whether Γ or Γ_s is considered, since these operators are related by a unitary transformation. Again there is a zero eigenvalue $\alpha^o_{o,o}$ corresponding to the eigenfunction $\psi^o_{oo} = f^{\frac{1}{2}}(\beta)$.

The correlation functions are then computed from eqs. (24) and (28) and we find

$$g^L_{pq}(t) = \sum_{n\neq o} \langle\psi^o_{oo}|D^{L*}_{p,q}|\psi^n_{pq}\rangle\langle\psi^n_{pq}|D^L_{p,q}|\psi^o_{oo}\rangle \exp(-\alpha^n_{pq}D_\perp t). \tag{44}$$

In the isotropic phase U=0, the Γ matrix is diagonal and the summation in eq. (44) reduces to

$$g^L_{pq}(t) = (2L+1)^{-1}\exp[-\{L(L+1)+(\kappa-1)q^2\}D_\perp t]. \tag{45}$$

Eq. (44) gives the correlation function as an infinite sum of decreasing exponentials. For computational purposes the solution must be obtained by truncating the expansion at some index n' such that terms with n>n' give a negligible contribution. The mathematical problem of the convergence of the series (44) has not been studied in detail; however in practice a twenty term expansion is sufficient to ensure convergence for $|\lambda|<6$ corresponding to a $\bar{P}_2$ value of 0.8. Although the correlation function appears to be given by a lengthy summation, it is observed that in many cases the first term gives the most important contribution, so that eq. (44) can be approximated by

$$g^L_{pq}(t) \simeq (\overline{|D^L_{p,q}|^2} - \delta_{pq}\delta_{qo}\bar{P}^2_L)\exp(-\alpha^L_{p,q}D_\perp t). \quad (46)$$

To check the validity of this approximation, the values of the zero frequency spectral densities calculated with the aid of eqs. (44) and (46) are compared in table 1. The spectral density is defined as the cosine Fourier transform of g(t),

$$j^L_{pq}(\omega) = \int_o^\infty dt\ g^L_{pq}(t)\cos\omega t \quad (47)$$

Table 1. *Spectral density functions $j^2_{pq}(0)$ in units of $(10D_\perp)^{-1}$*

(a) Exact values {Eq. (44)}

λ	$\bar{P}_2$	00	01	02	11	12	22
-1.5	.33	.384	.356	.141	.808	.208	.597
-3.0	.60	.172	.243	.042	1.956	.104	.845
-6.0	.82	.016	.086	.004	3.465	.030	1.062

(b) Approximate values {Eq. (46)}

λ	$\bar{P}_2$	00	01	02	11	12	22
-1.5	.33	.392	.362	.143	.363	.210	.598
-3.0	.60	.189	.255	.046	.343	.108	.850
-6.0	.82	.024	.090	.006	.233	.031	1.066

Orienting potential $U(\beta) = \lambda kT\ P_2(\cos\beta)$.
Rotational diffusion tensor anisotropy $\kappa = 1$.

and its value for $\omega=0$, which is the area of g(t), is a quantity of interest in magnetic relaxation measurements when the molecular motions are very fast. From table 1 we see that the one-term approximation works well in all cases except when p=q=1 and then two terms must be retained in this case.

The result given in eq. (46) has a remarkable resemblance to that derived from the strong collision model, but now the correlation times are different for different sets of the indices L,p,q,. The dependence of the time constants $\beta^L_{pq} = [\alpha^L_{pq}]^{-1}$ upon the orientational order $\bar{P}_2$ is displayed in figures 1 and 2 for $D_{||} = D_{\perp}$. In this case the eigenvalues

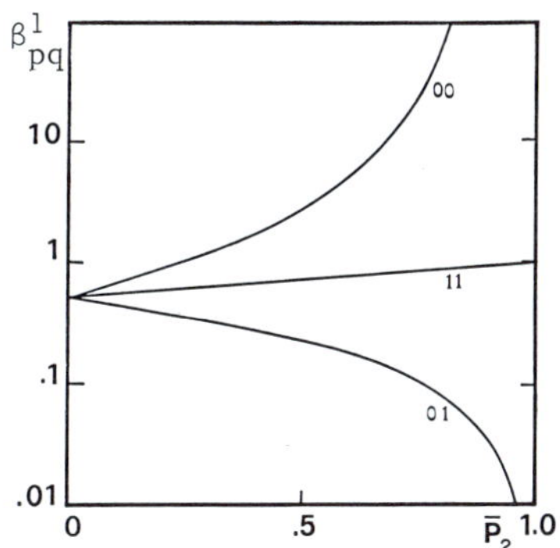

Fig.1 *Semilogarithmic plots of* β^1_{pq} *versus the orientational order parameter* $\bar{P}_2$.

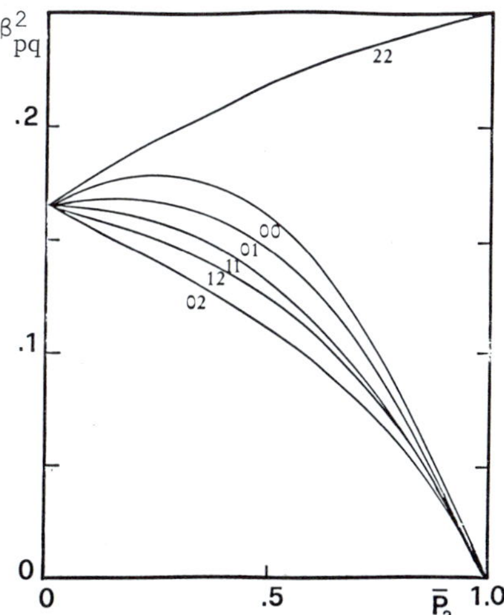

Fig.2 *Plots of* β^2_{pq} *versus the orientational order parameter* $\bar{P}_2$.

(and the correlation functions too) are symmetric with respect to interchange of p and q. For $\kappa \neq 1$, the eigenvalues α^L_{pq} are raised by an amount $(\kappa-1)q^2$. It is important to note that the variations of the effective rotational correlation times $\tau^L_{pq} = D^{-1}_{\perp}\beta^L_{pq}$ with ordering might be strongly influenced by the temperature dependence of the diffusion tensor components.

The values of the time constants for very highly ordered systems can be obtained by considering the asymptotic form of the diffusion operator. Under these conditions, the distribution function $f(\beta)$ is peaked about

$\beta=0$ and $\beta=\pi$, and large angular deviations from these limits are unlikely to occur. The diffusion operator Γ can then be expanded by keeping only terms to lowest order in β and $\pi-\beta$ respectively [8,11], and its eigenfunctions are obtained analytically. The resulting correlation functions, expressed in terms of the deviation from perfect alignment $\delta=1-\bar{P}_2$, assume the form [11],

$$g^L_{pq}(t) = C^L_{pq}\exp(-\alpha^L_{pq}D_\perp t). \tag{48}$$

The pre-exponential factors C^L_{pq} and the asymptotic rate constants α^L_{pq}, given in terms of the parameter δ for L=1 and 2, are shown in table 2.

Table 2. *Pre-exponential factors and rate constants in the asymptotic expressions for the correlation functions* $g^L_{pq}(t)$

L	p	q	C(δ)	α(δ)
1	0	0	$1-2\delta/3$	0
	0	1	$\delta/3$	$(3/\delta-1)+\kappa$
	1	0	$\delta/3$	$3/\delta$
	1	1	$\frac{1}{2}(1-\delta/3)$	κ
2	0	0	0	-
	0	1	δ	$(3/\delta-1)+\kappa$
	0	2	0	-
	1	0	δ	$3/\delta$
	1	1	$\frac{1}{2}(1-5\delta/3)$	κ
	1	2	$\delta/3$	$(3/\delta-2)+4\kappa$
	2	0	0	-
	2	1	$\delta/3$	$(3/\delta+1)+\kappa$
	2	2	$\frac{1}{2}(1-2\delta/3)$	4κ

We observe from the figures that all the time constants go to zero or remain finite with increasing order, except for β^1_{oo} which diverges as $\bar{P}_2$ approaches unit. The correlation function for the Z component of a molecular dipole moment is proportional to

$$\overline{\cos\beta(o)\cos\beta(t)} = g^1_{oo}(t). \tag{49}$$

When the orientational order of the mesophase increases, the motion of the axial component of the dipole slows down and eventually is completely frozen at $\bar{P}_2 = 1$, because no degree of freedom corresponding to rotations about the z axis can exist for a vectorial quantity. The large low-frequency shift of the dielectric dispersion is explained by this particular behaviour of the correlation time for the axial component of a molecular vector property [17].

Experimental Information on Orientational Time Correlation Functions

A complete description of the dynamical properties of a markovian stationary system is gained when the second-order joint probability function $W_2(\Omega_o;\Omega,t)$ is known. By analogy with the expansion of the orientational distribution function $f(\Omega)$ in a basis of Wigner functions, the joint probability can be written as

$$W_2(\Omega_o;\Omega,t) = \sum_{Lpq}\sum_{L'p'q'} (2L+1)(2L'+1)\overline{D^L_{p,q}(\Omega_o)D^{L'*}_{p',q'}(\Omega)}D^{L*}_{p,q}(\Omega_o)D^{L'}_{p',q'}(\Omega), \tag{50}$$

where eq. (23) has been used. In other Chapters it has been shown that the dynamics of molecular rotation can be probed by different spectroscopic techniques, and so we shall conclude with a brief summary of the information provided by the experimental methods which have been described.

Dielectric dispersion

In a dielectric relaxation experiment the real part of the complex dielectric permittivity $\varepsilon(\omega)$ is usually measured as a function of the frequency. The resulting dispersion curve has inflections at frequency values which can be related to the characteristic frequencies of the orientational correlation function for the molecular dipole component in the direction of the electric field,

$$\gamma_E(t) = \overline{\mu_E(0)\mu_E(t)}. \tag{51}$$

The problem of taking into account internal field effects, to relate $\varepsilon'(\omega)$ to $\gamma_E(t)$, is described elsewhere [22] and so will not be discussed here. For a uniaxial liquid crystal, experiments can be performed with the field parallel or perpendicular to the director (laboratory Z and X axis respectively). If we denote the components of the molecular electric dipole moment in the directions parallel and perpendicular to the long molecular axis by μ_ℓ and μ_t, then the relevant correlation functions are [17]

$$\overline{\mu_Z(0)\mu_Z(t)} = \mu_\ell^2 g^1_{00}(t)+\mu_t^2 g^1_{01}(t)$$

and

$$\overline{\mu_X(0)\mu_X(t)} = \mu_\ell^2 g^1_{10}(t)+\mu_t^2 g^1_{11}(t). \tag{52}$$

If any $g^1_{pq}(t)$ is characterized by a single decay time, the dispersion regions will be related to the relaxation frequencies $\alpha^1_{00},\alpha^1_{01}$ and $\alpha^1_{10},\alpha^1_{11}$ for the two experimental situations considered. The dependence of the time constants for first rank Wigner functions on the orientational order is shown in figure 1.

Neutron scattering

As we shall see in Chapter 20 the differential cross-section for incoherent quasielastic neutron scattering can be related to the intermediate scattering function $I_s(Q,t)$ which for a single proton is

$$I_s(Q,t) = \overline{\exp[i\mathbf{Q}.\{\mathbf{r}(t)-\mathbf{r}(0)\}]}. \tag{53}$$

Here Q is the momentum transfer vector and $\mathbf{r}(t)$ is the vector distance from the molecular centre of mass to the scattering nucleus, whose components in the laboratory frame vary in time because of molecular reorientation. We first introduce the Rayleigh expansion for a plane wave [1]

$$\exp(i\mathbf{Q}.\mathbf{R}) = 4\pi \sum_{L,m}(i)^L j_L(Qr)Y_{L,m}(\Omega_r)Y^*_{L,m}(\Omega_Q), \tag{54}$$

where j_L is a spherical Bessel function of rank L and $Y_{L,m}(\Omega)$ is a spherical harmonic. We now transform the $Y_{L,m}(\Omega_r)$ to a molecular axis system

$$Y_{Lm}(\Omega_r) = \sum_n D^{L*}_{m,n}(\Omega)Y_{L,n}(\Omega'), \tag{55}$$

where the angles Ω and Ω' define the orientation of the molecule relative to the reference system and of the nuclear vector relative to the molecular frame respectively. The angles Ω' and Ω_Q are determined by the geometry of the molecule and of the experimental arrangement, so only Ω is time dependent. In this way, we see that the scattering function $I_s(Q,t)$ is given as an infinite sum of correlation functions of Wigner rotation matrix components of any rank.

NMR and ESR

In the fast motion limit, magnetic relaxation is interpreted in terms of the quantum-mechanical correlation function of the perturbing hamiltonian, $H_1(t)$. Since this is expressed by second-rank tensor interactions, it would appear that magnetic spectroscopy can provide information only on the correlation functions of Wigner rotation matrix components of rank two (cf. eq. (50) of Chapter 19). However, this is true only in the fast motional regime, and the complete analysis of the spectral shape in conditions of relatively slow motion can yield more detailed information on the dynamics of molecular rotations; this can be seen in the following way [4,23]. The magnetic resonance spectrum is given by the Fourier transform of the spin autocorrelation function G(t) defined as

$$G(t) = \langle S_x(0)S_x(t)\rangle,$$
$$= \mathrm{Tr}\ \overline{S_x(0)S_x(t)}. \tag{56}$$

The equation of motion for $S_x(t)$ is

$$dS_x/dt = i[H_o+H_1\{\Omega(t)\},S_x(t)] = iL\{\Omega(t)\}S_x(t), \tag{57}$$

where the Liouville operator $L(t)$ has been introduced and $H_1\{\Omega(t)\}$ is the random perturbation resulting from the anisotropic parts of the spin hamiltonian. If the solution of the Heisenberg equation is introduced into eq. (56), and the average over the angular variables taken by assuming the motion of the liquid to be described by a markovian random process, we obtain finally

$$G(t) = \mathrm{Tr}\int\int d\Omega_o d\Omega S_x(0)f(\Omega_o)Q(\Omega_o|\Omega,t)S_x(0), \tag{58}$$

where

$$Q(\Omega_o|\Omega,t) = \sum_{n=o}^{\infty} Q_n(\Omega_o|\Omega,t) \tag{59}$$

and the Q_n are given by the recursion relations

$$Q_n(\Omega_o|\Omega,t) = \int_o^t dt'\int d\Omega' f(\Omega'|\Omega,t-t')iL(\Omega')Q_{n-1}(\Omega_o|\Omega',t') \tag{60}$$

and

$$Q_o(\Omega_o|\Omega,t) = f(\Omega_o|\Omega,t). \tag{61}$$

If the conditional probability $f(\Omega_o|\Omega t)$ is assumed to obey a differential equation like eq. (37) then $Q(\Omega_o|\Omega t)$ is found to satisfy the equation

$$\partial Q(\Omega_o|\Omega,t)/\partial t = \{iL(\Omega)-\Gamma\}Q(\Omega_o|\Omega,t), \tag{62}$$

which is a particular form of the stochastic Liouville equation, widely used in lineshape calculations [8,19]. By employing the expression of the joint probability density given in eq. (50) it can be shown that the fast motional results are recovered from the n=2 term of the expansion (59). However, correlation functions for Wigner components of rank higher than two appear when the molecular motions are slow and the expansion for $Q(\Omega_o|\Omega,t)$ must be taken to higher orders. A detailed analysis of some generalized perturbation schemes for lineshape calculations in the slow motion region, can be found in reference [24].

References

1. D.M. Brink and G.R. Satchler, "*Angular Momentum*"(Clarendon Press, 2nd Ed., Oxford 1968).

2. B.J. Berne in *"Physical Chemistry"*, vol. VIII B, H. Eyring, D.Henderson and W. Jost eds. (Academic Press, New York, 1971).
3. D.A. McQuarrie in *"Physical Chemistry"*, vol. XIB, H. Eyring, D. Henderson and W. Jost eds. (Academic Press, New York, 1975).
4. J.B. Pedersen in *"Electron Spin Relaxation in Liquids"* L.T. Muus and P.W. Atkins eds. (Plenum Press, New York, 1972).
5. K. Lakatos-Lindenberg and R.I. Cukier, *J. Chem. Phys.*, 62, 3271 (1975).
6. T.C. Farrar and E.D. Becker, *"Pulse and Fourier Transform NMR"* (Academic Press, New York, 1971).
7. G. Agostini and U. Segre, *Chem. Phys.* **35**, 407 (1978).
8. C.F. Polnaszek, G.V. Bruno and J.H. Freed, *J. Chem. Phys.*, 58, 3185 (1973).
9. R.I. Cukier, *J. Chem. Phys.*, 60, 734 (1974).
10. I. Oppenheim, G.H. Weiss and K.E. Shuler, *Adv. Mol. Relax. Process.*, **1**, 13 (1967).
11. P.L. Nordio and U. Segre, *J. Mag. Res.*, 27, 465 (1977).
12. W. Kauzmann, *Rev. Mod. Phys.*, 14, 12 (1942).
13. G.R. Luckhurst and A. Sanson, *Mol. Phys.*, 24, 1297 (1972).
14. L.D. Favro in *"Fluctuation Phenomena in Solids"* R.E. Burgess ed. (Ac. Press, N.Y.1965)
15. a) P.L. Nordio and P. Busolin, *J. Chem. Phys.*, 55, 5485 (1971);
    b) P.L. Nordio, G. Rigatti and U. Segre, *J. Chem. Phys.*, **56**, 2117 (1972).
16. L.P. Hwang and J.H. Freed, *J. Chem. Phys.*, 63, 118 (1975).
17. P.L. Nordio, G. Rigatti and U. Segre, *Mol. Phys.*, 25, 129 (1973).
18. P.L. Nordio and U. Segre, *Chem. Phys.*, **11**, 57 (1975).
19. C.F. Polnaszek and J.H. Freed, *J. Phys. Chem.*, 79, 2283 (1975).
20. L.D. Landau and E.M. Lifschitz, *"Quantum Mechanics"* (Pergamon Press, Oxford 1965), page 383.
21. a) R.L. Humphries, P.G. James and G.R. Luckhurst, *J. Chem. Soc., Faraday Trans II*, 68, 1031 (1972);
    b) G.R. Luckhurst, C. Zannoni, P.L. Nordio and U. Segre, *Mol. Phys.*, 30, 1345 (1975).
22. G.R. Luckhurst and C. Zannoni, *Proc. R. Soc.* **A343**, 389 (1975).
23. H. Sillescu and D. Kivelson, *J. Chem. Phys.*, 48, 3493 (1968).
24. B. Yoon, J.M. Deutch and J.H. Freed, *J. Chem. Phys.*, 62, 4687 (1975).

Chapter 19

MAGNETIC RESONANCE SPECTROSCOPY. DYNAMICAL ASPECTS

PIER LUIGI NORDIO and ULDERICO SEGRE

Istituto di Chimica Fisica dell'Universita di Padova, Italy

Relaxation and Lineshapes

The term relaxation is used to mean all the irreversible processes which bring a system back to equilibrium after it has been perturbed by some external force. For instance, if an electric field is applied to a fluid of polar molecules a polarization will be induced, but this will disappear some time after the field is removed, because of the randomization of the molecular orientations produced by brownian motions.

When a static magnetic field B_o is applied to a system of N molecules possessing spin angular momentum **S**, each individual spin processes about the field direction with an angular velocity ω_o equal to $g\mu_B B_o/\hbar$, its projection on this direction corresponds to one of the (2S+1) allowed orientations. At thermal equilibrium the distribution among the spin states is determined by Boltzmann's law, and a macroscopic magnetization in the direction of the field is created according to the Curie law,

$$M_o = NS(S+1)\beta g^2\mu_B^2 B_o/3, \tag{1}$$

where β is 1/kT. Quantum mechanically, the macroscopic observable M_o can be expressed in terms of the spin density matrix σ for the system[1],

$$\begin{aligned} M_o &= -g\mu_B \langle S_Z \rangle, \\ &= -g\mu_B \mathrm{Tr}\,\sigma_{eq} S_Z \end{aligned} \tag{2}$$

where

$$\sigma_{eq} = \exp(-\beta H_o)/\mathrm{Tr}\,\exp(-\beta H_o) \tag{3}$$

and

$$\begin{aligned} H_o &= g\mu_B B_o S_Z \\ &= \hbar\omega_o S_Z, \quad S_Z = \sum_i S_{Zi} \end{aligned} \tag{4}$$

where the angular brackets denote an ensemble average. The summation must be carried out over the N molecules composing the system. The equations are appropriate for a negative magnetic moment as in ESR; for NMR the sign of the nuclear g factor depends on the particular nucleus.

The equilibrium conditions may be temporarily disturbed, for example by the application of a pulse so as to alter the value of the magnetization in the main field direction (that we define to be the Z axis). Alternatively it may induce a component in a perpendicular direction; in each case we expect the magnetization components to decay after the pulse to the equilibrium value according to the rate equations,

$$dM_X(t)/dt = -\omega_o M_Y(t) - M_X(t)/T_2,$$

$$dM_Y(t)/dt = \omega_o M_X(t) - M_Y(t)/T_2$$

and

$$dM_Z(t)/dt = -(1/T_1)\{M_Z(t) - M_o\}; \tag{5}$$

or, in terms of $M_\pm = M_X \pm iM_Y$,

$$dM_\pm(t)/dt = \{\pm i\omega_o - (1/T_2)\}M_\pm(t). \tag{6}$$

These relations, derived for a single exponential decay of each component of the magnetization, are known as the phenomenological Bloch equations[1]. The rate constants T_1 and T_2 are called the longitudinal and transverse relaxation times, respectively. They are expected to be different on physical grounds, because the magnetic energy of the system, $U = -\mathbf{M}.\mathbf{B}_o$ depends only on the Z component of the magnetization, so that variations of M_Z imply an energy exchange with the motional degrees of freedom of the molecular environment (the lattice). For historical reasons the environment is referred to as the lattice and for this T_1 is also called the spin-lattice relaxation time. In a relaxation process involving M_Z, magnetic energy is irreversibly dissipated as thermal energy. Conversely, the transverse relaxation can be interpreted as a dephasing of the precessional motion of the spins, caused by the slightly different effective fields acting on the individual spins, so that energy is redistributed among the spins. These considerations support the hypothesis that different mechanisms govern the relaxation of M_Z and $M_\pm$. The relaxation times are indeed found to be different and T_2 is generally shorter than T_1.

Relaxation times of about one second are commonly encountered in NMR, and almost all modern spectrometers are able to observe the time dependence of the magnetization components and can be programmed to perform a variety of relaxation experiments [2]. For example, in the inversion-recovery measurements of nuclear T_1 the equilibrium magnetization is reversed instantaneously by means of a calibrated radio-frequency pulse, so that at t=0 an emission spectrum is observed. The time dependence of $M_Z(t)$ is followed by taking the NMR spectrum at various times, and is given by the solution of eq. (5) with the initial condition

$$M_Z(0) = -M_o$$

as

$$M_Z(t) = M_o\{1-2\exp(-t/T_1)\}. \quad (7)$$

ESR relaxation times range between $10^{-7}$and 10^{-5}s for normal radicals, and relaxation experiments cannot easily be performed on such a time scale. In this case, however, the width of the spectral lines provide information on the tranverse relaxation time T_2, as we shall see.

In fact, the question that we should answer is how the relaxation processes influence the spectra obtained under normal conditions, when a monochromatic magnetic field $B_1(t)$ is applied continuously to the sample in a direction perpendicular to the static magnetic field. The oscillating field will stimulate transitions between the spin levels created by the static field B_o, thus altering the Boltzmann distribution. This perturbation is opposed by the longitudinal relaxation mechanisms, which tends to bring the spin populations back to their equilibrium values. Moreover the oscillating field induces a time-dependent magnetization in the xy plane, which contrasts with the transverse relaxation mechanism, whose effect on the spectral shape can be understood within the framework of linear response theory [1,3]. This theory will now be outlined for a classical system and the results for a quantum mechanical system will be derived in the following section.

Let us suppose that a system is subjected to a time-dependent force F(t), whose effect is detected by measuring a physical property X(t) which constitutes the response of the system to the applied perturbation. As a basic hypothesis the response at time t is supposed to be linearly proportional to the strength of the perturbation for all previous times t',

$$\begin{aligned} X(t) &= \int_{-\infty}^{t} dt' R(t-t')F(t'), \\ &= \int_0^{\infty} d\tau R(\tau)F(t-\tau). \end{aligned} \quad (8)$$

R(t) is called the response function and is characteristic of the system. We now consider two particular ways of applying the perturbation:

i) step function perturbation $F(t) = F_o (t<0); = 0 (t>0)$.

In this case we may easily compute the time derivative of the response $X_S(t)$,

$$\begin{aligned} dX_S(t)/dt &= \int_0^{\infty} d\tau R(\tau)\{dF(t-\tau)/dt\}, \\ &= -F_o \int_0^{\infty} d\tau R(\tau)\delta(t-\tau), \\ &= -F_o R(t), \end{aligned} \quad (9)$$

since the derivative of the Heaviside step function is just the

Dirac delta function. It is convenient to introduce the normalized relaxation function,

$$\phi(t) = \{X_S(t)-X_S(\infty)\}/\{X_S(0)-X_S(\infty)\}; \tag{10}$$

ii) monochromatic perturbation $F(t) = \mathrm{Re}F_\omega \exp(i\omega t)$.

$$\begin{aligned} X_\omega(t) &= \mathrm{Re}F_\omega \int_0^\infty d\tau R(\tau)\exp\{i\omega(t-\tau)\}, \\ &= \mathrm{Re}\,\chi(\omega)F_\omega \exp(i\omega t), \\ &= \{\chi'(\omega)\cos\omega t-\chi''(\omega)\sin\omega t\}F_\omega . \end{aligned} \tag{11}$$

The complex susceptibility $\chi(\omega)$ is the one-sided Fourier transform of the response function, and can be expressed in terms of the relaxation function as,

$$\begin{aligned} \chi(\omega) &= \int_0^\infty dtR(t)\exp(-i\omega t), \\ &= \chi_0\{1-i\omega \int_0^\infty dt\phi(t)\exp(-i\omega t)\}. \end{aligned} \tag{12}$$

The expression in eq. (11) shows that the response $X_\omega(t)$ is given by an admixture of a term in-phase and another out-of-phase with $F(t)$. In other words, the same relaxation processes which cause the effect of a step perturbation to disappear some time after this is removed, prevent the response $X(t)$ from following an harmonic perturbation instantaneously. The connection between the two relaxation experiments (i) and (ii) is demonstrated by eq. (12), which shows that the frequency dependence of $\chi(\omega)$ can be calculated immediately if the time variation of $X(t)$, after the removal of a constant perturbation, is known.

In a magnetic resonance experiment, a linearly polarized oscillating field $B_X(t) = B_1\cos\omega t$ is applied in the X direction. It induces a magnetic moment $M_X(t)$ of the form given by eq. (11), where $\chi(\omega)$ is interpreted now as a complex magnetic susceptibility. The experiment measures the radio-frequency power absorbed by the spin system as a function of frequency. The rate at which energy is absorbed per unit volume by the sample from the B_X field is given by the average of the magnetic energy rate of change dU/dt over a cycle of the oscillating field [4],

$$\begin{aligned} I(\omega) &= (2\pi/\omega)^{-1}\int_0^{2\pi/\omega} dt(dU/dt), \\ &= -(2\pi/\omega)^{-1}\int_0^{2\pi/\omega} dt(M_X dB_X/dt), \\ &= -\tfrac{1}{2}\omega B_1^2\chi''(\omega), \\ &= \tfrac{1}{2}\omega^2 B_1^2\chi_0\int_0^\infty dt\phi(t)\cos\omega t. \end{aligned} \tag{13}$$

The NMR and ESR spectral lines are always relatively narrow, so that the variation of ω^2 within the linewidth can be neglected and the lineshape is determined essentially by the Fourier transform of the relaxation function $\phi(t)$ for $M_X(t)$. Now, for a system of particles of spin S, this function is obtained by solving eq. (6) with initial conditions $M_X(0)\neq 0$ and $M_Y(0) = 0$. The solution,

$$M_X(t) = M_X(0)\cos\omega_o t \exp(-t/T_2), \tag{14}$$

shows that after the removal of a weak perturbing field applied perpendicularly to the main field B_o, the spins perform a damped precession, in which the transverse component of the magnetization decays to zero with a characteristic time T_2. By taking the Fourier transform of the result in eq. (14), the spectrum is found to consist of a lorentzian line centred at ω_o, the width being determined by the parameter T_2^{-1}. Thus the normalized expression for the shape function is, in this case,

$$f(\omega) = T_2/\pi\{1+T_2^2(\omega-\omega_o)^2\}. \tag{15}$$

Another line at $\omega = -\omega_o$ is predicted, but for all practical purposes it has no effect on the spectrum, if T_2^{-1} is much less than ω_o, as implied in the previous discussion. Although the assumption of a single exponential decay is only an approximation to the relaxation behaviour for real systems, the conclusions concerning the relationship between relaxation in the time domain and the lineshape in the frequency domain are generally valid.

Theory of Relaxation

We shall derive first a quantum mechanical expression for the relaxation function $\phi(t)$ [5]. Let a constant perturbing force described by the spin hamiltonian $-g\mu_B B_\alpha S_\alpha$ be applied continuously to the spin system from $t = -\infty$ and removed at $t=0$. Its effect is observed by following the time dependence of the component $M_\alpha(t)$ of the macroscopic magnetization. If H_o is the hamiltonian of the system in the absence of the perturbation, as given in eq. (4), the total hamiltonian will be

$$\begin{aligned} H &= H_o - g\mu_B B_\alpha S_\alpha, && t < 0 \\ &= H_o, && t > 0. \end{aligned}$$

For the system prepared in this non-equilibrium condition at $t=0$, we can define an initial time density matrix $\sigma(0)$, according to eq. (3),

$$\sigma(0) = \exp\{-\beta(H_o - g\mu_B B_\alpha S_\alpha)\}/Z, \tag{16}$$

where Z is the partition function $\mathrm{Tr}\exp\{-\beta(H_o - g\mu_B B_\alpha S_\alpha)\}$. For positive times, the evolution of the physical observable $M_\alpha(t)$ is given by

$$M_\alpha(t) = -g\mu_B Tr\sigma(t)S_\alpha,$$
$$= -g\mu_B Tr\sigma(0)S_\alpha(t), \qquad (17)$$

where

$$S_\alpha(t) = \exp\{(i/\hbar)\mathcal{H}t\}S_\alpha\exp\{-(i/h)\mathcal{H}t\} \qquad (18)$$

and the second equality follows from the invariance of the trace when the operators are permuted cyclically. Expansion of $\sigma(0)$ in eq. (16) to terms linear in B_α and the assumption of high temperature approximation $\beta g\mu_B B_o \ll 1$, leads to the following expression for $M_\alpha(t)$ [5,6],

$$M_\alpha(t)-\langle M_\alpha\rangle = -\beta g^2\mu^2_B\{\langle S_\alpha(0)S_\alpha(t)\rangle-\langle S_\alpha\rangle^2\}B_\alpha, \qquad (19)$$

or

$$\{M_\alpha(t)-\langle M_\alpha\rangle\}/\{M_\alpha(0)-\langle M_\alpha\rangle\} = \{\langle S_\alpha(0)S_\alpha(t)\rangle-\langle S_\alpha\rangle^2\}/\{\langle S^2_\alpha\rangle-\langle S_\alpha\rangle^2\}. \qquad (20)$$

Comparison with eq. (10) gives

$$\phi(t) = \{\langle S_\alpha(0)S_\alpha(t)\rangle-\langle S_\alpha\rangle^2\}/\{\langle S^2_\alpha\rangle-\langle S_\alpha\rangle^2\}. \qquad (21)$$

The quantity $\langle S_\alpha(0)S_\alpha(t)\rangle$ is called the autocorrelation function of S_α.

An idealized experiment involving transverse relaxation would consist in the randomization of the X component of the magnetization after the removal of a small magnetic field B_1 applied along the X direction perpendicularly to the major static field B_o. According to eq. (13), the spectral lineshape in a continuous wave experiment will be proportional to

$$I(\omega) = Re\int dt\langle S_X(0)S_X(t)\rangle\exp(-i\omega t). \qquad (22)$$

From the discussion, it appears that two major routes can be followed in the theoretical interpretation of magnetic relaxation. Following the first path, the equation of motion of the density matrix $\sigma(t)$ is solved, and then the pertinent macroscopic observable $M_\alpha(t)$ calculated through the relation $Tr\sigma(t)S_\alpha$. The Redfield treatment [7] is the classic example of this procedure. Alternatively, kinetic equations can be set up directly for the macroscopic variables $M_\alpha(t)$ [8]. The result given in eq. (20) shows that the time dependence of $M_\alpha(t)$ and that of $\langle S_\alpha(0)S_\alpha(t)\rangle$ is indeed the same. We shall follow the second path, which leads quite naturally to the Bloch equations for the magnetization [9].

Relaxation Mechanisms in Liquids

If a magnetic interaction is modulated by some kind of motion, it acts as a time-dependent perturbation which causes transitions between the magnetic levels, and ultimately relaxation of the spin states. In liquids, the translational diffusion of the molecules modulates the intermolecular interactions, such as the dipole or the exchange coupling between spins located on pairs of molecules; the first effect is important in proton magnetic resonance, and the latter in relatively concentrated solutions of free radicals. We may note however that these effects can be minimized in NMR by suitable isotopic substitution, and in ESR by using dilute samples.

The anisotropic part of the intramolecular interactions, described in Chapter 16, fluctuates as a consequence of the rotational motion of the molecules. We shall ignore possible modulations of isotropic interactions resulting from internal motions, such as rotation of mobile groups or conformational changes. Instead we concentrate on the effects of molecular rotation, which provide the most relevant information on the molecular dynamics in ordered fluids.

We rewrite the spin hamiltonian given in eq. (2) of Chapter 16 in the form,

$$H = T^{(o,o)}A^{(o,o)} + \sum_{m,n} (-)^n T^{(2,-n)} D^2_{m,n}(\Omega) A^{(2,m)}, \tag{23}$$

where Ω denotes the instantaneous orientation of the molecule with respect to the laboratory frame. The molecular motions cause Ω to fluctuate randomly in time, and this causes a modulation of the energy levels and transition frequencies. The random fluctuations can be characterized through their amplitude and coherence. The amplitude is defined by the root-mean-square value Δ of the magnetic anisotropies, and the coherence is given by the correlation time τ_c, which characterises the molecular reorientations. For non-spherical molecules in anisotropic liquids the assumption of only one correlation time is a rather crude approximation, and we shall remove it shortly.

A random process is considered to be fast if $\Delta\tau_c$ is less than unity. In this case the spectrum consists of sharp lines, the resonance frequencies being determined by the static hamiltonian H_o obtained from the average of H in eq. (23) calculated with the orientational distribution function $f(\beta,\gamma)$ appropriate for a uniaxial liquid crystal. This gives

$$H_o = T^{(o,o)}A^{(o,o)} + \sum_m (-)^m T^{(2,-m)} \overline{D^2_{o,m}} A^{(2,o)}. \tag{24}$$

The difference $(H-H_o)$ is

$$H_1(t) = \sum_{m,n} (-)^n T^{(2,-n)} \{D^2_{m,n}(t) - \overline{D^2_{m,n}}\delta_{om}\} A^{(2,m)}, \tag{25}$$

which is a random function of time with average of zero, and can be treated as a perturbation which induces transitions among the eigenstates of H_o. We note that the time dependence of H_1 is contained entirely in the Wigner functions, if internal motions are ignored.

The lattice motions can be treated classically, and statistical methods must be employed because of their random nature. The processes we are interested in are stationary in character, which means that they are invariant under translation of the time axis. Fluctuations in a system in thermodynamic equilibrium are examples of stationary random processes, and it should be remembered that in the framework of the linear response theory the return to equilibrium after the application of a small perturbation is governed by the natural motions of the system, *i.e.* by the characteristic motions in the absence of the perturbation. The concept of correlation function plays a central role in the development of the theory. The autocorrelation function of $D^L_{m,n}(\Omega)$, when Ω is a random function of the time, is denoted as $\overline{D^{L*}_{m,n}(0)D^L_{m,n}(t)}$; the upper bar indicates an equilibrium average over the angular variables. This function takes the values $\overline{|D^L_{m,n}|^2}$ and $|\overline{D^L_{m,n}}|^2$ at times t=0 and t=∞ respectively, the time evolution depends upon the model assumed to describe the rotational motion, as we saw in Chapter 18.

For magnetic relaxation experiments in anisotropic liquids, it follows from eq. (25) that we are interested in the correlation functions for the fluctuations of the second rank components $D^2_{m,n}$ about their mean value $\overline{D^2_{m,n}}$. We shall denote these correlation functions by $g_{mn}(t)$,

$$g_{mn}(t) = \overline{D^{2*}_{m,n}(0)D^2_{m,n}(t)} - |\overline{D^2_{m,n}}|^2. \tag{26}$$

In fact, the relaxation times and spectral lineshapes are related to the spectral density functions $j_{mn}(\omega)$ which are the Fourier transforms of $g_{mn}(t)$,

$$j_{mn}(\omega) = (1/2)\int_{-\infty}^{+\infty} dt \exp(-i\omega t) g_{mn}(t). \tag{27}$$

We have seen that an approximate description of the correlation function can be given generally in terms of a single exponential decay with time constant τ_{mn}. As a consequence

$$g_{mn}(t) = \left(\overline{|D^2_{m,n}|^2} - |\overline{D^2_{m,n}}|^2\right)\exp(-t/\tau_{mn})$$

and

$$j_{mn}(\omega) = \left(\overline{|D^2_{m,n}|^2} - |\overline{D^2_{m,n}}|^2\right)\{\tau_{mn}/(1+\omega^2\tau^2_{mn})\} \tag{28}$$

Note that for isotropic liquids $\overline{D^2_{m,n}}$ vanishes and $\overline{|D^2_{m,n}|^2}$ is (1/5), because of the orthogonality properties of the Wigner functions.

In the discussion, cross correlation terms of the type $\overline{D^{2*}_{m,n}(0)D^2_{m',n'}(t)}$ have not been considered. Although the condition m=m' is dictated in uniaxial liquid crystals by the independence of the orientational potential on the Euler angle α, the restriction n=n' holds only for molecules of cylindrical symmetry for which the potential is a function of β alone. In this case,

$$\overline{D^2_{m,n}} = \delta_{mo}\delta_{no}\overline{D^2_{o,o}}.$$

To obtain a qualitative understanding of the effect of $H_1(t)$ on longitudinal and transverse relaxation (or lineshape) let us consider the example of an elongated molecule with an unpaired electron, partially oriented in a nematic solvent. We shall assume that the system is characterized by g-tensor anisotropy with axial symmetry in the direction of the long molecular axis, and that the orientational distribution function depends only upon the angle, β, between this axis and the director. Then

$$\begin{aligned} H_o &= \mu_B B_o\{g_{iso} + (2/3)(g_{\|} - g_{\perp})\overline{D^2_{o,o}}\}S_Z \\ &= \mu_B \bar{g} B_o S_Z \end{aligned} \tag{29}$$

where $g_{iso} = (g_{\|} + 2g_{\perp})/3$, and

$$H_1(t) = (2/3)\mu_B B_o(g_{\|} - g_{\perp})[\{D^2_{o,o}(t) - \overline{D^2_{o,o}}\}S_Z - (3/8)^{\frac{1}{2}}\{D^2_{1,o}(t)S_+ - D^2_{-1,o}(t)S_-\}], \tag{30}$$

or

$$H_1(t) = \Delta\{f_o(t)S_Z + f_1(t)S_+ + f_1^*(t)S_-\}. \tag{31}$$

The randomly fluctuating terms associated with the non-secular spin operators $S_\pm$ can clearly induce transitions between adjacent Zeeman levels with energy separation, $\hbar\omega_o$ equal to $\bar{g}\mu_B B_o$, thus contributing to longitudinal relaxation. These terms are in fact responsible for the equilibration of the populations among the spin levels. They also contribute to the lineshape since the transitions shorten the life time of the spin states and the consequent indeterminacy in the energy results in line broadening. The efficiency of this mechanism is expected to be high only if the lattice fluctuations have sufficient intensity at the frequency corresponding to the energy separation of the spin states. The contribution to the relaxation rates $1/T_1$ and $1/T_2$ is calculated to be of the order $\Delta^2 j_{10}(\omega_o)$ where the spectral densities are given by eq. (28). Conversely, the secular term $f_o(t)S_Z$ of $H_1(t)$ does not induce transitions between the eigenstates of H_o and so cannot contribute to longitudinal relaxation. The fluctuations of this term give rise to a spread of local magnetic fields in the Z direction or, equivalently, to a spread (or a dephasing) of the spin precession. Consequently it is a source of transverse relaxation, and an additional contribution to the linewidth. The magnitude of this secular contribution of $1/T_2$ is given approximately by $\Delta^2 j_{00}(o)$. When the condition $\omega_o^2\tau_c^2 < 1$ is satisfied for the correlation times of all Wigner components, in other words in the extreme narrowing condition, the secular and non-secular contributions become comparable.

In ESR, the electronic Zeeman levels are frequently split by the hyperfine coupling with nuclear spins. In this case, the anisotropic hyperfine interaction **I.A.S** contains spin operators $I_{\pm}S_Z$ which can induce transitions between the nuclear sublevels corresponding to the same electron spin state. Modulation of these terms gives rise to pseudo-secular contributions to the relaxation, expressed in terms of the spectral density function $j(\omega_{hf})$, where ω_{hf} is the separation between the hyperfine levels. Since ω_{hf} is much less than ω_o, the condition $\omega_{hf}^2\tau_c^2<1$ is likely to be satisfied.

Spin Dynamical Variables

In the following sections we derive equations for the time dependence of the relevant spectroscopic observables. The Bloch equations are inadequate to this purpose, because they are limited, by their phenomenological nature, to consider only the components of the magnetization vector, which cannot give a full description of a quantum system with spin angular momentum S. In fact, the energy of such a system is split into (2S+1) levels by the interaction with the magnetic field B_o. Consequently its spectroscopic behaviour needs a knowledge of the level populations and of the transition probabilities, to be completely identified. Any measurable quantity is the average of a suitable operator for the system, which in turn can be expressed as a linear combination of a complete set of linearly independent operators, rather like a state vector which can be represented as a linear combination of basis vectors. The number of independent operators needed to form a complete basis set is then $(2S+1)^2$ for a system with spin S, since any operator is represented by a (2S+1) x (2S+1) matrix. It follows therefore that the conventional choice of S_x, S_y, S_z (or S_+, S_-, S_z), supplemented by the scalar operator S^2, provides a complete set of operators only for $S=\frac{1}{2}$ systems.

It is convenient to choose an orthogonal operator basis set, which means that two basis operators Q_i and Q_j must obey the relationship [10],

$$\mathrm{Tr}(Q_iQ_j^+) = \sum_{MM'} \langle S,M|Q_i|S,M'\rangle\langle S,M'|Q_j^+|S,M\rangle,$$

$$= \delta_{ij}. \tag{32}$$

The orthogonality requirement is fulfilled when a spherical operator basis set $A^{(L,m)}$ is taken. The first rank components are

$$A^{(1,0)} = S_z,$$
$$A^{(1,\pm 1)} = \mp S_{\pm}/\sqrt{2} \tag{33}$$

and the other members of the set can be constructed with the aid of tensor algebra, thus

$$A^{(L+1,m)} = \sum_q C(1,L,L+1;\ q,m-q)A^{(1,q)}A^{(L,m-q)}. \tag{34}$$

The orthogonality property of these operators is easily demonstrated [11] by using the Wigner-Eckart theorem to compute the matrix elements in eq. (32). Finally, the normalized basis set is given by,

$$S^{(L,m)} = A^{(L,m)}\{2^L(2L+1)!(2S-L)!/(2S+L+1)!\}^{\frac{1}{2}}/L!. \tag{35}$$

Expressions for the normalized operators up to L=4 are given in ref.[12]. The components used in the following calculations are listed in table 1.

Table 1. *Explicit forms for the normalized operators* $S^{(L,m)}=N_{L,S}A^{(L,m)}$ *required in calculations with spin states S=1 and S=3/2* [12]

$A^{(1,0)}$	=	S_z
$A^{(1,1)}$	=	$-S_+/\sqrt{2}$
$N_{1,S}$	=	$\{3/S(S+1)(2S+1)\}^{\frac{1}{2}}$
$A^{(2,0)}$	=	$\{3S_z^2-S\ (S+1)\}/\sqrt{6}$
$A^{(2,1)}$	=	$-(S_zS_++S_+S_z)/2$
$N_{2,S}$	=	$\{30/S(S+1)(2S-1)(2S+1)(2S+3)\}^{\frac{1}{2}}$
$A^{(3,0)}$	=	$\{5S_z^3-3S_zS(S+1)+S_z\}/\sqrt{10}$
$A^{(3,1)}$	=	$-\{4(S_z^2S_++S_zS_+S_z+S_+S_z^2)-(S_+^2S_-+S_+S_-S_++S_-S_+^2)\}/\sqrt{120}$
$N_{3,S}$	=	$2\{35/(S-1)S(S+1)(S+2)(2S-1)(2S+1)(2S+3)\}^{\frac{1}{2}}$
$A^{(L,-m)}$	=	$(-)^mA^{(L,m)\dagger}$

Tensor operators up to rank L=2S must then be considered to meet the completeness requirement. We recall that the Wigner-Eckart theorem states that non-vanishing matrix elements $\langle S,M|S^{(L,m)}|S,M'\rangle$ require the values of L to be restricted by the triangular condition $0\leq L\leq 2S$. In terms of these operators, the physical observables of interest are given by the usual relationship

$$\mathcal{S}^{(L,m)}(t) = \mathrm{Tr}\rho(t)S^{(L,m)}, \tag{36}$$

where $\rho(t)$ is the density matrix for the total system of the spin plus the lattice. Italic type will be used to denote an instantaneous non-equilibrium ensemble average, while angular brackets will be reserved for equilibrium ensemble averages. In the previous sections, the expressions for the macroscopic magnetization components have been given with the implicit assumption that an average was taken over the lattice degrees of freedom. The reduced spin density matrix σ must then be written as

$$\sigma(t) = \mathrm{Tr}_{\ell}\rho(t). \tag{37}$$

The problem of calculating the time evolution of $S^{(L,m)}(t)$ is complex because it involves a knowledge of the lattice dynamics. In principle, the total hamiltonian for a spin system in contact with a thermal bath (or lattice) must be expressed as the sum of three terms, namely the spin, lattice, and spin-lattice interaction terms,

$$H = H_s + H_\ell + \lambda H_{s\ell}$$

where the parameter λ is introduced only to keep account of the perturbation expansions. The spin hamiltonian H_s and the perturbing term $H_{s\ell}$ are the same as H_0 and $H_1(t)$ given in the eqs. (24) and (25) respectively, the latter is chosen so that its average vanishes. The lattice hamiltonian H_ℓ describes the lattice states, but in practice the quantum description is restricted to the spin system, and no attempt is made to treat the lattice states explicitly. Rather, the time dependence induced by H_ℓ can be dealt with by making specific assumptions about the nature of the random processes which cause the fluctuations of $H_1(t)$, and ensemble averages over the lattice (liquid) variables are performed by means of appropriate distribution functions. This problem has been treated in Chapter 18, but it is helpful to reiterate some of the results here. The motion of the liquid is assumed to be specified completely by the equilibrium distribution $f(\Omega)$ and the conditional distribution $f(\Omega_0|\Omega,t)$ for the collection of lattice variables Ω. These functions satisfy the relations

$$df(\Omega_0|\Omega,t)/dt = -\Gamma_\Omega f(\Omega_0|\Omega,t),$$

$$f(\Omega_0|\Omega,0) = \delta(\Omega-\Omega_0)$$

and

$$\Gamma_\Omega f(\Omega) = 0, \tag{38}$$

where

$$f(\Omega) = \lim_{t\to\infty} f(\Omega_0|\Omega,t).$$

$f(\Omega_0|\Omega,t)$ is the conditional probability that the orientation is Ω at time t given the initial value Ω_o at time zero. Γ_Ω is a time-independent operator, whose explicit form depends upon the particular model chosen to describe the molecular motion. Accordingly the trace over orientational degrees of freedom, for any operator $A(\Omega)$, is replaced by the classical average

$$\bar{A} = \int d\Omega f(\Omega) A(\Omega) \tag{39}$$

and the time evolution is given by

$$\exp(-\Gamma_\Omega t) A(\Omega) = \int d\Omega_o f(\Omega_o | \Omega, t) A(\Omega_o). \tag{40}$$

The solution of the dynamical problem for the macroscopic physical observables is simplified considerably if a time-scale separation between the lattice and spin motions can be made. This situation obtains when the spin subsystem is weakly coupled to the lattice and the lattice motions are fast compared with the spin relaxation rate, in other words when the motional narrowing condition applies. As a consequence of this time-scale separation, the dynamical variables $S^{(L,m)}(t)$ form a complete set in the sense that they describe the relevant slow variables in the system. The effect of the rapidly varying variables can be treated as a fluctuating force and handled by means of perturbation procedures. To lowest order, linear laws for the relaxation of the magnetization, which are a generalization of the simple Bloch equations, are obtained [9].

It is important to note that the motional narrowing condition is generally satisfied in NMR relaxation experiments, but not necessarily in ESR, especially when viscous mesogens are studied. In this case one must resort to the stochastic Liouville equation [13] which provides a useful approach for non-perturbative solutions of the problem of spin dynamics, valid in the entire range of molecular mobility. The solutions of the Liouville equation, however, must be obtained numerically.

To illustrate the ideas we have developed, we shall consider molecular systems with total spin angular momentum S, so that the spin states are defined in terms of sharp quantum numbers S and M. For $S \geq 1$ the most important anisotropic interaction is given by a quadrupole type term **S.T.S**, and the spin hamiltonian of eqs. (24) and (25) can then be expressed in terms of the normalized spin operators, in frequency units, as

$$H_s = \omega_o S_z + N_{2S}^{-1} \overline{T^{(2,0)}} S^{(2,0)}, \tag{41a}$$

$$H_{s\ell} = N_{2S}^{-1} \sum_p (-)^p \{T^{(2,-p)}(t) - \overline{T^{(2,-p)}} \delta_{po}\} S^{(2,p)}, \tag{41b}$$

where the interaction tensor **T** is expressed in the laboratory frame and N_{2S} is the normalization factor from eq. (35).

Kinetic Equations for the Dynamical Variables

According to the previous arguments, and from the structure of the spin operators given in eq. (33) it follows that the longitudinal and transverse components of the magnetization are proportional, for a system of spin S, to $S^{(1,0)}(t)$ and $S^{(1,1)}(t)$ respectively. It is expected, however, that these terms can be computed only if the time dependence of all the members of the set $S^{(L,m)}(t)$ is known. The kinetic equations for the set of dynamical variables $S^{(L,m)}(t)$ are derived, under motional narrowing conditions, from a theoretical treatment developed originally by Mori. The theory, based on a projection operator technique which allows the extraction of the relevant slow variables from the entire set of variables necessary to specify a macroscopic system, was subsequently adapted by Hynes and Deutch [5] as well as by Kivelson and Ogan [9] to the spin relaxation problem. If the variables $S^{(L,m)}(t)$ and the corresponding operators $S^{(L,m)}$ are collected as the components of the vectors $\boldsymbol{S}(t)$ and $\mathbf{S}$ respectively, the relaxation equations are,

$$d\boldsymbol{S}(t)/dt = (i\Omega - R)\{\boldsymbol{S}(t) - \langle S\rangle\}, \tag{42}$$

where

$$\Omega = \langle [H_s, S]S^+\rangle, \tag{43a}$$

$$R = \int_0^\infty dt\mathbf{K}(t)\exp(-i\Omega t), \tag{43b}$$

$$\mathbf{K}(t) = \langle [H_{s\ell}(t), S(t)][H_{s\ell}, S]^+\rangle. \tag{43c}$$

The matrices Ω and R are called the frequency and relaxation matrix respectively. The eigenvalues of Ω are related to the eigen-frequencies of the static spin hamiltonian H_s. The relaxation matrix R is complex, the real part being related to the decay times and the imaginary part to the dynamic frequency shifts [9].

To order λ^2 in the spin-lattice interaction, the time dependence of the operators $H_{s\ell}$ and $S^{(L,m)}$ is given by [9]

$$H_{s\ell}(t) = \exp\{i(H_s + H_\ell)t\}H_{s\ell}\exp\{-i(H_s + H_\ell)t\}, \tag{44}$$

$$S^{(L,m)}(t) = \exp(iH_s t)S^{(L,m)}\exp(-iH_s t). \tag{45}$$

The last equation is obtained by noting that H_ℓ necessarily commutes with the spin operators. At the same level of approximation, the ensemble averages must be performed with the density operator

$$\rho_{eq} = \exp\{-\beta\hbar(H_s + H_\ell)\}/\mathrm{Tr}\,\exp\{-\beta\hbar(H_s + H_\ell)\}. \tag{46}$$

A quantum description is reserved again for the spin states, and the lattice motions are treated classically according to eqs. (39) and (40).

In taking the traces over spin states the high temperature expansion of the spin density operator is employed,

$$\begin{aligned}\sigma_{eq} &= \exp(-\beta\hbar H_s)/\mathrm{Tr}\,\exp(-\beta\hbar H_s),\\ &\simeq (1-\beta\hbar\omega_o S_Z)(2S+1)^{-1},\\ &\simeq (2S+1)^{-1}.\end{aligned} \tag{47}$$

If the second term on the right hand side of eq. (41a) is neglected in comparison with the Zeeman term when H_s is used in eq. (45), the time dependence of $S^{(L,m)}$ is simplified to

$$S^{(L,m)}(t) = \exp(im\omega_o t)S^{(L,m)}. \tag{48}$$

Finally, the following linear system of coupled differential equations is obtained, where the ensemble average is completely factorized in the lattice and spin variables,

$$\begin{aligned}dS^{(L,m)}(t)/dt &= \sum_{L',n} \{i\langle[H_s,S^{(L,m)}]S^{(L',n)+}\rangle\\ &-(N_{2S})^2 \sum_{p,q}(-1)^{p+q}\int_0^\infty dt\,\exp(ip\omega_o t)\{\overline{T^{(2,-p)}(t)T^{(2,-q)*}}-\overline{T^{(2,-p)}}\,\overline{T^{(2,-q)*}}\}\\ &\times\langle[S^{(2,p)},S^{(L,m)}][S^{(2,q)},S^{(L',n)}]^+\rangle\}(S^{(L',n)}(t)-\langle S^{(L',n)}\rangle).\end{aligned} \tag{49}$$

In isotropic liquids or in uniaxial mesophases, the condition for a non-vanishing average is p=q, and consequently m=n. We define,

$$\begin{aligned}G^{(p)}(t) &= \overline{T^{(2,p)}(t)T^{(2,p)*}}-\overline{T^{(2,p)}}\,\overline{T^{(2,p)*}},\\ &= \sum_q |T_M^{(2,q)}|^2 g_{pq}(t),\end{aligned} \tag{50}$$

where $g_{pq}(t)$ is the correlation function of the Wigner rotation matrix elements which transform the interaction tensor **T** from laboratory to molecular axes, defined in eq. (26):

$$J_p(p\omega_o) = \mathrm{Re}\int_0^\infty dt\,G^{(p)}(t)\exp(-ip\omega_o t), \tag{51}$$

$$K_p(p\omega_o) = \mathrm{Im}\int_0^\infty dt\,G^{(p)}(t)\exp(-ip\omega_o t) \tag{52}$$

and the differential equation for $S^{(L,m)}(t)$ may now be written as

$$\begin{aligned}dS^{(L,m)}(t)/dt &= \sum_{L'}\{i\langle[H_s,S^{(L,m)}]S^{(L',m)+}\rangle\\ &-(N_{2S})^2\sum_p(J_p+iK_p)\langle[S^{(2,p)},S^{(L,m)}][S^{(2,p)},S^{(L',m)}]^+\rangle\}(S^{(L'm)}(t)-\langle S^{(L',m)}\rangle),\\ &= \sum_{L'}(i\Omega_{LL'}-R^{(m)}_{LL'})(S^{(L'm)}(t)-\langle S^{(L'm)}\rangle).\end{aligned} \tag{53}$$

In conclusion, we see that only terms with the same m value are coupled together. It will also be easy to show that, because of the different parity, terms of even and odd rank can never mix. The commutators of tensor operators are evaluated by standard methods [11], and with the high temperature approximation the terms in the angular brackets are simply traces of spin operators. Their calculation provides the numerical coefficients for the spectral density functions J_p and K_p. The physical meaning of the quantities $S^{(L,m)}(t)$ which are sometimes referred to as statistical tensors [14] will become apparent from the specific examples discussed in the next section.

Application to S=1 and S=3/2 Spin Systems

We shall consider now the specific cases of S=1 and 3/2 to derive explicit expressions for the relaxation times. The first example may refer in ESR to a triplet (or biradical) state, and in NMR to the two equivalent *ortho* protons which are present in the benzene rings of almost all mesogenic molecules. In both cases the spins are relaxed mainly by dipolar interaction. The calculation is also applicable to any single deuterium or nitrogen nuclear spin relaxed by the quadrupole interaction. It follows in fact from the structure of the spin operators that the problem is isomathematical for all these different cases. The S=3/2 problem refers mainly to an electron spin quartet state when this is separated from the corresponding doublet states by an exchange interaction much larger than the dipole interactions between the spins. The solutions can also be applied to a system of three equivalent protons (such as a methyl or methoxy group) only if the cross-correlation terms between dipole interactions of different spin pairs are ignored.

We shall find it convenient to express the traces of some of the spin operators in terms of the relative intensities of the allowed transitions $|S,M-1\rangle \leftrightarrow |S,M\rangle$, defined as

$$\eta_{M-1,M} = (P_M - P_{M-1})|\langle S,M|S_+|S,M-1\rangle|^2,$$

$$= (\sigma_{MM} - \sigma_{M-1,M-1})\{S(S+1)-M(M-1)\}, \tag{54}$$

where P_M is the fractional population of the $|S,M\rangle$ spin state, which is given by the diagonal element σ_{MM} of the spin density matrix. For simplicity, an index $i=S-M+1$ $(M=S,S-1...,-S+1)$ will be used to label the relative intensity $\eta_i \equiv \eta_{M-1,M}$ corresponding to each of the 2S allowed transitions. Because of the selection rules, the only other quantities of interest are the off-diagonal spin density matrix elements $\sigma_{M-1,M} \equiv \chi_i$.

The expressions for the spin operators $S^{(L,m)}$, needed in the calculations are given in table 1. Recalling the definition

$$\mathcal{S}^{(L,m)} = \mathrm{Tr}\,\sigma S^{(L,m)},$$

$$= \sum_{M,M'} \sigma_{MM'}\langle M'|S^{(L,m)}|M\rangle, \tag{55}$$

we obtain the results shown in table 2. The frequency and relaxation matrices calculated on the basis of the normalized spin operators are given in the Appendix. We shall now comment briefly on the provisional results for the lineshapes and spin-lattice relaxation.

Table 2. *Statistical averages of the normalised spin operators.*

	S=1	S=3/2
$S^{(1,0)}$	$(\eta_1+\eta_2)/\sqrt{8}$	$(\eta_1+\eta_2+\eta_3)/\sqrt{20}$
$S^{(2,0)}$	$(\eta_1-\eta_2)/\sqrt{24}$	$(\eta_1-\eta_3)/6$
$S^{(3,0)}$	-	$(2\eta_1-3\eta_2+2\eta_3)/\sqrt{720}$
$S^{(1,1)}$	$-(\chi_1+\chi_2)/\sqrt{2}$	$-(\sqrt{3}\chi_1+2\chi_2+\sqrt{3}\chi_3)/\sqrt{10}$
$S^{(2,1)}$	$-(\chi_1-\chi_2)/\sqrt{2}$	$-(\chi_1-\chi_3)/\sqrt{2}$
$S^{(3,1)}$	-	$-(\chi_1-\sqrt{3}\chi_2+\chi_3)\sqrt{5}$

Lineshapes

If eq. (53) is solved for $S^{(\mu,1)}(t)$ with initial conditions $S^{(\mu,1)}(0)=\delta_{\mu 1}S^{(1,1)}(0)$, and the result for $S^{(1,1)}(t)$ Fourier transformed, the entire spectrum can be calculated. However, we usually prefer to deal with analytic expressions for the linewidth parameter T_2^{-1}. There are two situations in which eq. (53) offers simple solutions. The first refers to the case in which all the transitions occur at the same frequency. This condition is satisfied in the isotropic phase, or even in a mesophase if the director can be aligned at the magic angle with respect to the field since this makes $P_2(\cos\theta)$ equal to zero in eq. (19) of Chapter 16.

In both cases $T^{(2,0)}$ vanishes and if the small dynamic frequency shifts are neglected, the spectrum is a superposition of lorentzian lines centred at the same frequency. Their widths are related to the eigenvalues of the real part of the relaxation matrix. The solutions will then have the form

$$S^{(\mu,1)}(t) = \sum_{\nu,i} X_{\mu i}X_{\nu i}\exp\{(i\omega_o-\lambda_i)t\}S^{(\nu,1)}(0),$$

$$= S^{(1,1)}(0)\sum_i X_{\mu i}X_{1i}\exp\{(i\omega_o-2_i)t\}, \quad (56)$$

where

$$X^{-1}R^{(1)}X = \Lambda; \qquad \Lambda = \mathrm{diag}(\lambda_1\lambda_2\ldots\lambda_n). \quad (57)$$

The widths of the overlapping lorentzian lines in the spectrum will then be given by the eigenvalues λ_i which appear in the expression for $S^{(1,1)}(t)$ and their relative weights by $(X_{1i})^2$.

On the other hand, if **T** is the matrix which diagonalizes the frequency matrix **Ω**, then **T** relates the statistical tensors $S^{(\mu,1)}$ to the off-diagonal components χ_i of the density matrix,

$$\mathbf{T}^{-1}\mathbf{\Omega}\mathbf{T} = \boldsymbol{\omega},\ S = \mathbf{T}\boldsymbol{\chi}. \quad (58)$$

Thus, eq. (42) becomes:

$$d\boldsymbol{\chi}(t)/dt = \mathbf{T}^{-1}(i\Omega-R^{(1)})\mathbf{T}\chi(t),$$

$$= (i\omega-T^{-1}R^{(1)}T)\chi(t) \quad (59)$$

and approximate solutions for $\chi_i(t)$ are obtained if the off-diagonal terms of the transformed relaxation matrix are neglected. This approximation will be valid when the line spacings are much larger than the linewidths.

Spin-lattice relaxation

The general solution for $S^{(\mu,0)}(t)$ is given by:

$$S^{(\mu,0)}(t)-\langle S^{(\mu,0)}\rangle\delta_{\mu 1} = \sum_{\nu,i} Y_{\mu i}Y_{\nu i}\exp(-k_i t)\{S^{(\nu,0)}(0)-\langle S^{(\nu,0)}\rangle\delta_{\nu i}\}, \quad (60)$$

where $\langle S^{(1,0)}\rangle\propto\sum_i n_i^o, n_i^o$ is the equilibrium line intensity, and

$$\mathbf{Y}^{-1}\mathbf{R}^{(0)}\mathbf{Y} = \mathbf{K},\quad \mathbf{K} = \mathrm{diag}(k_1k_2\ldots k_n). \quad (61)$$

In NMR experiments the Z component of the magnetization can be inverted or annihilated by the effect of a radio-frequency pulse. The pulse is referred to as hard if it affects simultaneously all the lines of the spectrum, or soft if it selectively affects each of them [15]. For hard pulses, the proper initial conditions are given by

$$S^{(\mu,0)}(0) = \delta_{\mu 1}S^{(1,0)}(0),$$

$S^{(1,0)}(0) = -\langle S^{(1,0)}\rangle$ for 180°-τ-90° pulse experiment,

$= 0$ for a saturation experiment.

On the other hand, if a soft pulse is applied at the frequency of the i-th transition, the initial conditions become,

$$\eta_i(0) = -\eta_i^0 \ (180^\circ\text{-}\tau\text{-}\ 90^\circ) \text{ or zero (saturation)}$$

$$\eta_j(0) = \eta_j^0 \text{ for } j \neq i.$$

In general, under these conditions all the $S^{(\mu,0)}(0)$ values are different from zero. It follows that the decay to equilibrium of the intensity of each spectral line is strongly influenced by the experimental conditions, and one can take advantage of this to gain more information on the relaxation pathways.

These results are summarized below:

a) degenerate case, $\overline{T^{(2,0)}} = 0$

S = 1

Single lorentzian line of width $(3J_0+5J_1+2J_2)/2$. After a $180^\circ\text{-}\tau\text{-}90^\circ$ pulse, line intensity is recovered with an exponential law and time constant $1/T_1 = J_1+4J_2$.

S = 3/2

Line does not appear to be lorentzian: it is composed of two overlapping lorentzians of relative weight 3:2 and widths J_0+J_1 and J_1+J_2. Line intensity recovers according to

$$\eta = \eta^0[1 - \tfrac{2}{5}\{\exp(-2J_1t)+4\ \exp(-2J_2t)\}].$$

b) resolved spectrum $\overline{T^{(2,0)}} >> 1/T_2$.

S = 1

Two equal lorentzian lines separated by $\sqrt{6}\ \overline{T^{(2,0)}}$, with widths $(3J_0+3J_1+2J_2)/2$.

hard pulse: $\eta_1(t) = \eta_2(t) = \eta_1^0[1-2\ \exp\{-(J_1+4J_2)t\}]$

soft pulse on line 1

$$\eta_1(t) = \eta_1^0[1-\exp\{-(J_1+4J_2)t\}-\exp(-3J_1t)]$$

$$\eta_2(t) = \eta_2^0[1-\exp\{-(J_1+4J_2)t\}+\exp(-3J_1t)];$$

S = 3/2

Three lorentzian lines of relative intensity 3:4:3, separated by $\sqrt{6}\ T^{(2,0)}$. Outer lines have widths $(J_0+J_1+J_2)$, inner line has width (J_1+J_2).

hard pulse: $\eta_1(t) = \eta_3(t) = \eta_1^0\{1-2\exp(-2J_1t)\}$

$$\eta_2(t) = \eta_2^0\{1+2\exp(-2J_1t)-4\exp(-2J_2t)\}$$

soft pulse on line 1

$$n_1(t) = n_1^0[5-3\exp(-2J_1t)-5\exp\{-2(J_1+J_2)t\}-2\exp(-2J_2t)]/5$$

soft pulse on line 2

$$n_2(t) = n_2^0\{5-4\exp(-2J_1t)-6\exp(-2J_2t)\}/5.$$

Angular dependence of spin relaxation

According to the arguments developed in Chapter 16, if experiments can be performed by varying the alignment of the director with respect to the magnetic field, both the resonance frequencies and the relaxation effects are expected to change in a predictable manner. The theoretical expressions can be derived by starting from the hamiltonian given in eq. (18) of Chapter 16, and we find that the results presented in the previous section are still applicable provided $\overline{T^{(2,0)}}$ is reinterpreted as $d^2_{o,o}(\theta)\overline{T^{(2,0)}}$ and $J_p(\omega)$ as $J_p(\theta,\omega)$ which is given by

$$J_p(\theta,\omega) = \sum_q |d^2_{p,q}(\theta)|^2 J_q(\omega). \tag{62}$$

Determination of the Order Parameter $\bar{P}_4$

In Chapter 16 it has been shown that the line positions in NMR or ESR spectra provide information on the averages $\overline{D^2_{o,m}}$ only, because the magnetic interactions are all second rank. The dynamical features of the spectra can in principle give additional information on the orientational order. In fact the relaxation times are given in terms of the spectral density functions $j^2_{pq}(\omega)$, which can be related to higher terms of the expansion of the orientational distribution function $f(\beta)$.

According to the results of both the strong collision model and the diffusional model, given by eqs. (34) and (46) of Chapter 18, the spectral density functions which are defined in eq. (27) can be expressed as

$$j_{pq}(\omega) = (\overline{|D^2_{p,q}|^2} - \delta_{po}\delta_{qo}\bar{P}^2_2)\tau_{pq}/(1+\omega^2\tau^2_{pq}). \tag{63}$$

Here τ_{pq} is a rotational correlation time. Now the squares of the Wigner functions can be expressed in terms of the Legendre polynomials, $P_2(\cos\beta)$ and $P_4(\cos\beta)$, with the aid of the Clebsch-Gordan coefficients, so giving,

$$\overline{|D^2_{p,q}|^2} = \sum_L C(22L;\ p\text{-}p)C(22L;\ q\text{-}q)\bar{P}_L$$

where L is restricted to the values 0, 2 and 4. If sufficient experimental data are available, the value of $\bar{P}_4$ can be determined. This has been done by analysing the angular dependence of ESR linewidths of free radical probes [16,17] or the temperature dependence of the linewidths for a quartet state (S=3/2) [18].

Appendix

Here we give the frequency and relaxation matrices for S=1 and for S=3/2. The symbols J_p and K_p are used to denote $J_p(p\omega_o)$ and $K_p(p\omega_o)$ respectively.

S=1

$$\Omega = \begin{vmatrix} \omega_o & \sqrt{3/2}\,\overline{T^{(2,0)}} \\ \sqrt{3/2}\,\overline{T^{(2,0)}} & \omega_o \end{vmatrix}$$

eigenvalues: $\omega_o + \sqrt{3/2}\,\overline{T^{(2,0)}}$, $\omega_o - \sqrt{3/2}\,\overline{T^{(2,0)}}$

$$R^{(0)} = \begin{vmatrix} J_1+4J_2 & 0 \\ 0 & 3J_1 \end{vmatrix}$$

$$R^{(1)} = \begin{vmatrix} (3J_0+5J_1+2J_2)/2 + i(K_1+2K_2)/2 & 0 \\ 0 & (3J_0+J_1+2J_2)/2 + i(K_1+2K_2)/2 \end{vmatrix}$$

$$\mathrm{Re}(T^{-1}R^{(1)}T) = \begin{vmatrix} \tfrac{1}{2}(3J_0+3J_1+2J_2) & J_1 \\ J_1 & \tfrac{1}{2}(3J_0+3J_1+2J_2) \end{vmatrix}$$

$$n_1(t)-n_1^0 = \sqrt{2}\{S^{(1,0)}(t)-\langle S^{(1,0)}\rangle\}+\sqrt{6}S^{(2,0)}(t),$$

$$n_2(t)-n_2^0 = \sqrt{2}\{S^{(1,0)}(t)-\langle S^{(1,0)}\rangle\}-\sqrt{6}S^{(2,0)}(t),$$

$$\chi_1(t) = -\{S^{(1,1)}(t)+S^{(2,1)}(t)\}/\sqrt{2},$$

$$\chi_2(t) = -\{S^{(1,1)}(t)-S^{(2,1)}(t)\}/\sqrt{2}.$$

S = 3/2

$$\begin{vmatrix} \omega_o & 3\sqrt{2/5}\,\overline{T^{(2,0)}} & 0 \\ 3\sqrt{2/5}\,T^{(2,0)} & \omega_o & 2\sqrt{3/5}\,\overline{T^{(2,0)}} \\ 0 & 2\sqrt{3/5}\,\overline{T^{(2,0)}} & \omega_o \end{vmatrix}$$

eigenvalues: $\omega_o + \sqrt{6}\,\overline{T^{(2,0)}}$, ω_o, $\omega_o - \sqrt{6}\,\overline{T^{(2,0)}}$

$$\mathbf{R}^{(0)} = \begin{vmatrix} 2(J_1+4J_2)/5 & 0 & 4(J_1-J_2)/5 \\ 0 & 2(J_1+J_2) & 0 \\ 4(J_1-J_2)/5 & 0 & 2(4J_1+J_2)/5 \end{vmatrix}$$

eigenvalues: $2J_1$, $2(J_1+J_2)$, $2J_2$

$$\mathbf{R}^{(1)} \begin{vmatrix} (3J_0+5J_1+2J_2)/5 + i(K_1+2K_2)/5 & 0 & \sqrt{6}(J_0-J_2)/5 + i\sqrt{6}(2K_1-K_2)/5 \\ 0 & J_0+J_1+2J_2+iK_1 & 0 \\ \sqrt{6}(J_0-J_2)/5 + i\sqrt{6}(2K_1-K_2)/5 & 0 & (2J_0+5J_1+3J_2)/5 - i(K_1-3K_2)/5 \end{vmatrix}$$

eigenvalues (real part): J_1+J_2, $J_0+J_1+2J_2$, J_0+J_1.

$$\mathrm{Re}(T^{-1}R^{(1)}T) = \begin{vmatrix} J_0+J_1+J_2 & 0 & -J_2 \\ 0 & J_1+J_2 & 0 \\ -J_2 & 0 & J_0+J_1+J_2 \end{vmatrix}$$

$$\eta_1(t)-\eta_1^0=3\{S^{(1,0)}(t)-\langle S^{(1,0)}\rangle+\sqrt{5}S^{(2,0)}(t)+2S^{(3,0)}(t)\}/\sqrt{5},$$

$$\eta_2(t)-\eta_2^0=4\{S^{(1,0)}(t)-\langle S^{(1,0)}\rangle-3S^{(3,0)}(t)\}/\sqrt{5}.$$

$$\eta_3(t)-\eta_3^0=3\{S^{(1,0)}(t)-\langle S^{(1,0)}\rangle-\sqrt{5}S^{(2,0)}(t)+2S^{(3,0)}(t)\}/\sqrt{5}.$$

$$\chi_1(t) = -\{\sqrt{3}S^{(1,1)}(t)+\sqrt{5}S^{(2,1)}(t)+\sqrt{2}S^{(3,1)}(t)\}/\sqrt{10},$$

$$\chi_2(t) = -\{2S^{(1,1)}(t)-\sqrt{6}S^{(3,1)}(t)\}/\sqrt{10},$$

$$\chi_3(t) = -\{\sqrt{3}S^{(1,1)}(t)-\sqrt{5}S^{(2,1)}(t)+\sqrt{2}S^{(3,1)}(t)\}/\sqrt{10}.$$

References

1. A. Abragam, *"The Principles of Nuclear Magnetism"* (Clarendon Press, Oxford, 1961).
2. T.C. Farrar and E.D. Becker, *"Pulse and Fourier Transform NMR"* (Academic Press, New York, 1971).
3. C.P. Slichter, *"Principles of Magnetic Resonance"* (Harper and Row, New York, 1963).
4. L. Landau and E. Lifchitz, *"Statistical Physics"* (2nd Edition Pergamon Press, Oxford, 1967), page 286. See also ref. 1, page 40.
5. J.T. Hynes and J.M. Deutch in *"Physical Chemistry"* Vol. XIB, H. Eyring, D. Henderson and W. Jost eds. (Academic Press, New York, 1975).
6. D. Kivelson in *"Electron Spin Relaxation in Liquids"*, L.T. Muus and P.W. Atkins eds. (Plenum Press, New York, 1972).
7. A.G. Redfield, *Adv. Magn. Resonance* **1**, 1 (1965).
8. Ref 1, ch. VIII. See also P.W. Atkins, *Adv. Mol. Relaxation Processes*, **2**, 121 (1972).

9. D. Kivelson and K. Ogan, *Adv. Magn. Resonance*, **7**, 71 (1974).
10. L.T. Muus in *"Electron Spin Relaxation in Liquids"*, L.T. Muus and P.W. Atkins eds. (Plenum Press, New York, 1972).
11. M.E. Rose, *J. Math. Phys.* **3**, 409 (1962).
12. E. Ambler, J.C. Eisenstein and J.F. Schooley, *J. Math. Phys.*, **3**, 118 (1962).
13. J.H. Freed, in *"Spin Labelling"*, L.J. Berliner ed. (Academic Press, New York, 1976).
14. U. Fano, *Rev. Mod. Phys.*, **29**, 74 (1957).
15. J.M. Courtieu, C.L. Mayne and D.M. Grant, *J. Chem. Phys.*, **66**, 2669 (1977).
16. G.R. Luckhurst, M. Setaka and C. Zannoni, *Mol. Phys.*, **28**, 49 (1974).
17. M.A. Hemminga, *Chem. Phys.*, **6**, 87 (1974).
18. G. Kothe, A. Naujok and E. Ohmes, *Mol. Phys.*, **32**, 1215 (1976).

Chapter 20

INCOHERENT QUASIELASTIC NEUTRON SCATTERING

A.J. LEADBETTER AND R.M. RICHARDSON

Department of Chemistry, University of Exeter
Stocker Road, Exeter EX4 4QD, England.

General Aspects

A number of detailed discussions of the basic aspects of neutron scattering have appeared in recent years [1,2] and so we shall only summarise those results which are essential to this Chapter.

The most important property of neutrons, for our purposes, is that they combine a wavelength of the order of atomic or molecular dimensions (0.5 - 12Å) with a low energy (300 - 0.8 meV); n.b. 1 meV ≡ 8.07cm^{-1} ≡ 0.242 THz ≡ 1.52 x 10^{12}rad s^{-1}. This means that both the structure and dynamics of a system may be studied by the same experiment. The basic quantity which is measured is the double differential cross section $\partial^2\sigma(E_0,E_1,\phi)/\partial\Omega\partial E_1$ which gives the fraction of neutrons of incident energy E_0 scattered through an angle ϕ into an element of solid angle $d\Omega$ with an energy between E_1 and $E_1 + dE_1$. The natural variables, as functions of which the scattered neutron intensity is measured, are the energy exchanged between the neutron and the system

$$\hbar\omega = E_0 - E_s \qquad (1)$$

and the scattering vector Q (or momentum exchange $\hbar Q$) where

$$Q = k_0 - k_s \qquad (2)$$

and k_0 and k_s are, respectively, the incident and scattered wave vectors.

The scattering cross section is determined by the properties of the scattering system, is proportional to a flux factor k_s/k_0 and is related to the scattering lengths, b, of the different atoms in the system. Nuclei of the same element may have different values of b because of the existence of different isotopic species and different nuclear spin states. Since these are, in general, not correlated with nuclear positions this gives rise to an incoherent component

$$\sigma_{inc} = 4\pi(\langle b^2\rangle - \langle b\rangle^2), \qquad (3)$$

as well as a coherent component

$$\sigma_{coh} = 4\pi\langle b\rangle^2 \qquad (4)$$

of the total bound scattering cross section

$$\sigma_s = 4\pi\langle b^2\rangle,$$

$$= \sigma_{inc} + \sigma_{coh}. \qquad (5)$$

Here the angular brackets denote an average over all nuclei of the same element. Whereas σ_{coh} is responsible for the effects of interference between waves scattered from all the nuclei, σ_{inc} does not contribute to such effects except for those interference phenomena which are due to the scattering from one and the same nucleus. If we assume, for simplicity, that the scattering system contains only one kind of atom, then the double differential scattering cross section per atom is given by

$$(\partial^2\sigma)/(\partial\Omega\partial E) = (k_o/\hbar k_o)/\{\langle b\rangle^2 S_{coh}(Q,\omega) + (\langle b^2\rangle - \langle b\rangle^2)S_{inc}(Q,\omega)\}, \qquad (6)$$

where S_{coh} and S_{inc} are called the coherent and incoherent scattering functions (or scattering laws) respectively. The scattering laws depend only on the properties of the system and are independent of the experiment.

Most nuclei have coherent and incoherent cross sections between 0 and 10 barns (1 barn = 10^{-28} m^2). Among the common nuclei, hydrogen has an anomalously high incoherent cross section of 79.7 barns because the nuclear spin gives rise to two scattering lengths of opposite sign but with weighted average close to zero.

The coherent and incoherent scattering laws are related by Fourier transformation in space and time to the van Hove correlation functions,

$$S_{coh}(Q,\omega) = (1/2\pi)\int drdtG(r,t)\exp\{i(Q.r-\omega t)\} \qquad (7)$$

and

$$S_{inc}(Q,\omega) = (1/2\pi)\int drdtG_s(r,t)\exp\{i(Q.r-\omega t)\}. \qquad (8)$$

For liquid crystals the average de Broglie wavelength of the atoms in the molecules is small compared with the distances that they move through, and so it is safe and customary to use the classical interpretation of the correlation functions. If $R_i(t)$ represents the position of nucleus i at time t the correlation functions may be defined as follows:

(i) $G(r,t)$ is the ensemble averaged probability that if a nucleus is at R_o at an arbitrary zero time, there will be a nucleus at $R(t)$ (where $R = R_o + r$) at a later time t;

(ii) $G_s(r,t)$ is the ensemble averaged probability that if a nucleus is at R_o at an arbitrary zero time, the same nucleus will be at $R(t)$ at time t.

Thus by analogy with eq. (9) of Chapter 3

$$G(\mathbf{r},t) = \langle \sum_{i=j}^{N} \sum_{j=1}^{N} \delta(\mathbf{r}+R_{oi}-R_j) \rangle \tag{9}$$

and

$$G_s(\mathbf{r}.t) = \langle \sum_{i=j}^{N} \delta(\mathbf{r}+R_{oi}-R_i) \rangle, \tag{10}$$

where the sums are over the N atoms in the sample and the angular brackets indicate a thermal average.

For samples of condensed matter the most striking features of S_{coh} are Bragg peaks or broader maxima at particular values of Q and small $|\omega|$. These are the same values of Q as found by X-ray diffraction which measures $S_{coh}(Q)$,

$$S_{coh}(Q) = \int_{-\infty}^{+\infty} d\omega S_{coh}(Q,\omega). \tag{11}$$

The energy dependence of the coherent scattering law is not simple to predict but fortunately, if certain precautions are taken, one can regard the scattering by liquid crystals as completely incoherent, to a very good approximation. The precautions are as follows:

(i) the number of protons in the molecule should be large enough to make the total scattering cross section predominantly incoherent (typically, in the cases to be discussed the total scattering is more than 85% incoherent);

(ii) the total correlation function may be written as the sum of the self and distinct (or pair) correlation functions

$$G(\mathbf{r},t) = G_s(r,t)+G_d(r,t), \tag{12}$$

where $G_d(\mathbf{r},t)$ is the probability that if a nucleus is at R_o at time zero another nucleus will be at R(t) at time t. According to equations (7) and (8) the coherent scattering law will at least partly contain $G_s(r,t)$;

(iii) in fact the coherent scattered intensity is usually concentrated in a few Bragg peaks or broader maxima so that provided these regions in Q are avoided, the scattered intensity must indeed be almost entirely incoherent. This makes the interpretation of the neutron scattering experiments tractable since it is a very good approximation to consider only the motion of the protons and the theory of incoherent scattering may be used.

Theory of Incoherent Quasielastic Neutron Scattering

Introduction

In this section we will consider the incoherent scattering function

$S_{inc}(Q,\omega)$, or shorter $S_s(Q,\omega)$, of a proton; (if the system contains different kinds of protons which are dynamically inequivalent, then $S_s(Q,\omega)$ must be averaged over them). The scattering function will be calculated classically. The resulting symmetrical function $S_s(Q,\omega)$ does not fulfil the detailed balance condition. For the purpose of comparison with experimental data it must therefore be corrected with the detailed balance factor $\exp(\beta\hbar\omega/2)$, i.e. $S_s(Q,\omega)$ must be replaced by $\exp(\beta\hbar\omega/2)S_s(Q,\omega)$.

The scattering function is the time Fourier transform,

$$S_s(Q,\omega) = (1/2\pi)\int_{\infty}^{\infty} dt\ \exp(-i\omega t)I_s(Q,t), \tag{13}$$

of the intermediate scattering function $I_s(Q,t)$, which is given by

$$I_s(Q,t) = \langle\exp[iQ.\{R(t)-R_o\}]\rangle. \tag{14}$$

The angular brackets denote a thermal average of the exponential over all possible values of $R(t)$ and R_o weighted by the probability distribution for the initial position, $P(R_o)$, and by the conditional probability $P(R_o,R,t)$ for a nucleus which is at R_o at zero time being found at $R(t)$ at a later time t,

$$I_s(Q,t) = \int\int dR_o dR\ \exp\{iQ.(R-R_o)\}P(R_o,R,t)P(R_o). \tag{15}$$

Performing the ensemble average over the initial positions $\mathbf{R}_o$ gives the expression in terms of the van Hove correlation function,

$$I_s(Q,t) = \int dr\ \exp(iQ.r)G_s(\mathbf{r},t). \tag{16}$$

In other words one may say that $I_s(Q,t)$ is the spatial Fourier transform of $P(R_o,R,t)$ averaged over all initial positions. For the case of a polycrystal $I_s(Q\ t)$ must be averaged over all directions of Q.

G_s contains all the motions of the average proton but it is not feasible to attempt to write down a complete theory. In order to make the problem tractable we assume that the molecular motions may be divided into oscillatory and diffusion or random motions *which are independent*. These will give rise respectively to inelastic and quasielastic terms in $S_s(Q,\omega)$ and it is the latter which is especially interesting for liquid crystals.

Such a separation seems empirically justified if the experimental spectra are well separated into a quasielastic and an inelastic part. Theoretically the separation can be made under the following assumptions:
(a) the observed quasielastic scattering is due to the (diffusive) rotational motion of the molecule alone;
(b) the dynamical coupling between the centre of mass vibration of the

molecule, the torsional and intramolecular vibrations, and the diffusive rotational motion may be neglected. This assumption is necessary in order to make calculations feasible; it becomes probably less satisfactory in the large Q region, where the low energy part of the inelastic spectrum begins to blend into the quasielastic peak. In other words at large Q the experiment explores mainly small displacements and the distinction between rotational and vibrational displacements becomes more difficult.

The superposition of different molecular motions

The position vector **r** of a proton can be represented as

$$\mathbf{r} = \mathbf{a}+\mathbf{R}+\mathbf{u}, \tag{17}$$

where we now define **a** as the position vector of the centre of mass of the molecule, **R** as the radius vector of the proton relative to the centre of mass and **u** as the displacement of the proton from its equilibrium position due to vibrations. If we assume the translational, rotational and vibrational motions to be uncorrelated, eq. (14) can be factorized:

$$I_s(Q,t) = I^{trans}(Q,t)I^{rot}(Q,t)I^{vib}(Q,t). \tag{18}$$

The Fourier transformation of a product becomes a convolution (denoted by the symbol *) of the transforms and so the incoherent scattering law can be written:

$$S_s(Q,\omega) = S^T(Q,\omega) * S^R(Q,\omega) * S^V(Q,\omega), \tag{19}$$

where

$$f(x) * g(x) \equiv \int dx' f(x-x')g(x'). \tag{20}$$

Vibrations

The vibrational part of the scattering law $S^V(Q,\omega)$ is the transform, in space and time, of the self correlation function for the vibrational motion, $G_s(\mathbf{u},t)$. This can be written as:

$$G_s(\mathbf{u},t) = G_s(\mathbf{u},\infty)+\{G_s(\mathbf{u},t)-G_s(\mathbf{u},\infty)\}, \tag{21}$$

where ∞ represents a long time microscopically but a short one compared to the duration of the experiment. Performing the transformation we find

$$S^V(Q,\omega) = \delta(\omega)\int d\mathbf{u}\, \exp(i\mathbf{Q}.\mathbf{u})G_s(\mathbf{u},\infty)$$
$$+(1/2\pi)\int d\mathbf{u}\, dt\, \exp(i(\mathbf{Q}.\mathbf{u})-\omega t)\}\{G_s(\mathbf{u},t)-G_s(\mathbf{u},\infty)\}. \tag{22}$$

The first term represents elastic scattering and is sketched in figure 1a. The intensity of the elastic scattering can be represented by a Debye-Waller factor D(Q),

$$S^{V}(\mathbf{Q},\omega) = D(\mathbf{Q})\delta(\omega)+\{1-D(\mathbf{Q})\}f(\mathbf{Q},\omega), \tag{23}$$

where

$$D(Q) = \int du\ \exp(i\mathbf{Q}.\mathbf{u})G_s(\mathbf{u},\infty) \tag{24}$$

and $f(Q,\omega)$ represents the inelastic scattering and is normalized:

$$\int d\omega f(Q,\omega) = 1 \tag{25}$$

so that the $\{1-D(\mathbf{Q})\}$ term arises from the normalization condition of the scattering law,

$$\int d\omega S_s^V(Q,\omega) = 1. \tag{26}$$

If the vibrations are harmonic so that G_s is gaussian the Debye-Waller factor is

$$D(Q) = \exp(-2W)=\exp(-Q^2\langle u_Q^2\rangle), \tag{27}$$

where $\langle u_Q^2\rangle$ is the mean square displacement in the direction of Q.

It can be shown (see e.g. [1] Egelstaff p.30) that the inelastic scattering is proportional to the following factors:

$$f(Q,\omega)\alpha(Q^2\langle u_Q^2\rangle/2M)(n+\tfrac{1}{2}\pm\tfrac{1}{2})\exp(-2W)Z(\omega)/\omega, \tag{28}$$

where M is the mass of the scattering nucleus, $Z(\omega)$ is the phonon density of states and n is the Bose population number of a state with energy $\hbar\omega$:

$$n = \{\exp(-\hbar\omega\beta)-1\}^{-1}. \tag{29}$$

Provided that ω is low enough for there is be no dispersion of sound, the density of states is given by

$$Z(\omega) \sim \omega^2. \tag{30}$$

At low ω, n becomes very large and so for neutron energy loss and energy gain the inelastic scattering remains finite but independent of ω

$$f(\mathbf{Q},\omega)\alpha(Q^2\langle u_Q^2\rangle/2M)(\hbar\omega\beta)^{-1}(\omega^2/\omega),$$

$$= Q^2\langle u_Q^2\rangle x\ \text{constant}. \tag{31}$$

Thus in the quasielastic region the inelastic scattering merely contributes a Debye-Waller factor which affects the overall intensity and a flat background which inreases with Q^2, as we can see in figure 1a.

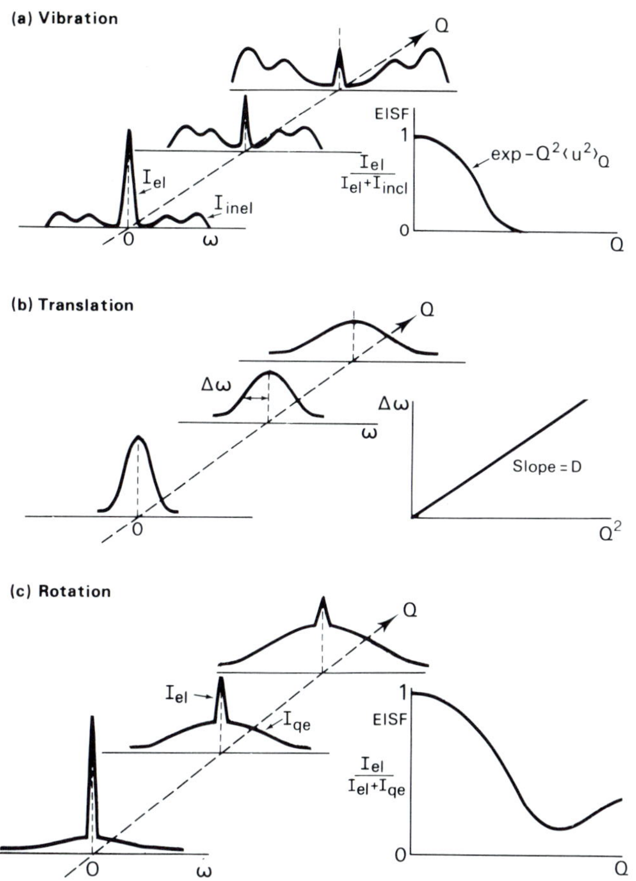

Fig 1. *Incoherent scattering law $S_s(Q,\omega)$ for various motions.*

From equation (19) the quasielastic incoherent scattering law is therefore a convolution of the rotational and translational parts multiplied by a Debye-Waller factor:

$$S_s^{qe}(\mathbf{Q},\omega) = e^{-Q^2\langle u_Q^2\rangle}\{S_s^T(\mathbf{Q},\omega) * S_s^R(\mathbf{Q},\omega)\} \qquad (32)$$

and

$$S_s^{tot}(\mathbf{Q},\omega) = S_s^{qe}(\mathbf{Q},\omega) + S_s^{inel}(\mathbf{Q},\omega). \qquad (33)$$

The scattering laws for some types of random molecular motion that could occur in liquid crystals will now be considered. These motions include translational diffusion, which may be anisotropic, reorientation about the long molecular axis and possibly slower tumbling about the short axes. These examples are discussed in terms of scattering from protons but, of course, the scattering laws apply to any incoherent scatterer.

Translational Diffusion

Isotropic translational diffusion is usually characterized by a self diffusion coefficient D. Solving the diffusion equation,

$$\frac{\partial G_s}{\partial t} = D \nabla^2 G_s \tag{34}$$

gives

$$G_s^T(a,t) = (4\pi D|t|)^{-3/2}\exp(-a^2/4D|t|)$$

and performing the Fourier transform (eqs. 15 and 13) gives a scattering law that is a single lorentzian function of full width at half maximum (FWHM) of $2DQ^2$ (cf. figure 1b):

$$S_s^T(Q,\omega) = (1/\pi)DQ^2/\{(DQ^2) + \omega^2\},$$

$$= (1/\pi)L(DQ^2), \tag{35}$$

where L(x) is a shorthand for a lorentzian function of half width at half height x. If a substance is not isotropic its diffusion must be characterized by a tensor but in a phase which is cylindrically symmetric on a microscopic level the translational diffusion can be characterized by two constants: $D_{||}$ for a diffusion parallel to the unique axes and $D_{\perp}$ for diffusion perpendicular to it. The scattering law is then still a lorentzian function

$$S_s^T(\mathbf{Q},\omega) = (1/\pi)L\{(D_{||}\cos^2\theta + D_{\perp}\sin^2\theta)Q^2\} \tag{36}$$

where θ is the angle between **Q** and the unique axis.

Rotational Scattering Functions

General For simplicity we now assume that all scattering nuclei are dynamically equivalent. When this is not the case, however, then the incoherent scattering function is simply the average of the different functions of the individual nuclei. To derive the rotational scattering function we start from eq. (15) but with **R** now representing the radius vector from the molecular centre of mass to the proton;

$$I_s(\mathbf{Q},t) = \iint d\mathbf{R}_0 d\mathbf{R} \exp\{iQ.(R-R_0)\}P^R(R_0,R,t)P^R(R_0). \tag{37}$$

The integrals are taken over the space available for the rotational motion of the nucleus. This space is restricted to part or all of the surface of a sphere and this restriction has a very important consequence for the nature of the experimental scattering law [2,3,4]. Thus if we consider the long time limit of $P^R(\mathbf{R}_0,\mathbf{R},t)$, this becomes independent of $\mathbf{R}_0$ and equal to the distribution function for the initial positions according to the ergodic theorem,

$$P^R(R_0,R,\infty) = P^R(R) = P\ (R_0).$$

The long time limit of the intermediate scattering function therefore reduces to

$$I_s^R(Q,\infty) = |\int d\mathbf{R}\ \exp(i\mathbf{Q}.\mathbf{R})\ P^R(\mathbf{R})|^2,$$
$$= |\ \langle\exp(i\mathbf{Q}.\mathbf{R})\rangle|^2, \qquad (38)$$

where the angular brackets indicate an average over all possible values of $\mathbf{R}$ (or $\mathbf{R}_o$). The rotational motion is confined to a restricted volume of space and so $P^R(\mathbf{R})$ and thus $I_s^R(Q,\infty)$ are finite. It is therefore useful to split the intermediate scattering function into this time independent and a time dependent term:

$$I_s^R(\mathbf{Q},t) = I_s^R(\mathbf{Q},\infty) + I_s^R(\mathbf{Q},t). \qquad (39)$$

Then, since the Fourier transform of a constant gives a δ function, we obtain

$$S_s^R(\mathbf{Q},\omega) = A_o(\mathbf{Q})\delta(\omega) + \{1-A_o(\mathbf{Q})\}f^{qe}(\mathbf{Q},\omega), \qquad (40)$$

where $A_o(\mathbf{Q}) = I_s^R(\mathbf{Q},\infty)$. This means (cf. figure 1c and compare eq. (23) for bound vibrational motion) that $S_s^R(\mathbf{Q},\omega)$ always contains a purely elastic component, $A_o(\mathbf{Q})\delta(\omega)$, superimposed on the quasielastic component $f^{qe}(\mathbf{Q},\omega)$. *$A_o(\mathbf{Q})$ is called the Elastic Incoherent Structure Factor (EISF).* It is simply the |scattering amplitude|^2 from the spatial distribution of the incoherently scattering particle (proton) taking account of interference between waves scattered from the particle at *different times* but not between different particles. Hence the EISF gives a direct measure of the time averaged spatial distribution of the proton which is very important information about the geometry of the rotational motions of the molecules, whereas the time evolution of the proton position is contained in $f^{qe}(\mathbf{Q},\omega)$. Note that $A_o(Q) \to 1$ as $Q \to 0$. The separation of the elastic and quasielastic components of the scattering in general requires a high resolution experiment.

We now consider various models which have been applied to specific cases; examples of the predicted forms of $A_o(\mathbf{Q})$ for some of these models are shown in figure 2. Further examples are included in the discussion of experimental results.

Rotational Jumps between a Finite Number of Allowed Orientations [2,5, 6,7] In this model it is assumed that the jump time is very much shorter than the time interval between jumps τ, i.e. instantaneous jumps. Then, the conditional probability of finding the proton at position $\mathbf{R}$ if it started at site i is

$$P^R(\mathbf{R}_o,\mathbf{R},t) = \sum_{j=1}^{m} W_{ji}(t)\delta(\mathbf{R}-\mathbf{R}_j), \qquad (41)$$

where m is the total number of available sites for a particular proton.

In terms of the motion of a single proton, $W_{ji}(t)$ is the probability of finding the proton at the allowed sites $\mathbf{R}_j$ if it was at the origin $\mathbf{R}_o = \mathbf{R}_i$ at t = 0. The probabilities W_{ji} are calculated from a system of simultaneous differential equations or rate equations:

$$dW_{ji}(t)/dt = (1/\tau)\{(1/n)\sum_{k=1}^{n} W_{ki}(t) - W_{ji}(t)\}, \tag{42}$$

where τ is the average time between two successive jumps of a proton, n is the number of proton sites that can be reached by a single rotational jump and equal probabilities are assumed for all n sites. The normalisation conditions are

$$\sum_{j=1}^{m} W_{ji}(t) = 1, \quad i = 1,2, \;.... \; m,$$

and the initial conditions are

$$W_{ii}(0) = 1, \; W_{ji}(0) = 0 \text{ for } j \neq i.$$

For illustration we will now calculate $P^R(\mathbf{R}_o,\mathbf{R},t)$ and $S^R_s(Q,\omega)$ for the most simple case of rotational jump diffusion. This is when there are only two sites, and corresponds, e.g. to a rotation by 180° such that a nucleus is transferred from one site to the other by this rotation. The sites are at $\mathbf{R}_1$ and $\mathbf{R}_2$ and the probabilities of finding the proton at each if it started at R_1 at t = 0 are $W_1(t)$ and $W_2(t)$. According to eq. (41) we have

$$dW_1/dt = (1/\tau)\{W_2(t) - W_1(t)\}. \tag{43}$$

In this simple case $W_2(t) = 1 - W_1(t)$, and for the initial conditions $W_1(0) = 1$ and $W_2(0) = 0$, the solution is

$$W_1(t) = \{1 + \exp(-2t/\tau)\}/2 \equiv W(t). \tag{44}$$

With the initial site $\mathbf{R}_o = \mathbf{R}_1$ and with $\mathbf{R}_2 = \mathbf{R}_1 + 2\mathbf{a}$ eq. (41) gives:

$$P^R(\mathbf{R}_1,\mathbf{R},t) = \delta(\mathbf{R}-\mathbf{R}_o)W(t) + \delta(\mathbf{R}-\mathbf{R}_o-2\mathbf{a})\{1-W(t)\}. \tag{45}$$

A similar result is obtained if the second site is chosen as the initial site (i.e. $\mathbf{R}_o = \mathbf{R}_2$) and as expected at infinite time these two distributions become independent of $\mathbf{R}_o$,

$$P^R(\mathbf{R}_1,\mathbf{R},\infty) = P\;(\mathbf{R}_2,\mathbf{R},\infty) = \tfrac{1}{2}\delta(\mathbf{R}-\mathbf{R}_1) + \tfrac{1}{2}\delta(\mathbf{R}-\mathbf{R}_2). \tag{46}$$

To obtain the orientational correlation function $P^R(\mathbf{R}_o,\mathbf{R},t)$ must be averaged over the two choices of $\mathbf{R}_o$:

$$G^R_S(\mathbf{r},t) = \delta(\mathbf{r})W(t) + \{\delta(\mathbf{r}-2\mathbf{a}) + \delta(\mathbf{r}+2a)\}\{1-W(t)\}/2, \tag{47}$$

where $\mathbf{r} = \mathbf{R}-\mathbf{R}_o$. The spatial Fourier transform then gives the intermediate scattering function:

$$I_S^R(\mathbf{Q},t) = W(t) + \{1-W(t)\}\cos 2\mathbf{Q}.\mathbf{a} \tag{48}$$

and finally time Fourier transformation leads to the incoherent scattering law

$$S_S^R(\mathbf{Q},\omega) = (1/2)\delta(\omega)(1 + \cos 2\,\mathbf{Q}.\mathbf{a}) + (1/2\pi)(1-\cos 2\,\mathbf{Q}.\mathbf{a})L(2/\tau) \tag{49}$$

where $L(2/\tau)$ is the lorentzian function $(2/\tau)/\{\omega^2+(2/\tau)^2\}$ (this has a half-width at half-maximum of $2/\tau$ rad s^{-1}.

The scattering law clearly depends on the relative orientations of $\mathbf{Q}$ and $\mathbf{a}$, and in particular, for $\mathbf{Q}$ perpendicular to $\mathbf{a}$

$$S_S^R(\mathbf{Q},\omega) = \delta(\omega) \tag{50}$$

and for $\mathbf{Q}$ parallel to $\mathbf{a}$ then

$$S_S^R(\mathbf{Q},\omega) = (1/2)\delta(\omega)(1+\cos 2Qa)+(1/2\pi)(1-\cos 2Q\;)L(2/\tau). \tag{51}$$

For the case of random orientation (i.e. a powder average) the result is

$$S_S^R(\mathbf{Q},\omega) = (1/2)\delta(\omega)\{1+(\sin 2Qa)/2Qa\}+(1/2\pi)\{1-(\sin 2Qa/2Qa\}L(2/\tau). \tag{52}$$

The elastic incoherent structure factor for this case, $A_o(\mathbf{Q}) = (1/2)\{1+(\sin 2Qa)/2Qa\}$, is shown in figure 2 as a function of Qa.

When the m allowed sites are equidistant points on a circle of radius a then, starting from eq. (42), the following general formula is obtained for the incoherent scattering function in a powder sample [8],

$$S_S^R(Q,\omega) = A_o(Q)\delta(\omega) + \pi^{-1}\sum_{k=1}^{m-1} A_k(Q)L(\tau_k^{-1}), \tag{53}$$

where

$$A_k(Q) = m\sum_{j=1}^{m} j_0(2\,Qa\,\sin\pi j/m)\cos(2\pi kj/m); \tag{54}$$

$j_0(x)$ is the zero order spherical Bessel function ($j_0(x) = x^{-1}\sin x$) and

$$\tau_k = \tau/2\sin^2(\pi k/m). \tag{55}$$

For the definition of the residence time τ see eq. (42). The EISF, $A_o(Q)$ of eq. (54), is shown in figure 2 as a function of Qa for m = 2 and for m = 4. It should be mentioned that for $m \to \infty$ eq. (53) - (55) are equivalent to continuous rotational diffusion on the circle with $D_r = 1/\tau_1$ {see also eq. (62)}.

Continuous Rotational Diffusion on a Sphere This is most usefully introduced as a special case of isotropic rotational motion in which on a time average a molecule has no preferred orientations in space. For such a case there is of course no difference in the scattering law between a single crystal and a powder since the motion of each individual molecule is already spatially isotropic. Sears [10] has shown that the intermediate scattering function for a proton moving on the surface of a sphere of radius a may be written

$$I_S^R(Q,t) = \sum_{L=0}^{\infty} (2L+1) j_L^{\,2}(Qa) F_L(t), \tag{56}$$

where j_L is a spherical Bessel function and

$$F_L(t) = \overline{P_L\{\cos\beta(t)\}} \tag{57}$$

is a rotational correlation function, P_L being a Legendre polynominal and $\beta(t)$ the angle through which some vector fixed in the molecule rotates in a time t. The $F_L(t)$ contain the details of the motion and for isotropic motions (e.g. rotational diffusion but not free rotation) all $F_L(t)$ decay to zero at long time except F_o which is unity. The general scattering law reads

$$S_S^R(Q,\omega) = j_0^{\,2}(Qa)\delta(\omega) + \sum_{L=1}^{\infty} (2L+1) j_L^{\,2}(Qa) S_L(\omega), \tag{58}$$

where $S_L(\omega)$ are the time Fourier transforms of the $F_L(t)$.

For rotational diffusion

$$F_L(t) = \exp(-L(L+1)D_r t) \tag{59}$$

and thus $S_L(\omega)$ is the lorentzian function $L\{L(L+1)D_r\}$ where D_r is a rotational diffusion constant.

It should be noted that in this case, as for any isotropic rotational motion, whatever the details of the motion at finite times, the EISF is given by

$$A_o(Q) = j_o^{\,2}(Qa); \tag{60}$$

this function is shown in figure 1.

Continuous rotational diffusion on a circle If the diffusional motion is confined to a circle of radius a, then it is necessary to consider the relative orientation of the circle and scattering vector and if the angle between **Q** and the rotation axis is θ then [9,11]

$$I_S^R(Q,t) = \sum_{m=-\infty}^{\infty} J_m^{\,2}(Qa\ \sin\theta)\exp(-m^2 D_r t) \tag{61}$$

and

$$S_S^R(Q,\omega) = J_0^{\,2}(Qa\ \sin\theta)\delta(\omega) + (2/\pi)\sum_{m=1}^{\infty} J_m^{\,2}(Qa\ \sin\theta)L(m^2 D_r), \tag{62}$$

where J_m is a Bessel function of the first kind.

The behaviour of $J_0^2(Qa \sin\theta)$ as a function of Qa is shown in figure 2 for $\theta = \pi/2$, i.e. for **Q** in the plane of the circle. For this configuration the scattering law is

$$S_{\perp}^{R}(Q,\omega) = J_0^2(Qa)\delta(\omega) + (2/\pi) \sum_{m=1}^{\infty} J_m^2(Qa)L(m^2 D_r). \qquad (63)$$

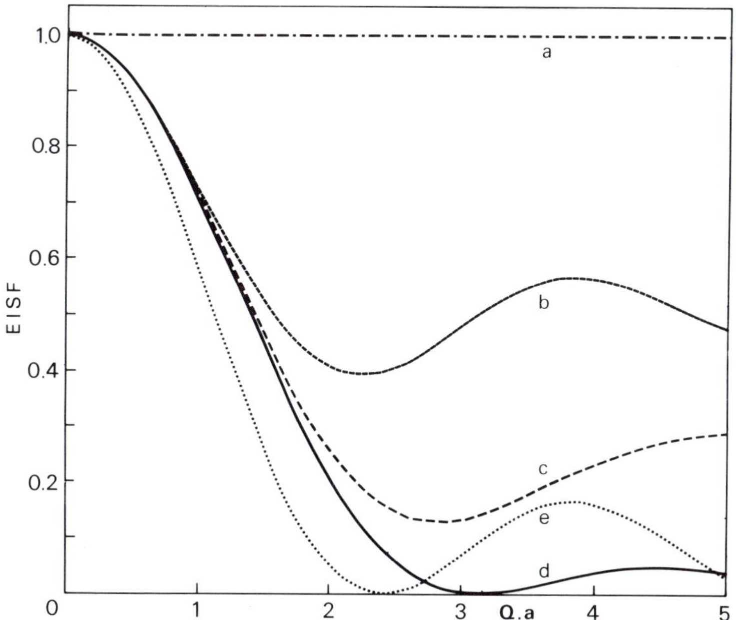

Fig. 2 *The Elastic Incoherent Structure Factor {EISF = $A_0(\mathbf{Q})$} for various models of rotational motion. (a) No motion along **Q** eqs. (50), (62) with θ=0}; (b) 2-fold jump between sites separated by 2a powder sample eqs.{(52), (53) and (54)};* (c) 4 *sites on circle, radius a, powder sample {eqs. (53) and (54)}; (d) any isotropic motion on surface of sphere, radius a {eq. (60), $j_0^2(Qa)$}; (e) continuous diffusion on circle radius a; Q in plane of circle {eq. (62) with θ = π }.*

At Qa < 2 this is dominated by the J_0 and J_1 terms so the scattering law is a purely elastic component and a quasielastic component of FWHM $2D_r$.

If the molecules are not perfectly aligned these scattering laws must be averaged over the various orientations of the molecules assumed to be at angle β to the average director (taken along the z axis). If the molecules have a *static* distribution f(β) such as the Maier-Saupe or an isotropic distribution the expression for the scattering law is

$$S\ (\mathbf{Q},\omega) = [\int d\Omega\{f(\beta)J_0^2(x)\delta(\omega)+(2/\pi) \sum_{m=1}^{\infty} J_m^2(x)L(m^2D_r)\}]/\int d\Omega f(\beta), \quad (64)$$

where

$$d\Omega = \sin\beta d\beta d\alpha.$$

When **Q⊥n** the argument, x, is

$$x = Qa(\cos^2\alpha\ \cos^2\beta + \sin^2\beta)^{\frac{1}{2}} \tag{65}$$

and when **Q** ||**n**

$$x = Qa\sin\beta. \tag{66}$$

When **Q** ||**n** the integration over α is trivial but all the other integrals must be evaluated numerically. Even if (β) is isotropic, no analytic expression can be found and it is preferable then to use eq. (53) for the powder-averaged jump model with m sufficiently large for the scattering law to be independent of further increase in m. At any given Q this occurs for finite values of m. (Remember that $1/\tau_1 \to D_r$ for $m \to \infty$).

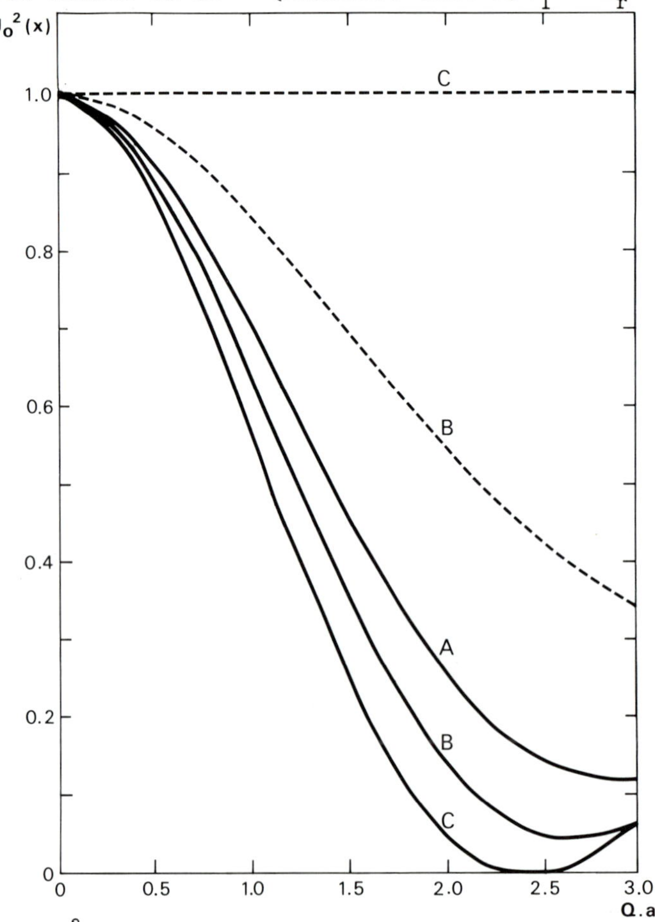

Fig. 3. *The value of $J_0^2(x)$ averaged over the Maier-Saupe distribution with order parameters; (A) 0, (B) 0.5 and (C) 1.0. x is defined by eq. (65) for Q⊥n (solid line) and by eq. (66) for* **Q**|| *n (dashed line).*

Figure 3 shows the average of $J_0^2(x)$ for both **Q**|| **n** and **Q⊥n** plotted against Qa for various values of the order parameter $\bar{P}_2$. $J_0^2(x)$ represents the proportion of purely elastic scattering in the quasielastic region of the scattering law (the EISF), and it can be seen that reduction of the order parameter from 1 affects this much more drastically when $\mathbf{Q} \parallel \mathbf{n}$ compared to $\mathbf{Q} \perp \mathbf{n}$

More complex Rotational Motions

These models for molecular rotations represent some simple extremes which cannot be expected, in general, to provide a full description of the motions in the case of liquid crystals. The models can be extended in a variety of ways, for example:

(a) motion between different sites in a real potential field rather than by an assumption of instantaneous jumps;

(b) successive (random) periods in which the molecules undergo different types of rotational motion; and

(c) simultaneous occurrence of more than one type of rotation.

We shall not consider these approaches in detail but note that it is often much simpler to determine the EISF for the more complex models than to solve the complete dynamics, since the EISF involves only the transform of the average spatial distribution {eq. (38)}. In this respect it is worth emphasising that the spatial extent of the projection along $\mathbf{Q}$ of this distribution (if not its detailed form) is revealed very simply by the rate of decrease of the EISF with Q (see figures 2 and 3).

Experimental Results

General Introduction

A schematic representation of the incoherent scattering law $S_s(\mathbf{Q},\omega)$ for the case where translational, rotational and vibrational contributions are all important is shown in figure 4. All liquid crystal in-

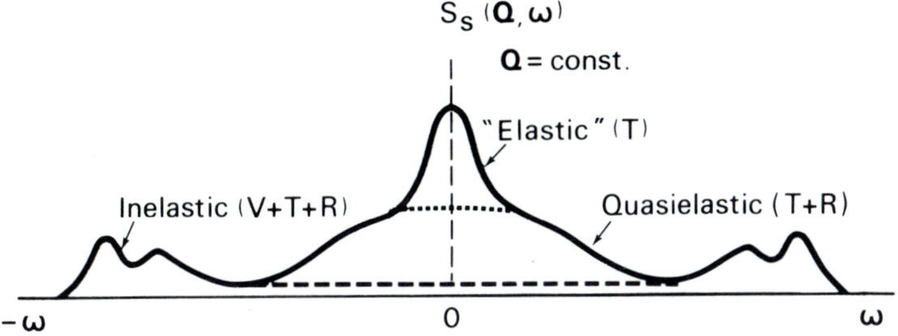

Fig. 4. *Schematic scattering law.*

investigations so far have used the assumption of separability of quasi-elastic and inelastic components of the spectrum. At least for the recent high resolution studies this seems to be well justified. Little detailed analysis has been attempted of the inelastic scattering if only because little detail has usually been observed, probably because of the many overlapping vibrational contributions, both intra and intermolecular in origin. We therefore restrict further discussion to the quasielastic region of the spectrum. When the translational diffusion coefficient is sufficiently small so that the translational broadening is negligible then provided that the rotation is fast enough to be observed (relative to the experimental resolution) the experimental determination of the EISF is in principle straightforward (cf. figures 7-12). The presence of a translational component produces additional complications in the data analysis. However, new instruments of high resolution make possible an experimental determination of the EISF. This obtains because of the general form of $S_s(\mathbf{Q},\omega)$

$$S_s(\mathbf{Q},\omega) = \exp(-Q^2 \langle u_Q^2 \rangle)[A_o(Q)L(DQ^2) + \{1-A_o(Q)\}f_s^{qe}(Q,\omega)], \tag{67}$$

with $A_o(\mathbf{Q}) \rightarrow 1$ as $Q \rightarrow 0$ (see figure 5).

Thus, if measurements of high enough resolution can be made at low enough Q then the translational components of the motion can be defined (figure 5). This value of the diffusion constant may then be used in

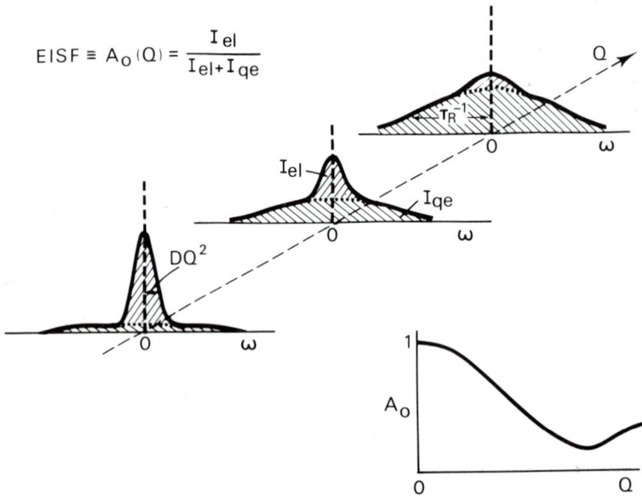

Fig.5 *Scattering law in quasielastic region.*

the analysis of measurements at higher Q so as to determine $A_o(\mathbf{Q})$, and $f_s^{qe}(\mathbf{Q},\omega)$. This of course depends upon the rotational motions being sufficiently rapid relative to the experimental resolution but it may well be possible even for simple liquids as illustrated in figure 6 for liquid cyclopropane [12]. The EISF is determined empirically from the ratio of the intensity of the central elastic (translationally broadened) component I_e to the total elastic plus quasielastic intensities, $I_e + I_{qe}$ (figure 5). The determination of the EISF is an important first step in the analysis of the data because it gives direct information about the geometrical aspects of the rotational motions and makes it easier to

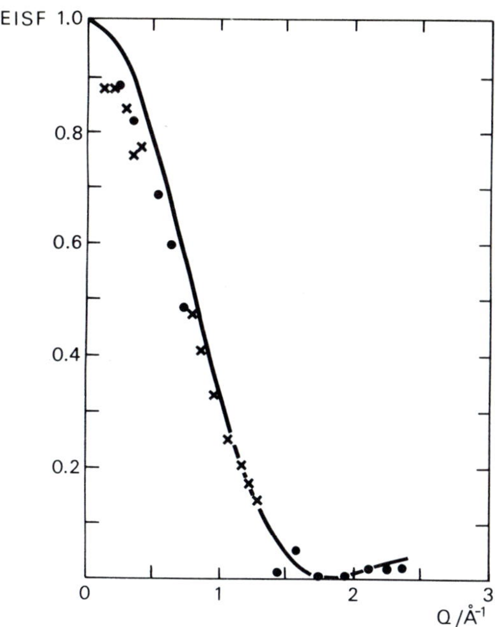

Fig.6 *EISF for cyclopropane at -118°C; the solid line is calculated from eq. (60).*

determine the dynamics. The width of the quasielastic component gives a direct measure of a correlation time, but a full model is required in order to specify properly the detailed dynamics.

Although the possible motions are rather complex for the relatively complicated liquid crystal molecules, neutron scattering provides two important means of simplifying the situation.

(i) Isotopic substitution: in particular the replacement of hydrogen by deuterium reduces the scattering cross section dramatically ($\sigma_H^{tot} \sim 80$ barns $\sigma_D^{tot} \sim 7$ barns) and, provided enough protons remain in different parts of the molecule, can render the motion of the substituted parts effectively invisible. This has been used extensively to determine or to eliminate the special motions of the long alkyl tails so as to study the motions of the molecular cores.

(ii) Because the scattering law depends on terms of the type, $\exp(i\mathbf{Q}.r)$ then only these components of the motion along **Q** are observed (cf. figures 2 and 3) so that by studying oriented specimens some of the different components of the motion can be separated experimentally.

The precise nature of the information derived from the EISF requires brief discussion at this stage. For infinitely good experimental resolution this is the time averaged (infinite time) self correlation function, $G_s(r,\infty)$, cf. eq. (38) and (40). In practice, however, the experimental resolution function Γ is of finite energy width so that any motions slower than say $\tau \gtrsim \hbar/0.1\ \Gamma$ cannot be observed, which means that this corresponds to infinite time. This will mean, of course, that the value of the EISF may be dependent on the experimental resolution but this will give useful information on the dynamics of the relevant motion. For example, in liquid crystals there is usually head-to-tail disorder of the molecular long axes and this must be dynamic so that in an infinite time each molecule will adopt both orientations. However, even if the average reorientation time were ∿ 1 μs this would be quite unobservable with present neutron experiments which have a maximum resolution ∿ 1 μeV corresponding to a maximum observable time scale of ∿ 10^{-8} s. Any motions slower than this simply act as a static distribution and the motions are therefore not observed. For this example the experimental $G_s(\mathbf{r},\infty)$ would then be a single-peaked distribution because $\infty \equiv 10^{-8}$ s! In this particular case the experiment is unable to distinguish between "up" and "down" but in general any static distribution will influence the observed scattering because of the effect of the relative orientation of the scattering vector **Q** and the rapid motions (e.g. rotation about the long axes) through the factors $\exp(i\mathbf{Q}.\mathbf{r})$; cf. figures 2 and 3.

We shall now consider a number of applications of the ideas outlined here to liquid crystals. Discussion will be restricted completely to relatively recent high resolution studies since earlier work with poorer resolution, although useful at the time, has been completely superceded. We will discuss first the rotation of end chains in the solid phase since this provides a very clear illustration of the application of the general theory. Then we will consider results on a series of phases from the most ordered smectics to the nematic, on different materials which respectively have phases where the molecules are tilted with respect to the layer normal and in which they are perpendicular to the layers. It will be shown that similar behaviour is found in a given phase type (e.g. S_E, both normal and tilted) and that the various phases are now becoming rather well characterised with respect to the molecular motions and that *all* show very rapid (∿ 10^{-11} s) rotational motions about the long axes.

End Group Rotation in Solid Phases of Liquid Crystalline Systems

Liquid crystals usually exhibit several different mesophases differing by the degree of randomness of the molecular motion. Neutron data on such systems are difficult to interpret because in general many different kinds of molecular motion are observed simultaneously and thus are superimposed on one-another in the experimental result. In particular this may include effects from random rotational motion of whole molecules and/or parts of them. In some cases internal molecular random rotation can already be observed in the solid phase, where no other rapid random motions are present. Such results may then be extrapolated to other phases existing at higher temperatures and facilitate the analysis of the more complex dynamical phenomena in these phases.

A very good example [13] is that of solid 4,4'-dimethoxyazoxybenzene (PAA) where two different partially deuteriated derivatives were studied:

PAA-ϕD_4

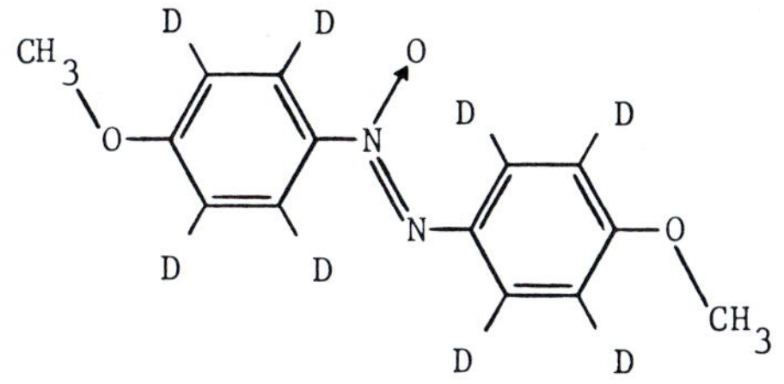

and PAA-CD_3

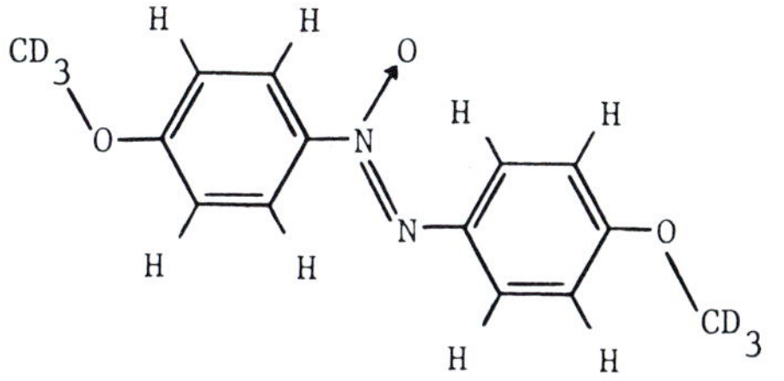

In this experiment use was made of the fact that, due to the dominant incoherent scattering cross-section of the proton, spectra of PAA-ϕD_4 mainly contain information on the motion of the two methyl groups. Thus a comparison with the spectra from PAA-CD_3 permits the separation of the methyl group motion from the motion of the whole molecule. This is clearly demonstrated in figure 7 which shows time-of-flight spectra of the two partially deuteriated PAA-derivatives. The PAA-ϕD_4 spectrum contains two dominant features which disappear completely when the methyl groups are deuteriated, the quasielastic scattering (underneath a well resolved elastic peak) and the strong inelastic peak near $\hbar\omega$ = 31 meV. This immediately shows that the quasielastic scattering is caused by a random rotational motion involving the methyl groups, but not the phenyl groups. One may imagine the following three different models: rotation of the methoxy group or of the methyl group alone or rotation of both simultaneously. A careful analysis of the EISF can help to distinguish between them.

Comparison of the EISFs for different models (figure 8) clearly shows that a rotation of the methoxy group must be excluded in favour of a three-fold rotation of the methyl group alone (model B). The small difference between model B and the experimental data is accounted for by a correction for incoherent elastic scattering due to (non-rotating)

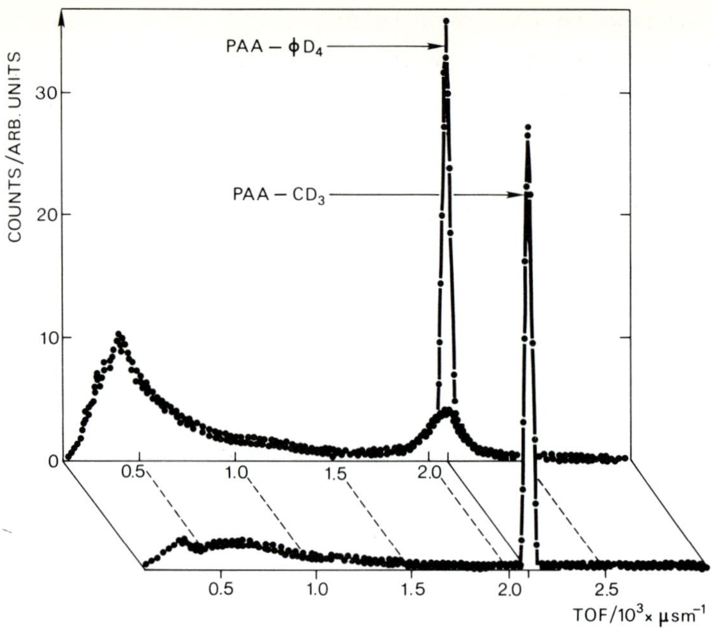

Fig. 7 *Time of flight spectra for two partially deuteriated derivatives of PAA at 100°C. Incident wavelength* λ_0 = *8.25Å and elastic momentum transfer* Q = *1.25Å*$^{-1}$.

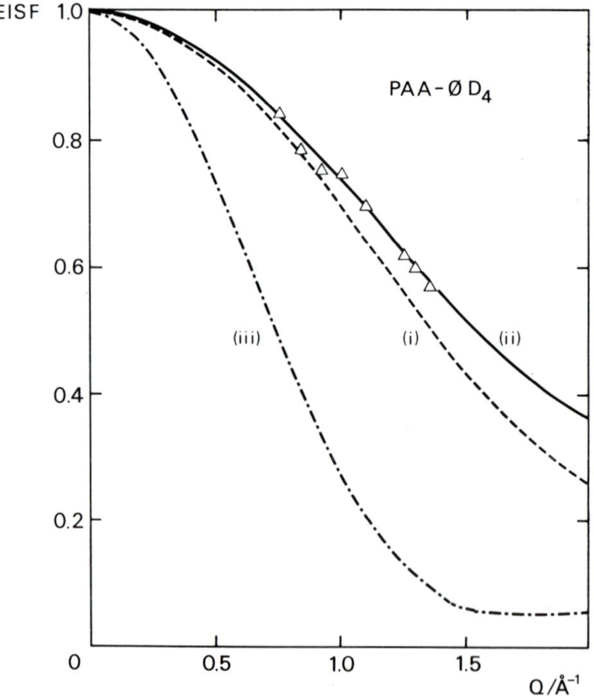

Fig. 8 *The elastic incoherent structure factor as a function of Q for PAA-*ϕD_4 *at 100°C (i) model B, rotation of methyl groups alone; (ii) model B corrected for incoherent elastic scattering and incompletely replaced hydrogen on the phenyl rings. Radius of gyration of methyl protons was 1.032Å; (iii) model C, simultaneous rotation of O-*CH_3 *and of* CH_3 *groups.*

deuterons and non-perfect deuteriation of the phenyl rings. The large difference between the model curves B and C is due to the appreciable difference in the rotation radii of the protons characterizing the different types of rotation. The threefold rotational jump model was fitted to the quasielastic data taken at different temperatures and examples of the fit are reproduced in figure 9. The width of the quasielastic peak

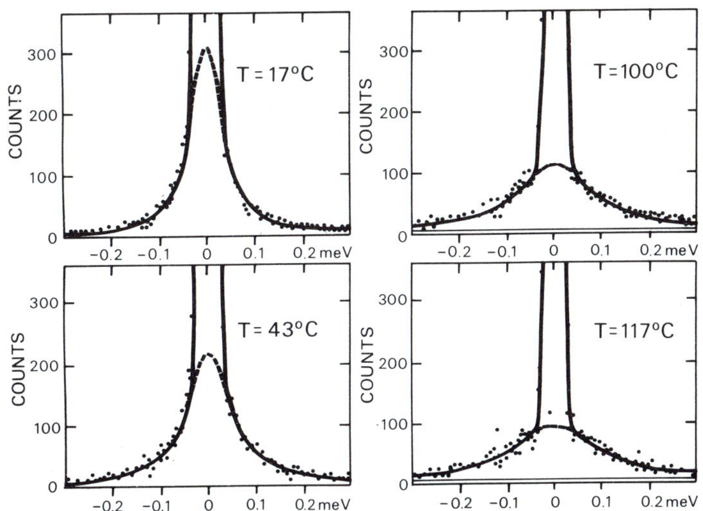

Fig. 9 *Neutron spectra versus energy transfer for PAA-ϕD_4 at four temperatures. Incident wavelength λ_0 = 9.45Å, elastic momentum transfer Q = 1.15Å^{-1}. The points are experimental, the full lines are calculated for the model of 3-fold jump reorientation (model B). The separate elastic and quasielastic components of the scattering are also shown.*

is closely related to the rotational correlation time τ. This was measured as a function of temperature and found to follow an Arrhenius law:

$$\tau = \tau_0 \exp(\beta E_a),$$

with

$$\tau_0 = (0.35 \pm 0.07) \times 10^{-12} \text{s}$$

and an activation energy

$$E_a = 10.6 \pm 0.5 \text{ kJ mol}^{-1}.$$

It should be mentioned that this value of the activation energy may be used to check calculations of the barrier height, assuming simple rotational potentials and using the position of inelastic peaks which are due to torsional excitations. For instance, the assumption of a threefold cosine potential leads to an activation energy which is definitely too large and which shows that the true potential shape deviates appreciably from the cosine form.

Similar experiments have been carried out on the solid phases of other liquid crystalline materials [14,15] in which it is found that the last methylene and/or the last methyl groups in the end chains rotate in the solid phase on a time scale of around 10^{-11} s.

Summary of Experiments on Liquid Crystalline Phases

The following substances have been studied by high resolution neutron scattering experiments.

(i) PAA;

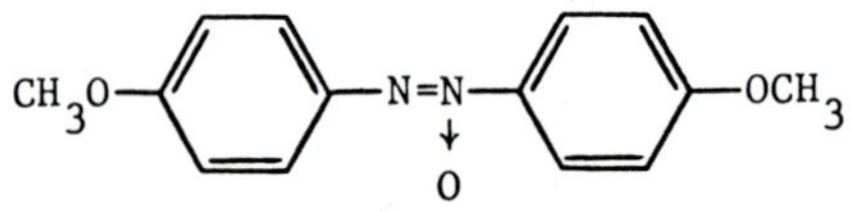

CH_3O— —N=N(→O)— —OCH_3

C 118°C N 135°C I [13,16,17]

(ii) MBCA;

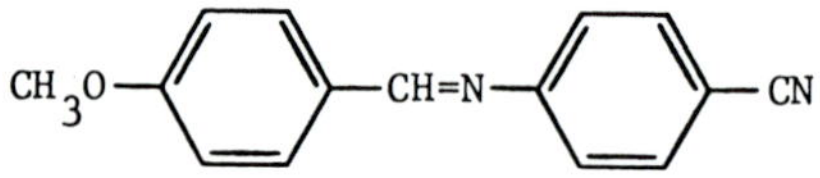

CH_3O— —CH=N— —CN

C 108°C N 119°C I [11,19]

(iii) 5CB;

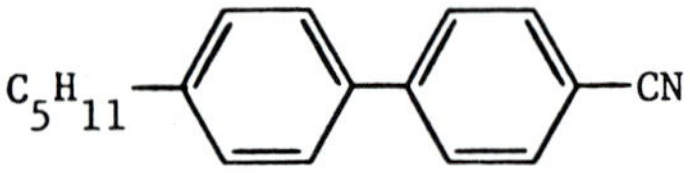

C_5H_{11}— — —CN

C 22.6°C N 35.1°C I [11,19]

(iv) EABAC;

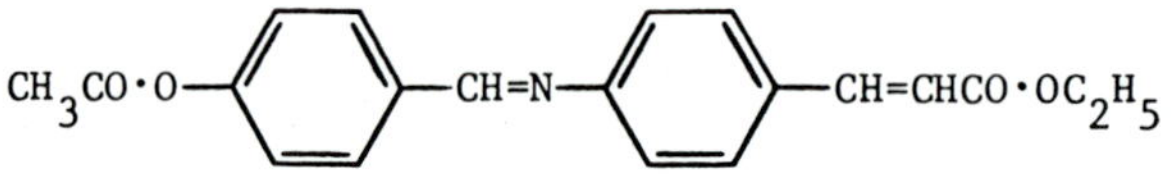

$CH_3CO{\cdot}O$— —CH=N— —$CH{=}CHCO{\cdot}OC_2H_5$

C 105.5°C S 125.0°C N 156.0°C I [15,20,21]

(v) TBBA

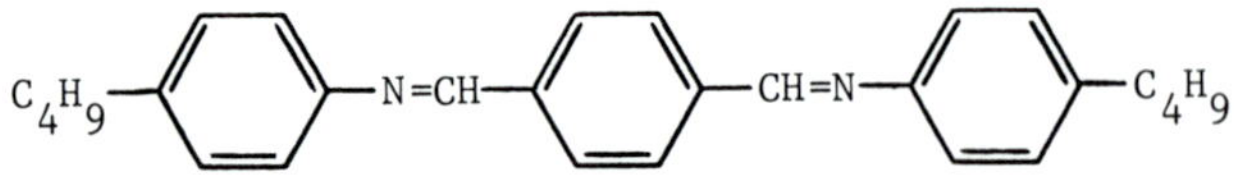

C_4H_9— —N=CH— —CH=N— —C_4H_9

C 113°C S_{B_t} 144°C S_C 172°C S_A 200°C N 238°C I

↑ ↓ 84°C

SVII ← S_{E_t}

68°C [14,22-29]

(vi) BPBAC;

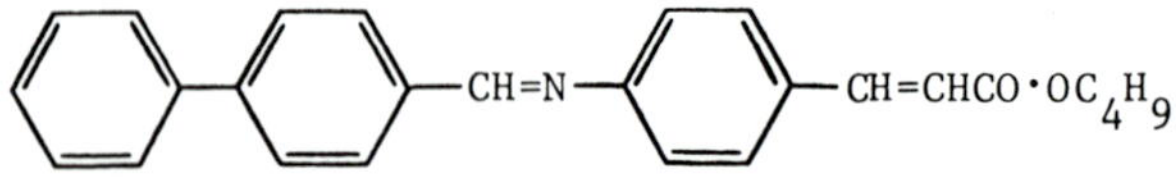

— —CH=N— —$CH{=}CHCO{\cdot}OC_4H_9$

C 77°C S_E 107°C S_B 172 C S_A 206°C I [30,31,15]

(vii) IBPBAC;

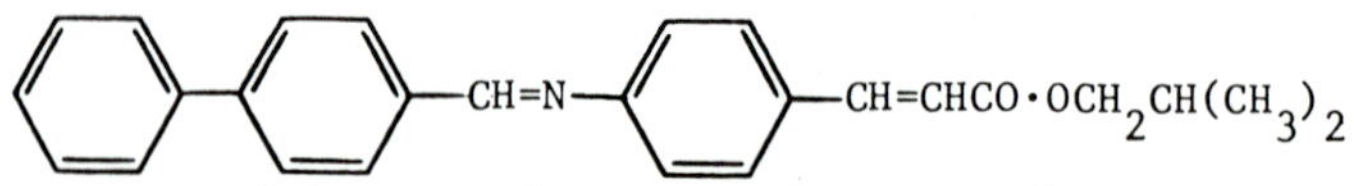

— —CH=N— —$CH{=}CHCO{\cdot}OCH_2CH(CH_3)_2$

C 86°C S_E 114°C S_B 162°C S_A 206°C N 214°C I [32]

In most cases the work has been carried out using the very high resolution instruments of the Institut Laue Langevin, Grenoble.

Some examples are briefly discussed in the next two sections and the final section summarizes the information that QENS has given about liquid crystals.

Translational diffusion

Eqs. (32), (36) and (40) show that the scattering law in the quasi-elastic region has the general form

$$S_S^{QE}(\mathbf{Q},\omega) = \exp(-Q^2\langle u_Q^2\rangle)\{A_o(\mathbf{Q})\pi^{-1}L[(D_{\parallel}\cos^2\theta+D_{\perp}\sin^2\theta)Q^2] + \sum_{n>0} A_n(\mathbf{Q})\pi^{-1}L[(D_{\parallel}\cos^2\theta+D_{\perp}\sin^2\theta)Q^2 + 1/\tau_n^R], \quad (68)$$

where $\sum_{n=0}^{\infty} A = 1$, A_n and τ_n^R will depend upon the nature of the rotational motion and $A_0(\mathbf{Q}) \to 1$ as $Q \to 0$. Translational diffusion in liquid crystals is usually slower than the rotational motion and can be observed by an experiment at very high resolution at low Q where the purely translational component of the scattering law is dominant. This can be achieved using the backscattering technique [18] which can measure the scattering law in the near elastic region ($|\hbar\omega| \lesssim 16$ μeV) with a very good resolution (∿ 1μeV FWHM). In this narrow energy range the broad rotational lorentzians approximate very well to a flat background. Typically, translational diffusion in a nematic phase gives a broadening of a few μeV for $Q \lesssim 0.5Å^{-1}$, as shown for PAA in figure 10. The spectra have been fitted

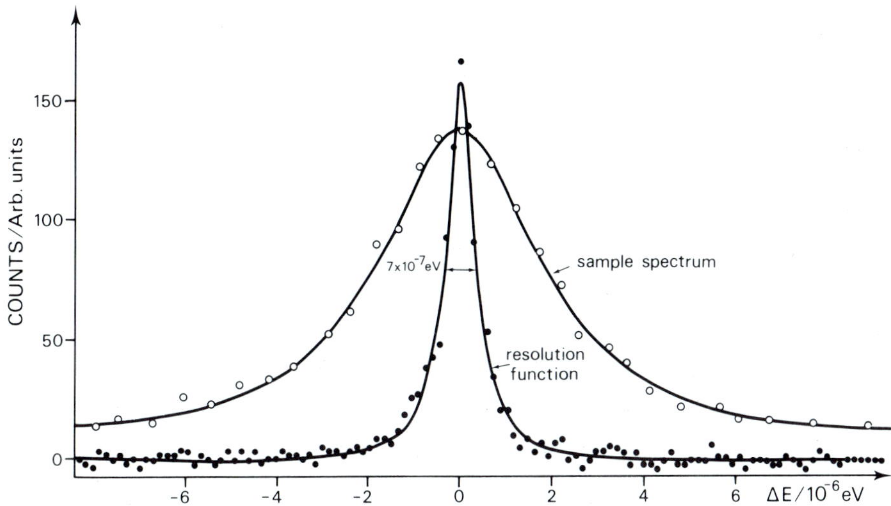

Fig. 10 *Typical ultra high resolution spectra.*

with a single lorentzian plus a flat background to represent the faster rotational components.

If an aligned liquid crystalline sample is used, the diffusion constants $D_{||}$ and $D_{\perp}$ may be measured by experiments with **Q**|| **n** and **Q**⊥**n**. The lorentzian broadening of the resolution function is proportional to Q^2 as shown in figure 11 and the slopes of the lines give the appropriate

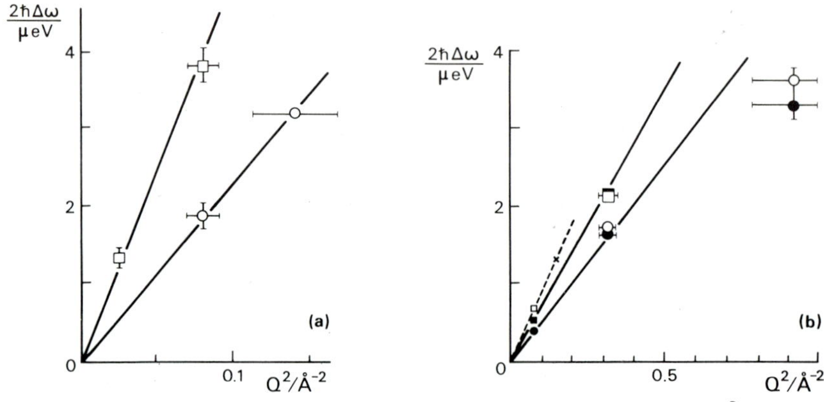

Fig. 11 *Lorentzian widths (fwhm) for (a) D-MBCA as functions of Q^2 in the nematic phase at 112°C. o: **Q⊥n**; □: **Q|| n** and (b) 5CB (o,□) and D-5CB (•,■) in the nematic phase at 23.5°C and in the isotropic liquid phase at 40°C(X). Circles: **Q⊥n**; squares: **Q|| n***

diffusion constants. The diffusion constants for nematic phases that have been measured by this method [11,16,21] are in the range $10^{-7} cm^2 s^{-1}$ to $5 \times 10^{-6} cm^2 s^{-1}$ but the temperature at which the nematic phase occurs is an important factor in determining their magnitude. The apparent anisotropies ($\gamma_{app} = D_{||}/D_{\perp}$) are between 1 and 2 but the real microscopic anisotropy may be greater than this because the distributions of orientations of the local director tends to reduce the observed anisotropy. The real anisotropy may be estimated using the approximate formula

$$\gamma_{real} = \{(2\bar{P}_2+1)\gamma_{app}+2(1-\bar{P}_2)\}/\{\gamma_{app}(1-\bar{P}_2)+\bar{P}_2+2\}, \tag{69}$$

where $\bar{P}_2$ is the order parameter.

Two measurements of the translational diffusion constants of aligned samples of the smectic A phase have also been carried out [21,26]. They have both yielded the surprising result that $D_{||} \approx D_{\perp} \approx 10^{-6} cm^2 s^{-1}$.

In a powder sample it is not possible to measure the anisotropy of translational diffusion but the scalar part of the diffusion tensor

$$\bar{D} = (2D_{\perp} + D_{||})/3, \tag{70}$$

can be measured if the anisotropy is not too large. This gives a useful estimate of the magnitude of the translational diffusion constant and studies have been carried out on powder samples of two compounds which exhibit smectic polymorphism [24,31]. In both cases, temperature dependent diffusion constants of $\sim 10^{-6} cm^2 s^{-1}$ have been found in the smectic phases with disordered layers (i.e. S_A and S_C for TBBA and S_A for BPBAC) but an apparently temperature independent value of $D \sim 5 \times 10^{-8}$ $cm^2\ s^{-1}$ in the smectic phases with ordered layers (i.e. S_H for TBBA and S_B, S_E for BPBAC).

Molecular rotations.

Several studies of molecular reorientation in liquid-crystalline phases using QENS have been made and usually samples with deuteriated end chains have been used so that only the motion of the cores of the molecules is seen. In general the rotational motion of liquid crystalline molecules about their long axes has a correlation time in the region 10^{-11} s to 10^{-10} s and so high resolution ($\lesssim$ 20 μeV FWHM) is still essential to see the scattering law clearly. Since translational diffusion is rather slower than rotation it is possible to separate the quasi-elastic scattering law into an elastic component (which may be slightly broadened by translational diffusion) and a quasielastic component whose width is dominated by the rotational motion, as shown in figure 5. An experimental value of the EISF is thus obtained which gives direct information on the geometry of all molecular motions which are fast enough to be seen (i.e. $\tau \lesssim 10^{-10}$ s) but remain localized on the same time scale.

Experiments on unaligned samples of TBBA and BPBAC have given qualitatively similar results [22,23,25,27-29; 15,30,31] although there are some differences in the detailed interpretation. Both compounds give quasielastic scattering (figures 12 and 13a,b,c) which may be separated into elastic and quasielastic parts. The half width of the quasielastic component (ΔE) gives the approximate reorientational correlation time ($\tau \sim \hbar/\Delta E$) which decreases continuously with temperature as shown in figures 14 and 15.

The experimental EISF decreases with increasing temperature as shown in figure 16 for BPBAC. In the S_E phase of BPBAC and the S_{VI} phases of TBBA the EISF is higher than that expected for uniaxial rotational diffusion or jump reorientation between six equal sites on a circle which implies that the rotation about the long axes is somewhat restricted in these phases. This has been interpreted as the result of orientational ordering of the molecules. If the molecules are rotationally diffusing about their long axes in an n fold cosine potential the expected EISF from a powder sample is [33]

$$\text{EISF} = \{1/\pi I_0^2(\beta') \int_0^\pi dx\, j_o(2Q\, a \sin x) I_0(2\beta' \cos nx). \qquad (71)$$

Here I_0 is a modified Bessel function of the first kind and β' is a parameter related to the orientational ordering parameter, β_n, which is given by

$$\beta_n = \overline{\cos n\gamma} = I_1(\beta')/I_0(\beta'),$$

where γ is the angle of rotation about the long molecular axis. For a real molecule eq. (71) should be averaged over the different gyration radii a.

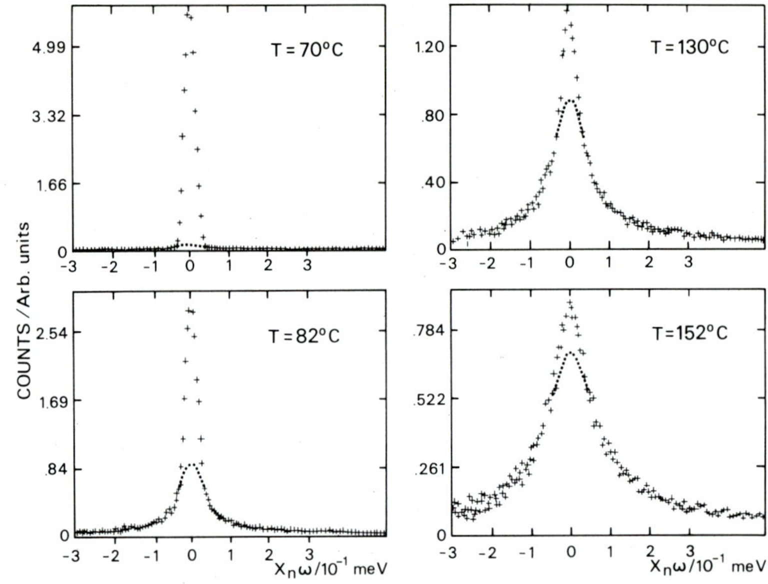

Fig. 12 *Scattering law at Q = 0.92Å^{-1} for D-TBBA, showing the separation into elastic and quasielastic components at four temperatures.*

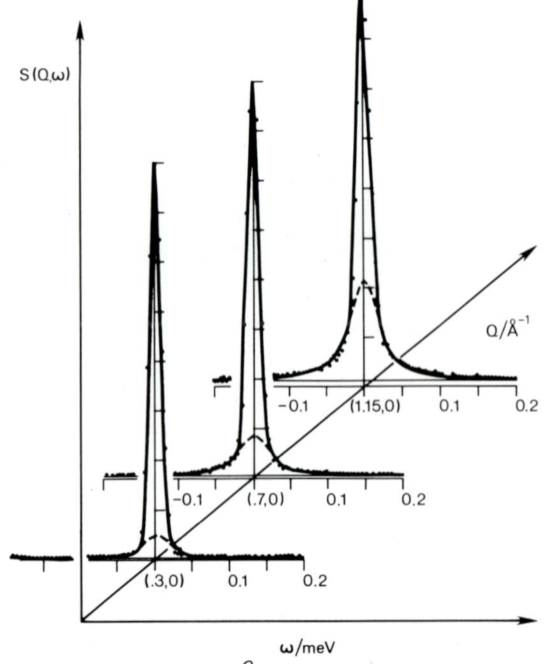

Fig. 13(a) *Scattering law for D-BBAC at 91°C in the smectic E phase showing the separation into elastic and quasielastic components and a flat, inelastic, background.*

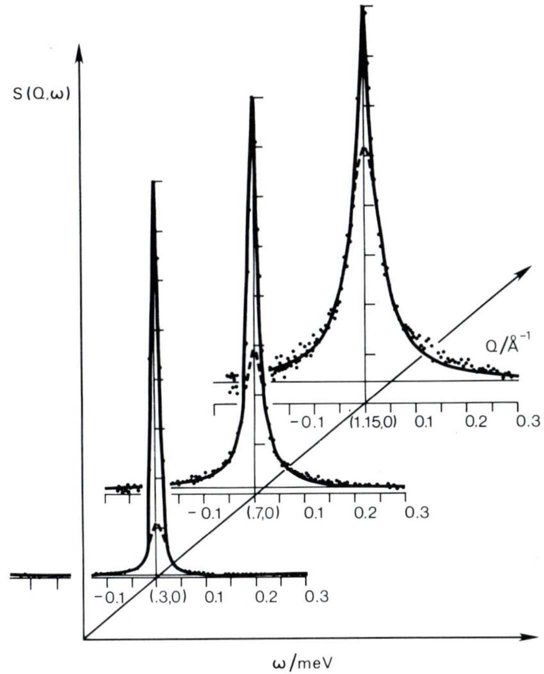

Fig. 13(b) *Scattering law for D-BPBAC at 166°C in the smectic B phase.*

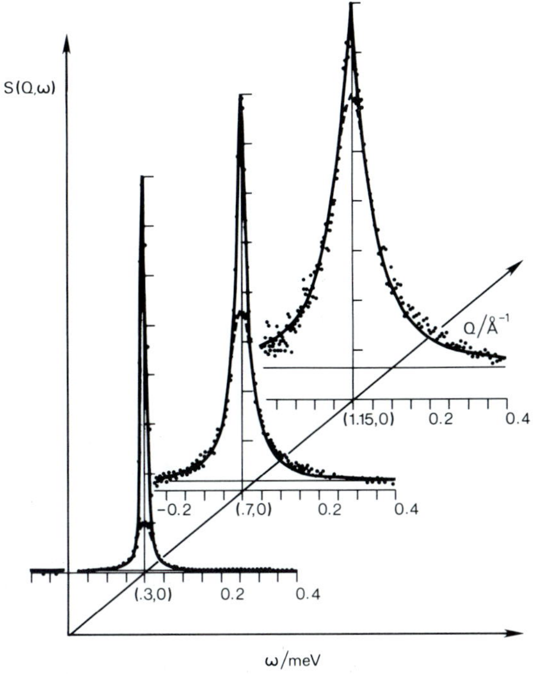

Fig. 13(c) *Scattering law for D-EABAC at 181°C in the smectic A phase.*

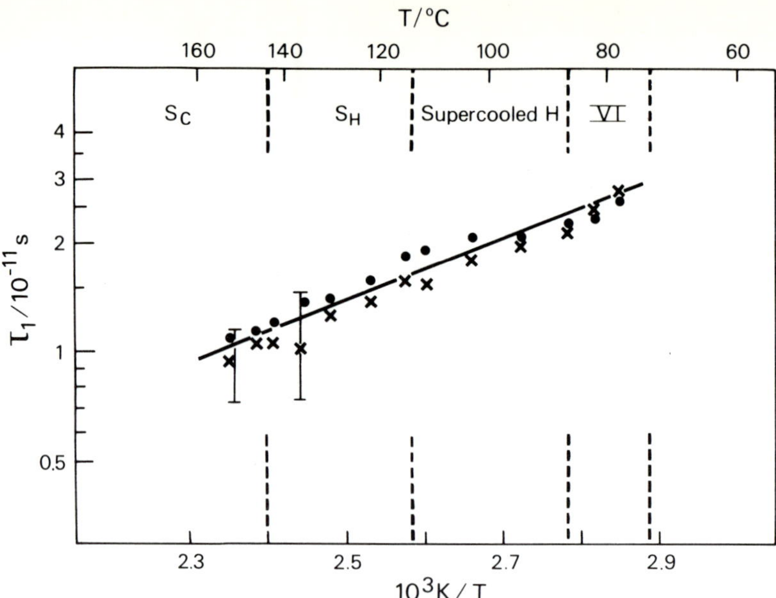

Fig. 14 *Arrhenius plot of the correlation time τ_1 for D-TBBA as a function of inverse temperature. τ_1 is defined by eq. (55) but to a good approximation $\tau_1 = l/D_r$ where D_r is the rotational diffusion constant. The slope corresponds to an average activation energy of 16.0 kJ mol^{-1}.*

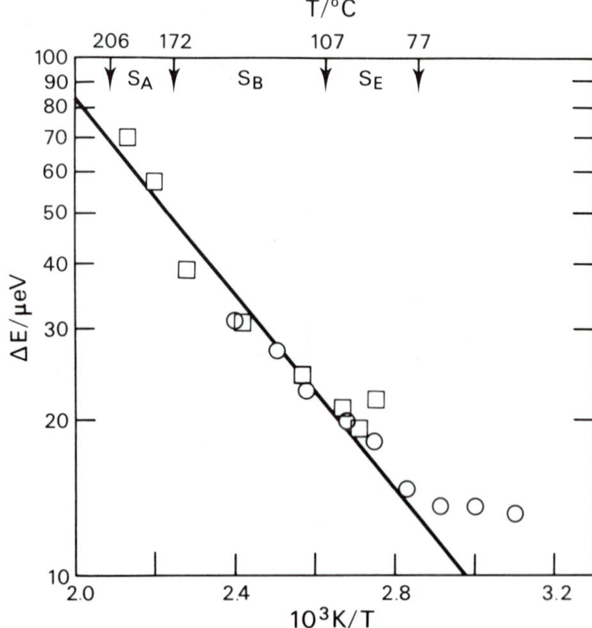

Fig. 15 *Arrhenius plot of the half width of the quasielastic component, ΔE, against the inverse of the absolute temperature for D-BPBAC. The slope of the line implies an effective activation energy of 18.0 kJ mol^{-1}.*

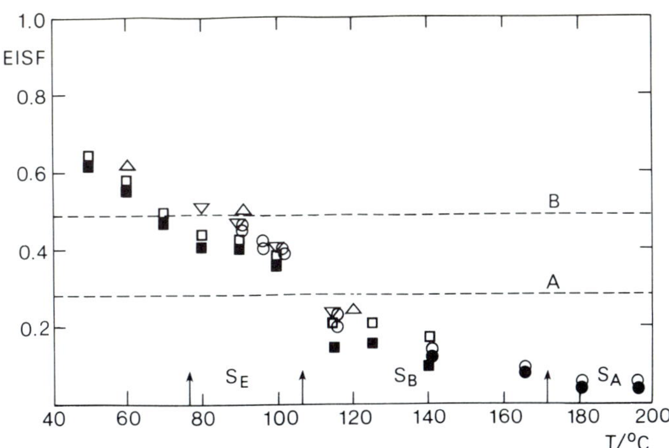

Fig. 16 *Experimental EISF at Q = 1.15Å^{-1} for D-BPBAC as a function at temperature, T. (A) is the EISF value expected from uniaxial rotational diffusion and (B) is the value expected for jump reorientation by π about the long molecular axis. The different types of point have been obtained by experiments at different times and using different techniques to extract the EISF. Their scatter gives an estimate of the accuracy of the experimental EISF.*

If there are two preferred orientations of the molecule (i.e.n = 2) the expected EISF can range from the value for jump reorientation by π about the long axes (when $\beta_2 = \overline{\cos 2\gamma} = 1$) to the value for uniaxial rotational diffusion (when $\beta_2 = 0$). However if the molecules can become localized at one site (n = 1) the EISF increases to unity when $\beta_1 = \overline{\cos\gamma} = 1$. Figure 16 shows that for BPBAC the EISF in the smectic E phase is consistent with two preferred sites separated by π but in the supercooled phase the EISF is higher than is possible for two equivalent sites which suggests that there is some further ordering (with one site preferred) taking place. This may be a precurser to the transition to the crystal.

In the smectic B phase of BPBAC the EISF is lower than expected for uniaxial rotational diffusion. This suggests that the molecules are undergoing an additional motion with an amplitude that increases with temperature. Two such motions have been proposed and these now are discussed in (i) and (ii).

(i) Although it is improbable that the molecules in a liquid crystal can reorient by π about their short axes on the neutron time scale, it is highly probable that the long axes can undergo fairly modest fluctuations of direction on this time scale. If the molecules are undergoing uniaxial diffusion plus a fluctuation of the long axes with an infinite time distribution:

$$g(\beta) = (\delta/4\sinh\delta)\,\exp(\delta\cos\beta), \tag{72}$$

where δ is a parameter that characterizes the distribution, then the EISF expected for a powder sample is

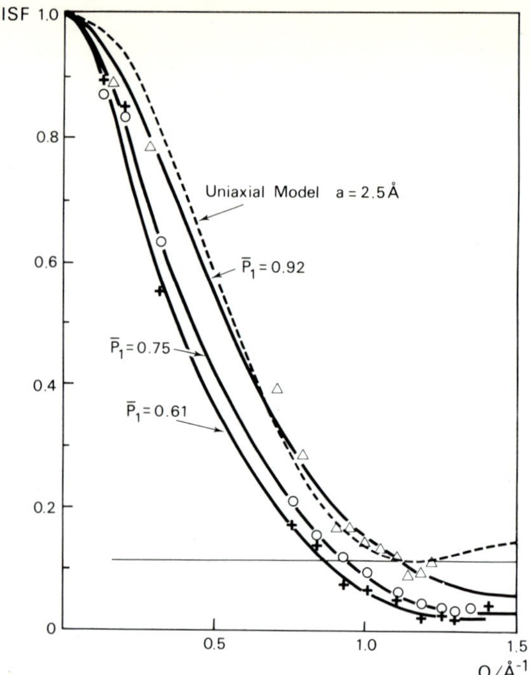

Fig. 17 *The points are the experimental EISF for D-TBBA at Δ 119°C, O 137°C, + 151°C and the dashed line is the best fit obtained to the 119°C data using the EISF calculated for a simple uniaxial rotational diffusion model. The solid lines are obtained using the real radii of the protons but allowing fluctuations of the long molecular axes as in eq. (74).*

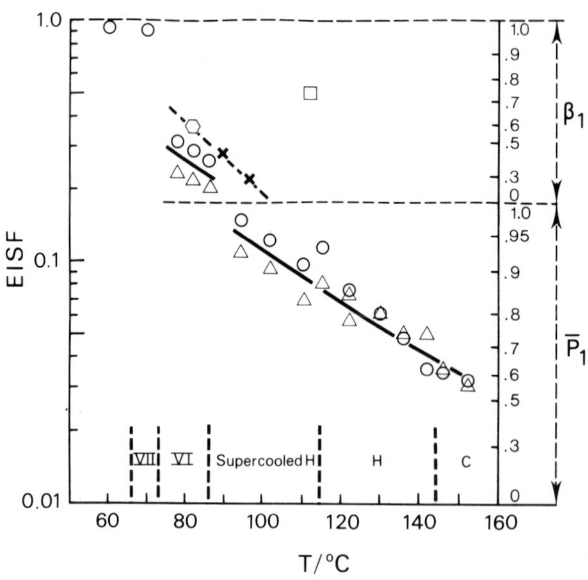

Fig. 18 *Experimental EISF of D-TBBA at Q = 1.18Å^{-1}. The scales on the right show the values of the order parameters for long axis fluctuations ($\overline{\cos\beta}$) and orientational ordering ($\overline{\cos\gamma}$) that correspond to the EISF data.*

$$\text{EISF} = \sum_{L=0}^{\infty} (2L+1) j_L^{\,2}(QR)\,\overline{P}_L^{\,2} P_L^{\,2}(\cos\mu), \tag{73}$$

where R is the radius from the centre of mass to the proton, μ is the angle between R and the long molecular axis **m**, $\overline{P}_L$ is the value of the Lth Legendre polynomial averaged over the distribution $g(\beta)$. For pure uniaxial rotational diffusion with no fluctuations we have $\overline{P}_1 = \overline{\cos\beta} = 1$ and for isotropic rotational diffusion $\overline{P}_1 = \overline{\cos\beta} = 0$.

As shown in figures 17 and 18 the EISF from the S_H phase of TBBA can be interpreted in terms of uniaxial rotational diffusion plus a fluctuation of the long molecular axes on the same time scale with $\overline{P}_1$ between 0.6 and 0.95.

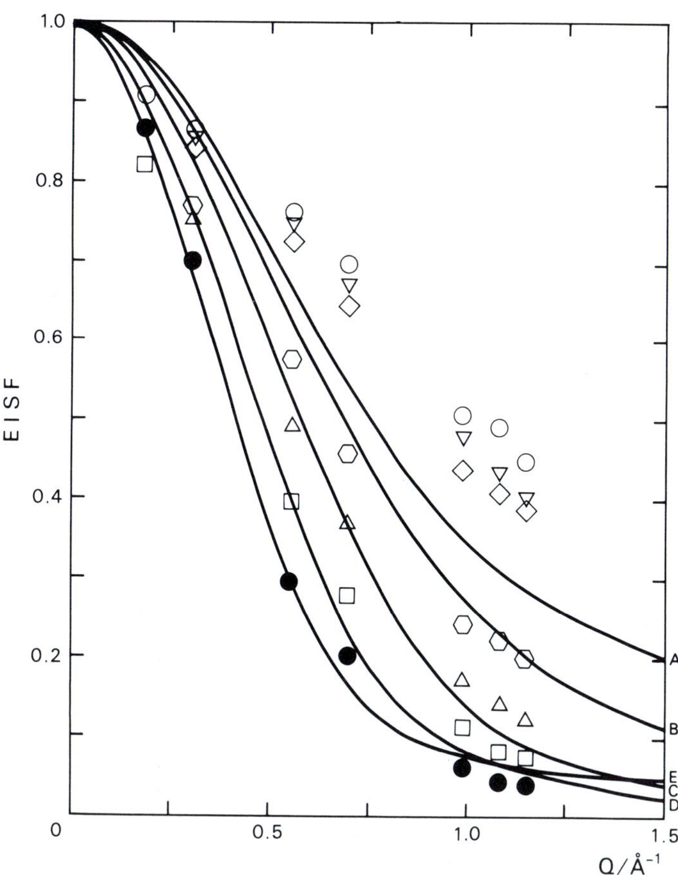

Fig. 19 *The points show the experimental EISF for D-BPBAC at O : 91°C, ∇: 96°C, : 101°C, : 116°C, Δ: 141°C, □: 166°C O: 181°C and 196°C. The lines show the EISF expected if the D-BPBAC molecules undergo uniaxial rotational diffusion plus an independent bound motion perpendicular to the layers of amplitude bÅ. The lines are not fitted to the points but are shown for the following values of b: (A) 0, (B) 1, (C) 2, (D) 3, (E) 4.*

(ii) A second possibility for an additional motion in the smectic A and B phases of BPBAC and TBBA is a bound motion perpendicular to the layers. This could be some overdamped oscillation of the molecules in the smectic layers which becomes softer with increasing temperature. Since the additional motion appears to be on a similar time scale to the rotations about the long axes this implies a coupling between the two motions and a simple model suggests a diffusive component perpendicular to the layers whose infinite time distribution would be the same as for an oscillatory motion. The EISF for a powder would then be given by:

$$\mathrm{EISF} = \int_0^{\pi/2} \sin\theta d\theta\; J_0^2(Qa \sin\theta) J_0^2(Qb \cos\theta), \qquad (74)$$

where a is the radius of gyration of the proton and b is the amplitude of the motion perpendicular to the layers. As shown in figure 19 quite modest values of b are able to explain the experimental EISF.

Both motion (i) and (ii) provide good explanations of the data from powder samples but results on aligned samples are needed to choose between them. Such experiments are currently in progress [15,31,32].

Conclusion

Quasielastic neutron scattering gives very direct information about the diffusive motion of the molecules because it is able to follow the motions of the protons themselves. The selection rules are very weak and so all sufficiently rapid motions of the protons are seen.

QENS has given the magnitude and anisotropies of translational diffusion in several phases and in particular it has shown that a major difference between the ordered and disordered smectic phases is the magnitude of the diffusion constant but that there is only a small difference between the smectic A and nematic phase.

In all liquid crystalline phases there is rapid reorientation about the long molecular axes which in the most ordered smectic phases (S_E and S_{VI}) becomes restricted. In the S_B, S_H, S_C and S_A phases there is at least one additional rapid motion ($\tau \sim 10^{-11}$s) whose nature has yet to be fully elucidated.

References

1. See for example: *Thermal Neutron Scattering*, Ed. P.A. Egelstaff, Academic Press, (1965). *Chemical Applications of Thermal Neutron Scattering*, Ed. B.T.M. Willis, Oxford University Press (1973). W. Marshall and S.W. Lovesey, *Theory of Neutron Scattering*, Oxford University Press, (1971). I.I. Gurevich and L.V. Tarasov, *Low Energy Neutron Physics*, North Holland, Amsterdam, (1968).
2. See for example: T. Springer, *Springer Tracts in Modern Physics*, **64**, (1972).
3. D.C. Champeney, *Fourier Transforms and their Physical Applications*, Academic Press, (1973).
4. A.J. Leadbetter and R.E. Lechner, Chapter 8, *The Plastic Crystalline State*, Ed. J.N. Sherwood, Wiley, (1979).

5. R. Stockmeyer and H. Stiller, *Phys. Stat. Sol.* **27**, 269, (1968).
6. K. Sköld, *J. Chem. Phys.* **49**, 2443, (1968).
7. R.E. Lechner, J.M. Rowe, K. Sköld and J.J. Rush, *Chem. Phys. Letts.* **4**, 444 (1969).
8. J.D. Barnes, *Neutron Inelastic Scattering,* p.287 IAEA, Vienna, (1972).
9. A.J. Dianoux, F. Volino and H. Hervet, *Mol. Phys.,* **30**, 1181 (1975).
10. V.F. Sears, *Canad. J. Phys.* **45**, 237 (1967).
11. A.J. Leadbetter, F.P. Temme, A. Heidemann and W.S. Howells, *Chem. Phys. Letts.* **34**, 363 (1975).
12. F. Volino and A.J. Dianoux, Proc. Euchem. Conf., *Organic Liquids: Structures, Dynamics and Chemical Properties,* Wiley (1979).
13. H. Hervet, A.J. Dianoux, R.E. Lechner and F. Volino, *J. Phys.* (Paris), **37**, 587 (1976).
14. F. Volino, A.J. Dianoux, R.E. Lechner and H. Hervet, *J. Phys.* (Paris), **36**, C1-83 (1975).
15. A.J. Leadbetter, R.M. Richardson and J.C. Frost, *J. Phys.* (Paris), in the press (1980); R.M. Richardson, Ph.D. Thesis, Bristol (1977).
16. J. Töpler, B. Alefeld and T. Springer, *Mol. Cryst. Liq. Cryst.*, **26**, 297 (1973).
17. F. Volino and A.J. Dianoux, unpublished work.
18. A. Alefeld, *Kerntechnik,* **14**, 15 (1972).
19. D.H. Bonsor, A.J. Leadbetter and F.P. Temme, *Mol. Phys.* **36**, 1805 (1978).
20. A.J. Leadbetter and R.M. Richardson, *Mol. Phys.* **35**, 1191 (1978).
21. A.J. Leadbetter, R.M. Richardson, B.A. Dasannacharya and W.S. Howells, *Chem. Phys. Lett.,* **39**, 501 (1976).
22. H. Hervet, F. Volino, A.J. Dianoux and R.E. Lechner, *J. Phys. Lett.* **35**, L-151 (1974).
23. A.J. Dianoux, F. Volino and H. Hervet, *Mol. Phys.,* **30**, 1181 (1975).
24. A.J. Dianoux, F. Volino, A. Heidemann and H. Hervet, *J. Phys. Lett.* **36**, L-275 (1975).
25. H. Hervet, F. Volino, A.J. Dianoux and R.E. Lechner, *Phys. Rev. Lett.* **34**, 451 (1975).
26. A.J. Dianoux, A. Heidemann, F. Volino and H. Hervet, *Mol. Phys.* **32**, 1521 (1976).
27. F. Volino, A.J. Dianoux and H. Hervet, *Solid State Comm.,* **18**, 453 (1976).
28. F. Volino, A.J. Dianoux and H. Hervet, *J. Phys.* (Paris), **37**, C3-55 (1976)
29. F. Volino, A.J. Dianoux and H. Hervet, *Mol. Cryst. Liq. Cryst.* **38**, 125 (1977).
30. R.M. Richardson, A.J. Leadbetter, C.J. Carlile and W.S. Howells, *Mol. Phys.,* **35**, 1697 (1978).
31. R.M. Richardson, A.J. Leadbetter and J.C. Frost, *Ann. Phys.* **3**, 177 (1978); A.J. Leadbetter, R.M. Richardson and C.J. Carlile, *J. Phys.* (Paris), **37**, C3-65 (1976).
32. A.J. Leadbetter, R.M. Richardson and J.C. Frost, *J. Phys.* (Paris), **40**, C3-125(1979).
33. A.J. Dianoux and F. Volino, *Mol. Phys.,* **34**, 1263 (1977).

SUBJECT INDEX

CHEMICAL INDEX